Airborne Particulate Matter
Sources, Atmospheric Processes and Health

ISSUES IN ENVIRONMENTAL SCIENCE AND TECHNOLOGY

SERIES EDITORS:

Professor R. E. Hester, University of York, UK
Professor R. M. Harrison, University of Birmingham, UK

EDITORIAL ADVISORY BOARD:

Professor P. Crutzen, Max-Planck-Institut für Chemie, Germany, **Professor S. J. de Mora**, Plymouth Marine Laboratory, UK, **Dr G. Eduljee**, SITA, UK, **Professor L. Heathwaite**, Lancaster University, UK, **Professor S. Holgate**, University of Southampton, UK, **Professor P. K. Hopke**, Clarkson University, USA, **Professor P. Leinster**, Environment Agency, UK, **Professor P. S. Liss**, School of Environmental Sciences, University of East Anglia, UK, **Professor D. Mackay**, Trent University, Canada, **Professor A. Proctor**, Food Science Department, University of Arkansas, USA, **Dr D. Taylor**, WCA Environmental Ltd, UK.

TITLES IN THE SERIES:

How to obtain future titles on publication

A subscription is available for this series. This will bring delivery of each new volume immediately on publication and also provide you with online access to each title *via* the Internet. For further information visit http://www.rsc.org/issues or write to the address below.

For further information please contact:
Sales and Customer Care, Royal Society of Chemistry, Thomas Graham House, Science Park, Milton Road, Cambridge, CB4 0WF, UK
Telephone: +44 (0)1223 432360, Fax: +44 (0)1223 426017, Email: booksales@rsc.org
Visit our website at www.rsc.org/books

ISSUES IN ENVIRONMENTAL SCIENCE AND TECHNOLOGY

EDITORS: R.E. HESTER, R.M. HARRISON AND X. QUEROL

42
Airborne Particulate Matter
Sources, Atmospheric Processes and Health

Issues in Environmental Science and Technology No. 42

Print ISBN: 978-1-78262-491-2
PDF eISBN: 978-1-78262-658-9
EPUB eISBN: 978-1-78262-886-6
ISSN 1350-7583

A catalogue record for this book is available from the British Library

Published by The Royal Society of Chemistry,
Thomas Graham House, Science Park, Milton Road,
Cambridge CB4 0WF, UK

Registered Charity Number 207890

For further information see our web site at www.rsc.org

Printed in the United Kingdom by CPI Group (UK) Ltd, Croydon, CR0 4YY, UK

Preface

Airborne particulate matter has been a highly topical subject for the past 20 years or so, although it has been a problem pollutant for very much longer. Historical records of problems with coal smoke pollution go back many centuries, but the first good quantitative records date back to the early 20th century when it was measured as "black smoke". An episode of severe pollution by black smoke and sulfur dioxide in December 1952 is believed to have caused around 4000 premature deaths in London alone, and comparable events occurred in other parts of the world. The recognised toxicity of black smoke and sulfur dioxide led to the development of mitigation policies, which greatly improved the quality of the atmosphere to the point that it was felt that there were no significant remaining effects on public health arising from these pollutants in most developed countries. However, the application of more advanced epidemiological methods in the 1990s demonstrated that significant public health impacts of exposure to airborne particles, now measured by the mass metrics $PM_{2.5}$ and PM_{10}, existed in North America and Europe and, by implication, in other parts of the world. Whilst most aspects of local air quality continued to improve in developed countries, this was a time when less developed countries were growing their economies rapidly, with an accompanying substantial increase in the concentration of airborne particles. As a result, the World Health Organisation's Global Burden of Disease Project ranks exposure to outdoor particles as the ninth largest preventable cause of disease in the global population.

This volume of the *Issues* series addresses airborne particulate matter from a global perspective. The first chapter, by Marc Guevara of the Barcelona Supercomputing Centre, gives quantitative information on the sources of particle emissions in a European context. Not only are particles emitted directly into the atmosphere, but they also form within the atmosphere from the oxidation of gases and undergo chemical transformations in the atmosphere. The latter secondary fraction typically makes the dominant contribution to $PM_{2.5}$ concentrations. In the second chapter, Neil Donahue of Carnegie Mellon University and co-authors pose the question: What are

Issues in Environmental Science and Technology No. 42
Airborne Particulate Matter: Sources, Atmospheric Processes and Health
Edited by R.E. Hester, R.M. Harrison and X. Querol
© The Royal Society of Chemistry 2016
Published by the Royal Society of Chemistry, www.rsc.org

the sources of particles? Their focus is on the secondary particles that are formed within the atmosphere and how the sources affect human exposure to airborne particles, both by mass and by number.

One of the key advances in the past 20 years has been the development of receptor modelling methods that take air quality data and use it to provide estimates of the quantitative contribution of different sources to measured concentrations. In the third chapter, John Watson and Judy Chow of the Desert Research Institute, Nevada, outline the principles of these methods, and in the fourth chapter, Philip Hopke of Clarkson University, New York State, provides further insights into methods and gives some case studies from North America. The following two chapters take data from two major European experimental studies to provide case studies of the application of receptor modelling methods to source apportionment. In doing so, they give powerful insights into the sources of particles in European cities. The AIRUSE LIFE+ team led by Xavier Querol of the Consejo Superior de Investigaciones Cientificas (CSIC) describes a study of cities in southern Europe. This is complemented by a study from northern Europe (the Joaquin Project) led by Edward Roekens of the Flanders Environment Agency, which has applied similar methods to five cities within northern Europe.

The subsequent two chapters provide insights into the apportionment of $PM_{2.5}$ (fine particles) and particulate matter of various size fractions in two of the world's hotspots. In Chapter 7, Mei Zheng and colleagues from Peking University consider the source apportionment of $PM_{2.5}$ within China, and in Chapter 8, Mukesh Khare and Isha Khanna of the Indian Institute of Technology, Delhi, review case studies of source apportionment from the Indian sub-continent.

No such volume would be complete without also considering the adverse health effects of airborne particles. Consequently, in Chapter 9, Frank Kelly of Kings College, London, reports on the health effects of airborne particles and how these relate to their composition, size and source.

We are delighted to have engaged such a distinguished group of authors, including several world leaders in their respective fields, to provide a truly authoritative and up-to-date volume describing many of the key aspects of airborne particulate matter. We believe that this will prove to be of widespread interest and be of great value to a range of communities, including policymakers, physical and life scientists, and students taking advanced courses in a range of environmental and health-related fields.

Ronald E. Hester
Roy M. Harrison
Xavier Querol

Contents

Issues in Environmental Science and Technology No. 42
Airborne Particulate Matter: Sources, Atmospheric Processes and Health
Edited by R.E. Hester, R.M. Harrison and X. Querol
© The Royal Society of Chemistry 2016
Published by the Royal Society of Chemistry, www.rsc.org

Case Studies of Source Apportionment and Suggested Measures at Southern European Cities   168

F. Amato, F. Lucarelli, S. Nava, G. Calzolai, A. Karanasiou, C. Colombi,
V. L. Gianelle, C. Alves, D. Custódio, K. Eleftheriadis, E. Diapouli, C. Reche,
A. Alastuey, M. C. Minguillón, M. Severi, S. Becagli, T. Nunes,
M. Cerqueira, C. Pio, M. Manousakas, T. Maggos, S. Vratolis,
R. M. Harrison and X. Querol

PM_{10} Source Apportionment in Five North Western European Cities—Outcome of the Joaquin Project   264

Dennis Mooibroek, Jeroen Staelens, Rebecca Cordell, Pavlos Panteliadis,
Tiphaine Delaunay, Ernie Weijers, Jordy Vercauteren, Ronald
Hoogerbrugge, Marieke Dijkema, Paul S. Monks and Edward Roekens

PM$_{2.5}$ Source Apportionment in China   293

Mei Zheng, Caiqing Yan and Xiaoying Li

Case Studies of Source Apportionment from the Indian Sub-continent   315

Mukesh Khare and Isha Khanna

Editors

Ronald E. Hester, BSc, DSc (London), PhD (Cornell), FRSC, CChem

Ronald E. Hester is now Emeritus Professor of Chemistry in the University of York. He was for short periods a research fellow in Cambridge and an assistant professor at Cornell before being appointed to a lectureship in chemistry in York in 1965. He was a full professor in York from 1983 to 2001. His more than 300 publications are mainly in the area of vibrational spectroscopy, latterly focusing on time-resolved studies of photoreaction intermediates and on biomolecular systems in solution. He is active in environmental chemistry and is a founder member and former chairman of the Environment Group of the Royal Society of Chemistry and editor of 'Industry and the Environment in Perspective' (RSC, 1983) and 'Understanding Our Environment' (RSC, 1986). As a member of the Council of the UK Science and Engineering Research Council and several of its sub-committees, panels and boards, he has been heavily involved in national science policy and administration. He was, from 1991 to 1993, a member of the UK Department of the Environment Advisory Committee on Hazardous Substances and from 1995 to 2000 was a member of the Publications and Information Board of the Royal Society of Chemistry.

Roy M. Harrison, BSc, PhD, DSc (Birmingham), FRSC, CChem, FRMetS, Hon MFPH, Hon FFOM, Hon MCIEH

Roy M. Harrison is Queen Elizabeth II Birmingham Centenary Professor of Environmental Health in the University of Birmingham. He was previously Lecturer in Environmental Sciences at the University of Lancaster and Reader and Director of the Institute of Aerosol Science at the University of Essex. His more than 400 publications are mainly in the field of environmental chemistry, although his current work includes studies of human health impacts of atmospheric pollutants as well as research into the chemistry of pollution phenomena. He is a past Chairman of the Environment Group of the Royal Society of Chemistry for whom he edited 'Pollution: Causes, Effects and Control' (RSC, 1983;

Fifth Edition 2014). He has also edited "An Introduction to Pollution Science", RSC, 2006 and "Principles of Environmental Chemistry", RSC, 2007. He has a close interest in scientific and policy aspects of air pollution, having been Chairman of the Department of Environment Quality of Urban Air Review Group and the DETR Atmospheric Particles Expert Group. He is currently a member of the DEFRA Air Quality Expert Group, the Department of Health Committee on the Medical Effects of Air Pollutants, and Committee on Toxicity.

List of Contributors

Peter J. Adams, Center for Atmospheric Particle Studies, Carnegie Mellon University, 5000 Forbes Avenue, Pittsburgh, USA.

A. Alastuey, Institute of Environmental Assessment and Water Research (IDAEA-CSIC), Barcelona, 08034, Spain.

C. Alves, Centre for Environmental & Marine Studies, Dep. of Environment, Univ. of Aveiro, 3810-193 Aveiro, Portugal.

F. Amato, Institute of Environmental Assessment and Water Research (IDAEA-CSIC), Barcelona, 08034, Spain. Email: fulvio.amato@idaea.csic.es

S. Becagli, Dep. of Chemistry, Università di Firenze, Sesto Fiorentino, 50019, Italy.

G. Calzolai, Dep. of Physics and Astronomy, Università di Firenze and INFN-Firenze, Sesto Fiorentino, 50019, Italy.

M. Cerqueira, Centre for Environmental & Marine Studies, Dep. of Environment, Univ. of Aveiro, 3810-193 Aveiro, Portugal.

L.-W. A. Chen, Desert Research Institute, Nevada System of Higher Education, Reno, NV 89512, USA. Graduate Faculty, University of Nevada, Reno, Nevada 89503, USA. Department of Environmental and Occupational Health, University of Nevada, Las Vegas 89154, USA.

J. C. Chow, Desert Research Institute, Nevada System of Higher Education, Reno, NV 89512, USA. Graduate Faculty, University of Nevada, Reno, Nevada 89503, USA. The State Key Laboratory of Loess and Quaternary Geology, Institute of Earth Environment, Chinese Academy of Sciences, Xi'an, Shaanxi, 710075, China.

C. Colombi, Environmental Monitoring Sector, Arpa Lombardia, Via Rosellini 17, I-20124 Milano, Italy.

Rebecca Cordell, University of Leicester, Department of Chemistry, LE1 7RH Leicester, United Kingdom.

D. Custódio, Centre for Environmental & Marine Studies, Dep. of Environment, Univ. of Aveiro, 3810-193 Aveiro, Portugal.

Tiphaine Delaunay, atmo Nord-Pas-de-Calais, 55 Place Rihour, 59044 Lille Cedex, France.

E. Diapouli, Institute of Nuclear and Radiological Science & Technology, Energy & Safety, N.C.S.R. Demokritos, 15341 Ag. Paraskevi, Attiki, Greece.

Marieke Dijkema, Public Health Service of Amsterdam, Department of Air Quality, Nieuwe Achtergracht 100, 1000 CE Amsterdam, the Netherlands.

Neil M. Donahue, Center for Atmospheric Particle Studies, Carnegie Mellon University, 5000 Forbes Avenue, Pittsburgh, USA. E-mail: nmd@andrew.cmu.edu

K. Eleftheriadis, Institute of Nuclear and Radiological Science & Technology, Energy & Safety, N.C.S.R. Demokritos, 15341 Ag. Paraskevi, Attiki, Greece.

G. Engling, Desert Research Institute, Nevada System of Higher Education, Reno, NV 89512, USA. Graduate Faculty, University of Nevada, Reno, Nevada 89503, USA.

Julia C Fussell, MRC-PHE Centre for Environment and Health, Facility of Life Sciences and Medicine, King's College London, 150 Stamford Street, London, SE1 9NH, UK.

V. L. Gianelle, Environmental Monitoring Sector, Arpa Lombardia, Via Rosellini 17, I-20124 Milano, Italy.

Marc Guevara, Barcelona Supercomputing Center – Centro Nacional de Supercomputación (BSC-CNS), Earth Sciences Department, Jordi Girona 29, Edificio Nexus II, 08034 Barcelona, Spain. E-mail: marc.guevara@bsc.es

R. M. Harrison, School of Geography, Earth & Environmental Sci., University of Birmingham. Edgbaston, Birmingham B15 2TT, UK. Also at: Department of Environmental Sciences/Center of Excellence in Environmental Studies, King Abdulaziz University, PO Box 80203, Jeddah, 21589, Saudi Arabia.

Ronald Hoogerbrugge, National Institute for Public Health and the Environment (RIVM), Antonie van Leeuwenhoeklaan 9, 3721 MA Bilthoven, the Netherlands.

Philip K. Hopke, Center for Air Resources Engineering and Science, Clarkson University, Potsdam, NY 13699, USA. Email: hopkepk@clarkson.edu

A. Karanasiou, Institute of Environmental Assessment and Water Research (IDAEA-CSIC), Barcelona, 08034, Spain.

Frank J Kelly, MRC-PHE Centre for Environment and Health, Facility of Life Sciences and Medicine, King's College London, 150 Stamford Street, London, SE1 9NH, UK. Email: frank.kelly@kcl.ac.uk

Isha Khanna, Indian Institute of Technology Delhi, Department of Civil Engineering, Room #220, Block IV, Hauz Khas, New Delhi – 110016, India.

Mukesh Khare, Indian Institute of Technology Delhi, Department of Civil Engineering, Room #220, Block IV, Hauz Khas, New Delhi – 110016, India. E-mail: kharemukesh@yahoo.co.in

Xiaoying Li, College of Environmental Sciences and Engineering, Peking University, 100871, Beijing, China.

Zhongju Li, Center for Atmospheric Particle Studies, Carnegie Mellon University, 5000 Forbes Avenue, Pittsburgh, USA.

F. Lucarelli, Dep. of Physics and Astronomy, Università di Firenze and INFN-Firenze, Sesto Fiorentino, 50019, Italy.

T. Maggos, Institute of Nuclear and Radiological Science & Technology, Energy & Safety, N.C.S.R. Demokritos, 15341 Ag. Paraskevi, Attiki, Greece.

M. Manousakas, Institute of Nuclear and Radiological Science & Technology, Energy & Safety, N.C.S.R. Demokritos, 15341 Ag. Paraskevi, Attiki, Greece.

M. C. Minguillón, Institute of Environmental Assessment and Water Research (IDAEA-CSIC), Barcelona, 08034, Spain.

Paul S. Monks, University of Leicester, Department of Chemistry, LE1 7RH Leicester, United Kingdom.

Dennis Mooibroek, National Institute for Public Health and the Environment (RIVM), Antonie van Leeuwenhoeklaan 9, 3721 MA Bilthoven, the Netherlands.

S. Nava, Dep. of Physics and Astronomy, Università di Firenze and INFN-Firenze, Sesto Fiorentino, 50019, Italy.

T. Nunes, Centre for Environmental & Marine Studies, Dep. of Environment, Univ. of Aveiro, 3810-193 Aveiro, Portugal.

Spyros N. Pandis, Center for Atmospheric Particle Studies, Carnegie Mellon University, 5000 Forbes Avenue, Pittsburgh, USA. Department of Chemical Engineering, University of Patras, Patras, Greece.

Pavlos Panteliadis, Public Health Service of Amsterdam, Department of Air Quality, Nieuwe Achtergracht 100, 1000 CE Amsterdam, the Netherlands.

C. Pio, Centre for Environmental & Marine Studies, Dep. of Environment, Univ. of Aveiro, 3810-193 Aveiro, Portugal.

Laura N. Posner, Center for Atmospheric Particle Studies, Carnegie Mellon University, 5000 Forbes Avenue, Pittsburgh, USA.

Albert A. Presto, Center for Atmospheric Particle Studies, Carnegie Mellon University, 5000 Forbes Avenue, Pittsburgh, USA.

X. Querol, Institute of Environmental Assessment and Water Research (IDAEA-CSIC), Barcelona, 08034, Spain.

C. Reche, Institute of Environmental Assessment and Water Research (IDAEA-CSIC), Barcelona, 08034, Spain.

Allen L. Robinson, Center for Atmospheric Particle Studies, Carnegie Mellon University, 5000 Forbes Avenue, Pittsburgh, USA.

Edward Roekens, Flanders Environment Agency (VMM), Department Air, Environment and Communication, Kronenburgstraat 45, 2000 Antwerp, Belgium.

M. Severi, Dep. of Chemistry, Università di Firenze, Sesto Fiorentino, 50019, Italy.

Manish Shrivastava, PNNL, USA.

Jeroen Staelens, Flanders Environment Agency (VMM), Department Air, Environment and Communication, Kronenburgstraat 45, 2000 Antwerp, Belgium. Email: j.staelens@vmm.be

Ryan C. Sullivan, Center for Atmospheric Particle Studies, Carnegie Mellon University, 5000 Forbes Avenue, Pittsburgh, USA.

Jordy Vercauteren, Flanders Environment Agency (VMM), Department Air, Environment and Communication, Kronenburgstraat 45, 2000 Antwerp, Belgium.

S. Vratolis, Institute of Nuclear and Radiological Science & Technology, Energy & Safety, N.C.S.R. Demokritos, 15341 Ag. Paraskevi, Attiki, Greece.

X. Wang, Desert Research Institute, Nevada System of Higher Education, Reno, NV 89512, USA. Graduate Faculty, University of Nevada, Reno, Nevada 89503, USA.

J. G. Watson, Desert Research Institute, Nevada System of Higher Education, Reno, NV 89512, USA. Graduate Faculty, University of Nevada, Reno, Nevada 89503, USA. The State Key Laboratory of Loess and Quaternary Geology, Institute of Earth Environment, Chinese Academy of Sciences, Xi'an, Shaanxi, 710075, China. E-mail: john.watson@dri.edu

Ernie Weijers, Energy research Centre of the Netherlands, Department of Air Quality, Westerduinweg 3, 1755 LE Petten, the Netherlands.

Daniel M. Westervelt, Center for Atmospheric Particle Studies, Carnegie Mellon University, 5000 Forbes Avenue, Pittsburgh, USA. Present address: Lamont-Doherty Earth Observatory, Columbia University, Palisades USA.

Caiqing Yan, College of Environmental Sciences and Engineering, Peking University, 100871, Beijing, China.

Mei Zheng, College of Environmental Sciences and Engineering, Peking University, 100871, Beijing, China. E-mail: mzheng@pku.edu.cn

Emissions of Primary Particulate Matter

M. GUEVARA

ABSTRACT

Particulate matter (PM) accounts for a complex group of air pollutants with properties and impacts that vary according to its composition and size. The emission rates, size and composition of primary PM emissions are challenging to determine since they depend not only on the sector considered, but also on the fuel properties, technology and other characteristics of the emission process. At the European level, fine carbonaceous particles are generally the dominant components of primary PM emissions, the most important sources of organic and black carbon being residential biomass combustion and diesel vehicle engines, respectively. On the other hand, soil particles generated by wind erosion processes, traffic resuspension, mining and construction operations, and agricultural land management activities are large contributors to the coarse fraction of primary PM emissions. European PM emissions are decreasing as a result of implemented EU legislation mainly focused on road transport and large point sources. Nevertheless, emissions released by residential solid fuel appliances have been increasing due to a lack of regulations, a tendency that is expected to change with the eco-design directive. The decrease of traffic PM exhaust emissions has also increased the importance of traffic non-exhaust emissions, a major source of metals in urban areas.

1 Introduction

Particulate matter (PM) is a generic term used to describe a mixture of solid particles and liquid droplets (aerosols) that vary in size and composition, depending on the location and time[1] (Table 1).

Issues in Environmental Science and Technology No. 42
Airborne Particulate Matter: Sources, Atmospheric Processes and Health
Edited by R.E. Hester, R.M. Harrison and X. Querol
© The Royal Society of Chemistry 2016
Published by the Royal Society of Chemistry, www.rsc.org

Table 1 Sources of origin and main components of coarse $PM_{10-2.5}$, fine $PM_{2.5}$ and ultrafine $PM_{0.1}$ primary particles.[a]

PM fraction	Sources of origin	Main components	Contribution
Coarse particles ($PM_{10-2.5}$)	Agricultural activities	Agricultural soil, OC	+++
	Traffic resuspension	Road dust	+++
	Windblown dust/ construction and mining activities/industrial resuspension	Si, Al, Ti, Fe	+++
	Tyre and brake wear	Cu, Zn	++
	Combustion in energy and manufacturing industries (coal, coke, heavy oil)	EC	++
	Wind-land fires and volcanoes	Volcanoes' ashes, burned OC	+
	Biological sources	Plant debris and fungal spores	+
	Ocean spray	Na, Cl, Mg	+
Fine ($PM_{2.5}$) and	Diesel-fuelled vehicle engines	BC	+++
Ultrafine particles ($PM_{0.1}$)	Biomass combustion	OC, PAHs	+++
	Maritime traffic	BC, OC, SO_4^{-2}	++
	Combustion in energy and manufacturing industries	Pb, Cd, As, Cr, V, Ni, Se, SO_4^{-2}	++
	Processes in non-metallic industries	Si, Al, Fe	+
	Metal processing activities	Pb, Cd, Cr, Zn	+

[a] +++ High contribution; ++ Medium contribution; + Low contribution.

PM is made up of a large number of components, including elemental or black carbon (BC) and organic carbon (OC) compounds, sulfate (SO_4^{-2}), nitrate (NO_3^-), trace metals, crustal material (*i.e.* soil particles) and sea salt.[2] PM also comes in a wide range of sizes and includes PM with diameter less than or equal to 10 μm (PM_{10}), PM with diameter less than or equal to 2.5 μm ($PM_{2.5}$), also denoted as fine particles, PM with diameter less than or equal to 0.1 μm ($PM_{0.1}$), also denoted as ultrafine particles (UFP), and PM with diameter less than or equal to 0.05 μm ($PM_{0.05}$), also denoted as nanoparticles.[3]

In terms of source of origin, PM can be directly emitted from anthropogenic (man-made) or natural sources (*i.e.* primary PM), or formed in the atmosphere from a series of gaseous combustion by-products such as volatile organic compounds (VOCs), ammonia (NH_3), oxides of sulfur (SO_x) and oxides of nitrogen (NO_x) (*i.e.* secondary PM). Primary PM originates predominantly from combustion (*e.g.* vehicle engines) and high-temperature processes (*e.g.* smelting and welding industrial operations),[4,5] as well as from mechanical disruption processes and man- or wind-induced events causing suspension of particles (*e.g.* traffic resuspension of street dust).[6,7]

On the other hand, secondary PM is formed by gas-to-particle conversion in the atmosphere and/or condensation of gaseous compounds on pre-existing aerosol particles, mainly involving NO_x, SO_x, NH_3 and VOCs, which may react with O_3, $^{\bullet}OH$ and other reactive molecules forming secondary inorganic aerosols (SIA) and secondary organic aerosols (SOA).[3]

Unlike other pollutants, such as SO_2 or NH_3, PM describes a complex group of air pollutants with properties and impacts that vary according to their composition and size. For instance, BC is linked to a range of climate impacts (*e.g.* increased temperatures) owing to its capability of directly absorbing light, reducing the albedo of snow and ice and interacting with clouds.[8] On the other hand, several European cohort studies have reported that short- and long-term exposure to $PM_{2.5}$ is associated with a number of health risks, such as lung cancer.[9] The results of these studies have formed the basis for the International Research Agency on Cancer (IARC) to classify PM as carcinogenic to human beings (Group 1).[10]

The main objective of the present chapter is to describe and analyse the main factors that characterize European primary PM emissions, including: main sources of origin, size distribution and chemical composition (speciation), current emission inventories, trends and regulations, and mitigation measures. Despite having a significant contribution to ambient particle concentrations,[11] secondary PM is not considered in the present chapter. The complexity of the atmospheric aerosol processes and other factors (*e.g.* precursor gases) influencing its formation suggest the need for treating it separately in a more extensive study.

Section 2 of this chapter lists and describes the main anthropogenic and natural emission sources that contribute to total PM emissions in Europe. In Sections 3 and 4 a thorough analysis of the size distribution and speciation of PM emissions is conducted, respectively. Section 5 describes the main European PM emission inventories currently used, while Section 6 performs an analysis of PM trends in Europe. Finally, Section 7 focuses on current regulations and mitigation measures that affect PM emissions.

2   Source Categories

Primary PM is derived from a wide range of sources (both natural and anthropogenic), the contribution of each one varying with the location, season and time of day[12] (Figure 1).

This section introduces and describes the sources that currently present the most significant contributions to European PM emissions.

2.1   Residential Combustion

Recently, interest has grown in biomass combustion as an environmentally friendly way of heating homes whilst at the same time reducing climate change impact and contributing to energy security. In this sense, the use of wood and other biomass in residential small combustion installations has

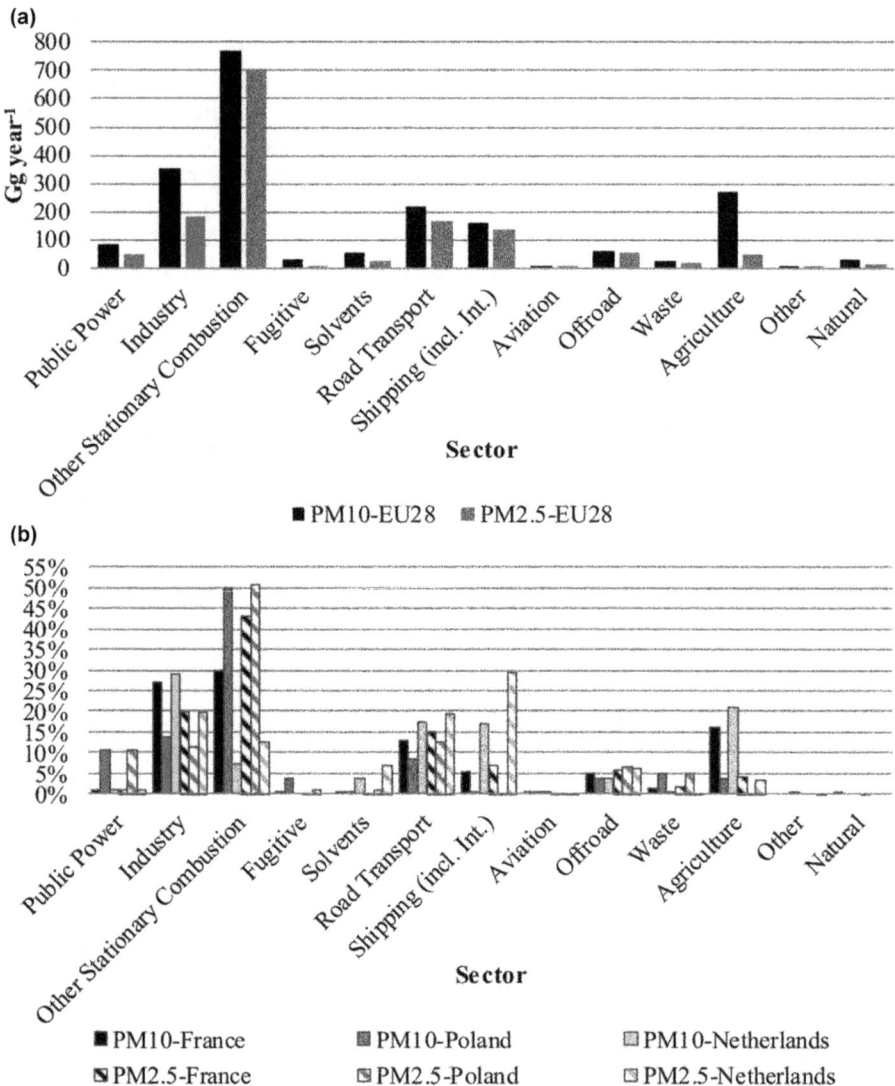

Figure 1 PM$_{10}$ and PM$_{2.5}$ annual emissions (Mg year^{-1}) per pollutant sector in the EU-28 region (a) and contribution (%) of each pollutant sector to total PM$_{10}$ and PM$_{2.5}$ emissions in France, Poland and The Netherlands in the year 2013 (b).[14]

been enhanced by several greenhouse gas strategies and targets for renewable energy. For instance, in 2014 the United Kingdom introduced the Domestic Renewable Heat Incentive (RHI), a financial support programme for renewable heat that offers payments to households for the installation of biomass heating systems to provide central heating and hot water. Moreover, the increase during the economic crisis of other fuel prices typically used in the residential sector (*e.g.* fuel oil) also triggered the use of biomass, which is less expensive, especially in those countries more affected by the recession, such as Greece.[13]

Despite being labelled as a renewable fuel that can contribute to mitigating climate change, the combustion of biomass in small heating combustion installations is currently a major source of primary PM emissions, especially in wintertime. In 2013, emissions released from small residential combustion appliances were reported as the largest source of PM_{10} (38%) and $PM_{2.5}$ (52%) in the EU-28 region.[14] The high contribution of residential wood combustion (RWC) is mainly owing to the fact that: (i) burning conditions are often inefficient (*i.e.* low combustion temperatures, which leads to incomplete combustion) and (ii) household appliances used for the combustion of biomass usually lack emission controls or regulations.

In each European country, the contribution of RWC towards the total PM_{10} and $PM_{2.5}$ emissions varies depending on its energy balance (*i.e.* relative amount of biomass consumed at the residential level with respect to other fuels) and the type of appliances used and fuels burned. The amount of appliances (*e.g.* fireplace, woodstove, pellet stove, pellet boilers) and biofuels (*e.g.* cork oak, pine, olive pit) used for RWC is very large and their use varies from one country to another. A survey conducted in Portugal reported that the main appliances used for RWC in the country are fireplaces (43%) and woodstoves or traditional ovens (24%).[15] On the other hand, in Finland the biggest portion of wood is burnt in masonry heaters and ovens (38%), log boilers (24%) and sauna stoves (15%), with fireplaces representing only 5% of the total combustion installations used.[16] Masonry heaters and ovens have qualities that allow better burning conditions, higher efficiency and subsequently lower PM emissions than conventional fireplaces. While in 2013 the amount of biomass used in the residential sector was 62% higher in Finland than in Portugal,[17] the amount of Finish $PM_{2.5}$ emissions reported for the same year and sector was just 28% larger than in Portugal.[14]

Several studies have shown that the amount of PM emitted varies widely with category of burning appliance and biomass type.[18–20] One of the most recent studies focusing on this topic was developed under the framework of the AIRUSE LIFE project. Emissions from different biofuels and appliances (those most prevalent in southern Europe) were analysed to obtain a detailed characterisation of emission profiles resulting from RWC. Results from this and previous studies shown that open fireplaces are the appliances that present the highest particulate emission factors (EFs; amount of pollutant emitted per activity unit) owing to low temperatures, which contribute to inefficient combustion. Compared to modern eco-labelled woodstove, $PM_{2.5}$ EF from traditional fireplaces can be up to 10–50 times higher.[18,19] Variations in the PM emissions can also be found within the same type of appliance owing to the operation conditions (*e.g.* air-staging settings and the thermal load).[21] On the other hand, the highest EF are observed for biomass fuels other than pellets (*e.g.* olive pit, shell of pine, nuts, almond shell), the variations being related to the different ash contents of the fuels.[22]

Apart from fuel parameters and operation conditions, the measurement protocol applied is another important factor that influences the variation of EF for the same appliance type. A detailed survey and review of the various RWC EF in use in Europe concluded that the most important type of measurement techniques are filter measurements, which measure only solid particles, and dilution tunnel measurements, which measure solid particles and condensates of semi-volatile organics.[23] The EF compiled by the study presented a high variation as a function of the technique used. For instance, EF for conventional wood stoves obtained with filter measurement ranged from 64 to 87 mg MJ^{-1}, while measurements in dilution tunnels showed results in the range from 340 to 544 mg MJ^{-1}. The choice to use a filter measurement- or dilution tunnel measurement-based EF can have a great impact when estimating PM RWC emissions and analysing the contribution of this source to primary organic aerosols (POA).[24]

2.2   Road Transport

Road transport is one of the main sources of PM in urban areas. In 2013, road transport alone was responsible for 12% and 13% of total primary PM_{10} and $PM_{2.5}$ emissions in the EU-28 region, respectively.[14] Nevertheless, at the city level these contributions can go up to 40–50%, thus constituting the main urban emission source.[25–27]

PM emissions from traffic are categorised according to the mode of formation.[28] The combustion of fuels, mainly gasoline and diesel, in internal combustion engines (exhaust emissions) is generally assumed as the principal mechanism by which PM is formed. On the other hand, road transport also involves the interactions between vehicles and the road surface and the use of brakes, which can result in the release of PM emissions. This category of emissions is known as non-exhaust emissions and includes: (i) tyre wear, (ii) brake wear, (iii) road surface wear and (iv) resuspension. The first three sources involve mechanical abrasion, grinding, crushing and corrosion processes, while the last one refers to the resuspension of the dust collected on the road surface owing to vehicle-generated turbulence.

The quantification of exhaust traffic emissions mainly depends on the engine type, engine age (i.e. Euro categories set up by European legislation), after-treatment technology, fuel properties (e.g. fuel sulfur content), level of maintenance of the vehicle, environmental conditions and driving conditions.[5] Exhaust emission rates from vehicles can be estimated from controlled conditions in laboratories (i.e. engine and chassis dynamometer studies) or real-world conditions (i.e. tunnel, remote sensing, on-road and on-board measurements).[29] The use of both approaches indicates that in general PM emission rates from diesel vehicles are significantly higher compared to those from gasoline ones, and that Heavy Duty Diesel Vehicles (HDDVs) are the highest emitters among the different diesel vehicle categories.[30–32] However, increasingly restrictive European diesel emission standards (Section 7) have resulted in a clear reduction of diesel PM emission levels by about

80–90%. In some cases vehicles equipped with diesel particle filters (DPF) (part of Euro 4 and all from Euro 5 and on) even show lower PM levels than gasoline vehicles.[33] The effect of speed on PM exhaust emissions is also increasingly reduced with the introduction of new Euro standards. Generally speaking, low-speed operations lead to higher emission rates. Nevertheless, and as shown by the two reference vehicle emission models in Europe (COPERT; COmputer Programme to calculate Emissions from Road Transport and HBEFA; Handbook of Emission Factors), the shape of emission rates *vs.* speed curves is flatter for the new emission standards than the old ones.[30–32] On the other hand, PM emission rates significantly increase during acceleration as well as with aggressive driving or heavy load conditions.[33]

Non-exhaust emissions are more difficult to quantify than exhaust emissions owing to the strong influence of not only the type of vehicle and traffic conditions, but also the material properties (*e.g.* tyre type, road pavement, grain size) and meteorological factors (*e.g.* temperature, road wetness).[6,34] Non-exhaust particles derived from resuspension processes seem dominant in terms of mass, although this can vary from one country to another owing to the effect of humidity, the use of studded tyres and the contribution from road sanding.[35] Resuspension PM_{10} emission rates estimated by roadside measurements on inner-city urban roads across Europe present a wide variation: UK (14–23 mg VKT^{-1}), Germany (57–109 mg VKT^{-1}), Denmark (46–108 mg VKT^{-1}), Finland (121 mg VKT^{-1}), Sweden (198 mg VKT^{-1}).[36–38] Several campaigns have pointed out a strong correlation between HDDVs and resuspension, the emission rate for this class of vehicle being up to 20 times higher than that for passenger cars.[39] Resuspension emissions in motorways tend to be lower than those in other types of roads (especially urban streets) since higher average vehicle speeds and traffic intensity lead to a lower on-road dust reservoir.[36] The large variation in the resuspension emission rates make them applicable only to the site of study or areas with similar characteristics. During recent years, different numerical approaches have been developed with the intention of reducing the dependency of non-exhaust emission quantification on local measurements.[40] One of the most recent models, the NORTRIP model, is capable of estimating non-exhaust traffic PM emissions based on the impact of surface wetness, the buildup of dust on the road surface, the surface moisture and the effects of applying traction maintenance measures (*e.g.* salting and sanding).[41]

Several studies across Europe have pointed out that the contribution of non-exhaust emissions to PM_{10} can be comparable or even higher than that of exhaust emissions, especially in Scandinavian and Mediterranean countries, owing to studded tyres and road sanding in the former and drier climates in the latter.[42] The contribution of non-exhaust emissions to total PM_{10} in urban areas is expected to grow during the coming years up to approximately 80–90% by 2020.[43] This increase is the result of a combination of several actions that are currently in place to reduce PM emissions from motor exhausts (both at legislative and technological levels) and a lack of abatement measures for non-exhaust emissions.

2.3   Energy and Manufacturing Industries

Emissions from energy (power plants and refineries) and manufacturing industries represent the second-largest source of primary PM_{10} (28%) and $PM_{2.5}$ (21%) in the EU-28 region.[14] However, and with the exception of certain cities close to industrial environments,[44] the contributions of these activities to primary PM in urban areas is less pronounced than that of road transport (around 10%).[25,45]

There are three main mechanisms by which industrial PM is formed. The first involves fuel combustion processes (*e.g.* coal, oil, coke) in conventional boilers, furnaces, gas turbines, reciprocating engines or other combustion devices. PM emissions can also arise from non-combustion processes, such as mechanical treatments of raw materials (non-metallic industries) or casting operations (iron and steel industries). Emissions derived from both combustion and non-combustion processes are usually channelled through ducts (*i.e.* stacks), which makes them more controllable. Finally, industrial PM emissions can also occur during the handling, transport and storage of dusty raw materials (*e.g.* clinker, cement). These emissions, referred to as "diffuse", are more complicated to quantify and control than the channeled ones, owing to the difficulties in determining their flux and location of occurrence inside the industrial areas.

PM emissions from combustion processes are mainly characterized by the type of fuel and technology used. Once released into the atmosphere, meteorological parameters (*e.g.* temperature, pressure) also play a key role in their vertical distribution and subsequent transportation.[46] Fuels with significant ash content (*i.e.* coal, oil and coke) have the highest potential to emit primary channeled PM. In the past, the Best Available Techniques (BAT) in large coal-fired power plants have been translated into abatement technologies, such as electrostatic precipitators (EP) or fabric filters (FF), which have allowed a great reduction of PM emissions. As a result, the current emission rate from a fluidised bed boiler ($\geq$300 MW) working with brown coal can be up to 4 times lower than the emissions derived from a gas oil reciprocating engine.[47] Combustion processes related to public electricity and heat production facilities alone presented a contribution of 4% to total PM_{10} and $PM_{2.5}$ primary emissions (in the EU-28 region in 2013.[14] These contributions largely vary from one country to another owing to the different energy generation systems. In Poland, where the production of electricity and heat mainly comes from coal-fired power plants, contribution from the public power sector goes up to 11% (PM_{10} and $PM_{2.5}$), while in France, where the main source of energy is nuclear, public power only accounts for 1% of total PM_{10} and $PM_{2.5}$ (Figure 1).[17]

Non-combustion channeled emissions are mainly associated with non-metallic mineral and iron and steel industries.[47] In the first case, PM emissions largely originate from pre- and after-treatments (*e.g.* milling processes in the cement industry), while in the iron and steel sector emissions are generated in sintering and pelletizing plants as well as in blast furnaces, used for the production of pig iron, and basic oxygen, open hearth and electric arc furnaces, used for the production of steel. Most of the time these emissions are conducted through stacks and subsequently controlled by efficient filters. Nevertheless, specific industrial processes, such as laser sintering of ceramic tiles, can entail non-controlled particle emissions, which can impact worker exposure.[4]

Cement, steel, ceramic and mining industries, in which bulk materials are usually stored, transported and handled in open air, are the facilities that present more potential for diffuse PM emissions.[48] These types of emissions are not only influenced by the characteristics of the industrial processes but also by meteorological factors (*e.g.* wind speed, precipitation) and material characteristics (*e.g.* raw material moisture content and particle size). The estimation of EF for this source has shown a large variation (from 7 to 400 g PM_{10} t^{-1} product) depending on the type of operation (*e.g.* transport of material on unpaved road) and control measure applied (*e.g.* enclosure and use of bag filters during handling operations).[49]

2.4 Maritime Traffic

Maritime traffic is a key component of the European economy. Compared to other modes of transport (*e.g.* trucks, trains) ship traffic is more fuel-efficient (*i.e.* fuel used per tonne-kilometre). The use of ships increased by more than 20% during the 1995–2012 period (with an average growth rate of 1% per year), and in 2012 the shipping sector was the second most used mode of freight transport in the EU-28 with 1401 billion tonne-kilometres (tkm), right after road transport (1692 billon tkm).[50] According to a recent report by the International Maritime Organization (IMO), it is expected that this form of transport will continue increasing in the future owing to globalization and the increase of global-scale trade.[51]

At the same time, maritime transport is considered an important contributor to primary PM in coastal areas[52] and subsequently to European coastal air quality degradation,[53] especially in the North Sea and the Mediterranean basin. Ship manoeuvring and hoteling operations (ships at berth), which occur in port areas usually located near cities, have been reported to contribute largely to primary PM emissions. In the Greek ports of Piraeus, Santorini, Mykonos, Corfu and Katakolo a total of 94.3 t year^{-1} PM emissions from cruise ships was estimated for the year 2013,[54] 85% of which was related to hoteling operations and 11.5% to the manoeuvring phase. On the other hand, ship hoteling in the port of Rotterdam (the Netherlands) was estimated to generate 248 t year^{-1} of PM_{10} in 2010.[55] At the European level, primary shipping emissions have been reported to influence atmospheric aerosol concentrations in coastal areas within about 1–7% of PM_{10} and 1–20% of $PM_{2.5}$.[53]

PM emissions from maritime traffic are mainly owing to combustion processes that take place in the ship engines. There are three main factors that control the total amount of emissions released by ships: engine load factor, engine type and fuel type.[52,56,57]

The powering of ships is delivered by their main engines (ME) and auxiliary engines (AE), which present different load factors (from 0 to 100%) depending on the operative profile of the ship. During the cruising and manoeuvring operations, the ME usually presents the highest load factor (50–75% during cruising and 10–30% during manoeuvring), while during the hoteling phase the AE is the main source of emissions (*i.e.* to cover the electricity requirements of the ship) and the MEs are switched off or running at low load (*e.g.* to provide power for pumps to load and unload liquid cargo). The dependence of PM emission factors on engine load may vary from ship to ship. Nevertheless, a recent review reported that at loads lower than 25%, emission rates can be significantly increased (up to 6.5 times).[58]

In terms of type of engine, ships can be equipped with marine diesel engines (slow-speed diesel engines, SSD; medium-speed diesel engines, MSD and high-speed diesel engines, HSD), steam turbines or gas turbines. SSD generate a greater fraction of hydrocarbons (HCs) than MSD and HSD, which may result in an increase in total PM emissions because of the formation of HC aerosols.[59]

The fuels used in maritime transport include marine heavy fuel oil (HFO), marine diesel oil (MDO), marine gasoline oil (MGO) and, more recently, liquefied natural gas (LNG). HFO is a residual product of the oil refinery process and its fuel sulfur content (FSC) can be up to 3.5%, while in the case of MDO/MGO the FSC is around 0.03%. The FSC has a crucial influence on PM emissions as primary sulfate is linearly dependent on it.[51] A review of published data from on-board studies on PM emissions from ships indicated ranges of emission rates between 0.18 and 0.48 g kWh^{-1} for MDO and 0.56 to 2.12 g kWh^{-1} for HFO.[57] Nevertheless, the same review study also indicated that the levels of fine and UF particle emissions are not necessarily reduced by this fuel shift.

PM emissions within port areas are not only produced owing to maritime traffic but also during loading/unloading operations of solid cargoes from ships (*e.g.* clinker, tapioca, phosphate). These operations generate dust that is firstly deposited in the dockside and later resuspended by the effects of port-related traffic or wind. The problematic issue relating to this emission source is similar to that found for diffuse industrial emissions (Section 2.3). In the framework of the LIFE project HADA (Automatic Tool for Environmental Diagnostic), average PM_{10} EF up to 140 ± 30 g min^{-1} were estimated for several operations and types of cargoes in Spanish harbours, which states the relevance of this source to port dust emissions.[60]

2.5 Agricultural Activities

During the last few years, agricultural activities, including fertilizer application, manure management and animal housing, have attracted scientific attention since they are the main European sources of NH_3 (their contribution is around 90%) and subsequently important contributors to secondary PM.[61] Nevertheless, agriculture also presents a notable contribution to primary PM_{10}, with a contribution of up to 14% in the EU-28 emission inventory in 2013.[14] The main activities that contribute to the formation of this pollutant include storage, handling and transport of agricultural products, manure management, agricultural waste burning, land preparation and harvesting.

Just as in the case of traffic resuspension (Section 2.2), emissions from land preparation and harvesting are not regulated by the Convention on Long-range Transboundary Air Pollution (CLRTAP) and are not included in the official emission inventories reported by the Member States (MS). Consequently, studies about the contribution of land management activities are currently scarce and a significant knowledge gap exists. Despite the small amount of available dedicated research, some studies have acknowledged that agricultural land operations (*e.g.* ploughing and harrowing) together with harvesting may create dust plumes, and although much of this dust is rather coarse-sized, significant amounts are carried in suspension over long distances, contributing to the background atmospheric dust load.[62] The contribution of these activities have been estimated to be around 5% of total primary PM_{10} in the EU-27 emission inventory,[63] but it can be more significant in countries and regions characterised by large agricultural regions ,such as The Netherlands.[64] Nevertheless, studies in which emission potentials were estimated for different land management activities present a wide range of values, showing EF variations of a factor of up to 50.[65]

2.6 Natural Sources

Natural sources, which involve no direct or indirect human activity, can present high contributions to total PM emissions. The sources included under this category are: (i) windblown (desert and local) dust, (ii) sea salt aerosols, (iii) volcanoes, (iv) primary biological aerosol particles and (v) wildland fires.[74,77,84]

Windblown dust defines the fugitive dust generated and transported by wind action. This occurs mainly in arid and semi-arid regions, although the process can also occur in surfaces covered by vegetation or man-made covers (*e.g.* roads, buildings). The major sources of dust are located in North Africa, the Saharan sources being considered as the most active ones in the world.[66] Recent estimates of the amount of dust exported annually from North Africa (usually referred to as desert dust or African dust) suggest that 400–2200 Tg year^{-1} is a plausible emission range.[67] A large fraction of the African dust is regularly transported from its source northwards across the Mediterranean to southern Europe,[68] and sometimes as far north as the United Kingdom.[69] Desert dust emissions have a significant impact on the background particle levels in the Mediterranean basin as they are responsible for a significant percentage (up to 70%) of PM_{10} daily level exceedances of the EC standard at background monitoring stations, especially in Spain and during the summertime.[70] On the other hand, in Europe there are also potentially erodible surfaces (local dust reservoirs) that can generate fugitive dust emissions. According to a study developed under the NatAir European project, the yearly amount of PM_{10} emitted by wind from the European territory is approximately in the range of 0.66–0.88 Tg year^{-1},[7] of which emissions from agricultural areas constitute an estimated 52%. Spain, together with France and Italy, are the European countries where the most important local dust reservoirs are located.[71] In the case of Spain, loamy soils in central Aragon (NE Spain) are often eroded by strong Cierzo winds, reporting observed dust events with vertical flux ranging from 0.4 to 70 $\mu g\ m^{-2}\ s^{-1}$.[72]

Sea salt aerosols under 10 μm in diameter are the dominant aerosols in marine surface air and can make a significant contribution to land-based PM levels, especially when surface wind speeds are high.[73] At the European level, the annual contribution that sea-salt emission makes to PM_{10} was estimated as 20 Tg year^{-1} for the year 2009,[74] the highest production of sea-salt found on the Atlantic Ocean during winter time, while in the Mediterranean Sea the highest emissions were estimated over the Aegean Sea during summer. A significant part of the variability in the emission estimation comes from the uncertainty associated with the parameterization of the sea-salt emission process, which mainly depends on surface wind speed as well as sea surface temperature, wave height and water salinity, among other parameters.[75]

Primary particles emitted by volcanic eruptions are formed through magma fragmentation and erosion of the vent walls. Volcanic aerosol emissions generally exhibit coarse size distribution and are mainly characterised by their plume height, mass eruption rate and their vertical distribution of mass (with the fine ashes concentred at the top).[76,77] Volcanic ash emission rates vary according to the eruptive style and the intensity and duration of the eruption. During the Eyjafjallajökull eruption, which took place in Iceland between April and May of 2010 and caused significant economic and social disruption in Europe, a total fine ash (diameter 2.8–28 μm) emission of 8 ± 4 Tg was found.[77] European volcanic activity is mainly limited to Iceland and the Mediterranean areas of Greece and Italy.[76,78] Nevertheless, volcanic particles can undergo long-range transport in the atmosphere since they have the potential to produce transient peaks in PM levels not only near the volcano area but also within distances of thousands of km.[79] Besides direct emissions, resuspension and dispersal of freshly deposited volcanic fine ash by wind also have a large impact on PM_{10} levels. Looking again at the example of the 2010 Eyjafjallajökull eruption, PM_{10} concentrations of up to 2000 $\mu g\ m^{-3}$ were registered in areas that were never hit directly by the eruptive plume owing to resuspended ash.[80]

Primary biological aerosol particles consist of material that derives from biological processes.[74,81] These types of aerosols are transferred into the atmosphere without any change in their chemical composition and they mainly include pollen, plant debris, fungal spores, bacteria and viruses. At the European level, the contribution of plant debris and fungal spores to

PM_{10} emissions has been estimated at 0.12 Tg year^{-1}.[74] However, there is currently a rather unsophisticated approach applied for the estimation of these emissions. The EF are not directly obtained but derived from a few sets of measured plant debris atmospheric concentrations that are compared to atmospheric concentrations of other compounds, for which the emission fluxes are known.[81] Moreover, emission rates are considered independent of the surface type or vegetation (excluding barren land and water area) and are temporally scaled (3 month periods) using observed seasonal cycles of plant debris and spore mass.[74] Hence, there is a need to better understand the release mechanisms associated with these primary biological aerosols (*e.g.* meteorological patterns that may influence the emission fluxes) and subsequently refine their emission estimates.

Wind-land fires, also referred as wildfires, are caused by burning forests, shrublands, grasslands and other vegetation (excluding agricultural waste burning). For the region of Europe, the global Fire INventory model (FINN) reported a total of 0.39 Tg year^{-1} and 0.22 Tg year^{-1} PM_{10} and $PM_{2.5}$ annual average emissions (2005–2009),[82] while the Global Fire Assimilation System (GFAS) estimated an average of 0.74 Tg year^{-1} and 0.46 Tg year^{-1} for annual PM_{10} and $PM_{2.5}$ emissions (2003–2011).[83] Wildfire emissions are especially relevant in forested Mediterranean countries, such as Spain, Portugal, France, Greece and Italy, where summers are drier and hotter than other European countries. These five southern MS present a combined average of 400 000 hectares of forestland burn every year and are estimated to be responsible for 0.17 Tg year^{-1} $PM_{2.5}$ average annual emissions (2003–2011).[84] Emissions from open vegetation fires basically depend on the land area burnt, the type of vegetation (*i.e.* fuel material), the amount of organic matter available, the properties and condition of the fuel material (*e.g.* dry, wet, decayed), and the combustion stage (*i.e.* flaming, smouldering).[82–84] Several laboratory studies and field campaigns indicate a wide variation in the emission factors associated with specific fuel types, most of them confirming that PM_{10} mass is dominated by $PM_{2.5}$ mass concentration.[85] As in the case of volcano emissions, the injection height of wildfire emissions is a critical parameter in the transport of the particles released to the atmosphere. Several factors, such as the energy released from the fire, fuel type and local meteorological conditions, determine the plume height, which can reach altitudes of up to 6.1–8.7 km above the surface.[86]

3 Particle Size Distribution

As previously stated, PM comes in a wide range of sizes according to its aerodynamic diameter, including: coarse particles ($PM_{2.5-10}$; diameter between 10 µm and 2.5 µm), fine particles ($PM_{2.5}$; diameter less than or equal to 2.5 µm), ultrafine particles (UFP) ($PM_{0.1}$; diameter less than or equal to 0.1 µm) and nanoparticles ($PM_{0.05}$; diameter less than or equal to 0.05 µm). The size of PM is directly linked to its potential for causing health problems since smaller particles penetrate further down the respiratory tract and even transfer to extrapulmonary organs, including the central nervous system.[87] While most severe adverse health effects have been typically associated with $PM_{2.5}$, other epidemiological studies suggest that PM_1 may have a greater potential for adverse health impacts.[88] The relative amounts of particles present in each size are expressed by mass concentration in the case of $PM_{2.5-10}$ and $PM_{2.5}$ and by number concentration (PNC) in the case of aerosols with diameters between 0.1 and 0.05 μm owing to their negligible mass.

Coarse particles are usually associated with mechanical disruption processes (*e.g.* crushing, grinding, and abrasion of surfaces) and the suspension of dust. Traffic non-exhaust emissions (wear processes and resuspension) are assumed to be dominated by the $PM_{2.5-10}$ fraction,[38] although in some cases particles in the fine particle range have also been found (approximately 15%).[89] Similarly, emissions derived from agricultural activities are mainly associated with the coarse size[62] as well as the diffuse emissions related to handling, transport and storage of dusty raw materials.[60] Regarding sea salt aerosols, approximately 95% of their total mass is in the coarse mode,[90] although in Atlantic zones its contribution to $PM_{2.5}$ can be up to 11%.[11] $PM_{2.5-10}$ tends to have a local impact (1 to 10s of km) and to settle on the ground through dry deposition processes (*e.g.* gravitational sedimentation) in a matter of hours. This is not the case for coarse particles related to wind-blown desert dust, which can be transported over thousands of km (Section 2.6).

Primary $PM_{2.5}$, UFP and nanoparticles are mainly formed from combustion and high-temperature processes, and industrial operations. Road transport, in particular diesel engines, is the major source of primary $PM_{0.1}$ and $PM_{0.05}$ emissions in urban environments,[91,92] with reported contributions of up to 97% of the total PNC.[93] Many of the PM produced by RWC as well as maritime traffic is also below 1 μm.[22,56] On the other hand, primary UFP and nanoparticle emissions from industrial processes such as tile sintering and laser ablation operations are also receiving increasing attention.[4] As opposed to coarse particles, PM in the accumulation mode (diameter between 0.1 and 2.5 μm) tend to have longer lifetimes (days to weeks) as they settle slowly and have low diffusivities, their travel distance being up to thousands of km.[3] On the other hand, UFP usually present lifetimes that go from minutes to hours owing to their tendency towards growth into the accumulation mode.

According to European official reported emissions in the year 2013, 32% of total primary PM_{10} emissions are considered to be in the $PM_{2.5-10}$ fraction and 68% in the $PM_{2.5}$ fraction.[14] In the coarse fraction, agricultural activities are the ones that present the largest contribution (36%), together with mining and construction activities (10%) and non-exhaust traffic emissions (9%). On the other hand, the fine fraction is mainly dominated by residential combustion (58%), energy and manufacturing industries (21%), and road transport (13%). Regarding UFP emissions, they can be indirectly obtained from primary particle number (PN) emission inventories (expressed as numbers of particles instead of mass) since PN emissions are dominated by UFP emissions and the difference between them is relatively small.[92] Recent primary PN emission inventories for Europe indicate a significant contribution from traffic as well as shipping emissions, especially in coastal urban areas such as Oslo, with total shares of 75% and 15%, respectively.[91] France, Spain, Germany, Italy, UK and Poland are reported as the major PN emitters in the EU-28 region, the sum of their traffic emissions representing approximately 72% of the total PN road transport emissions in EU-28.[93]

4 Speciation

Primary PM includes as principal components organic carbon (OC), black carbon (BC), trace metals, crustal material (*i.e.* soil particles), sea-salt and, to a lesser extent, sulfates (SO_4^{-2}). The chemical makeup of PM varies across Europe, depending on the emission source categories that characterize the region of study.

Carbonaceous particles (BC and OC) are generally the dominant components of primary PM emissions. The carbon fraction of PM is identified as having significant impacts on health, climate change, atmospheric photochemistry and aerosol–cloud interactions.[94] Primary BC and OC are mainly formed by incomplete combustion processes and are predominantly present in the fine and UF particle fractions.[67,92,95,96] BC is sometimes also defined using other terms, such as elemental carbon (EC), soot or graphitic carbon. While all the terms are used to denote light-absorbing carbon in atmospheric aerosol particles, each one of them identifies the specific instrument or measurement technique used to measure the quantity of the component. For instance, BC and EC are often used to indicate optical and thermal measurement methods, respectively. In this chapter, the term BC is generically used.

According to the EDGAR-HTAP_v2 global emissions inventory, the total amounts of primary anthropogenic (land-based, excluding ship emissions) BC and OC emissions released in Europe during 2010 were 0.38 Tg year^{-1} and 0.64 Tg year^{-1}, respectively, since the transport and residential sector is responsible for around 90% of the total emissions.[97] On the other hand, European official BC anthropogenic emissions report a total of 0.14 Tg year^{-1} released in the EU-28 region during 2013, the largest contributions being those of the residential (34%), traffic (32%), and national and international shipping (6%) sectors.[14] The contribution of emission sources to particulate carbonaceous emissions may vary according to PM size. The size-resolved emission inventory of carbonaceous particles addressed in the EUCAARI project indicates that the emission of OC in the fine fraction is dominated by the residential combustion of wood and coal, while the largest sources of EC in the UFP fraction are diesel transport and residential combustion.[95] On the other hand, and according to the TNO_MACC_II emission inventory, the most important sources of coarse OC and coarse BC in terms of total mass are not the transport and residential sectors but rather the agriculture (agricultural waste burning) and the power plants and industry sectors, respectively.[96]

Emission ratios of BC and OC to PM are critical to determine since they vary according to a large number of parameters, including fuel type, technology, combustion process efficiency, emission control and size of particles. Diesel engines are estimated as the largest contributors to primary BC emissions,[26,95] while gasoline engines are known to release a higher fraction of OC.[98] Nevertheless, some studies have pointed out that gasoline vehicle UFP emissions are dominated by the BC fraction.[99] In cases where advanced after-treatments are used (e.g. DPFs), a significant reduction of the BC fraction is also observed.[100] In the case of biomass combustion, OC generally dominates the PM emissions in small traditional appliances (e.g. fireplaces), while more efficient combustion installations show larger EC relative fractions owing to higher combustion temperatures and flaming combustion.[18] Fuel properties also influence the OC and BC contents in the particles emitted, with higher hydrocarbon (HC) emission rates contributing to higher OC contents.[20] The OC fraction released from biomass burning provides an important contribution to benzo[a]pyrene (BaP),[101] a polycyclic aromatic hydrocarbon (PAH) reported by the IARC as a probable carcinogen in humans.[102] In the case of maritime traffic, the OC fraction is typically larger than the BC one.[56] PM ship emissions of OC increase with the fuel sulfur content, whereas BC appears to have a significant dependence on the engine load and engine settings but not on the FSC.[56-58] SSD engines are found to generate a greater amount of OC fraction since they typically have a larger fraction of HCs coming through the engine.[58]

Crustal material includes soil particles generated by wind erosion processes (including desert dust contributions), traffic resuspension, handling, transport and storage of materials, and agricultural land management activities, among others. The main components that can be associated with crustal material include aluminium (Al), silicium (Si), calcium (Ca) and iron (Fe), which are usually associated with the coarse fraction ($PM_{2.5-10}$).[103] In Europe, soil particle emissions typically represent 5 to 20% of the ambient PM_{10} mass;[104] the contribution is higher in south-western and south-eastern Europe owing to the warmer and drier climate and the higher influence of African dust intrusions.

Tyre and brake wear emissions (as well as resuspension) are a major source of metals in urban areas.[28,103] In terms of heavy metals, brake wear is the most important source of emissions for copper (Cu), while for tyre wear the most important emission is zinc (Zn).[35] Tyre and brake wear sources represented 77% and 33% of total Cu and Zn emissions in the EU-28 region in 2013, respectively.[14] These two emission sources together also significantly contribute to total lead (Pb) emissions (10%), which in the past was dominated by gasoline exhaust emissions until the phasing out of leaded fuels in Europe.[105] Tyre and brake emissions also include other trace metals, such as arsenic (As), nickel (Ni), antimony (Sb), iron (Fe) and barium (Ba), the composition presenting a large variability across Europe owing to the dependence on the manufacturer and brand.[28,42] Outside urban areas, the metal concentrations of PM may partly originate from different sources such as energy and manufacturing combustion or industrial processes. Heavy metals are most abundant in high-temperature metal processing activities,[2,103] and hence the production of iron and steel is a key contributor to total European emissions of Pb, cadmium (Cd), mercury (Hg), chromium (Cr) and Zn.[14]

Primary sulfate aerosols usually present a residual fraction of total PM emissions (between 2 and 4%).[27,96] Nevertheless, sulfur emitted in the form of particles is important in combustion processes of high-sulfur fuels, which mainly occur in energy and manufacturing industries and in shipping. The fraction of primary sulfate to total PM is mainly influenced by the FSC of the fuel consumed; during coal combustion it has been reported to range from 10 to 45%,[106] while in the case of marine residual oil combustion (*i.e.* HFO) it can account for up to 80% of the weight of the emitted particles.[51,56] Sulfate is becoming less and less significant as a primary PM component owing to the general tendency to substitute coal for natural gas in the public power sector and HFO for MDO and LNG in the maritime sector (Section 7.4).

4.1   PM Speciation Source Profiles

PM speciation source profiles indicate the chemical species that comprise the PM emissions released from a specific source. These speciation profiles, commonly expressed as the mass ratio of each species to the total PM, are used to characterize the different components that are associated with individual pollutant sources. Currently there are different repositories of PM speciation source profiles freely accessible, with the objective of being used for different purposes, such as creating speciated PM emission inventories for photochemical air quality modelling[27] or providing input to the Chemical Mass Balance (CMB) receptor models (RM).[101] One of the best-known repositories is the United States EPA SPECIATE database, which has been publicly available since 1988, and it currently contains around 3000 entries.[107] Source profiles from this American repository are usually used in European emission and air quality modelling exercises owing to the scarcity of official and well-established European databases. With the objective of filling this gap, a new database of PM speciation source profiles in Europe has been recently developed (SPECIEUROPE).[2]

The SPECIEUROPE is a repository developed in the framework of the Forum for Air quality Modelling in Europe (FAIRMODE) that contains the chemical composition of PM emission sources reported in European scientific papers and official reports. Currently, SPECIEUROPE consists of 209 PM speciation profiles, combining measured, composite, calculated (from stoichiometric composition) and derived (results of source apportionment studies) profiles.

5   European PM Emission Inventories

Emission inventories are datasets used to estimate the amount of air pollutants being emitted to the atmosphere, caused by an anthropogenic or natural activity, at a certain geographical location for a given period of time. Emission inventories are generally recognized as key inputs to atmospheric modelling, especially when they are used to design effective control measures to mitigate the adverse impact of air pollution.[45] Statistical methods of source apportionment to indirectly assess pollutant sources from measurements have also proved the requirement of emission inventories as input data.[103] Therefore, during recent years a significant amount of emission datasets have been developed either for scientific or regulatory purposes.[108]

At the European level, the most used inventories to determine PM emissions and its impacts on air quality are: the Atmospheric Chemistry and Climate Model Intercomparison Project (ACCMIP),[109] the Emission Database for Global Atmospheric Research (EDGARv4.2),[110] the EDGAR-HTAP_v2,[97] the EMEP emission inventory,[111] the TNO-MACC-II emission inventory[96] and the Greenhouse gas and Air pollution INteractions and Synergies (GAINS) model.[112] Each of them presents different emission estimation methodologies, spatial resolutions, temporal coverages and applications (Table 2).

All of these European inventories focus their attention on the PM_{10} and $PM_{2.5}$ fractions (carbonaceous components included in some cases) giving no particular attention to UFP. This is a consequence of the fact that current European legislation on primary PM emissions is based on particle mass and not on particle number. However, the increasing evidence of the adverse health impacts related to UFP has also increased the attention on PN emission inventories. Numerous research studies and European projects, such as PARTICULATES or TRANSPHORM, have consolidated emission factor databases for constructing PN emission inventories in Europe. As a consequence, during recent years some of the aforementioned emission inventories have been revised in order to include PN emission estimations.[91,93,95] Nevertheless, the estimation of PN emissions is associated with a higher uncertainty than that linked with PM_{10} and $PM_{2.5}$ emission estimations. For instance, while the uncertainty of PM emissions from traffic sources has been reported to be between 10 and 20%,[113] the overall uncertainty of vehicular PN emissions can be up to 144–169% when aftertreatment device effects are included.[91] This increase of the uncertainty is mainly related to the set-ups of the measurements that define PN vehicle emission factors, including: (i) the consideration or not of volatile PN and (ii) the definition of the lower size cut-off used in the measurement. Considering that traffic is the most intensively studied source category for PN emissions, similar or higher uncertainty values can be assumed for other pollutant sectors.

Despite being well established and showing important improvements, European inventories are still not able to characterize primary PM emissions

Table 2 Summary of European emission inventories currently used in the scientific community for scientific and regulatory purposes.

Name	Source	Emission sources	Pollutants	Temporal resolution/coverage	Spatial resolution/coverage	Use	Approach used
ACCMIP	109	Anthropogenic biomass burning	SO_2, NO_x, CO, NMVOC, CH_4, NH_3, BC and OC	Decadal, 1850–2000	$0.5° \times 0.5°$, Global	Scientific	Combination of other inventories (RETRO, GAINS, EMEP)
EDGARv4.2	110	Anthropogenic biomass burning	SO_2, NO_x, CO, NMVOC, CH_4, NH_3, PM_{10}	Annual, 1970–2008	$0.1° \times 0.1°$, Global	Regulatory scientific	Combination of national AF with specific EF, disaggregated using different spatial proxies
EDGAR-HTAP_v2	97	Anthropogenic biomass burning	SO_2, NO_x, CO, NMVOC, CH_4, NH_3, PM_{10}, $PM_{2.5}$, BC and OC	Annual, 2008–2010	$0.1° \times 0.1°$, Global	Regulatory scientific	Compilation of different regional gridded inventories with EDGAR v4.2 spatial proxies
EMEP	111	Anthropogenic	SO_2, NO_x, CO, NMVOC, CH_4, NH_3, $PM_{2.5-10}$, $PM_{2.5}$	Annual, 1980–2013	$0.1° \times 0.1°$, European	Regulatory scientific	National emission inventories reported by parties and assigned to the EMEP grid
TNO-MACC-II	96	Anthropogenic	SO_2, NO_x, CO, NMVOC, CH_4, NH_3, PM_{10} and $PM_{2.5}$ (broken down into EC, OC, SO_4^{-2}, Na and other minerals)	Annual, 2003–2009	$1/8° \times 1/16°$, European	Scientific	Downscaling of National emission inventories through the use of specific spatial proxies
GAINS	112	Anthropogenic	SO_2, NO_x, CO, NMVOC, CH_4, NH_3, PM_{10}, $PM_{2.5}$, $PM_{0.1}$	Annual, 1990–2030	50 km $\times$ 50 km, European	Regulatory	Combination of national AF with specific EF and grid maps

to a satisfying level of detail. Reported emissions often present data gaps, missing sources and high uncertainties for the applied emission factors,[114] which entail high discrepancies between the different emission datasets.[115] This fact is especially relevant for fugitive emissions related to industrial and agricultural activities, where a problem of data gaps exists. On the other hand, the non-inclusion of key sources, such as traffic resuspension, in the national emission inventories reported under the CLRTAP, which are later used in well-established emission inventories such as EMEP or TNO_MACC-II, also entails large uncertainties.

Comparisons between European emission inventories with local emission inventories developed at the regional or urban scale have also pointed out significant discrepancies, especially in terms of allocation and total amount of PM residential biomass emissions.[116,117] These differences mainly come from the fact that emission inventories at European or national levels usually tend to rely to a larger degree on top-down approaches, while emission inventories developed for local and urban applications rely to a larger degree on bottom-up approaches. Both methods require information concerning activity factors (*e.g.* total amount of fuel consumed) and emission factors per activity (*e.g.* amount of pollutant emitted per activity unit). Nevertheless, emissions compiled through a bottom-up approach are based on specific information for each sector, such as housing units or number of vehicles per road link for domestic heating and traffic emissions, respectively. Alternately, top-down approaches are based on the disaggregation of variables defined at the regional or national level (*e.g.* fuel sold or consumed) in smaller areas based on auxiliary spatial surrogates that represent the activity (*e.g.* population density for wood burning emissions), thus achieving a higher spatial detail. Bottom-up approaches allow high spatial and temporal detail, although they also require a greater amount of data and thus more resources.

6   Long-term Trends in Europe

According to the European Union emission inventory report under the UNECE Convention on Long-range Transboundary Air Pollution (CLRTAP),[14] total emissions of primary PM_{10} have reduced by 19% across the EU-28 region between 2000 and 2013, driven by an 18% reduction in emissions of $PM_{2.5}$. On the other hand, BC emissions have seen a reduction of 35% over the same period.

The difference between the BC trend and that of PM_{10} and $PM_{2.5}$ is owing to significantly decreasing emissions in BC from road and off-road transport since 2000 (a decrease of 50% and 60%, respectively). The majority of the reduction in PM_{10} and $PM_{2.5}$ emissions has taken place in the anthropogenic sectors of public power (20% and 13%), industry (35% and 27%) and road transport (25% and 34%) owing to: (i) a fuel-switching from coal to natural gas for electricity generation, (ii) an introduction of after-treatment technologies in new vehicles, such as DPF (driven by the legislative Euro standards) and (iii) an implementation of BATs in the industrial sector, including improvements in the performance of pollution abatement equipment. Moreover, a marked decrease has been recorded since 2008 in the hardest-hit countries by the economic crisis (*i.e.* Italy, Portugal and Spain). The influence of the economic recession on PM has also been reported by different European interpretation trend studies.[118]

During 2013 the contribution of residential combustion to total PM emissions significantly increased in comparison to 2000 (by 13% in PM_{10} and 17% in both $PM_{2.5}$ and BC) and it is the only sector in which emissions have risen between 2000 and 2013 (by 11% for PM_{10}, 13% for $PM_{2.5}$ and 12% for BC) (Figure 2). This evolution can partly be explained by the increase of

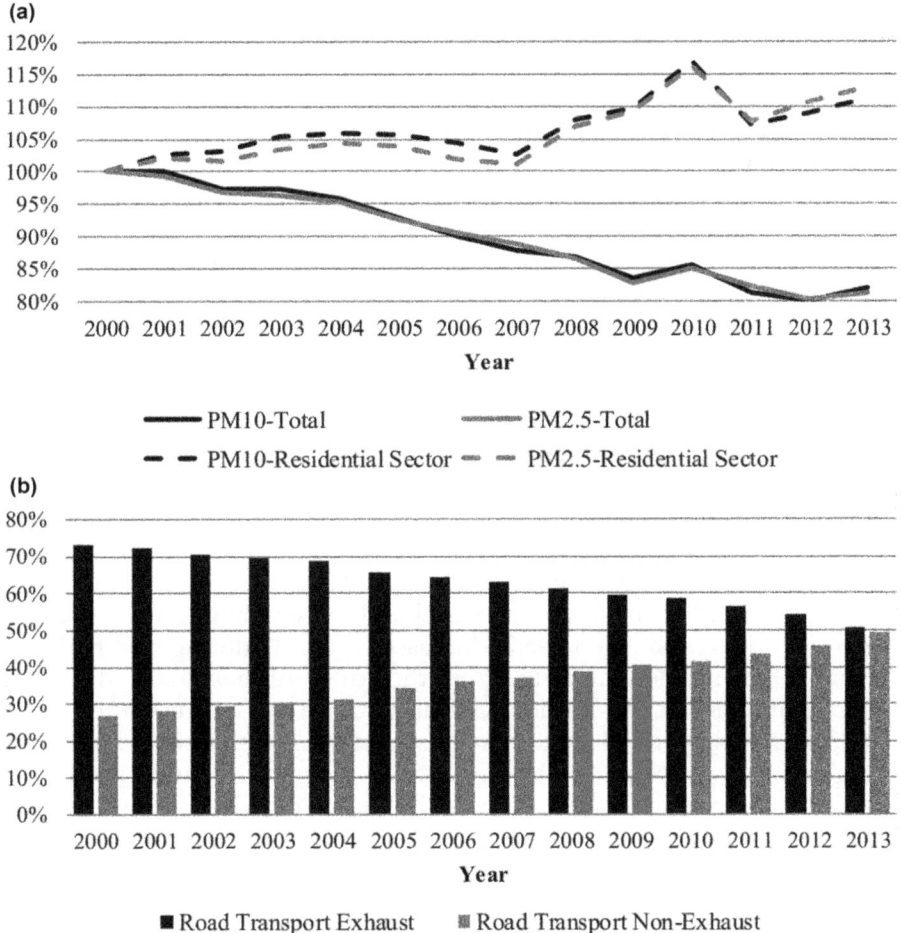

Figure 2 Trend of PM_{10} and $PM_{2.5}$ emissions (%) from total pollutant sources and residential combustion sources (a) and trend of the contribution (%) of exhaust and non-exhaust emissions in total road transport PM_{10} emission in the EU-28 region (b).[14]

biomass burning at the residential level, especially in Eastern European countries (see section below). Moreover, during this period, European efforts have been especially focused on exhaust diesel PM emissions control strategies, which has caused an important decrease in the exhaust road transport's contribution to total BC (from 43% to 33%). The decrease of traffic PM exhaust emissions has increased the importance of non-exhaust emissions in the coarse fraction; the relative contribution from non-exhaust emissions in road transport has increased from 27% to 49% for PM_{10} from 2000 to 2013 (Figure 2). For these results it is important to note that traffic resuspension emissions are not included in the official MS inventories and subsequently road transport non-exhaust contributions may be underestimated.

Looking at the variations between countries, the largest reductions of PM_{10} for 2000–2013 have been reported by Cyprus (66%), France (35%) and Hungary (35%). In the case of $PM_{2.5}$, Cyprus and France are also among those countries that have shown the greatest reduction (73% and 42%, respectively) together with The Netherlands (50%). BC emission reductions are led by The Netherlands (61%), the United Kingdom (53%) and France (44%). The large reduction of BC observed in the case of The Netherlands is partly explained by the increase of the market share of hybrid-electric vehicles during this period (9.7%), the largest of the whole EU-28 region.[119] However, despite all the reductions observed, France was the MS with the largest contribution to total PM_{10}, $PM_{2.5}$ and BC emissions in 2013 (14.4, 14.2 and 26.7%, respectively), which was also the case back in 2000. These results can be explained by the patterns observed in the road transport and residential combustion sectors: (i) diesel dominated the French passenger car market with a 66% of the total share[119]—the EU-28 share was 53% in 2013—and (ii) biomass was, after natural gas, the fuel most used in French households (28%) and represented 19% of the total biomass consumed in the EU-28 region.[17]

In contrast to the aforementioned countries, PM emissions have increased in some countries since 2000; the greatest increases have been reported by far in Romania for all PM emissions (PM_{10}, 21%; $PM_{2.5}$, 31% and BC, 43%). The explanation for this lies in the fact that the use of biofuels and waste in the residential combustion sector increased by 27% during this period.[17] With this increase, Romania rose from the 7th to 4th position in the list of top contributors to $PM_{2.5}$ and BC emissions in EU-28 (just after Italy). Similarly, Bulgaria has also increased its levels of PM emissions during the same period owing to a rise in residential biomass consumption (53%). Although the emissions have dropped by 11%, Poland is another Eastern European country in which residential emissions have a significant impact (50% of total $PM_{2.5}$). In this case, the main fuel consumed is coal, which represented 68% of the total coal consumed at the residential level in the EU-28 region in 2013.[17]

Emissions of primary PM_{10}, $PM_{2.5}$ and BC are expected to decrease across the EU-28 region in the coming years as vehicle technologies are further improved and stationary fuel combustion emissions are controlled through abatement techniques or the use of low-sulfur fuels (natural gas). However, it is possible that Eastern European countries experience a different evolution if the tendency towards biomass and coal combustion in the residential sector continues to increase. In this sense, regulations and mitigation measures targeting solid fuel appliance and their energy efficiency are necessary (Section 7). Moreover, the impact of international energy markets and human-induced global warming can entail changes in PM trends. For instance, and according to the Spanish Electric System statistics, the use of coal in the Spanish energy production system has increased from 14.6% to 21.3% between 2013 and 2015. This increase has been mainly triggered by two elements: (i) the extreme dry and warm conditions that affected the country during this year, which entailed a reduction of hydroelectric power (from 14.2% to 11.2%) and (ii) the reduction of the use of coal in the United States (owing to an increase of fracking technology to extract natural gas) and the subsequent reduction of coal prices.

7 Regulations and Mitigation Measures

There are currently three air pollutant emission reporting obligations in the EU: the 1979 CLRTAP, the National Emission Ceilings Directive 2001/81/EC (NECD) and the EU Greenhouse Gas Monitoring Mechanism. All of them require MS and other Parties to annually report their national atmospheric emission inventories. Additionally, the NECD sets pollutant-specific emission ceilings for each country, which were to be met by 2010 as well as in future years. Nevertheless, NECD emission targets are currently focused on emissions of the secondary PM precursors (*i.e.* NO_x, SO_2, NMVOCs and NH_3) and do not include primary PM. The proposal to amend the NECD is currently still under preparation and should set emission ceilings to be respected by 2020 and 2030 for the four already regulated substances and for PM2.5 and CH_4 as well.[120]

Although EU emission targets for primary PM are currently non-existent, there are several directives and regulations that affect the emissions of this pollutant. The following sections describe some of the most important European regulations and mitigation measures that are currently being applied with the objective of reducing PM emissions.

7.1 Eco-design Directive

Whereas emissions from large point sources have decreased during the last decade as a result of several EU legislations (Large Combustion Plant Directive 2001/80/EC and Industrial Emissions Directive 2010/75/EU), emissions released by small residential solid fuel appliances have been increasing, especially in Eastern European countries, owing to a lack of regulation at EU level (Section 6). In order to tackle this problem, the European Commission launched the Directive 2009/125/EC (Eco-design Directive) to establish a framework for the setting of eco-design requirements for energy-related products. The Eco-design directive provides a framework and tools for improving the environmental performance of residential combustion appliances.

According to studies conducted to estimate the impact of implementing the European Eco-design Directive, the introduction of the Eco-design standards without requirements for improved energy efficiencies would imply a decrease of total European $PM_{2.5}$ emissions from residential small combustion sources of 38% in 2020, 70% in 2030 and 83% in 2050.[121] Similarly, BC emissions would be reduced by 25% in 2020 and by 75% in 2050. If Eco-design standards include all the requirements for improved energy efficiencies, reduction of emissions would be larger. However, the study concludes that this scenario of application is very unfeasible since it assumes a very fast replacement of the existing inefficient devices by new equipment and an unlimited availability of pellets.

In parallel to the development of the Eco-design Directive, some European countries have also issued national emission standards for small residential heating installations, which are already in effect. This is the case in countries such as Germany and Sweden, for example, which have introduced voluntary eco-labelling of stoves with standards for efficiency and emissions.[94]

7.2 Diesel Vehicle Emission Standards and Abatement Technologies

The process of dieselization that European vehicle fleets have suffered during the last decades,[119] combined with the significant contribution of diesel-fuelled vehicles to BC emissions[95] and the classification of diesel engine exhaust as carcinogenic to humans (Group 1),[102] has resulted in more stringent emission regulations and new abatement technologies.

In particular, the adoption of DPF has contributed greatly to reducing emissions, with mass basis PM reductions typically cited as between 85 and 95% when compared to direct engine emissions.[5] The application of DPF has been indirectly incentivized by the increasing restrictions of the European emission standard PM limits for diesel Euro 5/V (applied from 2009) and 6/VI (applied from 2014) vehicle categories. In the case of passenger diesel cars, the standard has been reduced from 140 mg km^{-1} (Euro 1) to 5 mg km^{-1} (Euro 5 and 6), while in the case of HDDV the limits have changed from 0.612 g kWh^{-1} (Euro I) to 0.02 g kWh^{-1} (Euro V) and 0.01 g kWh^{-1} (Euro VI). These emission standards, however, have proved to be largely ineffective in reducing diesel NO_x emissions; several real-world measurement campaigns showed that real-world diesel NO_x emissions exceed certification limits.[33]

Regarding HDDV, the utilisation of natural gas in diesel engines has also become a promising and highly attractive alternative in the transportation sector. Emissions in natural gas and diesel dual-fuelled engines have been largely investigated, the results indicating that there is a significant decrease in PM emission under dual fuel mode.[122]

7.3   Urban Access Regulations

Besides the application of emission standards and abatement technologies, another way to reduce traffic PM emissions in urban areas is through the reduction of traffic congestion and traffic jams. With this objective in mind, in recent years several European cities have set up regulations for vehicles entering their area, including: (i) urban road tolls (entrance to an area is subject to payment), (ii) low emission zones (LEZs; entrance to an area is regulated by vehicle emission standards) and (iii) Key Access Regulation Schemes (Key-ARS; entrance to an area is regulated by other elements, such as special permits or time of the day). While urban road tolls and Key-ARS have a clear impact on the amount of vehicles that circulates throughout the area of application, LEZs also contribute to renewing and modernizing the vehicle fleet of the city (*i.e.* citizens buy newer vehicle or retrofit the ones they have in order to reduce their emissions and comply with the restrictions of the LEZs). Several studies have shown how the application of urban access regulations has had a positive impact on urban traffic emissions, with PM_{10} primary emission reductions increasing to almost 20%.[123] Nevertheless, there is currently a debate about the effects of these measures on PM urban concentration levels. A review on the efficacy of LEZs to improve urban air quality in five EU countries (Denmark, Germany, Netherlands, Italy and UK) concludes that there have been mixed results. While German cities have reported reductions in annual mean PM_{10} concentrations up to 7%, no clear effects have been observed in other urban areas.[124] These results may be related to different causes, including the types of vehicles restricted in the LEZs (German LEZs restrict HDDVs and passengers cars, while in the rest of the cities the restriction is limited to HDDVs). The reduction of PM_{10} concentration by only a few percent may be also related to the fact that LEZs do not impact on non-exhaust traffic PM_{10} emissions, which present a significant contribution to total urban road transport primary emissions (Figure 2).

7.4   MARPOL Convention

IMO is an international agency of the United Nations (UN) that addresses maritime traffic air pollution through the International Convention for the Prevention of Pollution from Ships (MARPOL) and its Annexes, which limits ship emissions for SO_x, NO_x, O_3-depleting substances and VOCs.

Although the regulation does not include explicit PM emission limits, there are caps on FSC that directly control SO_x and indirectly primary sulfate compounds. MARPOL designates emission control areas (ECAs; the Baltic Sea area and the North Sea area in Europe) where ships have to use on-board fuel oil with an FSC of no more than 0.10% (m/m) from 1 January 2015, while outside these ECAs (in which the Mediterranean Sea area is included) the current limit is 3.50%, falling to 0.50% after 1 January 2020. In line with this regulation, the EC published the Marine Fuel Oils Directive 2005/33/EC with the obligation on ships to use 0.1% FSC while at berth (*i.e.* securely moored or anchored in the port while loading, unloading or hoteling) from 1 January 2010.

Both the ECA regulation and the EC Directive do not prohibit the use of HFO; its use is allowed as long as the fuel meets the applicable FSC limit or if approved emission abatement technologies are used in the ship to limit SO_x emissions. Nevertheless, one of the main goals of these regulations is to force maritime transport to the use of cleaner maritime fuels. Traditionally, LNG has been basically used in LNG carriers. However, the use of this fuel is currently becoming more interesting for liner service ships (*i.e.* ferries) and other merchant ships. In this sense, LNG is increasingly being adopted as a marine fuel with solutions that include a dual-fuel engine that can run on either LNG or HFO, with LNG being expected to take 11% of the market share in 2030.[51]

References

1. http://www3.epa.gov/pm/index.html (last accessed January 2016).
2. D. Pernigotti, C. A. Belis and L. Span, *Atmos. Pollut. Res.*, 2015, **7**(2), 307.
3. M. Shepherd, R. McClellan, S. Pandis, G. Hidy, T. Pace, F. Fehsenfeld, D. Hastie, P. Solomon, J. Chow, C. Blanchard, J. Brook, E. Vega, J. Watson, C. Seigneur, M. Moran, I. Tombach, K. McDonald, J. Vickery and P. McMurry, *Particulate Matter Science for Policy Makers: A NARSTO Assessment*, ed. P. McMurry, M. Shepherd and J. Vickery, Cambridge University Press, Cambridge, 2004, p. 103.
4. A. S. Fonseca, M. Viana, X. Querol, N. Moreno, I. de Francisco, C. Estepa and G. F. de la Fuente, *J. Aerosol Sci.*, 2015, **88**, 48.
5. M. Matti Maricq, *Aerosol Sci.*, 2007, **38**, 1079.
6. F. Amato, F. R. Cassee, H. A. C. Denier van der Gon, R. Gehrig, M. Gustafsson, W. Hafner, R. M. Harrison, M. Jozwicka, F. J. Kelly, T. Moreno, A. S. H. Prevot, M. Schaap, J. Sunyer and X. Querol, *J. Hazard. Mater.*, 2014, **275**, 31.
7. M. Korcz, J. Fuda and C. Kli, *Atmos. Environ.*, 2009, **43**, 1410.
8. B. Booth and N. Bellouin, *Nature*, 2015, **519**, 167.
9. O. Raaschou-Nielsen, Z. J. Andersen, R. Beelen, E. Samoli, M. Stafoggia, G. Weinmayr, B. Hoffmann, P. Fischer, M. J. Nieuwenhuijsen, B. Brunekreef, W. W. Xun, K. Katsouyanni, K. Dimakopoulou, J. Sommar, B. Forsberg, L. Modig, A. Oudin, B. Oftedal, P. E. Schwarze, P. Nafstad, U. de Faire, N. L. Pedersen, C. G. Ostenson, L. Fratiglioni, J. Penell, M. Korek, G. Pershagen, K. T. Eriksen, M. Sorensen, A. Tjonneland, T. Ellermann, M. Eeftens, P. H. Peeters, K. Meliefste, M. Wang, B. Bueno-de-Mesquita, T. J. Key, K. de Hoogh, H. Concin, G. Nagel, A. Vilier, S. Grioni, V. Krogh, M. Y. Tsai, F. Ricceri, C. Sacerdote, C. Galassi, E. Migliore, A. Ranzi, G. Cesaroni, C. Badaloni,

F. Forastiere, I. Tamayo, P. Amiano, M. Dorronsoro, A. Trichopoulou, C. Bamia, P. Vineis and G. Hoek, *Lancet Oncol.*, 2013, **14**, 813.

10. D. Loomis, Y. Grosse, B. Lauby-Secretan, F. El Ghissassi, V. Bouvard, L. Benbrahim-Tallaa, N. Guha, R. Baan, H. Mattock and K. Straif, *Lancet Oncol.*, 2013, **14**, 1262.

11. J.-P. Putaud, R. Van Dingenen, A. Alastuey, H. Bauer, W. Birmili, J. Cyrys, H. Flentje, S. Fuzzi, R. Gehrig, H. C. Hansson, R. M. Harrison, H. Herrmann, R. Hitzenberger, C. Hüglin, A. M. Jones, A. Kasper-Giebl, G. Kiss, A. Kousa, T. A. J. Kuhlbusch, G. Löschau, W. Maenhaut, A. Molnar, T. Moreno, J. Pekkanen, C. Perrino, M. Pitz, H. Puxbaum, X. Querol, S. Rodriguez, I. Salma, J. Schwarz, J. Smolik, J. Schneider, G. Spindler, H. ten Brink, J. Tursic, M. Viana, A. Wiedensohler and F. Raes, *Atmos. Environ.*, 2010, **44**, 1308.

12. F. Karagulian, C. A. Belis, C. Francisco, C. Dora, A. M. Prüss-Ustün, S. Bonjour, H. Adair-Rohani and M. Amann, *Atmos. Environ.*, 2015, **120**, 475.

13. A. Saffari, N. Daher, C. Samara, D. Voutsa, A. Kouras, E. Manoli, O. Karagkiozidou, C. Vlachokostas, N. Moussiopoulos, M. M. Shafer, J. J. Schauer and C. Sioutas, *Environ. Sci. Technol.*, 2013, **47**, 13313.

14. EEA, 2015, *European Union emission inventory report 1990–2013 under the UNECE Convention on Long-range Transboundary Air Pollution (LRTAP)*, EEA Technical report No 8/2015, 2015, p. 130.

15. C. Gonçalves, C. Alves and C. Pio, *Atmos. Environ.*, 2012, **50**, 297.

16. V.-V. Paunu, Master's Thesis, Aalto University, 2012.

17. http://www.iea.org/ (last accessed January 2016).

18. A. P. Fernandes, C. A. Alves, C. Gonçalves, L. Tarelho, C. Pio, C. Schimdl and H. Bauer, *J. Environ. Monit.*, 2011, **13**, 3196.

19. C. Gonçalves, C. Alves, M. Evtyugina, F. Mirante, C. Pio, A. Caseiro, C. Schmidl, H. Bauer and F. Carvalho, *Atmos. Environ.*, 2010, **44**, 4474.

20. M. Obaidullah, S. Bram, V. K. Verma and J. De Ruyck, *Int. J. Renewable Energy Res.*, 2012, **2**, 147.

21. H. Lamberg, O. Sippula, J. Tissari and J. Jokiniemi, *Energy Fuels*, 2011, **25**, 4952.

22. E. D. Vicente, M. A. Duarte, A. I. Calvo, T. F. Nunes, L. Tarelho and C. A. Alves, *Fuel Process. Technol.*, 2015, **131**, 182.

23. T. Nussbaumer, C. Czasch, N. Klippel, L. Johansson and C. Tullin, *Particulate emissions from biomass combustion in IEA countries. Survey on Measurements and Emission Factors*, IEA Bioenergy Task 32, Zurich, 2008.

24. H. A. C. Denier van der Gon, R. Bergström, C. Fountoukis, C. Johansson, S. N. Pandis, D. Simpson and A. Visschedijk, *Atmos. Chem. Phys. Discuss.*, 2014, **14**, 31719–31765.

25. R. Borge, J. Lumbreras, J. Pérez, D. de la Paz, M. Vedrenne, J. M. de Andrés and M. E. Rodríguez, *Sci. Total Environ.*, 2014, **466–467**, 809.

26. R. N. Colvile, E. J. Hutchinson, J. S. Mindell and R. F. Warren, *Atmos. Environ.*, 2001, **35**, 1537.
27. M. Guevara, F. Martínez, G. Arévalo, S. Gassó and J. M. Baldasano, *Atmos. Environ.*, 2013, **81**, 209.
28. P. Pant and R. M. Harrison, *Atmos. Environ.*, 2013, **77**, 78.
29. V. Franco, M. Kousoulidou, M. Muntean, L. Ntziachristos, S. Hausberger and P. Dilara, *Atmos. Environ.*, 2013, **70**, 84.
30. D. Gkatzoflias, C. Kouridis, L. Ntziachristos and Z. Samaras. *COPERT IV. Computer programme to calculate emissions from road transport User manual* (version 5.0); 2007.
31. S. Hausberger, M. Rexeis, M. Zallinger and R. Luz. *Emission factors from the model PHEM for the HBEFA version 3.* Institute for internal combustion engines and thermodynamics Report Nr. I-20a/2009 Haus-Em 33a/08/679. Graz University of Technology, 2009.
32. M. Kousoulidou, G. Fontaras, L. Ntziachristos, P. Bonnel, Z. Samaras and P. Dilara, *Atmos. Environ.*, 2013, **64**, 329.
33. G. Fontaras, V. Franco, P. Dilara, G. Martini and U. Manfredi, *Sci. Total Environ.*, 2014, **468–469**, 1034.
34. A. J. Thorpe and R. M. Harrison, *Sci. Total Environ.*, 2008, **400**, 270.
35. R. M. Harrison, A. M. Jones, J. Gietl, J. Yin and D. C. Green, *Environ. Sci. Technol.*, 2012, **46**, 6523.
36. R. Gehrig, M. Hill, B. Buchmann, D. Imhof, E. Weingarter and U. Baltensprenger, *Int. J. Environ. Pollut.*, 2004, **22**, 312.
37. M. Ketzel, G. Omstedt, C. Johansson, I. Düring, M. Pohjola, D. Oettl, J. Gidhagen, P. Wåhlin, A. Lohmeyer, M. Haakana and R. Berkowicz, *Atmos. Environ.*, 2007, **41**, 9370.
38. A. J. Thorpe, R. M. Harrison, P. G. Boulter and I. S. McCrae, *Atmos. Environ.*, 2007, **41**(37), 8007.
39. F. Amato, A. Karanasiou, T. Moreno, A. Alastuey, J. A. G. Orza, J. Lumbreras, R. Borge, E. Boldo, C. Linares and X. Querol, *Atmos. Environ.*, 2012, **61**, 580.
40. M. T. Pay, P. Jimenez-Guerrero and J. M. Baldasano, *Atmos. Environ.*, 2011, **45**, 802.
41. B. R. Denby, I. Sundvor, C. Johansson, L. Pirjola, M. Ketzel, M. Norman, K. Kupiainen, M. Gustafsson, G. Blomqvist and G. Omstedt, *Atmos. Environ.*, 2013, **77**, 283.
42. H. A. C. Denier van der Gon, M. E. Gerlofs-Nijland, R. Gehrig, M. Gustafsson, N. Janssen, R. M. Harrison, J. Hulskotte, C. Johansson, M. Jozwicka, M. Keuken, K. Krijgsheld, L. Ntziachristos, M. Riediker and F. R. Cassee, *J. Air Waste Manage.*, 2013, **63**(2), 136.
43. M. Mathissen, V. Scheer, U. Kirchner, R. Vogt and T. Benter, *Atmos. Environ.*, 2012, **59**, 232.
44. J. M. Baldasano, A. Soret, M. Guevara, F. Martínez and S. Gassó, *Sci. Total Environ.*, 2014, **473–474**, 576.
45. A. Soret, M. Guevara and J. M. Baldasano, *Atmos. Environ.*, 2014 **99**, 51.

46. M. Guevara, A. Soret, G. Arévalo, F. Martínez and J. M. Baldasano, *Atmos. Environ.*, 2014, **99**, 618.
47. http://www.eea.europa.eu/publications/emep-eea-guidebook-2013 (last accessed January 2016).
48. M. Santacatalina, C. Reche, M. C. Minguillón, A. Escrig, V. Sanfelix, A. Carratalá, J. F. Nicolás, E. Yubero, J. Crespo, A. Alastuey, E. Monfort, J. V. Miró and X. Querol, *Sci. Total Environ.*, 2010, **408**(21), 4999.
49. E. Monfort, V. Sanfélix, I. Celades, S. Gomar, F. Martín, B. Aceña and A. Pascual, *Atmos. Environ.*, 2011, **45**, 7286.
50. http://ec.europa.eu/transport/facts-fundings/statistics/pocketbook-2014_en.htm (last accessed January 2016).
51. T. W. P. Smith, J. P. Jalkanen, B. A. Anderson, J. J. Corbett, J. Faber, S. Hanayama, E. O'Keeffe, S. Parker, L. Johansson, L. Aldous, C. Raucci, M. Traut, S. Ettinger, D. Nelissen, D. S. Lee, S. Ng, A. Agrawal, J. J. Winebrake, M. Hoen, S. Chesworth and A. Pandey, *Third IMO GHG Study 2014*, Tech. rep., International Maritime Organization (IMO), London, UK, 2014, p. 327.
52. F. Di Natale and C. Carotenuto, *Transp. Res. Part D: Transp. Environ.*, 2015, **40**, 166–191.
53. M. Viana, P. Hammingh, A. Colette, X. Querol, B. Degraeuwe, I. de Vlieger and J. van Aardenne, *Atmos. Environ.*, 2014, **90**, 96.
54. A. Maragkogianni and S. Papaefthimiou, *Transp. Res. Part D: Transp. Environ.*, 2015, **36**, 10.
55. J. H. J. Hulskotte and H. A. C. Denier van der Gon, *Atmos. Environ.*, 2010, **44**(9), 9.
56. L. Mueller, G. Jakobi, H. Czech, B. Stengel, J. Orasche, J. M. Arteaga-Salas, E. Karg, M. Elsasser, O. Sippula, T. Streibel, J. G. Slowik, A. S. H. Prevot, J. Jokiniemi, R. Rabe, H. Harndorf, B. Michalke, J. Schnelle-Kreis and R. Zimmermann, *Appl. Energy*, 2015, **155**, 204.
57. H. Winnes and E. Fridell, *J. Air Waste Manage. Assoc.*, 2009, **59**(12), 1391.
58. D. A. Lack and J. J. Corbett, *Atmos. Chem. Phys.*, 2012, **12**, 3985.
59. D. A. Cooper and T. Gustafsson, *Methodology for Calculating Emissions from Ships. 1. Update of Emission Factors*, Swedish Meteorological Hydrological Institute, Norrköping, Sweden, 2004, p. 47.
60. F. Martín, M. Pujadas, B. Artiñano, F. Gómez-Moreno, I. Palomino, N. Moreno, A. Alastuey, X. Querol, J. Basora, J. A. Luaces and A. Guerra, *Atmos. Environ.*, 2007, **41**, 6356.
61. S. López-Aparicio, C. Guerreiro, M. Viana, C. Reche and X. Querol, *Contribution of agriculture to Air Quality problems in cities and in rural areas in Europe*. ETC/ACM Technical Paper 2013/10, 2013, p. 29.
62. K. D. Moore, M. D. Wojcik, R. S. Martin, C. C. Marchant, D. S. Jones, W. J. Bradford, G. E. Bingham, R. L. Pfeiffer, J. H. Prueger and J. L. Hatfield, *J. Appl. Remote Sens.*, 2015, **9**, 1.
63. O. Oenema, G. Velthof, Z. Klimont and W. Winiwarter, *Emissions From Agriculture and Their Control Potentials*, ed. M. Amann, IIASA, 2012, p. 27.

64. C. Hendriks, R. Kranenburg, J. Kuenen, R. van Gijlswijk, R. W. Kruit, A. Segers, H. A. C. Denier van der Gon and M. Schaap, *Atmos. Environ.*, 2013, **69**, 289.

65. H. A. C. Denier van der Gon, M. Jozwicka, E. Hendriks, M. Gondwe and M. Schaap, *Mineral Dust as a component of Particulate Matter*. Netherlands Research Program on Particulate Matter Report 500099003, 2010, p. 28.

66. P. Ginoux, J. M. Prospero, O. Torres and M. Chin, *Environ. Model. Software*, 2004, **19**, 113.

67. N. Huneeus, M. Schulz, Y. Balkanski, J. Griesfeller, J. Prospero, S. Kinne, S. Bauer, O. Boucher, M. Chin, F. Dentener, T. Diehl, R. Easter, D. Fillmore, S. Ghan, P. Ginoux, A. Grini, L. Horowitz, D. Koch, M. C. Krol, W. Landing, X. Liu, N. Mahowald, R. Miller, J.-J. Morcrette, G. Myhre, J. Penner, J. Perlwitz, P. Stier, T. Takemura and C. S. Zender, *Atmos. Chem. Phys.*, 2011, **11**, 7781.

68. S. Basart, M. T. Pay, O. Jorba, C. Perez, P. Jimenez-Guerrero, M. Schulz and J. M. Baldasano, *Atmos. Chem. Phys.*, 2012, **12**, 3363.

69. M. Dall'Osto, R. M. Harrison, E. J. Highwood, C. O'Dowd, D. Ceburnis, X. Querol and E. P. Achterberg, *Atmos. Environ.*, 2010, **44**, 3135.

70. X. Querol, J. Pey, M. Pandolfi, A. Alastuey, M. Cusack, N. Pérez, T. Moreno, M. Viana, N. Mihalopoulos, G. Kallos and S. Kleanthous, *Atmos. Environ.*, 2009, **43**, 4266.

71. P. Borrelli, P. Panagos and L. Montanarella, *Sustainability*, 2015, 7(7), 8823.

72. L. Gomes, J. L. Arrué, M. V. López, G. Sterk, D. Richard, R. Gracia, M. Sabre, A. Gaudichet and J. P. Frangi, *Catena*, 2003, **52**, 235.

73. M. Sofiev, J. Soares, M. Prank, G. de Leeuw and J. Kukkonen, *J. Geophys. Res.*, 2011, **116**, D21302.

74. N. Liora, K. Markakis, A. Poupkou, T. M. Giannaros and D. Melas, *Atmos. Environ.*, 2015, **122**, 493.

75. M. Spada, O. Jorba, C. Perez García-Pando, Z. Janjic and J. M. Baldasano, *Atmos. Chem. Phys.*, 2013, **13**, 11735.

76. F. Sigmundsson, S. Hreinsdóttir, A. Hooper, T. Árnadóttir, R. Pedersen, M. J. Roberts, N. Óskarsson, A. Auriac, J. Decriem, P. Einarsson, H. Geirsson, M. Hensch, B. G. Ófeigsson, E. Sturkell, H. Sveinbjörnsson and K. L. Feigl, *Nature*, 2010, **468**, 426.

77. A. Stohl, A. J. Prata, S. Eckhardt, L. Clarisse, A. Durant, S. Henne, N. I. Kristiansen, A. Minikin, U. Schumann, P. Seibert, K. Stebel, H. E. Thomas, T. Thorsteinsson, K. Tørseth and B. Weinzierl, *Atmos. Chem. Phys.*, 2011, **11**, 4333.

78. W. D'Alessandro, A. Aiuppaa, S. Bellomo, L. Brusca, S. Calabrese, K. Kyriakopoulos, M. Liotta and M. Longo, *J. Geochem. Explor.*, 2013, **131**, 1.

79. S. Sandrini, L. Giulianelli, S. Decesari, S. Fuzzi, P. Cristofanelli, A. Marinoni, P. Bonasoni, M. Chiari, G. Calzolai, S. Canepari, C. Perrino and M. C. Facchini, *Atmos. Chem. Phys.*, 2014, **14**, 1075.

80. T. Thorsteinsson, T. Jóhannsson, A. Stohl and N. I. Kristiansen, *J. Geophys. Res.*, 2012, **117**, B00C0.
81. W. Winiwarter, H. Bauer, A. Caseiro and H. Puxbaum, *Atmos. Environ.*, 2009, **43**, 1403.
82. C. Wiedinmyer, S. K. Akagi, R. J. Yokelson, L. K. Emmons, J. A. Al-Saadi, J. J. Orlando and A. J. Soja, *Geosci. Model Dev.*, 2011, **4**, 625.
83. J. W. Kaiser, A. Heil, M. O. Andreae, A. Benedetti, N. Chubarova, L. Jones, J.-J. Morcrette, M. Razinger, M. G. Schultz, M. Suttie and G. R. van der Werf, *Biogeosciences*, 2012, **9**, 527.
84. S. Turquety, L. Menut, B. Bessagnet, A. Anav, N. Viovy, F. Maignan and M. Wooster, *Geosci. Model Dev.*, 2014, **7**, 587.
85. M. O. Andreae and P. Merlet, *Global Biogeochem. Cycles*, 2001, **15**, 995.
86. A. Veira, S. Kloster, S. Wilkenskjeld and S. Remy, *Atmos. Chem. Phys.*, 2015, **15**, 7155.
87. M. R. Heal, P. Kumar and R. M. Harrison, *Chem. Soc. Rev.*, 2012, **41**(19), 6606.
88. S. Weichenthal, *Environ. Res.*, 2012, **115**, 26.
89. A. Dahl, A. Gharibi, E. Swietlicki, A. Gudmundsson, M. Bohgard, A. Ljungman, G. Blomqvist and M. Gustafsson, *Atmos. Environ.*, 2006, **40**, 1314.
90. J. H. Seinfeld and S. N. Pandis, *Atmospheric Chemistry and Physics – From Air Pollution to Climate Change*, 2nd edn, John Wiley & Sons, New York, USA, 2006, p. 1232.
91. J. Kukkonen, M. Karl, M. P. Keuken, H. A. C. Denier van der Gon, B. R. Denby, V. Singh, J. Douros, A. Manders, Z. Samaras, N. Moussiopoulos, S. Jonkers, M. Aarnio, A. Karppinen, L. Kangas, S. Lützenkirchen, T. Petäjä, I. Vouitsis and R. S. Sokhi, *Geosci. Model Dev. Discuss.*, 2015, **8**, 5873.
92. P. Kumar, L. Morawska, W. Birmili, P. Paasonen, M. Hug, M. Kulmala, R. M. Harrison, L. Norford and R. Britter, *Environ. Int.*, 2014, **66**, 1.
93. P. Paasonen, A. Asmi, T. Petäjä, M. K. Kajos, M. Äijälä, H. Junninen, T. Holst, J. P. D. Abbatt, A. Arneth, W. Birmili, H. A. C. Denier van der Gon, A. Hamed, A. Hoffer, L. Laakso, A. Laaksonen, W. R. Leaitch, C. Plass-Dülmer, S. C. Pryor, P. Räisänen, E. Swietlicki, A. Wiedensohler, D. R. Worsnop, V.-M. Kerminen and M. Kulmala, *Nat. Geosci.*, 2013, **6**, 438.
94. T. C. Bond, S. J. Doherty, D. W. Fahey, P. M. Forster, T. Berntsen, B. J. DeAngelo, M. G. Flanner, S. Ghan, B. Kärcher, D. Koch, S. Kinne, Y. Kondo, P. K. Quinn, M. C. Sarofim, M. G. Schultz, M. Schulz, C. Venkataraman, H. Zhang, S. Zhang, N. Bellouin, S. K. Guttikunda, P. K. Hopke, M. Z. Jacobson, J. W. Kaiser, Z. Klimont, U. Lohmann, J. P. Schwarz, D. Shindell, T. Storelvmo, S. G. Warren and C. S. Zender, *J. Geophys. Res.: Atmos.*, 2013, **118**, 5380.
95. M. Kulmala, A. Asmi, H. K. Lappalainen, U. Baltensperger, J.-L. Brenguier, M. C. Facchini, H.-C. Hansson, Ø. Hov, C. D. O'Dowd, U. Poschl, A. Wiedensohler, R. Boers, O. Boucher, G. de Leeuw,

H. A. C. Denier van der Gon, J. Feichter, R. Krejci, P. Laj, H. Lihavainen, U. Lohmann, G. McFiggans, T. Mentel, C. Pilinis, I. Riipinen, M. Schulz, A. Stohl, E. Swietlicki, E. Vignati, C. Alves, M. Amann, S. Arabas, P. Artaxo, H. Baars, D. C. S. Beddows, R. Bergstrom, J. P. Beukes, M. Bilde, J. F. Burkhart, F. Canonaco, S. L. Clegg, H. Coe, S. Crumeyrolle, B. D'Anna, S. Decesari, S. Gilardoni, M. Fischer, A. M. Fjaeraa, C. Fountoukis, C. George, L. Gomes, P. Halloran, T. Hamburger, R. M. Harrison, H. Herrmann, T. Hoffmann, C. Hoose, M. Hu, A. Hyvarinen, U. Horrak, Y. Iinuma, T. Iversen, M. Josipovic, M. Kanakidou, A. Kiendler-Scharr, A. Kirkevag, G. Kiss, Z. Klimont, P. Kolmonen, M. Komppula, J.-E. Kristjansson, L. Laakso, A. Laaksonen, L. Labonnote, V. A. Lanz, K. E. J. Lehtinen, L. V. Rizzo, R. Makkonen, H. E. Manninen, G. McMeeking, J. Merikanto, A. Minikin, S. Mirme, W. T. Morgan, E. Nemitz, D. O'Donnell, T. S. Panwar, H. Pawlowska, A. Petzold, J. J. Pienaar, C. Pio, C. Plass-Duelmer, A. S. H. Prévôt, S. Pryor, C. L. Reddington, G. Roberts, D. Rosenfeld, J. Schwarz, Ø. Seland, K. Sellegri, X. J. Shen, M. Shiraiwa, H. Siebert, B. Sierau, D. Simpson, J. Y. Sun, D. Topping, P. Tunved, P. Vaattovaara, V. Vakkari, J. P. Veefkind, A. Visschedijk, H. Vuollekoski, R. Vuolo, B. Wehner, J. Wildt, S. Woodward, D. R. Worsnop, G.-J. van Zadelhoff, A. A. Zardini, K. Zhang, P. G. van Zyl, V.-M. Kerminen, K. S. Carslaw and S. N. Pandis, *Atmos. Chem. Phys.*, 2011, **11**, 13061.

96. J. Kuenen, A. Visschedijk, M. Jozwicka and H. A. C. Denier van der Gon, *Atmos. Chem. Phys. Discuss.*, 2014, **14**, 5837.

97. G. Janssens-Maenhout, M. Crippa, D. Guizzardi, F. Dentener, M. Muntean, G. Pouliot, T. Keating, Q. Zhang, J. Kurokawa, R. Wankmüller, H. A. C. Denier van der Gon, J. J. P. Kuenen, Z. Klimont, G. Frost, S. Darras, B. Koffi and M. Li, *Atmos. Chem. Phys.*, 2015, **15**, 11411.

98. L. Ntziachristos, Z. Ning, M. D. Geller, R. J. Sheesley, J. J. Schauer and C. Sioutas, *Atmos. Environ.*, 2007, **41**, 5864.

99. D. A. Sodeman, S. M. Toner and K. A. Prather, *Environ. Sci. Technol.*, 2005, **39**(12), 4569.

100. L. Ntziachristos, G. Mellios, G. Fontaras, S. Gkeivanidis, M. Kousoulidou, D. Gkatzoflias, T. Papageorgiou and C. Kouridis, *Updates of the Guidebook Chapter on Road Transport*, LAT Report No 0706, 2007, p. 63.

101. C. A. Belis, F. Karagulian, B. R. Larsen and P. K. Hopke, *Atmos. Environ.*, 2013, **69**, 94.

102. http://monographs.iarc.fr/ENG/Classification (last accessed January 2016).

103. M. Viana, T. A. J. Kuhlbusch, X. Querol, A. Alastuey, R. M. Harrison, P. K. Hopk, W. Winiwarter, M. Vallius, S. Szidat, A. S. H. Prévôt, C. Hueglin, H. Bloemen, P. Wåhlin, R. Vecchi, A. I. Miranda, A. Kasper-Giebl, W. Maenhaut and R. Hitzenberge, *Aerosol Sci.*, 2008, **39**, 827.

104. X. Querol, A. Alastuey, C. R. Ruiz, B. Artiñano, H. C. Hansson, R. M. Harrison, E. Buringh, H. M. ten Brink, M. Lutz, P. Bruckmann, P. Straehl and J. Schneider, *Atmos. Environ.*, 2004, **38**, 6547.
105. H. A. C. Denier van der Gon and W. Appelman, *Sci. Total Environ.*, 2009, **407**, 5367.
106. Z. Guo, Z. Li, J. Farquhar, A. J. Kaufman, N. Wu, C. Li, R. R. Dickerson and P. Wang, *J. Geophys. Res.*, 2010, **115**, D00K07.
107. H. Simon, L. Beck, P. V. Bhave, F. Divita, Y. Hsu, D. Luecken, J. D. Mobley, G. A. Pouliot, A. Reff, G. Sarwar and M. Strum, *Atmos. Pollut. Res.*, 2010, **1**, 196.
108. S. Darras, C. Granier, C. Liousse, A. Mieville, D. Boulanger, G. Brissebrat, M. Paulin and H. Richard, *Geophys. Res. Abstr.*, 2014, **16**, EGU2014-6705-1.
109. J.-F. Lamarque, T. C. Bond, V. Eyring, C. Granier, A. Heil, Z. Klimont, D. Lee, C. Liousse, A. Mieville, B. Owen, M. G. Schultz, D. Shindell, S. J. Smith, E. Stehfest, J. van Aardenne, O. R. Cooper, M. Kainuma, N. Mahowald, J. R. McConnell, V. Naik, K. Riahi and D. P. van Vuuren, *Atmos. Chem. Phys.*, 2010, **10**, 7017.
110. G. Janssens-Maenhout, A. M. R. Petrescu, M. Muntean and V. Blujdea, *Greenhouse Gas Meas. Manage.*, 2011, **1**(2), 132.
111. http://www.ceip.at/ (last accessed January 2016).
112. M. Amann, I. Bertok, J. Borken-Kleefeld, J. Cofala, C. Heyes, L. Höglund-Isaksson, Z. Klimont, B. Nguyen, M. Posch, P. Rafaj, R. Sandler, W. Schöpp, F. Wagner and W. Winiwarter, *Environ. Model. Software*, 2011, **26**, 1489.
113. I. Kioutsioukis, C. Kouridis, D. Gkatzoflias, P. Dilara and L. Ntziachristos, *Procedia Soc. Behav. Sci.*, 2010, **2**, 7690.
114. W. Winiwarter, T. A. J. Kuhlbusch, M. Viana and R. Hitzenberger, *Atmos. Environ.*, 2009, **43**, 3819.
115. C. Granier, B. Bessagnet, T. C. Bond, A. D'Angiola, H. A. C. Denier van der Gon, G. J. Frost, A. Heil, J. W. Kaiser, S. Kinne, Z. Klimont, S. Kloster, J.-F. Lamarque, C. Liousse, T. Masui, F. Meleux, A. Mieville, T. Ohara, J.-C. Raut, K. Riahi, M. G. Schultz, S. J. Smith, A. Thompson, J. van Aardenne, G. R. van der Werf and D. P. van Vuuren, *Clim. Change*, 2011, **109**, 163.
116. M. Guevara, M. T. Pay, F. Martínez, A. Soret, H. A. C. Denier van der Gon and J. M. Baldasano, *Atmos. Environ.*, 2014, **98**, 134.
117. R. M. A. Timmermans, H. A. C. Denier van der Gon, J. J. P. Kuenen, A. J. Segers, C. Honoré, O. Perrussel, P. J. H. Builtjes and M. Schaap, *Urban Climate*, 2013, **6**, 44.
118. X. Querol, A. Alastuey, M. Pandolfi, C. Reche, N. Pérez, M. C. Minguillón, T. Moreno, M. Viana, M. Escudero, A. Orio, M. Pallarés and F. Reina, *Sci. Total Environ.*, 2014, **490**, 957.
119. http://eupocketbook.theicct.org/ (last accessed January 2016).
120. http://ec.europa.eu/environment/air/pollutants/rev_nec_dir.htm (last accessed January 2016).

121. J. Cofala and Z. Klimont, *Emissions From Households and Other Small Combustion Sources and Their Reduction Potential*, ed. M. Amman, IIASA, 2012, p. 42.

122. L. Wei and P. Geng, *Fuel Process. Technol.*, 2016, **142**, 264.

123. http://urbanaccessregulations.eu/ (last accessed January 2016).

124. C. Holman, R. Harrison and X. Querol, *Atmos. Environ.*, 2015, **111**, 161.

Where Did This Particle Come From? Sources of Particle Number and Mass for Human Exposure Estimates

NEIL M. DONAHUE,* LAURA N. POSNER, DANIEL M. WESTERVELT,
ZHONGJU LI, MANISH SHRIVASTAVA, ALBERT A. PRESTO,
RYAN C. SULLIVAN, PETER J. ADAMS, SPYROS N. PANDIS AND
ALLEN L. ROBINSON

ABSTRACT

Atmospheric chemistry dominates the size distribution and composition of most fine particles inhaled by humans. However, it is important to distinguish between secondary particles—new particles formed in the atmosphere—and secondary mass—molecules formed in the atmosphere that condense to existing particles. In many ways the life stories of particles viewed from the perspectives of particle number concentrations and particle mass concentrations are distinct. Individual particle cores can often be said to have an individual source, while the mass on individual particles comes from myriad sources. This, plus the aforementioned chemical processing in the atmosphere, must be kept in mind when considering the health effects of fine particles.

1 Introduction

Fine particles ($PM_{2.5}$, or particles smaller than 2.5 micrometers in diameter) induce negative human health effects, including mortality.[1,2] The current

*Corresponding author.

Issues in Environmental Science and Technology No. 42
Airborne Particulate Matter: Sources, Atmospheric Processes and Health
Edited by R.E. Hester, R.M. Harrison and X. Querol
© The Royal Society of Chemistry 2016
Published by the Royal Society of Chemistry, www.rsc.org

best estimates are that inhalation of $PM_{2.5}$ causes between 2 and 3 million deaths per year world wide[3,4] and roughly 75 million Disability Adjusted Life Years (DALYs).[3] The health effects appear worst in vulnerable populations, including children and the elderly, and measurable effects, including mortality, extend to very low ambient concentrations.[1,5]

The evidence for these effects is strong, yet we lack mechanistic under-standing connecting particle exposures to health effects. Toxicologists and epidemiologists agree that that both particle size and composition influence the $PM_{2.5}$ dose response.[6,7] However, the strongest epidemiological associ-ation is with the $PM_{2.5}$ mass concentration, and so the health standard remains a simple mass standard. The common statutory size cutoff at 2.5 micrometers is also a complication; from the perspective of aerosol science alone, a cutoff at 1 micrometer (PM_1) would be more sensible, as the sources and composition of the submicron fraction differ in many ways from the supermicron fraction, even for $PM_{2.5}$, where particles between 1 and 2.5 micrometers are most often the small tail of the coarse particle mode.[8,9] Here we shall for the most part address only PM_1.

There has also been a growing interest in connecting PM health effects directly with sources. However, the relationship between individual sources and PM concentrations faces two significant complications. First, "the source" of an individual particle can often be meaningfully defined when considering PM number concentrations, though particle coagulation causes some ambiguity; however, "the" source of the mass of most particles is not meaningful. This is because a large majority of the mass of most individual particles comes from myriad sources and arrives *via* condensation. To the extent that health effects are driven by PM mass, "a diesel particle" may not exist very often. Second, much of the PM_1 mass is secondary, having been chemically transformed in the atmosphere; and most submicrometer particles are secondary (from new-particle formation). However, most of the mass on submicrometer particles inhaled by people away from urban centers probably resides on primary particles (though the mass is largely secondary). The situation in urban centers and very near roadways is uncertain.

Chemical processing is especially extensive for organics,[10] which may turn out to be potent drivers of negative health effects; even in urban centers, oxidized organic mass typically exceeds reduced organic mass by 2 : 1 during the summer.[11,12] Because of this, individual particles being inhaled, even in urban areas, often contain much more mass from condensation, repre-senting the complete mix of local and regional sources, than within any core that can be associated with a single source. An additional consequence of having vapor condensation dominate fine-particle mass is that atmospheric chemistry can and does play a major role. Thus, while many specific sources certainly provide the raw material for fine-particle mass, chemistry in the at-mosphere may be the dominant source of the specific compounds responsible for human health effects.

Here we shall explore what we now know (and do not know) about the different stories of particle-number and particle-mass life cycles, and discuss how recent developments have transformed our understanding and may influence future health-effects research. For most of the case studies we shall draw on our own work in the eastern United States, but the conclusions are generalizable after considering regional differences in emissions sources; the central point here is not which sources predominate but rather the ubiquity of oxidation chemistry in the process. Recognition of the crucial role played by atmospheric chemistry is essential for designing both control strategies and health-effects studies. Nowhere is this more significant than with organic PM, where the aged organic compounds may have dramatically different effects than the more reduced precursors (including primary organic aerosols) directly emitted from sources.

2   Background

Particulate matter is a uniquely complicated pollutant. This is because "particles" comprise a huge ensemble of objects defined solely by being a condensed phase suspended in the atmosphere. Even if we restrict ourselves to PM_1, particles exist over a size range from approximately 1 nm diameter to 1 μm diameter—from just larger than clusters containing a handful of molecules to objects containing more than 10^9 molecules—with an equally enormous range of composition and physical properties, such as hygroscopicity and viscosity. Particles and constituents can range from soluble to insoluble, solid to liquid,[13,14] and spherical to highly structured.[15]

A reasonable picture for most fine particles follows a core–shell model (Figure 1), where a source-specific core is covered with a coating dominating the mass that has a similar composition on most particles. Many cores are directly emitted from primary sources. High-temperature sources dominate urban settings (combustion, cooking, *etc.*), but sea-spray and biological particles can also contribute to sub-micrometer primary particle emissions. Different measurement techniques are more or less sensitive to the shell (which is typically non-refractory, relatively uniform, and dominant in mass) and the core (which may be refractory, heterogeneous, and of modest mass). We must emphasize that the "core" and "shell" may sometimes mix freely and that the shell may include more than one separate phase (for example, an inorganic and an organic phase). In addition, in some cases the "core" is an almost abstract concept, as with new-particle formation, where it simply consists of the very lowest vapor pressure condensates and instead refers to the source (*in situ* formation) more than to the content of the core.

Because condensation is central to both new-particle formation and condensational growth, it deserves special attention. Particle growth rates are directly proportional to the gas-phase concentration of condensable vapors (the "condensation driving force").[16,17] However, condensation to particles, defined by the condensation sink (the suspended surface area concentration multiplied by a deposition velocity) is also generally the major sink of those vapors, and so the balance between production of condensable vapors and the condensation sink will determine the steady-state

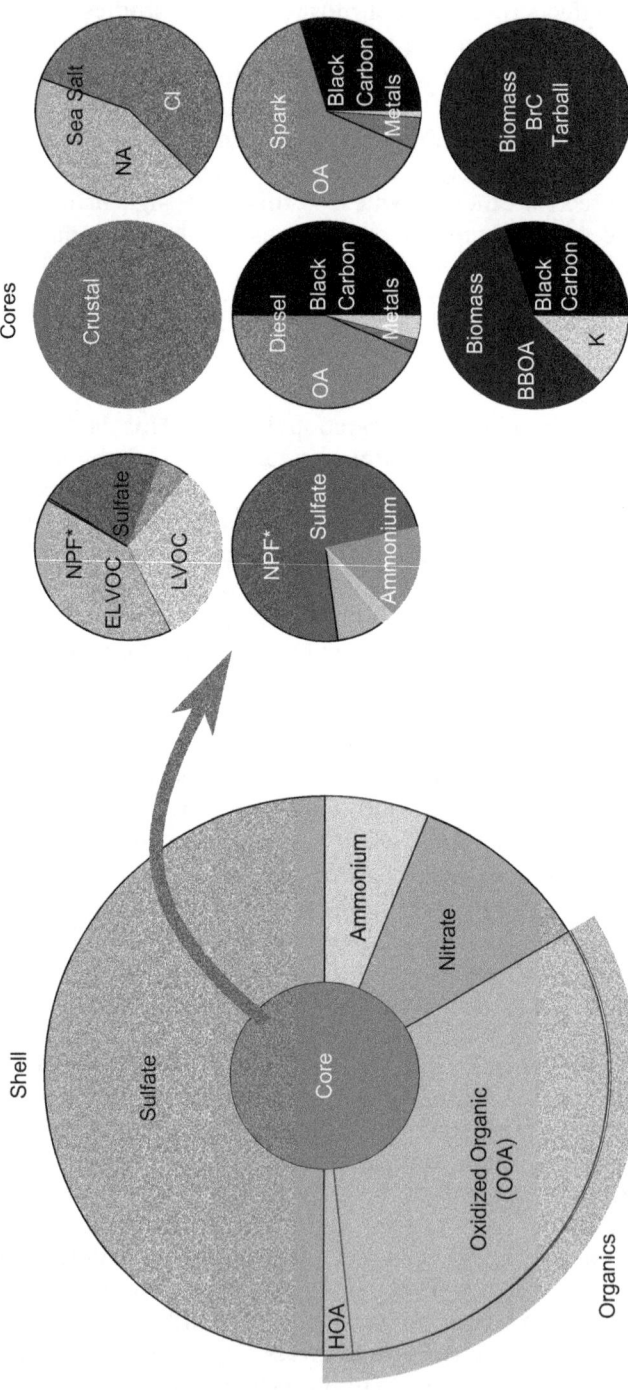

Figure 1 A core–shell model of fine-particle composition. Particle mass is dominated by a relatively uniform shell of constituents that arrive *via* condensation (indicated on the left in the outer ring). These include sulfate, nitrate, ammonia, and organics (some reduced hydrocarbon like organic aerosol, HOA; most oxidized organic aerosol, OOA), and they are measured with various bulk mass measurements including filter assays and online techniques such as Aerosol Mass Spectrometry (AMS). Particle number is dominated by a diverse array of cores with numerous origins and unique signatures (indicated on the right). Cores have source-specific origins and composition, as indicated. Some cores arise from new-particle formation (NPF) and are distinguishable from the shells only by composition; these particles are completely secondary. Cores can often be measured with single-particle mass-spectrometer (SP-MS) techniques.

concentration of those vapors.[18] That in turn will govern both the new-particle formation rate and the particle growth rate.

While it is easy to imagine that new-particle formation might be higher in relatively pristine areas because of the low condensation sink and conversely that growth rates might be higher in polluted areas because of the high production rates, this is not necessarily the case. New-particle formation rates and growth rates are positively correlated and both are buffered by the positive correlation between the production rate of condensable vapors and the total particle mass (and thus surface area) loading. Figure 2 shows the cumulative probability distributions of the condensation sink for a remote site (Hyytiälä, Finland) and a polluted site (San Pietro Capofiume, Italy), based on observations originally presented by Westervelt *et al.*[19] Aside from a clear bump above the 90th percentile associated with extremely hazy events in the Po Valley, the two distributions are simply shifted by roughly 1 order of magnitude, with a median value near 6 h^{-1} in Hyytiälä and near 60 h^{-1} in San Pietro Capofiume. The condensation sink is the uptake frequency of vapors with particles (here for sulfuric acid, assuming that each collision results in uptake); vapors thus typically collide with particles once every 10 minutes in Hyytiälä and once per minute in San Pietro Capofiume. This also defines the equilibration timescale for the aerosol suspension,[20] if mass transfer within particles is sufficiently rapid.[21] In addition, in each location,

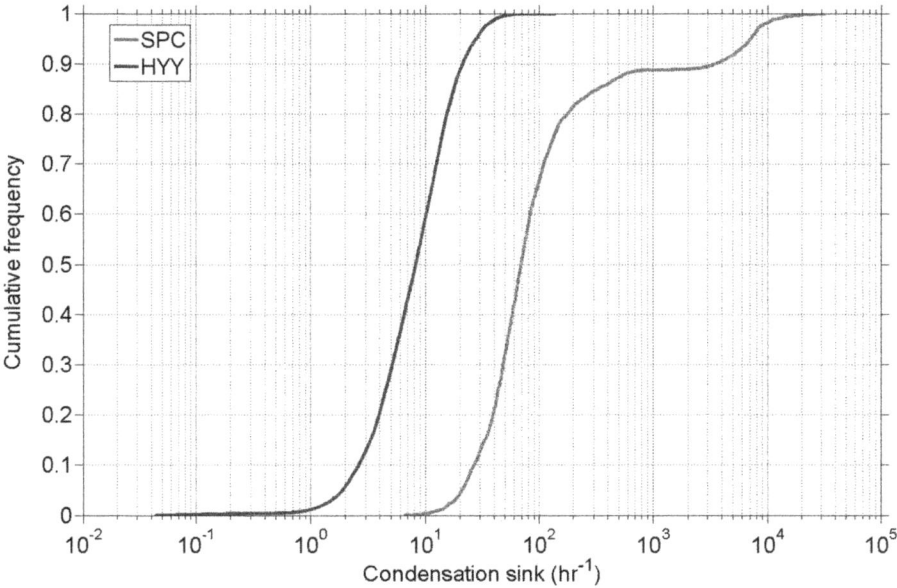

Figure 2 Cumulative probability distributions of the condensation sink (in h^{-1}) from a remote location (the boreal-forest site of Hyytiälä, Finland, "HYY", left) and a frequently polluted industrial region (the Po Valley regional background site, San Pietro Capofiume, Italy, "SPC", right).

the enormous majority of cases fall within about one order of magnitude (the slope of the probability distributions is $\simeq 1.0$).

3 Particle Mass Concentrations

Particle mass is frequently decoupled from particle number because most mass on every particle in the atmosphere arrives either *via* condensation of low vapor-pressure compounds formed by gas-phase chemical reactions or *via* condensation of more volatile compounds followed by a condensed-phase chemical reaction that fixes the material in the condensed phase. This applies to sulfate, nitrate, ammonium, and most organics—even a substantial fraction of organics traditionally regarded as primary organic aerosol (POA) are semi-volatile.[22] In total, these constituents usually comprise the large majority of the bulk particle mass[11] and even most of the mass on most individual particles.[23] What is left is an often relatively refractory core consisting of elemental carbon, metals, perhaps sodium chloride, some crustal material, and other constituents,[24] which is typically unique and source specific, in contrast to the relative homogeneity of the particle mass.

The evidence that condensation dominates fine PM mass comes from field measurements, lab studies, and theory. The simplest generic piece of evidence is that the large majority of fine-particulate mass is found on accumulation-mode particles between 100 and 1000 nanometers in diameter, where condensation is by far the dominant growth mechanism, and where particles appear to be generally internally mixed.[8] We shall go through more specific evidence for the major constituents in turn.

Sulfate is an almost trivial example as oxidation of SO_2 has long been recognized as the dominant source.[8] Both gas-phase oxidation by OH radicals and aqueous oxidation within cloud droplets by HOOH play a major role, but each is "condensation" under a broad definition as each consists of a secondary chemical process that adds to the particle mass by bringing vapors to the particle (in the case of aqueous sulfate, the sulfate condenses on the cloud condensation nucleus if and when the droplet evaporates). Furthermore, on a regional scale sulfate tends to be dominated by coal combustion in areas where PM mass is a health concern,[25] and thus control of SO_2 emissions at intense point sources remains the clear control strategy. Near roadways the sulfur content of (diesel) fuel can be a substantial contributor to ultrafine particles.[26] On the other hand, because of emissions controls driven by concerns over acid deposition (and because the respiratory tract contains abundant ammonia), sulfate is often substantially neutralized by ammonia; this may limit negative health effects.

Nitrate is broadly recognized as a semivolatile constituent of fine PM, typically found as ammonium nitrate and only present when air masses have excess ammonia. As a semivolatile, it is by definition a condensation product. Nitrate forms *via* condensation of nitric acid followed by reaction with a base (typically ammonia), and nitric acid in turn is formed principally by gas-phase oxidation of NO_2. NO_2 has a more diverse set of sources than SO_2, though NO_x sources are mostly anthropogenic. Typically, intense emissions from power plants and area emissions from mobile sources are the largest sources, with the mobile sources dominating in urban centers. As with ammonium sulfate, there is little toxicological evidence for negative effects from modest ammonium nitrate exposure. However, for both there is epidemiological evidence that adverse health effects may indeed exist.[6,7,27]

Ammonia is the most common base reacting with both sulfuric and nitric acid—highly volatile on its own it will remain in particles as part of a salt. Sources are mostly agricultural, and once again condensation is the dominant source of PM.[28] Because ammonia is very light, ammonium is usually a relatively small constituent of fine-particle mass, though ammonia availability can play a disproportionate role in PM composition.[29]

Depending on location, organics will comprise from one-third to more than 90% of the fine-particle mass.[30] Where the story of inorganic mass is fairly simple, the story of the organics is complex. However, a significant finding in the past decade is that the large majority of organic mass is found in an oxidized form, which collects in a mixture that has been named "oxidized organic aerosol" (OOA) in the community of researchers employing the Aerodyne aerosol mass spectrometer (AMS) for *in situ* measurements.[11,12,31] In spite of the complexity of the organic fraction, various forms of factor analysis have shown that ambient AMS data can be described by a relatively small set of factors associated with different sources and processes. The OOA factors appear to be associated with atmospheric oxidation chemistry, broadly secondary organic aerosol (SOA), while various factors appear to be associated with different primary organic aerosol (POA) sources.

Little is known about the chemical composition of OOA, only that in aggregate OOA tends to produce similar, highly fragmented mass spectra upon electron ionization and that OOA appears to evolve from a "fresh" form known as semi-volatile OOA (SV-OOA) to an older, more oxidized form called low-volatility OOA (LV-OOA).[12,30,32] Both forms of OOA are highly oxidized: SV-OOA has an oxygen-to-carbon ratio (O : C) between 0.4 : 1 and 0.6 : 1, while LV-OOA has $0.6 < O : C < 1$.[32-35] In general all forms of OOA are more oxidized in the summer than in the winter.[32,36,37] In addition, thermodenuder measurements suggest that SV-OOA is relatively volatile while LV-OOA is significantly less volatile (thus the names), but in each case the quite broad temperature range of thermal vaporization suggests a relatively broad array of constituents spanning several orders of magnitude in vapor pressure.[38-40]

In remote settings OOA dominates OA mass,[11] but even in urban centers OOA often makes up more than two-thirds of the OA mass, especially during the summer.[11,12] The rest of the OA mass consists of several fractions associated with primary emissions. These factors include "hydrocarbon-like organic aerosol" (HOA), which strongly resembles lubricating oil and other emissions associated with vehicles, "cooking organic aerosol" (COA), which shows a temporal pattern strongly correlated with mealtimes,[41,42] and "biomass burning organic aerosol" (BBOA), which appears to come from biomass burning.[12,43,44] It is now accepted that OOA largely derives from condensation. In remote settings, the ratio of HOA to elemental carbon is somewhat less than that observed in urban centers,[45] while OOA:EC grows substantially. Even very near urban centers, first SV-OOA then LV-OOA grow rapidly onto fine particles.[35]

Laboratory evidence strongly supports this conclusion. All major primary OC emission sources investigated so far are significantly semi volatile,[46–48] with 75–90% of the organic mass evaporating from particles as they are diluted from concentrations typical of source-testing environments to concentrations typical of the urban atmosphere. For diesel engines,[48–51] gasoline automobile engines,[52–55] scooters,[56] gas turbine and jet engines,[57] and wood burning,[58,59] the vapors evolved from these emissions can be rapidly oxidized in the gas phase by OH radicals to produce a sharp increase in OA mass consisting of material that strongly resembles ambient OOA according to AMS analysis.[59–61] A quick jump in OOA soon after the maximum of primary emissions from mobile sources is exactly what is observed in Mexico City, for example,[30,35] and this "non-traditional" SOA plays a major role in explaining that jump in model simulations.[49,62] Thus, far from being mostly non-volatile and non-reactive, the vast majority of organic emissions undergo a complex cycle of evaporation, oxidation, and condensation, as depicted in Figure 3.[63,64]

The OOA factors are specific to AMS analysis, but they appear to be highly correlated with another bulk measure—"water soluble organic carbon" (WSOC), which is also measured continuously and online.[65] Once again, WSOC is operationally defined and often dominates the total OC mass. Downwind of urban centers, WSOC mass rises with respect to CO (a roughly conserved tracer of mobile sources),[45,66] again suggesting strong gas-phase formation of condensable organic material.

Another tool for OA source attribution is Fourier Transform Infrared (FTIR) analysis of filter samples.[67] Because FTIR is sensitive to specific functional groups on organic compounds (*e.g.* aliphatic *vs.* aromatic carbon), FTIR analysis can provide constraints where the high degree of fragmentation in AMS mass spectra loses information. Residual functional groups can retain source-specific features during atmospheric transport and aging and thus help to constrain OA sources longer than bulk mass spectra. Key molecular markers can also be critical to source attribution, especially in conjunction with secondary molecular markers and bulk measurements, such as AMS spectra.[68] Complementary measurements can be made with NMR, which also provides quantitative measurements of total functional group abundance from OA filter samples.[69,70] Factor analysis of the NMR spectra can, for example, quantify the fraction of biogenic SOA in filter samples [71] as well as the primary to secondary ratio in BBOA.[72]

The only refractory constituent that consistently appears in accumulation mode particles with a significant mass fraction is elemental carbon (EC). However, EC on average comprises only about 10% of the $PM_{2.5}$ mass—in the Eastern US an OC:EC of 5:1 is typical, and it is typical for organics to make up about half of the $PM_{2.5}$ mass and inorganics the other half. Smaller

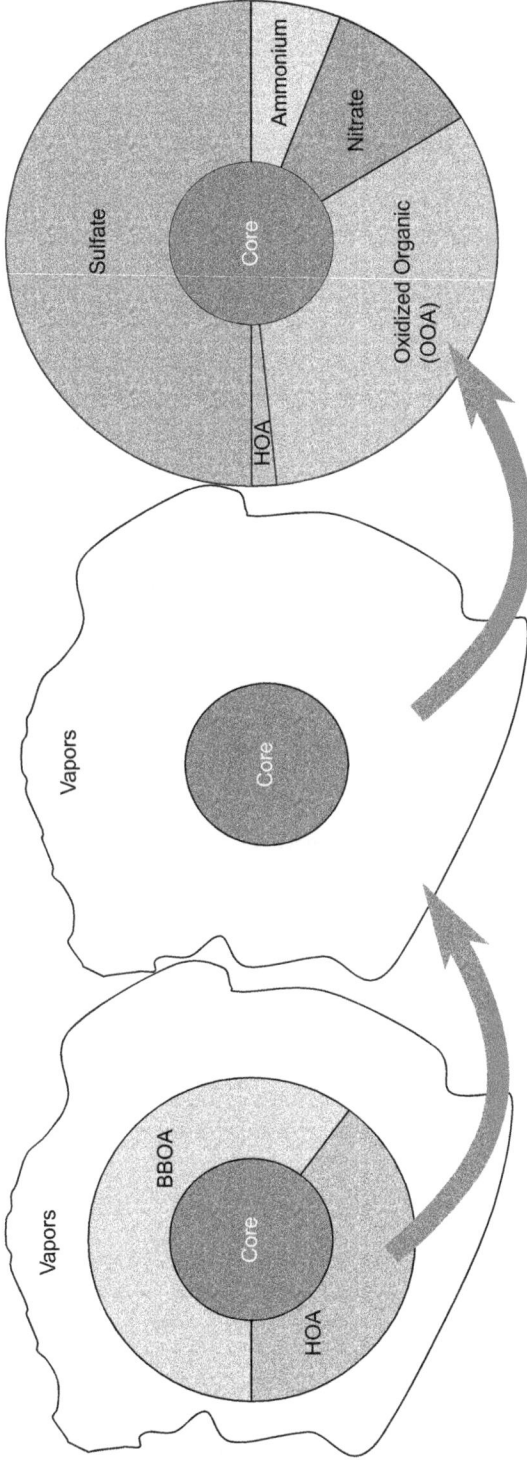

Figure 3 The life cycle of organics associated with organic aerosol emissions. Primary emissions typically contain a very large fraction of semi-volatile constituents (labeled "BBOA" and "HOA" in this bulk representation) on a core including black carbon and the very lowest vapor pressure organics). The semi-volatile constituents evaporate and either get oxidized in the gas phase (becoming OOA) or simply re-condense on accumulation mode particles (becoming "condensation" HOA). At the same time, inorganic components condense on the same accumulation-mode particles to form internally mixed aggregate fine particles.

contributions from, for example, crustal material[73] also occur mostly at the high end of the $PM_{2.5}$ size distribution, so this conclusion is even stronger for PM_1. Consequently, the effective (spherical) core diameter is typically about half of the particle diameter, but most of the mass is in the condensational shell.

Here we must emphasize a crucial point. Any source attribution based on some unique attribute of a single particle—for example clear biological structure or a unique combination of elemental carbon and trace metals—applies only to that core. It is easy to misattribute all of the mass on that particle to the core source—for example to a biological or a diesel particle—and this can lead to a significant over estimation of the significance of particular sources with respect to mass. However, source attribution based on mass composition—either with an AMS or with chemical mass balance (CMB) methods[74,75]—is intrinsically less vulnerable to "core bias" but may suffer when CMB tracers (especially organics) are oxidized.[76]

4   Particle Number Concentrations

The fine-particle number budget is far less well understood than the fine-particle mass budget, both locally and globally.[19,77–79] As with mass, the major questions are the split between anthropogenic and biogenic sources, and the split between primary emissions (sea spray, fires, fine dust) and secondary formation (new-particle formation). Furthermore, ultrafine particles are extremely short lived (they either grow or are lost to coagulation) and so questions of scale in both measurements and models can strongly influence conclusions.

Now that we have established that particle mass arises largely *via* condensation, it follows that particle number is to first order divorced from particle mass: each particle must come from somewhere. There are two significant linkages in spite of this: new-particle formation is a specialized form of condensation, and particle surface area (and thus particle number lifetime) is also strongly influenced by condensation. In some sense particle mass devours particle number *via* coagulation; while this has a second-order effect on individual particle mass (and none on bulk mass), it has a first-order effect on number. This means that the sources of particle number need not have much to do with particle mass, though they might. Even within a class, such as organics, organic aerosol mass and the organic contributors (if any) to new-particle formation need not be the same, nor even correlated.

Because most of the mass on most particles arrives *via* condensation, it follows that most particles grow substantially during their residence in the atmosphere. The largest source of particles on a global scale is new-particle formation, at the smallest possible particle size ($\simeq 1$ nm), though as we shall discuss urban-scale exposure may paint a different picture.

The next largest source is ultrafine emissions, mostly from combustion, over a size range peaking between 10 and 50 nm. Most emissions measurements, however, constrain the mass emissions far more accurately than the number emissions, and a shift in the modal size by 50%, conserving mass, implies more than a factor of three change in number. The largest sink of particle number is coagulation. Here by convention larger particles devour smaller ones, and most particle number losses involve the smallest particles (freshly nucleated or emitted) colliding with particles near the surface-area modal size, which is typically 100–300 nm diameter. However, those larger particles themselves will for the most part have grown up *via* condensation, and so it is far less obvious whether any core in the larger particle will contain more mass than the subsumed smaller particle.

Individual particle measurements can be very informative about sources of particle number, especially because the methods used to analyze single particle data often involve clustering algorithms,[80–82] and these particle clusters can then be compared to measurements of known sources for ambient particle source identification.[83–87] Most single-particle analysis techniques—including laser-ablation mass spectrometry, and electron and x-ray microscopies—lack the sensitivity to analyze ultrafine particles ($d_p < 100$ nm).[88–91] This is unfortunately where the majority of particle number exists, and this hampers our ability to determine the sources of most particles by number. Growing mobility-size selected ultrafine particles by water condensation can facilitate their analysis by single-particle mass spectrometry. This revealed the composition of individual particles measured in La Jolla, CA, USA to be complex mixtures of combustion-derived "cores" (from cars and trucks, biomass burning, and also ships) and (presumably secondary) organic and inorganic material, and a small contribution from metallic cores. Smaller particles, down to 60 nm, contained relatively less secondary material such as ammonium, nitrate, and amines, compared to larger particles.[92]

Other advanced mass spectrometry techniques that determine the composition of small collections of size-selected ultrafine particles have also found these to be complex mixtures of ammonium sulfate, nitrate/nitrogen, organics, and a small contribution from metals.[93,94] Particles containing oxidized carbonaceous matter were generally neutralized, whereas particles containing unoxidized carbonaceous matter or no carbon at all were acidic.[95] Measurements of nucleation-mode particles with $d_p = 8$–10 nm found these contained mixtures of amines and aminium salts, oxidized organics, and sulfate.[96] The much larger amount of organics compared to sulfate typically measured in these nucleation-mode particles suggests that the organics play an important and even dominant role in the initial rapid growth of newly nucleated particles that is required for them to survive against death by coagulation.[97]

The measurements of ultrafine and nucleation-mode particle composition we do have indicate that even these young and very small particles quickly become mixtures of primary and secondary inorganic and organic components. Any particle in the upper end of the ultrafine size mode will have already coagulated with several other particles and swallowed their cores in the process, while also accumulating secondary material *via* condensation and heterogeneous reactions. Thus thinking about a particle as having one "core" that indicates the original source of that particle may be too simplistic as one moves away from the source. Focusing on determining a particle's refractory core is useful for distinguishing between primary and secondary contributions to particle number and mass, as all secondary components are non-refractory (they evaporate at ~500 °C when analyzed) but many primary components (combustion-derived soot, coal fly ash, and some inorganics; mechanically generated crustal, sea spray, and metallic particles) are refractory. As the sources, production mechanisms, and likely the toxicity of primary refractory particles are quite different from secondary particle components, this distinction is valuable for designing effective air pollution control strategies, and motivating PM health effects studies that target the most hazardous sources and components.

Individual particle sources and their compositions can also control the rate and type of secondary components a particle accumulates. Vanadium emitted along with combustion aerosol from ships may catalyze the oxidation of SO_2 to sulfate.[98] Oxalate, a major component of secondary organic aerosol, was found enriched in vanadium-rich particles measured in Mexico City.[99] Chelation of particulate metals by organic acids can alter the solubility and toxicity of transition metals, and also enhance the accumulation of secondary organics. The vanadium and other complex metal mixtures such as zinc chlorides found in PM were attributed to refuse burning in Mexico City.[24] Refuse burning and industrial and construction activities can emit significant levels of metallic particles, exposing residents and workers in urban areas to unusually high levels of particulate metals. Oxalic acid was also found enriched in Asian mineral dust particles, attributed to acid–base neutralization by alkaline carbonate minerals in the natural dust particles.[100] This demonstrates the potential feedbacks between "natural" and anthropogenic pollutants, where natural particle cores provide surfaces for condensation, and also for heterogeneous reactions that can increase the amount of secondary aerosol produced beyond that formed by condensation alone.

4.1   New-particle Formation

New-particle formation occurs whenever the gas-phase concentration of condensable vapors C_i^v exceeds a critical value. Specifically, the saturation ratio $S_i^v = C_i^v/C_i^o$ must be sufficiently large. There is strong evidence that new-particle formation in the boundary layer almost always involves multiple species,[101,102] usually including sulfuric acid,[103] but also some mixture of bases (ammonia and amines)[104] and oxidized organics.[105–107]

New-particle formation is surprisingly common in urban areas,[108–111] even when they are heavily polluted. Even though the condensation sink of vapors to particles is usually high in polluted areas, the production rate of condensable vapors can also be high, and the end result can be high steady-state saturation ratios of condensable (nucleating) vapors. The interplay between production and condensational loss is evident in Figure 2. If at the same time the production rate of condensable species (sulfuric acid and highly oxidized organics, specifically) is 10 times higher in the Po Valley than in the boreal forest of Finland (which is roughly true), then the steady-state concentration (and excess saturation) of the potential nucleating species will be the same in the two locations.

Many identified new-particle formation events involve well-defined formation and growth patterns known as "bananas" based on their characteristic appearance in particle size distribution plots. An example is shown in Figure 4a for an urban background site in Pittsburgh, USA. Particle size distributions are typically measured with a scanning mobility particle sizer (SMPS), in this case with a lower size limit near 3 nm.[112] Size distributions are plotted as a surface plot over the course of a day, typically with a logarithmic scale. The new-particle formation and growth event thus appears as a curving banana-like feature, typically starting when photochemical activity is high (either near noon or some hours after daybreak).[102,113] The growth rate of the feature is defined as the rate of growth of the feature mode, dn_N^{max}/dt.[16,113] The banana is curved because of the logarithmic y (diameter) axis—the actual growth rate of the feature is generally linear.

These banana-type events often last for 6–12 hours, during which air can travel 100 km or more. This suggests that many new-particle formation events are regional, and coordinated observations of new-particle formation events have confirmed this.[109,112] However, there may be some sampling bias, as bananas are easy to identify and are strongly associated with new-particle formation. New-particle formation events in urban areas are not always be associated with "banana" events.[114,115]

Under most circumstances, growth rates vary between 1 and 10 nm h^{-1}.[16,104] New-particle formation and growth are not easily separated, and so measured growth rates are a key diagnostic. This is partly because the same species responsible for new-particle formation may be important for growth as well, and the excess saturation of vapors thus drives both processes.

Sulfuric acid vapor measurements play an important role, both because sulfuric acid is thought to participate in new-particle formation in almost all cases and because sulfuric acid is thought to condense onto particles irreversibly. Consequently, the measured growth rate can be compared to the growth rate predicted by condensation of sulfuric acid, $\Gamma = GR^{obs}/GR^{SA}$, with or without small corrections for co-condensation of ammonia.[104] Sometimes sulfuric acid is sufficient to explain the observed growth, so $\Gamma \simeq 1$.

The events in Pittsburgh typified by Figure 4 probably fall into this category. Though the Pittsburgh Air Quality Study did not include sulfuric acid vapor measurements, larger particles ($d_p > 50$ nm) measured with an Aerosol Mass Spectrometer were almost pure ammonium sulfate,[116] and SO_2 measurements could be used to model new-particle formation events. Those models reproduced the observed growth rates well using only sulfuric

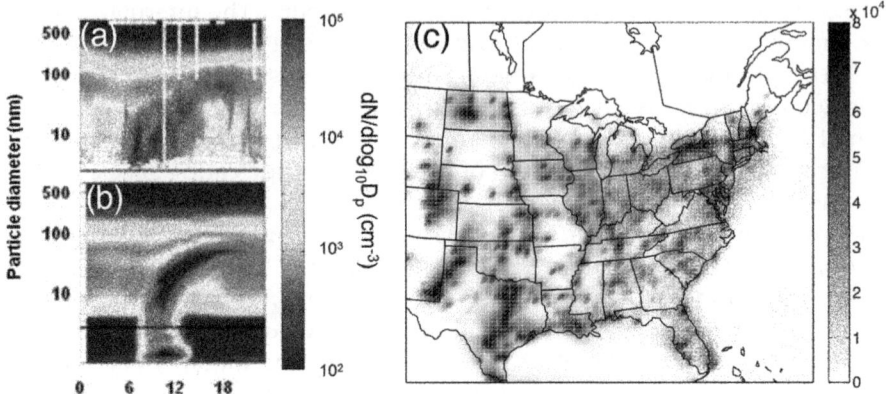

Figure 4   (a) New-particle formation event in Pittsburgh (27 July 2001). The *x*-axis is local time over one day, while the (logarithmic) *y*-axis is particle mobility diameter, d_p. Plotted hue indicates the differential size distribution of the particle number concentration ($\overset{\circ}{n}_N$) on a log scale. (b) Model simulation of the same event for Pittsburgh using a ternary H_2SO_4-NH_3-H_2O parameterization, (c) Model average number concentrations of particles smaller than 10 nm at the surface in the eastern US for 14–28 July 2001 (on a linear scale). While the range of hues suggests variability in the overall intensity of new-particle formation, all shades indicate a widespread new-particle formation episode.

acid condensation.[117] However, more often $\Gamma > 10$,[104,118] and this excess growth has been associated with condensation of organics, partly by the process of elimination.[16,118]

Three-dimensional chemical transport models can simulate new-particle formation with reasonable success. The Eastern United States has relatively high sulfuric acid vapor concentrations because of heavy reliance on coal combustion for electrical power generation.[74] In this case, it is likely that ternary nucleation involving sulfuric acid, ammonia, and water vapor is sufficient to explain new-particle formation.[119,120] In Figure 4b we show a simulation of the data from Figure 4a taken from a 3D-CTM PMCAMx-UF simulation described by Jung *et al.*[77] using the ternary nucleation scheme of Napari *et al.*[119] tempered by an empirical scaling factor of 10^{-5} to reduce the predicted new-particle formation rate. This factor is now understood to arise because the smallest clusters of ammonia and sulfuric acid do not chemically dissociate into ammonium and bisulfate, rendering them much less stable and thus less likely to nucleate.[121,122] The model simulations had high skill predicting new-particle formation over a 17 day period (82% predictive success) and generally predicted growth rates in addition to the presence or absence of new-particle formation with good fidelity while somewhat over predicting the number concentrations (by a factor of 2.5).[77] The overall spatial pattern for the smallest particles in the simulation ($N_{<10}$) is shown in Figure 4c.

Recent field observations suggest that a stronger base such as dimethyl amine may be required to quantitatively explain observed new-particle formation in areas dominated by inorganic (sulfuric acid + base) nucleation,[101] while laboratory measurements also suggest that nucleation rates for the ammonia + sulfuric acid system are too slow at a given sulfuric acid concentration to explain the observed new-particle formation.[123,124] It is thus possible that the empirical scaling factor used to obtain Figure 4a and b is itself a proxy for organic bases that either supplement or replace ammonia. Elsewhere in more remote regions there are strong indications that new-particle formation may involve stabilization of sulfuric acid by highly oxidized organic vapors,[105,106] and there are at least indications from laboratory experiments that oxidation products of aromatic hydrocarbons can contribute to ternary new-particle formation as well.[125]

4.2 Primary Particle Number Emissions

Primary emissions in the eastern US are dominated by gasoline vehicles and industrial sources, with broad emissions size distributions that peak in the 10–50 nm size range.[126] Together they constitute slightly more than 70% of the total primary particle number emissions. We show spatial maps in Figure 5a. In each case we derived the number emissions estimates by

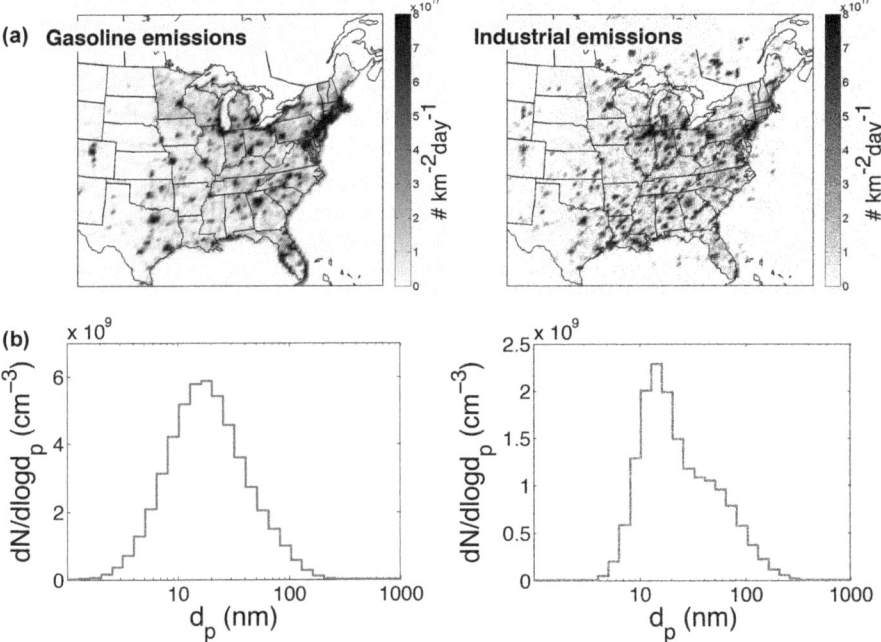

Figure 5 (a) Main primary particle emissions over the eastern United States (in July 2001), in particles km^{-2} day^{-1} (red is 8×10^{17}). Gasoline vehicles and industrial emissions are the major primary sources, reflecting major urban centers and the industrial midwest. (b) Size distributions assumed for gasoline vehicle and industrial emissions to convert data from mass flux inventories to number flux emissions estimates.

converting relatively well constrained mass flux inventories into number fluxes by using literature data on primary particle emissions size distributions.[127] Those constraints are uncertain, and thus the primary number flux estimates are also highly uncertain. However, it is clear that various combustion sources dominate the primary particle emissions (both number and mass) for sub-micrometer particles, and it is also clear that the mode for those primary emissions generally falls between 10 and 50 nm.

In most urban areas, primary particle emissions are dominated by vehicles. Typically, vehicular emissions have a bimodal distribution. A "soot mode" rich in elemental carbon has a peak in the number distribution between 50 and 100 nm, consisting of particles that form from incomplete combustion in the engine itself.[128,129] An additional "nucleation" mode, composed primarily of organics, is formed as exhaust vapors are cooled with a number mode around 15–30 nm.[128,129] Both gasoline and diesel vehicles emit both modes, but the soot mode is more pronounced in diesel emissions. The smaller mode dominates the number of emitted particles. Overall, gasoline vehicles probably contribute more than diesel to the total number of emissions, but the diesel contribution is non-negligible for the size range larger than 50 nm.

Other sources of primary particles that may be important in some regions or episodically include cooking, industrial combustion, biomass burning, and sea spray. Biomass-burning particles tend to be larger, with a number mode around 100 nm[130,131] and a composition that is richer in organics or elemental carbon for smoldering and flaming combustion, respectively. Emissions of ultrafine sea spray are attested in several measurements,[132,133] but the fluxes and sizes are poorly characterized.

In general, the relative magnitudes of these fluxes are not well constrained. As mentioned earlier, most number emissions are estimated from mass emissions inventories combined with assumptions about size distributions. The size distribution measurements are more sparse and sometimes potentially inconsistent with the mass emissions, *e.g.* if the number size distribution measurements do not cover the full size range contributing to particle mass or if a lognormal fit to the number distribution is made that does not represent well the "tail" of the distribution that dominates the mass emissions. The emerging emphasis on ultrafine particles and number concentrations demands more systematic and careful characterization of the number and sizes of primary particles from various sources. Nevertheless, the importance of vehicle emissions is well established. The organic-rich vehicle nucleation mode dominates number size distributions in curbside measurements and also in the urban background, where it is seen to correlate with weekday rush hour traffic patterns.[134] Tunnel studies have provided very valuable information about vehicle emissions including number fluxes.[127,135,136] Source apportionment studies for $PM_{0.1}$ have pointed to gasoline and diesel vehicles, cooking, residential wood burning, and rail.[137,138]

The behavior of vehicle emissions is highly dynamic with the potential for semi-volatile organics to both condense and evaporate, shifting the particle size distribution in the first few minutes after emission.[139–141] Recent vehicle emissions controls mostly target emissions of the larger "soot mode" particles that dominate the mass emissions even in gasoline vehicles. This has caused some to worry that a mass-focused emissions approach may inadvertently increase the number of emitted particles overall by increasing the number of organic-rich particles forming in the vehicle "nucleation mode". This is physically plausible and observed to occur in some cases, but the actual effects depend upon the detailed technology deployed and its operation.[142] Tunnel studies that resolve individual vehicle emission factors indicate that there is a class of trucks that are super-emitters for mass emissions and a separate class that were super-emitters for number emissions.[136] Exposure to ultrafine particles from vehicle emissions is one hypothesized explanation for epidemiological data showing adverse health effects in the near-roadway (100–200 m) environment.[143]

4.3 Primary Emissions vs. New-particle Formation

For simulations of the eastern United States during July 2001 using PMCAMx-UF, new-particle formation dominated the total particle number source term and surface concentration as shown in Figure 6a. The $N_{<100}$ concentration is almost identical to the $N_{<10}$ concentration plotted in Figure 4c. However, Figure 6b shows that in the accumulation mode (N_{100}), where essentially all of the PM$_1$ mass resides, there is almost no spatial correlation between the number distribution and either $N_{<100}$ or $N_{<10}$. Figure 6c and d show the fraction of particles formed *via* new-particle formation for both $N_{<100}$ and N_{100}.

Even though new-particle formation dominates the ultrafine budget, and even though most particles larger than 100 nm "grew up" from particles in the $N_{<100}$ range, primary emissions for the most part contribute more than half of the total number concentration above 100 nm. This seeming paradox is resolved by the survival probability, which is an extremely strong function of size.[144] The particles that actually survive to 100 nm size are heavily weighted toward the high end of the 1–100 nm size distribution, which is much more robustly represented by primary emissions.

Concentration or composition maps such as the new-particle fractions shown in Figure 6 convey the geographic distribution of a complex and variable field, but they can be misleading when considering human exposure. Many of the grid cells represented by the maps are sparsely populated, while a few are densely populated; urban centers are typically a single grid cell in these maps. One can even see in the lower panel showing the $N_{<100}$ nucleated fraction small lighter spots amid the broad field. Those are major urban centers such as Chicago, Detroit, and New York City. To overcome this bias, we integrated the US census data for 2000 onto the PMCAMx grid used in these simulations.[145] This allows us to calculate population-weighted concentration statistics relevant to human exposure. In Figure 7 we show the cumulative distribution of the nucleated fraction for both $N_{<100}$

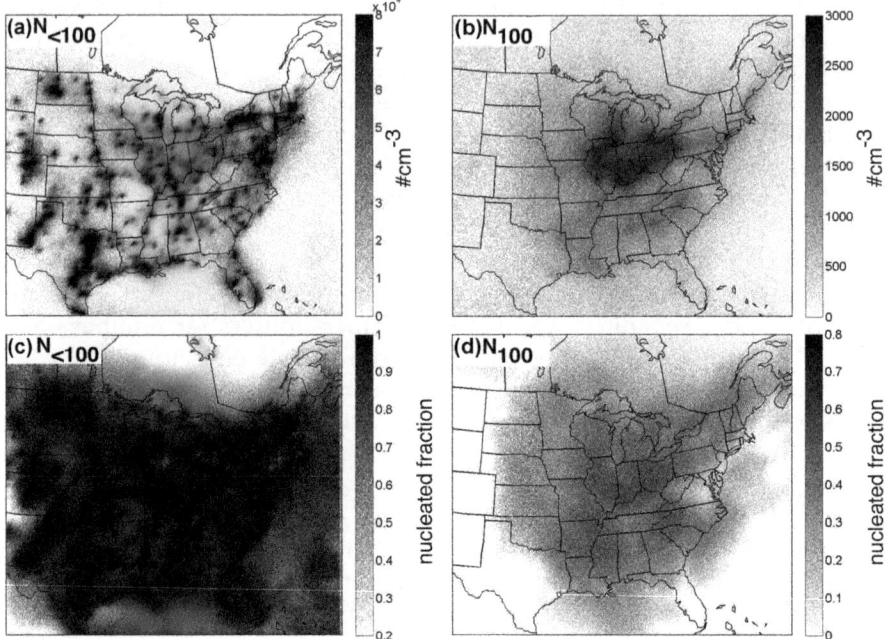

Figure 6   (Top) Simulated surface particle number concentrations (# cm^{-3}) for (a) particles smaller than 100 nm ($N_{<100}$) and (b) particles larger than 100 nm (N_{100}). The spatial patterns are completely different, and $N_{<100}$ exceeds N_{100} by more than a factor of 100 in most cases. (Bottom, c and d) Fraction of particles ($N_{<100}$ and N_{100}) from new-particle formation: most sub-100 nm particles are from new-particle formation, and most larger particles are not.

and N_{100} in July 2001 based on census data that has been mapped onto the same 36×36 km grid used in the PMCAMx-UF model. The result is qualitatively similar to the maps—a large fraction of ultrafine particles being inhaled by residents of the eastern US (in July) are secondary particles, but the fraction is not as high as the maps might suggest. Furthermore, very few of the accumulation mode particles (N_{100}) inhaled by humans are secondary.

4.4   Issues of Scale

Another issue important to considering primary *vs.* secondary ultrafine particles is the question of scale. The simulation results we have shown so far are for a relatively coarse-grid model, with 36×36 km grid spacing. A recent empirical source-attribution study in Leipzig, Germany reported the estimated fraction of secondary particles at three sites: the town center, an urban background site (the IFT labs) and a regional background site (Melpitz, well known for new-particle formation studies). Not surprisingly, the authors found that over half of the ultrafine particles in the regional background site were local secondary particles, especially in summer.

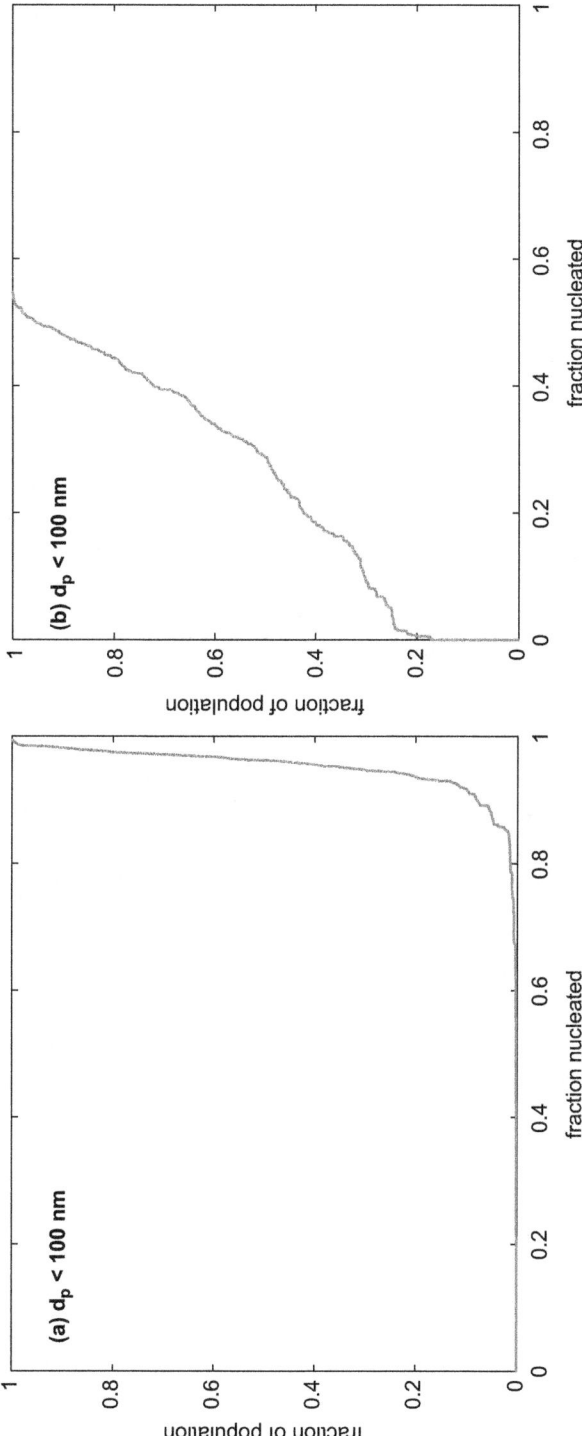

Figure 7 Cumulative distributions of exposure to secondary particles (formed *via* new-particle formation) *vs.* primary particles, based on July 2001 population in the eastern US and the nucleation fractions shown in Figure 6. (a) Ultrafine particles ($N_{<100}$) and (b) accumulation-mode particles (N_{100}).

However, in the urban background, and especially the metropolitan center sites, the primary fraction rose steadily until it dominated the ultrafine number concentration even during midsummer.[146] Likewise, in urban areas with "Mediterranean" climates, a cluster analysis of ultrafine particle observations suggests that new-particle formation can be a significant thought not usually dominant source for ultrafine particle number in urban centers.[114]

We can turn to urban-scale EC observations as a tracer for primary particle distributions. One example of EC concentrations and OC:EC values on an urban scale is shown in Figure 8 for Pittsburgh and Allegheny County, USA.[147,148] Observations at a succession of 70 sampling sites reached with an instrumented van during 2011–2014 provided the input to a land-use regression model for extrapolation over the county at a nominal spatial resolution of 100 m. Three of the explanatory variables in the land-use regression model were proximity to major highways, proximity to major industrial sources (especially two metallurgical coke works) and elevation (a proxy for river valleys in a region with sharp topography). The land-use regression suggests strong gradients tied to all three buffers, with coincidence (traffic and industrial sites in river valleys) showing by far the highest EC levels. Figure 8b shows OC:EC values, which are largely anti-correlated with the EC pattern because the EC pattern is primary and diminishes with dispersion.

Figure 8 shows large (order of magnitude) variations in both EC concentration and OC:EC at sub-kilometer spatial scales. This can have substantial impacts on (1) population exposures, especially if PM health effects are a function not only of particle mass, but of composition as well, and

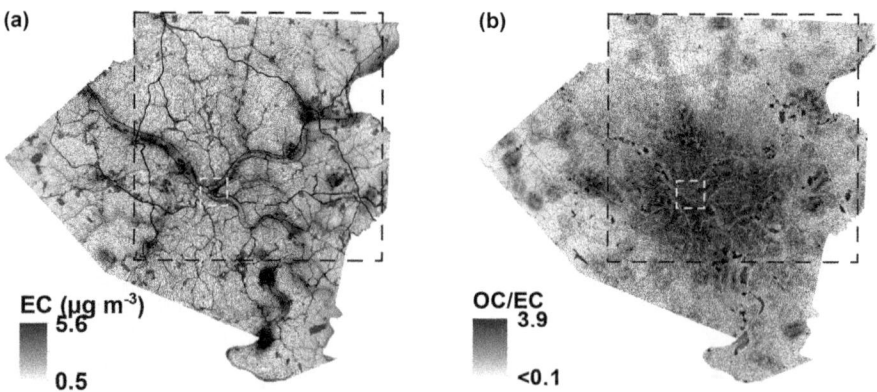

Figure 8   (a) Elemental carbon (EC) concentrations in Pittsburgh, Pennsylvania and surrounding Allegheny County during 2011–2014 based on mobile observations at 70 sites and interpolation *via* land-use regression based on proximity to roadways, elevation (a measure of enclosure in river valleys) and distance from major industrial sources. (b) Elemental carbon to organic carbon ratios (OC/EC) for the same dataset. The dashed boxes in both images show a 36 × 36 km grid box and a 4 × 4 km grid box.

(2) evolution of particle number, mass, and composition in near-source *versus* background locations. The EC emitted along roadways and near industrial areas, as shown above, are emitted at small sizes ($d_p < 100$ nm) and thus at high number concentrations. These particles can serve as cores (condensation sinks) that are subsequently coated with organics and other secondary species as they are transported away from the source areas. The sharp gradient in EC concentration in Figure 8a is driven primarily by dilution of fresh emissions, whereas the reverse gradient of OC:EC in Figure 8b is a consequence of dilution of fresh emissions, mixing with the regional background, and secondary aerosol condensation.

The entire county domain in Figure 8 is approximately one grid box in the regional model simulations discussed above, which is indicated with large dashed rectangles in the figure. Model resolution is known to influence model performance, especially resolving primary emissions with large gradients in their emissions patterns.[149,150] However, even a 4×4 km fine grid centered over the greater Paris region (like the smaller squares in Figure 8) only resolved BC in the urban center as a single unstructured plume extending over ~10 km,[149] while the results shown in Figure 8 suggest much higher resolution variability. The population distribution within the county is also highly non-uniform, with dense urban clusters and a great deal of open space. We do not yet have proxies for the secondary fraction of ultrafine particles for this area, but presuming that the results from Leipzig apply at least qualitatively, it is likely that the cumulative distributions shown in Figure 7 would show a substantially higher fraction of primary ultrafine particles, especially for individuals living near major roadways or industrial sites.

5 Implications for Human Exposure

The bottom line of the discussion above for human exposure is that atmospheric chemistry drives a significant fraction of human exposure to both particle number and fine particle mass. First, concerning particle number, particle number concentrations globally are driven by new-particle formation; however, primary emissions can be an important source of particle number in urban areas. Therefore, it is clear that both mechanisms (primary emissions and new-particle formation) contribute significantly to human exposures, but few studies have derived quantitative estimates. The distributions shown in Figure 7 overestimate the contribution of new-particle formation to particle number exposure because of the relatively coarse model grid. Simply decreasing the grid spacing in regional chemical transport models is unlikely to resolve gradients such as those shown in Figure 8. A key challenge is these strong spatial gradients in particle number near sources. For example, near-road studies show greatly elevated concentrations (factor of five or more) to particle number near roads,[151] consistent with the strong gradients in elemental carbon shown in Figure 8. More research is needed to quantify the fraction of human exposure to particle number driven by new-particle formation.

The picture of condensation and secondary aerosols is much clearer with respect to human exposure to fine-particle mass than to particle number. The large majority of the PM_1 particle mass inhaled under most circumstances (even in urban areas) is formed from the condensation of secondary products formed by atmospheric chemistry. Even near sources, the contribution of direct emissions is generally modest. For example, within a few hundred meters of a roadway the relative increase in fine PM mass is relatively modest, less than 30% over background and much less than the enhancements in particle number.[151] However, roughly half of those inhaled PM_1 particles will contain some sort of primary core, which in many cases will not be water soluble. Because most particles are lost to coagulation (to 100–1000 nm particles forming most of the condensation sink shown in Figure 2), it also follows that most of those particles will actually contain more than one core. Once again, the secondary coating on those particles will include highly oxidized organic compounds, whose health effects are not well studied.

Organic aerosol is a major component of fine PM mass. OA is unique in that it has substantial primary and secondary sources. Over the past decade our understanding of OA sources has dramatically changed from one based on a largely non-volatile, primary dominated aerosol to one that includes a more dynamic aerosol dominated by secondary formation.[49,152] This evolving understanding of has significant implications for human exposure. These are summarized in Figure 9: compared with traditional models of nonvolatile primary organic emissions, we expect a significant reduction in OA mass near sources as the primary emissions evaporate, followed by progressive increases in OA mass downwind as the vapors are oxidized to form SOA. This is demonstrated in model calculations with a relatively coarse grid,[49,145] but to date has not been implemented in high-resolution urban dispersion models. Broadly, evaporation of the primary emissions smooths out urban:regional gradients, consistent with ambient data from urban:rural pairs in the EPA IMPROVE and EPA STN network.[49]

The dynamic nature of organic emissions is predicted to lead to fairly strong seasonal and latitudinal effects. This is because the vapor pressures are temperature (and thus season) dependent and because oxidant levels (specifically OH) are also believed to have substantial seasonal trends.[153-155] For this reason, the seasonality of the primary : secondary ratio in OA may be especially strong in the northeastern US and in other locations with strong seasonal temperature fluctuations. Model simulations for the eastern US suggest a strong seasonal variation to the extent of evaporation and aging of OA owing to a combination of reduced saturation concentrations and lower OH radical concentrations in the winter compared to the summer.[145]

However, the most dramatic consequence of incorporating the cycle of evaporation, oxidation, and re-condensation into chemical transport models is a dramatic shift in the composition of organics that people are predicted

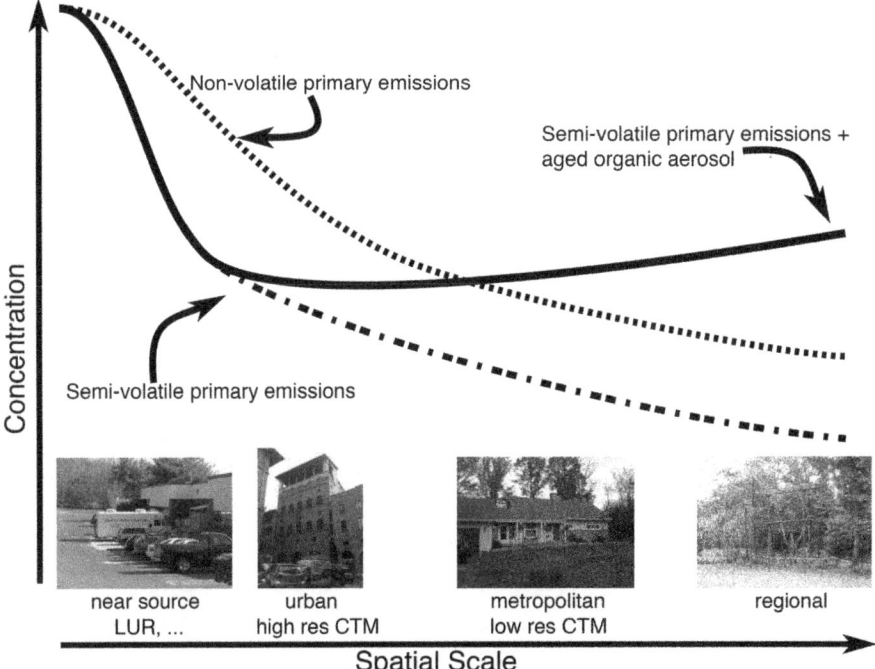

Figure 9 Changes to organic aerosol levels at various distances from urban centers based on evaporation and aging of primary emissions, compared to a traditional "non-volatile" picture of primary organic aerosol emissions. Near urban centers, the OA levels will generally be lower than earlier expected because primary emissions will mostly evaporate instead of remaining on their source particle. However, downwind of the source location, the resulting organic vapors will rapidly oxidize and recondense on different, accumulation-mode particles. Because vapors co-emitted with the primary emissions also lead to SOA formation, the downwind particle mass may exceed the downwind mass in the traditional framework. Different scales can be resolved with different model types: metropolitan and larger scales with relatively low-resolution chemical transport models (CTM, 36×36 km); urban scale with high-resolution CTMs (4×4 km); and near source exposure with land-use regression (LUR) and other methods.

to inhale. In Figure 10 we show cumulative exposure distributions to primary and secondary OA based on PMCAMx simulations for the eastern US in July 2001.[49,145] Note that these are *separate* CPDs for POA and SOA, so one cannot simply add the exposures. This figure reveals the really dramatic change in predicted human exposure between the traditional view of relatively non-volatile and inert primary emissions and our revised view that the organic emissions are semivolatile and chemically very active. The traditional view, shown in the left-hand panel of Figure 10, predicts that people everywhere will be exposed to more primary than secondary OA and that this difference will be greatest in urban centers (where the highest exposures are found).

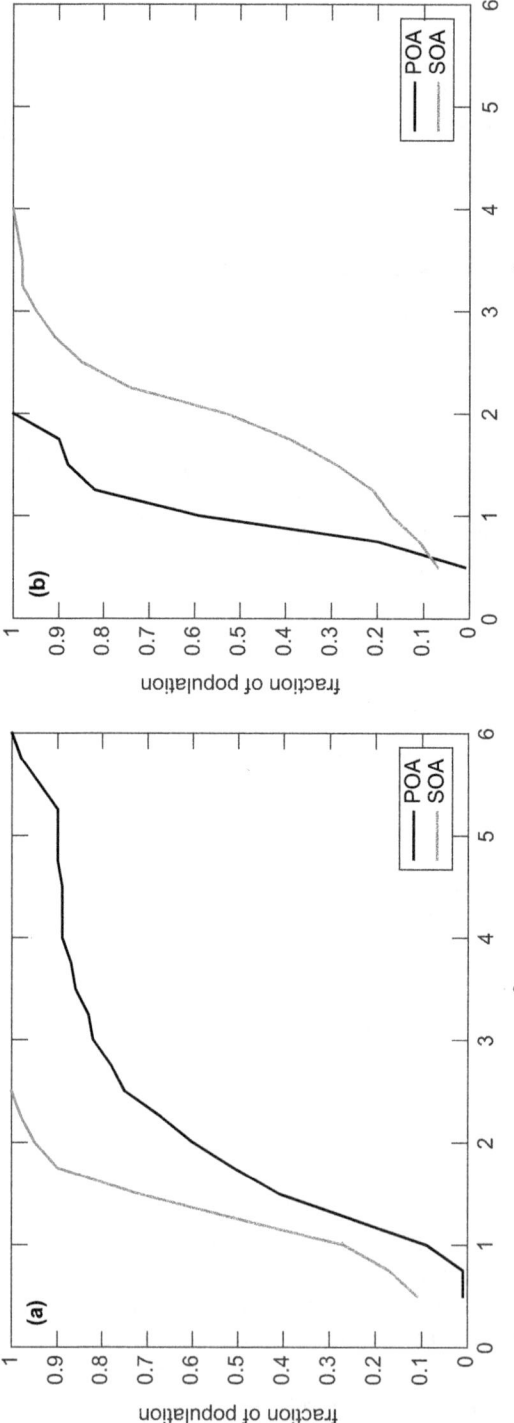

Figure 10   Cumulative probability distribution of human exposure in the eastern US for (a) a "traditional" organic aerosol framework dominated by non-volatile OA emissions, and (b) the revised framework including evaporation, oxidation, and recondensation of most primary emissions.

About 15% of the population is exposed to unusually high POA levels in this traditional simulation. In contrast, the revised simulation shown in Figure 10 shows that SOA dominates over POA, at almost all exposure levels, and that the peak exposure is considerably muted—only a few percent of the population is exposed to the highest SOA levels and the spike is less than a microgram above the upper quartile. If we hypothesize a different health effect from POA and SOA, consistent with toxicity studies performed in chamber aerosols,[156,157] we thus conclude that the revised simulation predicts a slightly lower dose of potentially much more potent material.

It is also not at all clear whether SOA derived from different sources will have different health effects, or to what extent the highly processed OOA in ambient AMS observations has similar health effects to less oxidized SOA typically formed in smog chambers.[158] Quite distinct primary sources (for example, diesel emissions and wood burning) can be aged to yield material with very similar spectra (each strongly resembling OOA) in an AMS.[32,60,61] Traditional SOA sources including α-pinene SOA and toluene SOA can also ultimately resemble OOA,[159–161] though often the fresh SOA is a poor analogue for OOA.[159,162] Unfortunately, at present we know of no epidemiological data that are suitable to test any hypothesized difference between HOA and OOA exposure.

Although there are spatial gradients in particle mass near sources, they are substantially smaller than the gradients in particle number.[151] Therefore, while there is likely some bias towards secondary organic aerosol in the distributions shown in Figure 10, it is substantially less than for particle number (Figure 7). However, it is estimated that 30% to 45% of people living in large cities live within 300 to 500 m from a highway or a major road, which is the area most highly affected by traffic emissions.[143] More research is needed to better quantify the contribution of HOA and OOA to human exposure in these sorts of near-source regions.

It is important to stress that the difference between the old and new paradigms for organic aerosol is not confined to some regional background haze; data from Mexico City and elsewhere show that the oxidized form of organic aerosol builds up very rapidly, and that even in areas with very high organic aerosol concentrations (such as these megacities), the OOA constituents dominate the organic aerosol within a few hours on a photo-chemically active day.[62] This establishes conclusively that the local chemistry is a source (you can not build up high concentrations through dispersion—maxima unambiguously identify source regions). It is only very near sources—for example, very near roadways—that a significant gradient in very fresh *versus* "aged" (meaning hours, not days) particles is likely to be evident. It is therefore crucial that future human exposure studies take this chemistry into account. Epidemiological studies need to include the significant oxidation of organics likely included in most human exposure, and toxicological studies need to address the processed nature of the organics as well.

Acknowledgements

This work was supported by grant 1447056 from the US National Science Foundation.

References

1. C. A. Pope, M. Ezzati and D. W. Dockery, *N. Engl. J. Med.*, 2009, **360**, 376–386.
2. D. W. Dockery, C. A. Pope, X. Xu, J. D. Spengler, J. H. Ware, M. E. Fay, B. G. Ferris and F. E. Speizer, *N. Engl. J. Med.*, 1993, **329**, 1753–1759.
3. S. S. Lim, T. Vos, A. D. Flaxman, G. Danaei, K. Shibuya, H. Adair-Rohani, M. A. AlMazroa, M. Amann, H. R. Anderson, K. G. Andrews, M. Aryee, C. Atkinson, L. J. Bacchus, A. N. Bahalim, K. Balakrishnan, J. Balmes, S. Barker-Collo, A. Baxter, M. L. Bell, J. D. Blore, F. Blyth, C. Bonner, G. Borges, R. Bourne, M. Boussinesq, M. Brauer, P. Brooks, N. G. Bruce, B. Brunekreef, C. Bryan-Hancock, C. Bucello, R. Buchbinder, F. Bull, R. T. Burnett, T. E. Byers, B. Calabria, J. Carapetis, E. Carnahan, Z. Chafe, F. Charlson, H. Chen, J. S. Chen, A. T.-A. Cheng, J. C. Child, A. Cohen, K. E. Colson, B. C. Cowie, S. Darby, S. Darling, A. Davis, L. Degenhardt, F. Dentener, D. C. Des Jarlais, K. Devries, M. Dherani, E. L. Ding, E. R. Dorsey, T. Driscoll, K. Edmond, S. E. Ali, R. E. Engell, P. J. Erwin, S. Fahimi, G. Falder, F. Farzadfar, A. Ferrari, M. M. Finucane, S. Flaxman, F. G. R. Fowkes, G. Freedman, M. K. Freeman, E. Gakidou, S. Ghosh, E. Giovannucci, G. Gmel, K. Graham, R. Grainger, B. Grant, D. Gunnell, H. R. Gutierrez, W. Hall, H. W. Hoek, A. Hogan, I. Hosgood, H. Dean, D. Hoy, H. Hu, B. J. Hubbell, S. J. Hutchings, S. E. Ibeanusi, G. L. Jacklyn, R. Jasrasaria, J. B. Jonas, H. Kan, J. A. Kanis, N. Kassebaum, N. Kawakami, Y.-H. Khang, S. Khatibzadeh, J.-P. Khoo, C. Kok, F. Laden, R. Lalloo, Q. Lan, T. Lathlean, J. L. Leasher, J. Leigh, Y. Li, J. K. Lin, S. E. Lipshultz, S. London, R. Lozano, Y. Lu, J. Mak, R. Malekzadeh, L. Mallinger, W. Marcenes, L. March, R. Marks, R. Martin, P. McGale, J. McGrath, S. Mehta, Z. A. Memish, G. A. Mensah, T. R. Merriman, R. Micha, C. Michaud, V. Mishra, K. M. Hanafiah, A. A. Mokdad, L. Morawska, D. Mozaffarian, T. Murphy, M. Naghavi, B. Neal, P. K. Nelson, J. M. Nolla, R. Norman, C. Olives, S. B. Omer, J. Orchard, R. Osborne, B. Ostro, A. Page, K. D. Pandey, C. D. Parry, E. Passmore, J. Patra, N. Pearce, P. M. Pelizzari, M. Petzold, M. R. Phillips, D. Pope, I. Pope, C. Arden, J. Powles, M. Rao, H. Razavi, E. A. Rehfuess, J. T. Rehm, B. Ritz, F. P. Rivara, T. Roberts, C. Robinson, J. A. Rodriguez-Portales, I. Romieu, R. Room, L. C. Rosenfeld, A. Roy, L. Rushton, J. A. Salomon, U. Sampson, L. Sanchez-Riera, E. Sanman, A. Sapkota, S. Seedat, P. Shi, K. Shield, R. Shivakoti, G. M. Singh, D. A. Sleet, E. Smith, K. R. Smith, N. J. Stapelberg, K. Steenland, H. Stöckl, L. J. Stovner, K. Straif, L. Straney, G. D. Thurston, J. H. Tran,

R. Van Dingenen, A. van Donkelaar, J. L. Veerman, L. Vijayakumar, R. Weintraub, M. M. Weissman, R. A. White, H. Whiteford, S. T. Wiersma, J. D. Wilkinson, H. C. Williams, W. Williams, N. Wilson, A. D. Woolf, P. Yip, J. M. Zielinski, A. D. Lopez, C. J. Murray and M. Ezzati, *Lancet*, 2012, **380**, 2224–2260.

4. R. Burnett, C. I. Pope, M. Ezzati, C. Olives, S. Lim, S. Mehta, H. Shin, G. Singh, B. Hubbell, M. Brauer, H. Anderson, K. Smith, J. Balmes, N. Bruce, H. Kan, F. Laden, A. Prüss-Ustün, M. Turner, S. Gapstur, W. Diver and A. Cohen, *Environ. Health Perspect.*, 2014, **122**, 397–403.

5. D. Crouse, P. Peters, A. van Donkelaar, M. Goldberg, P. Villeneuve, O. Brion, S. Khan, D. Atari, M. Jerrett, C. Pope, M. Brauer, J. Brook, R. Martin, D. Stieb and R. Burnett, *Environ. Health Perspect.*, 2012, **120**, 708–714.

6. R. D. Peng, F. Dominici, R. Pastor-Barriuso, S. L. Zeger and J. M. Samet, *Am. J. Epidemiol.*, 2005, **161**, 585–594.

7. M. L. Bell, K. Ebisu, R. D. Peng, J. Walker, J. M. Samet, S. L. Zeger and F. Dominici, *Am. J. Epidemiol.*, 2008, **168**, 1301–1310.

8. J. H. Seinfeld and S. N. Pandis, *Atmospheric Chemistry and Physics*, John Wiley & Sons, Hoboken, New Jersey, 2nd edn, 2006.

9. S. Fuzzi, U. Baltensperger, K. Carslaw, S. Decesari, H. Denier van der Gon, M. C. Facchini, D. Fowler, I. Koren, B. Langford, U. Lohmann, E. Nemitz, S. Pandis, I. Riipinen, Y. Rudich, M. Schaap, J. G. Slowik, D. V. Spracklen, E. Vignati, M. Wild, M. Williams and S. Gilardoni, *Atmos. Chem. Phys.*, 2015, **15**, 8217–8299.

10. J. H. Kroll, N. M. Donahue, J. L. Jimenez, S. Kessler, M. R. Canagaratna, K. Wilson, K. E. Alteri, L. R. Mazzoleni, A. S. Wozniak, H. Bluhm, E. R. Mysak, J. D. Smith, C. E. Kolb and D. R. Worsnop, *Nat. Chem.*, 2011, **3**, 133–139.

11. Q. Zhang, J. L. Jimenez, M. R. Canagaratna, J. D. Allan, H. Coe, I. Ulbrich, M. R. Alfarra, A. Takami, A. M. Middlebrook, Y. L. Sun, K. Dzepina, E. Dunlea, K. Docherty, P. F. De-Carlo, D. Salcedo, T. Onasch, J. T. Jayne, T. Miyoshi, A. Shimono, S. Hatakeyama, N. Takegawa, Y. Kondo, J. Schneider, F. Drewnick, S. Borrmann, S. Weimer, K. Demerjian, P. Williams, K. Bower, R. Bahreini, L. Cottrell, R. J. Griffin, J. Rautiainen, J. Y. Sun, Y. M. Zhang and D. R. Worsnop, *Geophys. Res. Lett.*, 2007, **34**, L13801.

12. V. A. Lanz, M. R. Alfarra, U. Baltensperger, B. Buchmann, C. Hueglin and A. S. H. Prévôt, *Atmos. Chem. Phys.*, 2007, **7**, 1503–1522.

13. T. Koop, J. Bookhold, M. Shiraiwa and U. Poeschl, *Phys. Chem. Chem. Phys.*, 2011, **13**, 19238–19255.

14. A. Virtanen, J. Joutsensaari, T. Koop, J. Kannosto, P. Yli-Pirila, J. Leskinen, J. M. Makela, J. K. Holopainen, U. Poeschl, M. Kulmala, D. R. Worsnop and A. Laaksonen, *Nature*, 2010, **467**, 824–827.

15. S. Maria, L. Russell, M. Gilles and S. Myneni, *Science*, 2004, **306**, 1921–1924.

16. I. Riipinen, T. Yli-Juuti, J. R. Pierce, T. Petäjä, D. R. Worsnop, M. Kulmala and N. M. Donahue, *Nat. Geosci.*, 2012, **5**, 453–458.

17. S. A. K. Häkkinen, H. E. Manninen, T. Yli-Juuti, J. Merikanto, M. K. Kajos, T. Nieminen, S. D. D'Andrea, A. Asmi, J. R. Pierce, M. Kulmala and I. Riipinen, *Atmos. Chem. Phys.*, 2013, **13**, 7665–7682.
18. N. M. Donahue, E. R. Trump, I. Riipinen and J. R. Pierce, *Geophys. Res. Lett.*, 2011, **38**, L16801.
19. D. M. Westervelt, J. R. Pierce, I. Riipinen, W. Trivitayanurak, A. Hamed, M. Kulmala, A. Laaksonen, S. Decesari and P. J. Adams, *Atmos. Chem. Phys. Discuss.*, 2013, **13**, 8333–8386.
20. R. Saleh, N. M. Donahue and A. L. Robinson, *Environ. Sci. Technol.*, 2013, **47**, 5588–5594.
21. M. Shiraiwa, M. Ammann, T. Koop and U. Poeschl, *Proc. Natl. Acad. Sci.*, 2011, **108**, 11003–11008.
22. A. L. Robinson, A. P. Grieshop, N. M. Donahue and S. W. Hunt, *J. Air Waste Manage. Assoc.*, 2010, **60**, 1204–1222.
23. D. M. Murphy, D. J. Cziczo, K. D. Froyd, P. K. Hudson, B. M. Matthew, A. M. Middlebrook, R. E. Peltier, A. Sullivan, D. S. Thomson and R. J. Weber, *J. Geophys. Res.: Atmos.*, 2006, **111**, D23S32.
24. R. C. Moffet, Y. Desyaterik, R. J. Hopkins, A. V. Tivanski, M. K. Gilles, Y. Wang, V. Shutthanandan, L. T. Molina, R. G. Abraham, K. S. Johnson, V. Mugica, M. J. Molina, A. Laskin and K. A. Prather, *Environ. Sci. Technol.*, 2008, **42**, 7091–7097.
25. K. M. Wagstrom and S. N. Pandis, *Atmos. Environ.*, 2011, **45**, 347–356.
26. A. M. Jones, R. M. Harrison, B. Barratt and G. Fuller, *Atmos. Environ.*, 2012, **50**, 129–138.
27. M. Strak, G. Hoek, K. J. Godri, I. Gosens, I. S. Mudway, R. v. Oerle, H. M. H. Spronk, F. R. Cassee, E. Lebret, F. J. Kelly, R. M. Harrison, B. Brunekreef, M. Steenhof and N. A. H. Janssen, *PLoS One*, 2013, **8**, e58944.
28. R. W. Pinder, P. J. Adams, S. N. Pandis and A. B. Gilliland, *J. Geophys. Res.: Atmos.*, 2006, **111**, D16310.
29. R. W. Pinder, P. J. Adams and S. N. Pandis, *Environ. Sci. Technol.*, 2007, **41**, 380–386.
30. J. L. Jimenez, M. R. Canagaratna, N. M. Donahue, A. S. H. Prévôt, Q. Zhang, J. H. Kroll, P. F. DeCarlo, J. Allan, H. Coe, N. L. Ng, A. C. Aiken, K. D. Docherty, I. M. Ulbrich, A. P. Grieshop, A. L. Robinson, J. Duplissy, J. D. Smith, K. R. Wilson, V. A. Lanz, C. Hueglin, Y. L. Sun, A. Laaksonen, T. Raatikainen, J. Rautiainen, P. Vaattovaara, M. Ehn, M. Kulmala, J. M. Tomlinson, D. R. Collins, M. J. Cubison, E. J. Dunlea, J. A. Huffman, T. B. Onasch, M. R. Alfarra, P. I. Williams, K. Bower, Y. Kondo, J. Schneider, F. Drewnick, S. Borrmann, S. Weimer, K. Demerjian, D. Salcedo, L. Cottrell, R. Griffin, A. Takami, T. Miyoshi, S. Hatakeyama, A. Shimono, J. Y. Sun, Y. M. Zhang, K. Dzepina, J. R. Kimmel, D. Sueper, J. T. Jayne, S. C. Herndon, A. M. Trimborn, L. R. Williams, E. C. Wood, C. E. Kolb, U. Baltensperger and D. R. Worsnop, *Science*, 2009, **326**, 1525–1529.
31. Q. Zhang, D. R. Worsnop, M. R. Canagaratna and J.-L. Jimenez, *Atmos. Chem. Phys.*, 2005, **5**, 3289.

32. N. L. Ng, M. R. Canagaratna, Q. Zhang, J. L. Jimenez, J. Tian, I. M. Ulbrich, J. H. Kroll, K. S. Docherty, P. S. Chhabra, R. Bahreini, S. M. Murphy, J. H. Seinfeld, L. Hildebrandt, N. M. Donahue, P. F. DeCarlo, V. A. Lanz, A. S. H. Prevot, E. Dinar, Y. Rudich and D. R. Worsnop, *Atmos. Chem. Phys.*, 2010, **9**, 4625–4641.

33. M. Canagaratna, J. L. Jimenez, J. Kroll, Q. Chen, S. Kessler, P. Massoli, L. Hildebrandt Ruiz, E. Fortner, L. Williams, K. Wilson, J. Surratt, N. M. Donahue, J. Jayne and D. Worsnop, *Atmos. Chem. Phys.*, 2015, **15**, 253–272.

34. A. C. Aiken, P. F. DeCarlo, J. H. Kroll, D. R. Worsnop, J. A. Huffman, K. Docherty, I. M. Ulbrich, C. Mohr, J. R. Kimmel, D. Sueper, Q. Zhang, Y. Sun, A. Trimborn, M. Northway, P. J. Ziemann, M. R. Canagaratna, R. Alfarra, A. S. Prevot, J. Dommen, J. Duplissy, A. Metzger, U. Baltensperger and J. L. Jimenez, *Environ. Sci. Technol.*, 2008, **42**, 4478–4485.

35. P. F. DeCarlo, I. M. Ulbrich, J. Crounse, B. de Foy, E. J. Dunlea, A. C. Aiken, D. Knapp, A. J. Weinheimer, T. Campos, P. O. Wennberg and J. L. Jimenez, *Atmos. Chem. Phys.*, 2010, **10**, 5257–5280.

36. L. Hildebrandt, G. J. Englehardt, C. Mohr, E. Kostenidou, V. A. Lanz, A. Bougiatioti, P. F. DeCarlo, A. S. H. Prévôt, U. Baltensperger, N. Mihalopoulos, N. M. Donahue and S. N. Pandis, *Atmos. Chem. Phys.*, 2010, **10**, 4167–4186.

37. L. Hildebrandt, E. Kostenidou, N. Mihalopoulos, D. R. Worsnop, N. M. Donahue and S. N. Pandis, *Geophys. Res. Lett.*, 2010, **37**, L23801.

38. C. D. Cappa and J. L. Jimenez, *Atmos. Chem. Phys. Discuss.*, 2010, **10**, 1901–1938.

39. E. Kostenidou, B.-H. Lee, G. J. Engelhart, J. R. Pierce and S. N. Pandis, *Environ. Sci. Technol.*, 2009, **43**, 4884–4889.

40. A. Paciga, E. Karnezi, E. Kostenidou, L. Hildebrandt, M. Psichoudaki, G. J. Engelhart, B.-H. Lee, M. Crippa, A. S. H. Prévôt, U. Baltensperger and S. N. Pandis, *Atmos. Chem. Phys. Discuss.*, 2015, **15**, 22263–22289.

41. Y.-L. Sun, Q. Zhang, J. J. Schwab, K. L. Demerjian, W.-N. Chen, M.-S. Bae, H.-M. Hung, O. Hogrefe, B. Frank, O. V. Rattigan and Y.-C. Lin, *Atmos. Chem. Phys.*, 2011, **11**, 1581–1602.

42. J. Yin, S. A. Cumberland, R. M. Harrison, J. Allan, D. E. Young, P. I. Williams and H. Coe, *Atmos. Chem. Phys.*, 2015, **15**, 2139–2158.

43. J.-E. Petit, O. Favez, J. Sciare, F. Canonaco, P. Croteau, G. Močnik, J. Jayne, D. Worsnop and E. Leoz-Garziandia, *Atmos. Chem. Phys.*, 2014, **14**, 13773–13787.

44. M. Crippa, F. Canonaco, V. A. Lanz, M. Äijälä, J. D. Allan, S. Carbone, G. Capes, D. Ceburnis, M. Dall'Osto, D. A. Day, P. F. DeCarlo, M. Ehn, A. Eriksson, E. Freney, L. Hildebrandt Ruiz, R. Hillamo, J. L. Jimenez, H. Junninen, A. Kiendler-Scharr, A.-M. Kortelainen, M. Kulmala, A. Laaksonen, A. A. Mensah, C. Mohr, E. Nemitz, C. O'Dowd, J. Ovadnevaite, S. N. Pandis, T. Petäjä, L. Poulain, S. Saarikoski,

K. Sellegri, E. Swietlicki, P. Tiitta, D. R. Worsnop, U. Baltensperger and A. S. H. Prévôt, *Atmos. Chem. Phys.*, 2014, **14**, 6159–6176.

45. J. A. de Gouw, A. M. Middlebrook, C. Warneke, P. D. Goldan, W. C. Kuster, J. M. Roberts, F. C. Fehsenfeld, D. R. Worsnop, M. R. Canagaratna, A. A. P. Pszenny, W. C. Keene, M. Marchewka, S. B. Bertman and T. S. Bates, *J. Geophys. Res.*, 2005, **110**, D16305.
46. E. M. Lipsky and A. L. Robinson, *Environ. Sci. Technol.*, 2006, **40**, 155–162.
47. M. K. Shrivastava, E. M. Lipsky, C. O. Stanier and A. L. Robinson, *Environ. Sci. Technol.*, 2006, **40**, 2671–2677.
48. R. Chirico, P. F. DeCarlo, M. F. Heringa, T. Tritscher, R. Richter, A. S. H. Prévôt, J. Dommen, E. Weingartner, G. Wehrle, M. Gysel, M. Laborde and U. Baltensperger, *Atmos. Chem. Phys.*, 2010, **10**, 11545–11563.
49. A. L. Robinson, N. M. Donahue, M. K. Shrivastava, A. M. Sage, E. A. Weitkamp, A. P. Grieshop, T. E. Lane, J. R. Pierce and S. N. Pandis, *Science*, 2007, **315**, 1259–1263.
50. E. A. Weitkamp, A. T. Lambe, N. M. Donahue and A. L. Robinson, *Environ. Sci. Technol.*, 2008, **42**, 7950–7956.
51. S. Nakao, M. Shrivastava, A. Nguyen, H. Jung and D. Cocker III, *Aerosol Sci. Technol.*, 2011, **45**, 964–972.
52. T. D. Gordon, A. A. Presto, A. A. May, N. T. Nguyen, E. M. Lipsky, N. M. Donahue, A. Gutierrez, M. Zhang, C. Maddox, P. Rieger, S. Chattopadhyay, H. Maldonado, M. M. Maricq and A. L. Robinson, *Atmos. Chem. Phys.*, 2014, **14**, 4661–4678.
53. S. H. Jathar, T. D. Gordon, C. J. Hennigan, H. O. T. Pye, G. Pouliot, P. J. Adams, N. M. Donahue and A. L. Robinson, *Proc. Natl. Acad. Sci.*, 2014, **111**, 10473–10478.
54. S. M. Platt, I. El Haddad, A. A. Zardini, M. Clairotte, C. Astorga, R. Wolf, J. G. Slowik, B. Temime-Roussel, N. Marchand, I. Ježek, L. Drinovec, G. Močnik, O. Möhler, R. Richter, P. Barmet, F. Bianchi, U. Baltensperger and A. S. H. Prévôt, *Atmos. Chem. Phys.*, 2013, **13**, 9141–9158.
55. E. Z. Nordin, A. C. Eriksson, P. Roldin, P. T. Nilsson, J. E. Carlsson, M. K. Kajos, H. Hellén, C. Wittbom, J. Rissler, J. Löndahl, E. Swietlicki, B. Svenningsson, M. Bohgard, M. Kulmala, M. Hallquist and J. H. Pagels, *Atmos. Chem. Phys.*, 2013, **13**, 6101–6116.
56. S. M. Platt, I. E. Haddad, S. M. Pieber, R. J. Huang, A. A. Zardini, M. Clairotte, R. Suarez-Bertoa, P. Barmet, L. Pfaffenberger, R. Wolf, J. G. Slowik, S. J. Fuller, M. Kalberer, R. Chirico, J. Dommen, C. Astorga, R. Zimmermann, N. Marchand, S. Hellebust, B. Temime-Roussel, U. Baltensperger and A. S. H. Prévôt, *Nat. Commun.*, 2014, **5**, 4749.
57. M. A. Miracolo, C. J. Hennigan, M. Ranjan, N. T. Nguyen, T. D. Gordon, E. M. Lipsky, A. A. Presto, N. M. Donahue and A. L. Robinson, *Atmos. Chem. Phys.*, 2011, **11**, 4135–4147.
58. A. P. Grieshop, J. M. Logue, N. M. Donahue and A. L. Robinson, *Atmos. Chem. Phys.*, 2009, **8**, 1263–1277.

59. M. F. Heringa, P. F. DeCarlo, R. Chirico, T. Tritscher, J. Dommen, E. Weingartner, R. Richter, G. Wehrle, A. S. H. Prévôt and U. Baltensperger, *Atmos. Chem. Phys.*, 2011, **11**, 5945–5957.

60. A. M. Sage, E. A. Weitkamp, A. L. Robinson and N. M. Donahue, *Atmos. Chem. Phys.*, 2008, **8**, 1139–1152.

61. A. P. Grieshop, N. M. Donahue and A. L. Robinson, *Atmos. Chem. Phys.*, 2009, **8**, 2227–2240.

62. K. Dzepina, R. M. Volkamer, S. Madronich, P. Tulet, I. M. Ulbrich, Q. Zhang, C. D. Cappa, P. J. Ziemann and J. L. Jimenez, *Atmos. Chem. Phys. Discuss.*, 2009, **9**, 4417–4488.

63. N. M. Donahue, A. L. Robinson and S. N. Pandis, *Atmos. Environ.*, 2009, **43**, 94–106.

64. R. M. Harrison, A. M. Jones, D. C. Beddows, M. Dall'Osto and I. Nikolova, *Atmos. Environ.*, 2016, **125**(Part A), 1–7.

65. A. P. Sullivan, R. J. Weber, A. L. Clements, J. R. Turner, M. S. Bae and J. J. Schauer, *Geophys. Res. Lett.*, 2004, **31**, L13105.

66. A. P. Sullivan, R. E. Peltier, C. A. Brock, J. A. de Gouw, J. S. Holloway, C. Warneke, A. G. Wollny and R. J. Weber, *J. Geophys. Res.: Atmos.*, 2006, **111**, D23S46.

67. L. M. Russell, R. Bahadur and P. J. Ziemann, *Proc. Natl. Acad. Sci. U. S. A.*, 2011, **108**, 3516–3521.

68. Y. Zhao, N. M. Kreisberg, D. R. Worton, G. Isaacman, D. R. Gentner, A. W. H. Chan, R. J. Weber, S. Liu, D. A. Day, L. M. Russell, S. V. Hering and A. H. Goldstein, *J. Geophys. Res.: Atmos.*, 2013, **118**, 11,388–11,398.

69. P. Sannigrahi, A. P. Sullivan, R. J. Weber and E. D. Ingall, *Environ. Sci. Technol.*, 2006, **40**, 666–672.

70. F. Moretti, E. Tagliavini, S. Decesari, M. C. Facchini, M. Rinaldi and S. Fuzzi, *Environ. Sci. Technol.*, 2008, **42**, 4844–4849.

71. E. Finessi, S. Decesari, M. Paglione, L. Giulianelli, C. Carbone, S. Gilardoni, S. Fuzzi, S. Saarikoski, T. Raatikainen, R. Hillamo, J. Allan, T. F. Mentel, P. Tiitta, A. Laaksonen, T. Petäjä, M. Kulmala, D. R. Worsnop and M. C. Facchini, *Atmos. Chem. Phys.*, 2012, **12**, 941–959.

72. M. Paglione, S. Saarikoski, S. Carbone, R. Hillamo, M. C. Facchini, E. Finessi, L. Giulianelli, C. Carbone, S. Fuzzi, F. Moretti, E. Tagliavini, E. Swietlicki, K. Eriksson Stenström, A. S. H. Prévôt, P. Massoli, M. Canaragatna, D. Worsnop and S. Decesari, *Atmos. Chem. Phys.*, 2014, **14**, 5089–5110.

73. N. J. Pekney, C. I. Davidson, L. M. Zhou and P. K. Hopke, *Aerosol Sci. Technol.*, 2006, **40**, 952–961.

74. Y. Zeng and P. Hopke, *Atmos. Environ.*, 1989, **23**, 1499–1509.

75. R. Subramanian, N. M. Donahue, A. Bernardo-Bricker, W. F. Rogge and A. L. Robinson, *Atmos. Environ.*, 2006, **40**, 8002–8019.

76. A. L. Robinson, N. M. Donahue and W. F. Rogge, *J. Geophys. Res.: Atmos.*, 2006, **111**, D03302.

77. J. Jung, C. Fountoukis, P. J. Adams and S. N. Pandis, *J. Geophys. Res.: Atmos.*, 2010, **115**, D03203.

78. J. R. Pierce and P. J. Adams, *J. Geophys. Res.: Atmos.*, 2006, **111**, D06203.
79. K. S. Carslaw, L. A. Lee, C. L. Reddington, K. J. Pringle, A. Rap, P. M. Forster, G. W. Mann, D. V. Spracklen, M. T. Woodhouse, L. A. Regayre and J. R. Pierce, *Nature*, 2013, **503**, 67–71.
80. A. Zelenyuk, D. Imre, E. J. Nam, Y. Han and K. Mueller, *Int. J. Mass Spectrom.*, 2008, **275**, 1–10.
81. P. J. G. Rehbein, C.-H. Jeong, M. L. McGuire and G. J. Evans, *Aerosol Sci. Technol.*, 2012, **46**, 584–595.
82. T. P. Rebotier and K. A. Prather, *Anal. Chim. Acta*, 2007, **585**, 38–54.
83. M. Dall'Osto and R. M. Harrison, *Atmos. Chem. Phys.*, 2012, **12**, 4127–4142.
84. A. P. Ault, T. M. Peters, E. J. Sawvel, G. S. Casuccio, R. D. Willis, G. A. Norris and V. H. Grassian, *Environ. Sci. Technol.*, 2012, 4331–4339.
85. M. P. Tolocka, D. A. Lake, M. V. Johnston and A. S. Wexler, *J. Geophys. Res.: Atmos.*, 2005, **110**, D07S04.
86. S. M. Toner, L. G. Shields, D. A. Sodeman and K. A. Prather, *Atmos. Environ.*, 2008, **42**, 568–581.
87. C. Giorio, A. Tapparo, M. Dall'Osto, D. C. Beddows, J. Esser-Gietl, R. M. Healy and R. M. Harrison, *Environ. Sci. Technol.*, 2015, 150219170910008.
88. R. C. Sullivan and K. A. Prather, *Anal. Chem.*, 2005, **77**, 3861–3885.
89. K. A. Pratt and K. A. Prather, *Mass Spectrom. Rev.*, 2012, **31**, 17–48.
90. A. Laskin, J. Laskin and S. A. Nizkorodov, *Environ. Chem.*, 2012, **9**, 163.
91. A. Laskin, J. P. Cowin and M. J. Iedema, *J. Electron Spectrosc. Relat. Phenom.*, 2006, **150**, 260–274.
92. M. D. Zauscher, M. J. K. Moore, G. S. Lewis, S. V. Hering and K. A. Prather, *Anal. Chem.*, 2011, **83**, 2271–2278.
93. B. R. Bzdek, C. A. Zordan, G. W. Luther and M. V. Johnston, *Aerosol Sci. Technol.*, 2011, **45**, 1041–1048.
94. A. Held, G. J. Rathbone and J. N. Smith, *Aerosol Sci. Technol.*, 2009, **43**, 264–272.
95. J. P. Klems, C. a. Zordan, M. R. Pennington and M. V. Johnston, *Anal. Chem.*, 2012, **84**, 2253–2259.
96. J. N. Smith, K. C. Barsanti, H. R. Friedli, M. Ehn, M. Kulmala, D. R. Collins, J. H. Scheckman, B. J. Williams and P. H. McMurry, *Proc. Natl. Acad. Sci. U. S. A.*, 2010, **107**, 6634–6639.
97. J. N. Smith, M. J. Dunn, T. M. VanReken, K. Iida, M. R. Stolzenburg, P. H. McMurry and L. G. Huey, *Geophys. Res. Lett.*, 2008, **35**, L04808.
98. A. P. Ault, C. J. Gaston, Y. Wang, G. Dominguez, M. H. Thiemens and K. A. Prather, *Environ. Sci. Technol.*, 2010, **44**, 1954–1961.
99. R. C. Moffet, B. de Foy, L. T. Molina, M. J. Molina and K. A. Prather, *Atmos. Chem. Phys.*, 2008, **8**, 4499–4516.
100. R. C. Sullivan and K. A. Prather, *Environ. Sci. Technol.*, 2007, **41**, 8062–8069.
101. M. Chen, M. Titcombe, J. Jiang, C. Jen, C. Kuang, M. L. Fischer, F. L. Eisele, J. I. Siepmann, D. R. Hanson, J. Zhao and P. H. McMurry, *Proc. Natl. Acad. Sci.*, 2012, **109**, 18713–18718.

102. M. Kulmala, J. Kontkanen, H. Junninen, K. Lehtipalo, H. E. Manninen, T. Nieminen, T. Petäjä, M. Sipilä, S. Schobesberger, P. Rantala, A. Franchin, T. Jokinen, E. Järvinen, M. Äijälä, J. Kangasluoma, J. Hakala, P. P. Aalto, P. Paasonen, J. Mikkilä, J. Vanhanen, J. Aalto, H. Hakola, U. Makkonen, T. Ruuskanen, R. L. Mauldin, J. Duplissy, H. Vehkamäki, J. Bäck, A. Kortelainen, I. Riipinen, T. Kurtén, M. V. Johnston, J. N. Smith, M. Ehn, T. F. Mentel, K. E. J. Lehtinen, A. Laaksonen, V.-M. Kerminen and D. R. Worsnop, *Science*, 2013, **339**, 943–946.

103. P. H. McMurry, *J. Colloid Interface Sci.*, 1980, **78**, 513–527.

104. C. Kuang, M. Chen, J. Zhao, J. Smith, P. H. McMurry and J. Wang, *Atmos. Chem. Phys.*, 2012, **12**, 3573–3589.

105. S. Schobesberger, H. Junninen, F. Bianchi, G. Lönn, M. Ehn, K. Lehtipalo, J. Dommen, S. Ehrhart, I. K. Ortega, A. Franchin, T. Nieminen, F. Riccobono, M. Hutterli, J. Duplissy, J. Almeida, A. Amorim, M. Breitenlechner, A. J. Downard, E. M. Dunne, R. C. Flagan, M. Kajos, H. Keskinen, J. Kirkby, A. Kupc, A. Kürten, T. Kurtén, A. Laaksonen, S. Mathot, A. Onnela, A. P. P. L. Rondo, F. D. Santos, S. Schallhart, R. Schnitzhofer, M. Sipilä, A. Tomé, G. Tsagkogeorgas, H. Vehkamäki, D. Wimmer, U. Baltensperger, K. S. Carslaw, J. Curtius, A. Hansel, T. Petäjä, M. Kulmala, N. M. Donahue and D. R. Worsnop, *Proc. Natl. Acad. Sci.*, 2013, **110**, 17223–17228.

106. F. Riccobono, S. Schobesberger, C. E. Scott, J. Dommen, I. K. Ortega, L. Rondo, J. Almeida, A. Amorim, F. Bianchi, M. Breitenlechner, A. David, A. Downard, E. Dunne, J. Duplissy, S. Ehrhart, R. C. Flagan, A. Franchin, A. Hansel, H. Junninen, M. Kajos, H. Keskinen, A. Kupc, O. Kupiainen, A. Kürten, T. Kurtén, A. N. Kvashin, A. Laaksonen, K. Lehtipalo, V. Makhmutov, S. Mathot, T. Nieminen, T. Olenius, A. Onnela, T. Petäjä, A. P. Praplan, F. D. Santos, S. Schallhart, J. H. Seinfeld, M. Sipilä, D. V. Spracklen, Y. Stozhkov, F. Stratmann, A. Tomé, G. Tsagkogeorgas, P. Vaattovaara, H. Vehkamäki, Y. Viisanen, A. Vrtala, P. E. Wagner, E. Weingartner, H. Wex, D. Wimmer, K. S. Carslaw, J. Curtius, N. M. Donahue, J. Kirkby, M. Kulmala, D. R. Worsnop and U. Baltensperger, *Science*, 2014, **344**, 717–721.

107. R. Zhang, L. Wang, A. F. Khalizov, J. Zhao, J. Zheng, R. L. McGraw and L. T. Molina, *Proc. Natl. Acad. Sci.*, 2009, **106**, 17650–17654.

108. C. O. Stanier, A. Y. Khlystov and S. N. Pandis, *Aerosol Sci. Technol.*, 2004, **38**, 253–264.

109. T. Hussein, J. Martikainen, H. Junninen, L. Sogacheva, R. Wagner, M. Dal Maso, I. Riipinen, P. P. Aalto and M. Kulmala, *Tellus, Ser. B*, 2008, **60**, 509–521.

110. S. Guo, M. Hu, M. L. Zamora, J. Peng, D. Shang, J. Zheng, Z. Du, Z. Wu, M. Shao, L. Zeng, M. J. Molina and R. Zhang, *Proc. Natl. Acad. Sci.*, 2014, **111**, 17373–17378.

111. A. Hamed, J. Joutsensaari, S. Mikkonen, L. Sogacheva, M. Dal Maso, M. Kulmala, F. Cavalli, S. Fuzzi, M. C. Facchini, S. Decesari, M. Mircea,

K. E. J. Lehtinen and A. Laaksonen, *Atmos. Chem. Phys.*, 2007, 7, 355–376.

112. C. O. Stanier, A. Y. Khlystov, W. R. Chan, M. Mandiro and S. N. Pandis, *Aerosol Sci. Technol.*, 2004, **38**, 215–228.

113. M. Dal Maso, M. Kulmala, I. Riipinen, R. Wagner, T. Hussein, P. Aalto and K. Lehtinen, *Boreal Environ. Res.*, 2005, **10**, 323–336.

114. M. Brines, M. Dall'Osto, D. C. S. Beddows, R. M. Harrison, F. Gómez-Moreno, L. Núñez, B. Artíñano, F. Costabile, G. P. Gobbi, F. Salimi, L. Morawska, C. Sioutas and X. Querol, *Atmos. Chem. Phys.*, 2015, **15**, 5929–5945.

115. M. Minguillón, M. Brines, N. Pérez, C. Reche, M. Pandolfi, A. Fonseca, F. Amato, A. Alastuey, A. Lyasota, B. Codina, H.-K. Lee, H.-R. Eun, K.-H. Ahn and X. Querol, *Atmos. Res.*, 2015, **164–165**, 118–130.

116. Q. Zhang, C. O. Stanier, M. R. Canagaratna, J. T. Jayne, D. R. Worsnop, S. N. Pandis and J. L. Jimenez, *Environ. Sci. Technol.*, 2004, **38**, 4797–4809.

117. J. G. Jung, P. J. Adams and S. N. Pandis, *Atmos. Environ.*, 2006, **40**, 2248–2259.

118. I. Riipinen, J. R. Pierce, T. Yli-Juuti, T. Nieminen, S. Häkkinen, M. Ehn, H. Junninen, K. Lehtipalo, T. Petäjä, J. Slowik, R. Chang, N. C. Shantz, J. P. D. Abbatt, W. R. Leaitch, V.-M. Kerminen, D. R. Worsnop, S. N. Pandis, N. M. Donahue and M. Kulmala, *Atmos. Chem. Phys.*, 2011, **11**, 3865–3878.

119. I. Napari, M. Noppel, H. Vehkamäki and M. Kulmala, *J. Geophys. Res.: Atmos.*, 2002, **107**, AAC 6-1–AAC 6-6.

120. J. Kirkby, J. Curtius, J. Almeida, E. Dunne, J. Duplissy, S. Ehrhart, A. Franchin, S. Gagne, L. Ickes, A. Kuerten, A. Kupc, A. Metzger, F. Riccobono, L. Rondo, S. Schobesberger, G. Tsagkogeorgas, D. Wimmer, A. Amorim, F. Bianchi, M. Breitenlechner, A. David, J. Dommen, A. Downard, M. Ehn, R. C. Flagan, S. Haider, A. Hansel, D. Hauser, W. Jud, H. Junninen, F. Kreissl, A. Kvashin, A. Laaksonen, K. Lehtipalo, J. Lima, E. R. Lovejoy, V. Makhmutov, S. Mathot, J. Mikkila, P. Minginette, S. Mogo, T. Nieminen, A. Onnela, P. Pereira, T. Petaja, R. Schnitzhofer, J. H. Seinfeld, M. Sipila, Y. Stozhkov, F. Stratmann, A. Tome, J. Vanhanen, Y. Viisanen, A. Vrtala, P. E. Wagner, H. Walther, E. Weingartner, H. Wex, P. M. Winkler, K. S. Carslaw, D. R. Worsnop, U. Baltensperger and M. Kulmala, *Nature*, 2011, **476**, 429–432.

121. T. Kurtén, L. Torpo, C.-G. Ding, H. Vehkamäki, M. R. Sundberg, K. Laasonen and M. Kulmala, *J. Geophys. Res.: Atmos.*, 2007, **112**, D04210.

122. I. K. Ortega, T. Kurtén, H. Vehkamäki and M. Kulmala, *Atmos. Chem. Phys.*, 2008, **8**, 2859–2867.

123. J. Almeida, S. Schobesberger, A. Kürten, I. K. Ortega, O. Kupiainen, A. P. Praplan, A. Amorim, F. Bianchi, M. Breitenlechner, A. David, J. Dommen, N. M. Donahue, A. Downard, E. Dunne, J. Duplissy,

S. Ehrhart, R. C. Flagan, A. Franchin, R. Guida, A. Hansel, H. Junninen, M. Kajos, H. Keskinen, A. Kupc, T. Kurtén, A. N. Kvashin, A. Laaksonen, K. Lehtipalo, J. Leppä, V. Loukonen, V. Makhmutov, S. Mathot, M. J. McGrath, T. Nieminen, T. Olenius, A. Onnela, T. Petäjä, F. Riccobono, I. Riipinen, L. Rondo, F. D. Santos, S. Schallhart, R. Schnitzhofer, J. H. Seinfeld, M. Sipilä, Y. Stozhkov, F. Stratmann, A. Tomé, G. Tsagkogeorgas, Y. Viisanen, A. Vrtala, P. E. Wagner, E. Weingartner, H. Wex, D. Wimmer, P. Ye, T. Yli-Juuti, K. S. Carslaw, M. Kulmala, J. Curtius, U. Baltensperger, D. R. Worsnop, H. Vehkamäki and J. Kirkby, *Nature*, 2013, **502**, 359–363.

124. C. N. Jen, P. H. McMurry and D. R. Hanson, *J. Geophys. Res.: Atmos.*, 2014, **119**, 7502–7514.

125. A. Metzger, B. Verheggen, J. Dommen, J. Duplissy, A. S. H. Prevot, E. Weingartner, I. Riipinen, M. Kulmala, D. V. Spracklen, K. S. Carslaw and U. Baltensperger, *Proc. Natl. Acad. Sci.*, 2010, **107**, 6646–6651.

126. L. N. Posner and S. N. Pandis, *Atmos. Environ.*, 2015, **111**, 103–112.

127. G. A. Ban-Weiss, M. M. Lunden, T. W. Kirchstetter and R. A. Harley, *J. Aerosol Sci.*, 2010, **41**, 5–12.

128. D. Kittelson, W. Watts and J. Johnson, *J. Aerosol Sci.*, 2006, **37**, 913–930.

129. D. Kittelson, W. Watts, J. Johnson, J. Schauer and D. Lawson, *J. Aerosol Sci.*, 2006, **37**, 931–949.

130. J. Rissler, E. Swietlicki, J. Zhou, G. Roberts, M. O. Andreae, L. V. Gatti and P. Artaxo, *Atmos. Chem. Phys.*, 2004, **4**, 2119–2143.

131. J. Rissler, A. Vestin, E. Swietlicki, G. Fisch, J. Zhou, P. Artaxo and M. O. Andreae, *Atmos. Chem. Phys.*, 2006, **6**, 471–491.

132. A. D. Clarke, S. R. Owens and J. Zhou, *J. Geophys. Res.: Atmos.*, 2006, **111**, D06202.

133. K. A. Prather, T. H. Bertram, V. H. Grassian, G. B. Deane, M. D. Stokes, P. J. DeMott, L. I. Aluwihare, B. P. Palenik, F. Azam, J. H. Seinfeld, R. C. Moffet, M. J. Molina, C. D. Cappa, F. M. Geiger, G. C. Roberts, L. M. Russell, A. P. Ault, J. Baltrusaitis, D. B. Collins, C. E. Corrigan, L. A. Cuadra-Rodriguez, C. J. Ebben, S. D. Forestieri, T. L. Guasco, S. P. Hersey, M. J. Kim, W. F. Lambert, R. L. Modini, W. Mui, B. E. Pedler, M. J. Ruppel, O. S. Ryder, N. G. Schoepp, R. C. Sullivan and D. Zhao, *Proc. Natl. Acad. Sci.*, 2013, **110**, 7550–7555.

134. C. O. Stanier, A. Y. Khlystov and S. N. Pandis, *Atmos. Environ.*, 2004, **38**, 3275–3284.

135. G. A. Ban-Weiss, J. P. McLaughlin, R. A. Harley, M. M. Lunden, T. W. Kirchstetter, A. J. Kean, A. W. Strawa, E. D. Stevenson and G. R. Kendall, *Atmos. Environ.*, 2008, **42**, 220–232.

136. G. A. Ban-Weiss, M. M. Lunden, T. W. Kirchstetter and R. A. Harley, *Environ. Sci. Technol.*, 2009, **43**, 1419–1424.

137. T. Kuwayama, C. R. Ruehl and M. J. Kleeman, *Environ. Sci. Technol.*, 2013, **47**, 13957–13966.

138. J. Hu, H. Zhang, S. Chen, Q. Ying, C. Wiedinmyer, F. Vandenberghe and M. J. Kleeman, *Environ. Sci. Technol.*, 2014, **48**, 4980–4990.

139. K. M. Zhang and A. S. Wexler, *Atmos. Environ.*, 2004, **38**, 6643–6653.
140. K. Zhang, A. S. Wexler, Y. F. Zhu, W. C. Hinds and C. Sioutas, *Atmos. Environ.*, 2004, **38**, 6655–6665.
141. K. M. Zhang, A. S. Wexler, D. A. Niemeier, Y. F. Zhu, W. C. Hinds and C. Sioutas, *Atmos. Environ.*, 2005, **39**, 4155–4166.
142. S. Biswas, S. Hu, V. Verma, J. D. Herner, W. H. Robertson, A. Ayala and C. Sioutas, *Atmos. Environ.*, 2008, **42**, 5622–5634.
143. H. E. Institute, *Traffic-Related Air Pollution: A Critical Review of the Literature on Emissions, Exposure, and Health Effects*, Special Report 17, 2010.
144. J. R. Pierce and P. J. Adams, *Atmos. Chem. Phys.*, 2007, 7, 1367–1379.
145. M. K. Shrivastava, T. E. Lane, N. M. Donahue, S. N. Pandis and A. L. Robinson, *J. Geophys. Res.: Atmos.*, 2008, **113**, D18301.
146. N. Ma and W. Birmili, *Sci. Total Environ.*, 2015, **512–513**, 154–166.
147. Y. Tan, E. M. Lipsky, R. Saleh, A. L. Robinson and A. A. Presto, *Environ. Sci. Technol.*, 2014, **48**, 14186–14194.
148. Y. Tan, T. R. Dallmann, A. L. Robinson and A. A. Presto, *Atmos. Environ.*, 2016, **134**, 51–60, 2016.
149. C. Fountoukis, D. Koraj, H. D. van der Gon, P. Charalampidis, C. Pilinis and S. Pandis, *Atmos. Environ.*, 2013, **68**, 24–32.
150. C. A. Stroud, P. A. Makar, M. D. Moran, W. Gong, S. Gong, J. Zhang, K. Hayden, C. Mihele, J. R. Brook, J. P. D. Abbatt and J. G. Slowik, *Atmos. Chem. Phys.*, 2011, **11**, 3107–3118.
151. A. A. Karner, D. S. Eisinger and D. A. Niemeier, *Environ. Sci. Technol.*, 2010, **44**, 5334–5344.
152. N. M. Donahue, W. K. Chuang, S. A. Epstein, J. H. Kroll, D. R. Worsnop, A. L. Robinson, P. J. Adams and S. N. Pandis, *Environ. Chem.*, 2013, **10**, 151–157.
153. A. H. Goldstein, S. C. Wofsy and C. M. Spivakovsky, *J. Geophys. Res.: Atmos.*, 1995, **100**, 21023–21033.
154. T. Canty and K. Minschwaner, *J. Geophys. Res.: Atmos.*, 2002, **107**, ACH 1-1–ACH 1-6.
155. S. Vaughan, T. Ingham, L. K. Whalley, D. Stone, M. J. Evans, K. A. Read, J. D. Lee, S. J. Moller, L. J. Carpenter, A. C. Lewis, Z. L. Fleming and D. E. Heard, *Atmos. Chem. Phys.*, 2012, **12**, 2149–2172.
156. W. Rattanavaraha, E. Rosen, H. Zhang, Q. Li, K. Pantong and R. M. Kamens, *Atmos. Environ.*, 2011, **45**, 3848–3855.
157. L. Künzi, M. Krapf, N. Daher, J. Dommen, N. Jeannet, S. Schneider, S. Platt, J. G. Slowik, N. Baumlin, M. Salathe, A. H. Prévôt, M. Kalberer, C. Strähl, L. Dümbgen, C. Sioutas, U. Baltensperger and M. Geiser, *Sci. Rep.*, 2015, **5**, 11801 EP.
158. N. M. Donahue, K. M. Henry, T. F. Mentel, A. K. Scharr, C. Spindler, B. Bohn, T. Brauers, H. P. Dorn, H. Fuchs, R. Tillmann, A. Wahner, H. Saathoff, K. H. Naumann, O. Möhler, T. Leisner, L. Müller, M.-C. Reinnig, T. Hoffmann, K. Salow, M. Hallquist, M. Frosch, M. Bilde, T. Tritscher, P. Barmet, A. P. Praplan, P. F. DeCarlo,

J. Dommen, A. S. H. Prévôt and U. Baltensperger, *Proc. Natl. Acad. Sci.*, 2012, **109**, 13503–13508.

159. U. Baltensperger, M. Kalberer, J. Dommen, D. Paulsen, M. Alfarra, H. Coe, R. Fisseha, A. Gascho, M. Gysel, S. Nyeki, M. Sax, M. Steinbacher, A. Prevot, S. Sjogren, E. Weingartner and R. Zenobi, *Faraday Discuss.*, 2005, **130**, 265–278.

160. J. E. Shilling, Q. Chen, S. M. King, T. Rosenoern, J. H. Kroll, D. R. Worsnop, P. F. DeCarlo, A. C. Aiken, D. Sueper, J. L. Jimenez and S. T. Martin, *Atmos. Chem. Phys.*, 2009, **9**, 771–782.

161. L. Hildebrandt Ruiz, A. Paciga, K. Cerully, A. Nenes, N. M. Donahue and S. N. Pandis, *Atmos. Chem. Phys.*, 2015, **15**, 8301–8313.

162. N. M. Donahue, S. A. Epstein, S. N. Pandis and A. L. Robinson, *Atmos. Chem. Phys.*, 2011, **11**, 3303–3318.

Source Apportionment: Principles and Methods

J. G. WATSON,* J. C. CHOW, L.-W. A. CHEN, G. ENGLING AND X. L. WANG

ABSTRACT

Receptor model source apportionment has been facilitated by the availability of particulate matter (PM) speciation networks that measure elements, ions, and carbon fractions, and the availability of effective variance (EV)- and positive matrix factorization (PMF)-chemical mass balance (CMB) solutions to identify and quantify source contributions. However, receptor modeling software is too often applied without a thorough evaluation of the results. Quantitative source contribution estimates derived from these solutions must be challenged as part of a larger modeling and data analysis effort that supplies a "weight of evidence" for the major contributors. PMF-derived source factors should be compared with measured source profiles to identify potential source mixing within a factor and collinearities among factors. EV-CMB solutions should justify the use of measured profiles from other areas as representing those in the study area. Cost-effective methods exist to obtain more relevant source profiles that better represent the potential contributors. As pollution controls reduce primary emissions, elemental source markers and elemental carbon are becoming less useful for distinguishing among source types. Much more information can be obtained from speciation network filters at minimal additional cost to provide more specific markers related to important source types, such as solid fuel

*Corresponding author.

Issues in Environmental Science and Technology No. 42
Airborne Particulate Matter: Sources, Atmospheric Processes and Health
Edited by R.E. Hester, R.M. Harrison and X. Querol
© The Royal Society of Chemistry 2016
Published by the Royal Society of Chemistry, www.rsc.org

combustion for heating and cooking and secondary organic aerosol contributions. Receptor models have been productive for identifying sources, quantifying their contributions, and justifying regulations for residential wood combustion and cooking emission reduction strategies. When used as complements to source-oriented models and emission inventory development, air quality management practices can more accurately allocate pollution control resources.

1   Introduction

Receptor-oriented source apportionment models intend to identify and quantify contributions from different source types to pollutant concentrations measured at receptors. The receptors are usually neighborhood-, urban-, and regional-scale air quality monitoring stations established to determine compliance with ambient air quality standards (AAQS). The most common objective of a source apportionment study is to determine which source types require additional emission reductions to bring ambient concentrations into AAQS compliance. High quantitative precision is not needed for this objective, as long as source contributions can be classified as: (1) dominant, which would make control measures a top priority; (2) moderate, which would orient efforts toward those sources that are most practical or cost-effective to control; or (3) small compared to other contributors, and therefore of lower control priority. For suspended particulate matter (PM), AAQS apply to 24-hour and annual averages of mass concentration in the $PM_{2.5}$ and PM_{10} (particles with aerodynamic diameters <2.5 and <10 µm, respectively) size fractions,[3] and source apportionment is applied to these sizes and time periods.

Previous reviews[4–18] provide details on the guidance, mathematics, measurements, applications, evaluation tests, and results for receptor modeling. Since the early 1970s, more than 2500 journal articles and reports related to receptor model source apportionment have been published, many with credible source contributions, but there is a growing number with dubious source identifications and contribution estimates. While the availability of PM composition measurements from speciation networks coupled with receptor modeling software supplied by the US Environmental Protection Agency (EPA)[20] has made it easy to produce source apportionment outputs, these advances have not necessarily added to, and have often obscured, understanding of how PM air quality can be improved.

This chapter is directed at the consumers of source apportionment results, those charged with managing air quality through emission reduction strategies. These consumers should use receptor model results as part of a broader "weight of evidence" approach[21,22] that challenges their accuracy in light of other information, such as that derived from applying multiple solution methods, evaluating performance measures, obtaining more source-specific PM measurements, conducting detailed case studies, and reconciling with emission inventories and source-oriented models. Modern air quality management also requires more specific source identification than "mobile sources," "biomass burning," "industry," "fugitive dust," "sea salt," and "secondary sulfates and nitrates," which are often the categories indicated by a receptor model study. Contributions from these sources are always found in most urban and many regional settings, and control measures have been devised and implemented.[24] As these well-documented sources are controlled through changes in operating conditions, fuels and feedstocks, hardware, and emission treatment devices, it becomes more important to identify less obvious sources and "high-emitters" within the broader categories. As primary emissions decrease, the importance of secondary organic aerosol (SOA) increases.[28]

Refining source contributions is seldom possible when using only the mass, elements, water-soluble inorganic ions, and carbon fractions available from widely-used speciation networks such as the US National Park Service's Interagency Monitoring of PROtected Visual Environments (IMPROVE)[29] and the US EPA's Chemical Speciation Network (CSN).[30] The value of such long-term networks has been recognized in other countries, with adaptations currently being implemented in Canada,[31] Europe,[32] mainland China,[33] Hong Kong,[34,35] and India,[36] with shorter-term campaigns in many other countries. However, more information can, and should, be obtained from existing compliance and speciation network samples that can address these challenges.

2 Diurnal, Spatial, and Chemical Patterns Indicate PM Origins

Source apportionment models are based on solutions to the Chemical Mass Balance (CMB) equations outlined in Box 1. A more descriptive term might be "Property Mass Balance," since source markers are not restricted to PM chemical composition; they may also include light absorption, morphological, and size observables. Varying patterns of these properties in time and space indicate the nature of the contributors. Figure 1 compares wintertime hourly changes in PM mass from two nearby sites in a mountain valley. This example illustrates the utility of monitors with different zones of representation (ZOR):[37] (1) micro-scale (10–100 m); (2) middle-scale (100–500 m); (3) neighborhood-scale (500 m–4 km); (4) urban-scale (4–100 km); (5) regional-scale (100–1000 km); (6) continental scale (1000–10 000 km); and (7) global-scale (>10 000 km). Samples from each of the smaller ZORs include increments in source contributions over those from the larger ZORs. Most compliance and speciation sites are neighborhood- to urban-scale ZORs, although there are a growing number of networks with micro- to middle-scale ZORs, especially near roadsides.[38] The ZOR may change from day to day, or even within a day, as evidenced by the convergence of $PM_{2.5}$ and PM_{10} levels at mid-day (*e.g.*, Figure 1) when atmospheric mixing improves.

Box 1 The Chemical Mass Balance (CMB) equations.[1]

$$C_{iklm} = \sum_{j=1}^{J} F_{ijm}S_{jklm}$$

$i =$ one of I PM chemical components or physical properties selected because they are important components of mass (>1%), source markers, or have adverse effects on health, visibility, climate, or ecosystem damage.

$j =$ one of J potentially contributing source types, emissions with similar contents and mass proportions of selected properties i.

$k =$ one of K PM sampling intervals at the same location.

$l =$ one of L sampling locations.

$m =$ one of M PM size fraction, usually $PM_{2.5}$ and PM_{10}. The coarse fraction ($PM_{10-2.5}$) is the difference between the two.

$C_{iklm} =$ concentration ($\mu g\ m^{-3}$) of PM component i measured for time period k at location l in size range m.

$F_{ijm} =$ mass fraction of component i of source type j in size range m as it would be measured at the receptor.

$S_{jklm} =$ contribution ($\mu g\ m^{-3}$) of source type j for time period k at location l in size range m.

The utility of different particle size fractions is evident in comparing Figure 1a–b, as the coarse fraction is nearly always dominated by fugitive dust, much of which deposits near the point of origin.[39] Figure 1 also illustrates the need to understand the context of available measurements, as the emissions interact with the meteorology and with sporadic activities (such as road de-icing) that accompany certain meteorological conditions. More species-specific time-resolved aerosol patterns can be incorporated into CMB solutions with appropriate instrumentation.[40,41] Short-duration samples (*e.g.*, 5 min averages) often show sharp peaks superimposed on a slowly-varying background, which can be de-convoluted to separate nearby contributions from those having a larger zone of influence.[42]

Figure 2 illustrates the value of inexpensive short-term studies to confirm the source and spatial extent of the nighttime increase in Figure 1a. Brown carbon (BrC) is a biomass burning indicator[43] available from multi-wavelength aethalometers.[44] BrC can also be measured on filter samples[45] and will soon be reported with thermal/optical carbon fractions in the IMPROVE network.[46,47] Multi-wavelength light absorption measurements have been used in receptor models to separate wintertime wood smoke from vehicle engine exhaust emissions.[48–50]

Figure 3 compares portions of three source profiles representing different categories (stationary industrial *vs.* area-wide fugitive dust) and different sub-sources within the fugitive dust category. The *n*-alkanes are chosen

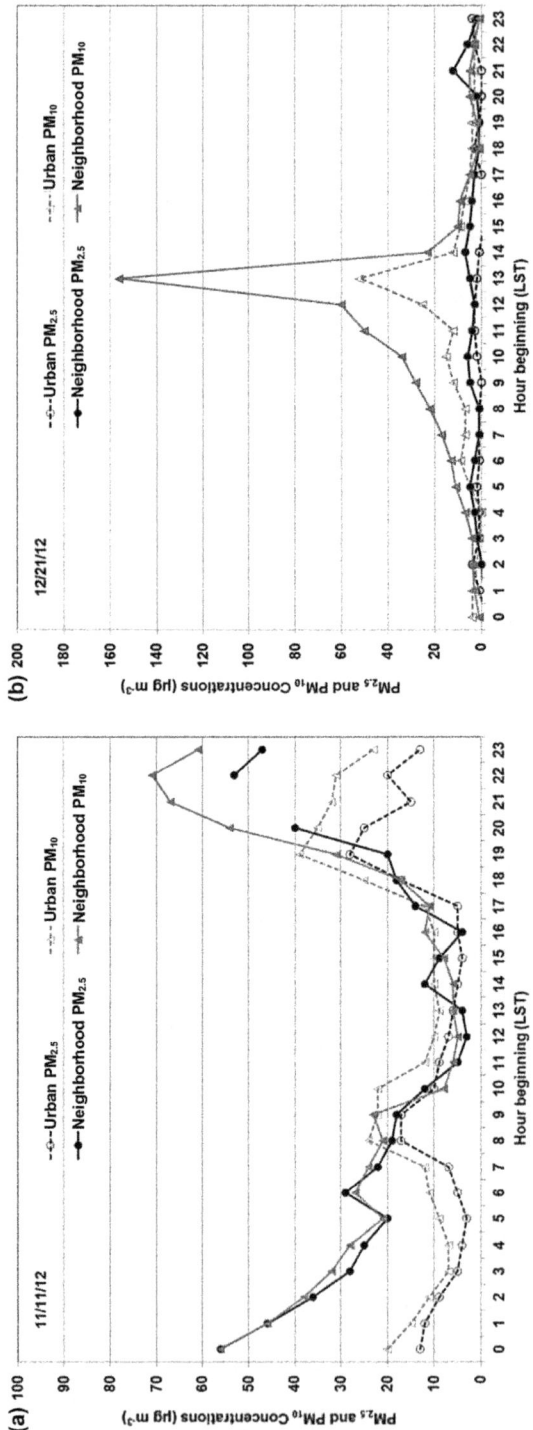

Figure 1 Hourly PM$_{2.5}$ and PM$_{10}$ patterns measured with a Beta Attenuation Monitor (BAM) at neighborhood- and urban-scale compliance monitors separated by <5 km in a US mountain valley during winter. Pattern (a), which occurred during a cold, stagnant high pressure system, shows evidence of a morning engine exhaust contribution from 0700 to 1000 local standard time (LST) until the shallow surface layer couples to the valley-wide layer. The rapid increases starting at ~1800 LST indicate a combination of engine exhaust and home heating, especially at the neighborhood-scale site that is surrounded by residences known to use wood for heating. The small PM$_{10}$ increment indicates that fugitive dust contributions, typically from paved roads given the site environs, are small. Pattern (b) was measured during a snowstorm accompanied by an unstable atmosphere that disperses emissions. The large PM$_{10-2.5}$ (PM$_{10}$ minus PM$_{2.5}$) increase at mid-day corresponds to the application of de-icing material on the roadways. Long-term chemical speciation measurements at the urban-scale site revealed increased secondary nitrate formation during snow-cover,[114] which appears to be a common feature of populated mountain valleys during snowy winters.[115] When the BAM tape is regularly marked with a time-stamp, it is possible to derive elemental and microscopic markers from the deposited sample spots.[116]

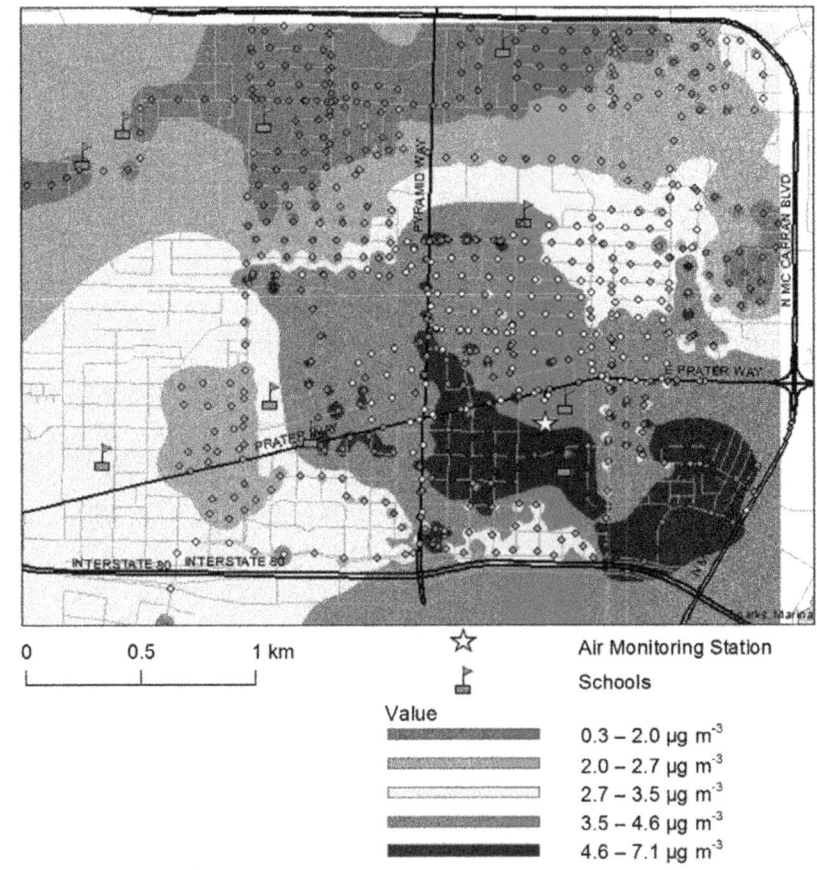

Figure 2 Spatial distribution of brown carbon (BrC, in μg m^{-3}), a marker for the smoldering phase of biomass burning, measured with a 2 wavelength (370 nm and 880 nm) microaethalometer[117] around the neighborhood-scale site (star) in Figure 1 from ∼2000 to next day ∼0200 LST. Data were spatially interpolated by kriging from average values assigned to evenly spaced grid squares. Wood is used for heating with older appliances in this older residential area. The BrC cloud was not detected during forecast burning prohibitions, indicating that residents complied with the ban.

because the patterns are most different among the >200 chemical species quantified for these profiles. These are composites of multiple samples, and the standard deviations indicated by the error bars show large variabilities, which is often the case for organic compounds. Each compound constitutes a small fraction of the organic carbon (OC), so there is room for abundance variations of more than an order of magnitude for a given source type. Such variability is not possible for PM properties that constitute an important fraction (>1%) of the PM mass. In spite of this variability, there are clear distinctions in the patterns of the emitted compounds, and these distinctions are exploited by solutions to the CMB model.

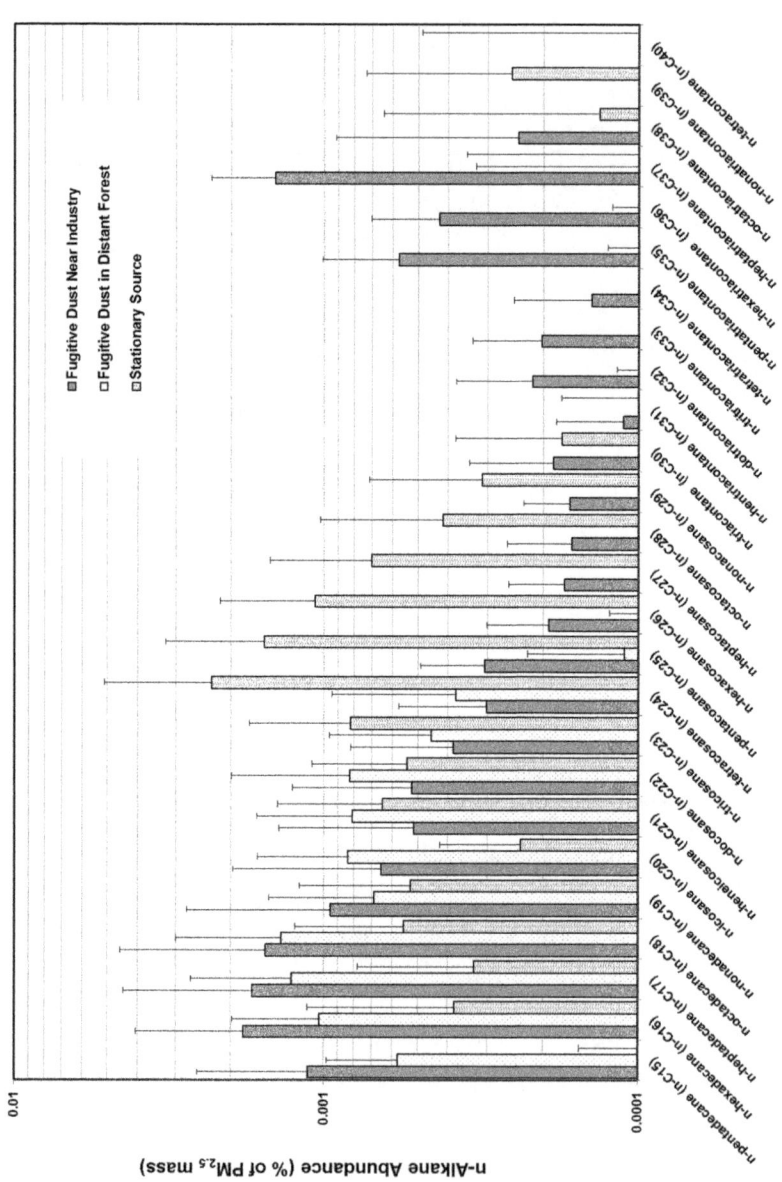

n-Alkanes

Figure 3 Source profile for *n*-alkanes comparing abundances in fugitive dust sources close to and distant from oil sands facilities in northern Alberta, Canada. [118,119] Error bars indicate standard deviations of multiple tests. Fugitive dusts include those from paved and unpaved roads, tailings ponds, mining overburden, and storage piles. The stationary source profile was obtained by dilution sampling of stack effluent from an upgrading facility. Despite the variable abundances, there are clear differences in the chemical patterns among these source types.

3 Solutions to the CMB Equations

Box 2 outlines the commonly applied solutions to the CMB equations; less-common variations and alternatives are discussed in the reviews cited above. The tracer (TR) and reconstructed mass (RM) solutions are easily applied without software, while the effective variance (EV)[51] and positive matrix

Box 2 Common solutions to the CMB equations.

Tracer (TR) solution:[1] Assumes a unique marker i for each source type j with a constant abundance for that source type. $S_{jklm} = C_{iklm}/F_{ijm}$, where i corresponds to the marker for source j.

Reconstructed Mass (RM) solution:[2] Divides PM into its major chemical components of ammonium sulfate $((NH_4)_2SO_4)$, ammonium nitrate (NH_4NO_3), organic matter (OM, organic carbon [OC] × a multiplier), elemental carbon (EC), minerals, and salt, derived from network measurements of mass, sulfate ion $(SO_4^{=})$ or sulfur (S), nitrate ion (NO_3^{-}), ammonium ion (NH_4^{+}), OC, EC, and major mineral components of aluminum (Al), silicon (Si), calcium (Ca), iron (Fe), and titanium (Ti). The IMPROVE RM formula[19] is most often used as a starting point: $(NH_4)_2SO_4 = 1.375 \times SO_4^{=}$ or $4.125 \times S$; $NH_4NO_3 = 1.29 \times NO_3^{-}$; $OM = 1.8OC$; $EC = EC$; $Minerals = 2.2 \times Al + 2.49 \times Si + 1.63 \times Ca + 2.42 \times Fe + 1.94 \times Ti$; and $Salt = 1.8Cl^{-}$. The multipliers intend to compensate for unmeasured species, such as hydrogen (H) and oxygen (O) in organic compounds and O in mineral oxides, and for water-soluble inorganic cations (*e.g.*, sodium [Na^{+}] and NH_4^{+}) that are not measured in the IMPROVE network but are available from the Chemical Speciation Network (CSN).

Effective Variance (EV) solution:[23] Minimizes the difference between measured and calculated C_{iklm} weighted by uncertainties in the measured marker abundances of source profiles and ambient concentrations such that:

$$\sum_{i}^{I} \frac{\left(C_{iklm} - \sum_{j=1}^{J} F_{ijm} S_{jklm} \right)^2}{\sigma_{C_{iklm}}^2 + \sum_{j=1}^{J} \sigma_{F_{ijm}}^2 S_{jklm}^2} = \text{Minimum}$$

where $\sigma_{C_{iklm}}$ are uncertainties of the measured concentrations of sample species i for time period k at location l in size range m and $\sigma_{F_{ijm}}$ are uncertainties of the profile abundances. Since source contribution estimates, S_{jklm}, are unknown, the EV solution initializes them to zero, then iterates until they converge on single values for these source contribution estimates. A byproduct of the EV solution, $\sigma_{S_{jklm}}$, is a propagated uncertainty of S_{jklm} that defines the precision of the source contribution estimates.

Positive Matrix Factorization (PMF) solution:[25-27] Uses samples over multiple time periods or multiple locations to minimize the differences between measured and calculated C_{iklm}, where:

$$\sum_{l=1}^{L}\sum_{k=1}^{K}\sum_{i=1}^{I}\left\{\left(C_{iklm}-\sum_{j=1}^{J}F_{ijm}S_{jklm}\right)\bigg/\sigma_{C_{iklm}}\right\}^{2}=\text{Minimum}$$

This is subject to constraints of non-negativity for F_{ijm} and S_{jklm}, which are iteratively calculated. The F_{ijm} are source factors that must be associated with a measured source profile to identify the source type. Additional constraints can be placed on F_{ijm} using PMF5.0 software to nudge them toward real-world chemical patterns.

factorization (PMF)[52] solutions require multiple iterations implemented in computer programs. The software should not be confused with the CMB model, despite the US EPA software designations as CMB8.2, which implements the EV-CMB solution, and PMF5.0, which implements the PMF-CMB solution. Information is gained by applying each of these solutions and comparing the results. The TR-CMB solution is limited in that there are few unique properties for a given source type, which is why the term "marker" is a more accurate term than "tracer" when referring to indicators of the presence or absence of a source contribution. For example, the carbon 14 (^{14}C) isotope is unique to PM of biogenic origin (*e.g.*, biomass burning, bioaerosols, or conversion of volatile organic compounds [VOC] emissions to PM), although its abundances vary with the age of the burned biomass.[53,54] Vanadium (V) and nickel (Ni) are often dominated by, but are not unique to, residual-oil combustion, and are now more commonly derived from large marine engines than from power stations.[55] A number of the more source-specific markers that might be used in the tracer solution are listed in Table 1, with a more complete list in Table 2. A variation of the tracer solution is the "diagnostic ratio" in which ratios among markers are used as "tracers" for different source types.[56] These ratios often do not take into account the variability within a source type, as indicated in Figure 3, which would add uncertainty to the distinction. The excess OC/EC (organic to elemental carbon) ratio[57,58] is another variation of the TR solution, in which a ratio of 2 to 3 is assumed for primary aerosol, with the excess OC ascribed to SOA formation. This method depends on the presence of large and relatively constant EC emissions, which have been decreasing with modern gasoline and diesel engines.[59,60] OC/EC ratios have always been variable in biomass burning emissions, and they are becoming more variable with the introduction of clean diesel engines.[61] Many other source markers, especially elements (*e.g.*, lead [Pb] in gasoline), have been eliminated or reduced due to pollution controls (several elemental markers were

Table 1 Chemical markers for different sources (complements Table 20.1 of ref. 4).

Major source type	Source sub-type	Specific markers[a]
Traffic-related emissions	Road dust	Al, Si, Fe, Cu, Sb, Ba, styrenebutadiene rubber (SBR), benzothiazole (tire wear), and asphaltenes (pavement wear)
	Gasoline engine exhaust	17 α(H), 21 β(H)-hopane, 17 α(H)-diastigmastane, other hopanes, steranes, and PAH diagnostic ratios
	Diesel engine exhaust	High molecular weight hydroxycarbonyls, hopanes, steranes, and PAH diagnostic ratios
Fossil fuel combustion	Fuel oil combustion (*e.g.*, heating oil and kerosene in externally-fired boilers)	PAH diagnostic ratios
	Residual-oil combustion (bunker fuel)	Ni and V
	Uncontrolled coal combustion (*e.g.*, domestic heating and cooking)	Zn, As, Se, Cd, Hg, Pb, S, $SO_4^=$, picene, and PAH diagnostic ratios
	Controlled coal combustion (*e.g.*, power plants)	As, Se, S, $SO_4^=$, picene, and PAH diagnostic ratios
Anthropogenic combustion	Tobacco smoke	Iso/anteiso alkanes
	Meat cooking	Cholesterol, palmitic acid, palmitoleic acid, stearic acid, and oleic acid
	Trash/plastic burning	1,3,5-Triphenylbenzene and tris(2,4-di-*tert*-butylphenyl)phosphate
Biomass burning	Softwood	Resin acids, guaiacol derivatives, retene, and levoglucosan/mannosan ratio (3–6)
	Hardwood	Syringol derivatives and levoglucosan/mannosan ratio (15–25)
	Straw and grasses	Levoglucosan/mannosan ratio (>30)
	Peat	Levoglucosan/mannosan ratio ($\sim$10)
	Biomass burning SOA	3-Methyl-5-nitrocatechol; 3-methyl-6-nitrocatechol; 4-methyl-5-nitrocatechol; 6-nitroguaiacol; 4,6-dinitroguaiacol; and HULIS (humic-like substances)
Natural sources	Mineral dust	CO_3^{2-}, Al, Si, and Ca
	Sea salt	Na^+ and Cl^-
	Bioaerosol—fungi	Arabitol, mannitol, and ergosterol
	Bioaerosol—bacteria	Hydroxy fatty acids

Table 1   Continued

Major source type	Source sub-type	Specific markers[a]
Secondary organic aerosol (SOA)	Anthropogenic SOA	2,3-Dihydroxy-4-oxopentanoic acid, *o*-phthalic acid, 2,4,6-trimethylphenol, oxy-PAHs, oxalic acid, and excess OC/EC ratios
	Isoprene-derived biogenic SOA	2-Methylthreitol, 2-methylerythritol, 2-methylglyceric acid, *cis*-2-methyl-1,3,4-trihydroxy-1-butene, 3-methyl-2,3,4-trihydroxy-1-butene, *trans*-2-methyl-1,3,4-trihydroxy-1-butene, and excess OC/EC ratios
	Monoterpene-derived biogenic SOA	*cis*-Pinic acid, *cis*-pinonic acid, *trans*-norpinic acid, *cis*-caric acid, limonic acid, ketolimononic acid, 3-hydroxyglutaric acid, 3-methyl-1,2,3-butanetricarboxylic acid, and excess OC/EC ratios

[a]Documented in Supplemental Table 2.

deemed toxic), so TR apportionments are limited with the commonly available PM properties. Still, an examination of some of the additional markers listed in Table 2 with the TR solution is a good starting point to limit the number of contributing source types.

The RM solution is also a pre-requisite, as it points out the major PM components. Many source apportionment studies jump straight to the more complex EV-CMB and/or PMF-CMB solutions and "discover" secondary sulfates and nitrates, fugitive dust, and salt (*e.g.*, from marine aerosol, windblown playas, and de-icing materials). Enough studies have been done to be confident that nearly all of the measured ammonium (NH_4^+), sulfate (SO_4^{2-}), and nitrate (NO_3^-) is of secondary origin and regionally distributed. This can usually be confirmed with ion balances comparing measured NH_4^+ with that predicted by assuming SO_4^{2-} as sulfuric acid (H_2SO_4), ammonium bisulfate (NH_4HSO_4), and ammonium sulfate (($NH_4)_2SO_4$) plus ammonium nitrate (NH_4NO_3). More acidic SO_4^{2-} forms indicate a closer source of precursors that have not yet encountered sufficient ammonia (NH_3) to become neutralized.

The big unknown in the RM solution regards sources of the organic matter (OM) fraction. Some indication of this is given by the OC multiplier, which is usually 1.2–1.4 for nearby sources, and 1.8–2.2 for SOA and biomass burning emissions.[2,62] When the sum of the RM components exceeds the measured mass, the most probable cause is a high OC multiplier. Schauer *et al.*[63] advocate applying the more complex solutions only to the carbonaceous fraction to minimize interference with the other components of the RM solution.

The EV- and PMF-CMB solutions intend to obtain more source specificity and reduce uncertainty. Rather than using simple markers or binary ratios, they use the entire pattern of source and receptor measurements. They also contain diagnostics, described by Watson *et al.*[64] for the EV and by Reff *et al.*[13] for the PMF, that are used to identify deviations from CMB assumptions and to evaluate uncertainties in the source contribution estimates. While the TR solution is sensitive to a single marker, the EV and PMF solutions use multiple markers, so if one marker's abundance is overestimated in a measured profile or derived source factor, it is likely that another marker's abundance is underestimated, thereby pulling the source contribution estimate toward a more accurate value. As with many environmental models, however, there is a temptation to accept their outputs as reality without considering effects of the simplifications and assumptions needed to simulate that reality.

4   CMB Model Assumptions and Effects of Deviations

The CMB model assumptions are:

1. Source profile abundances are constant and represent the profile that arrives at the receptor.
2. Aerosol properties add linearly.
3. Source compositions are linearly independent of each other.
4. The number and types of sources are known.
5. The number of source types is less than or equal to the number of measured properties.
6. Input data uncertainties are random, uncorrelated, and normally distributed.

These apply to all of the solutions in Box 2. There are additional constraints for each solution method described in their software manuals and several of the review articles cited above. A useful receptor modeling study must examine the effects of deviations from these assumptions on the source contribution estimates.

With regard to Assumption 1, it is apparent from Figure 3 and many published source profiles that these abundances are not constant. Profile variations are both systematic and random, caused by: (1) transformation and deposition between emissions points and the receptor; (2) differences in fuel type and operating processes between similar sources or the same source at a different time; and (3) uncertainties of or differences between the source profile measurement methods. For the EV solution, which seeks the best fit of source profile combinations to the ambient data, various profiles representing the same source type can be applied, which may also indicate a source subtype. The PMF solution requires that the profile is the same across all of the samples, and the data included should be divided such that this is the case. Ambient samples corresponding to special events, such as the Indian Diwali festival,[65] that are dominated by fireworks, should be treated separately. Profiles may vary seasonally, owing to different heating methods and cold engine starts during winter, so these groups might be treated separately. Source profiles have been compiled in several databases,[66–70] but these are often dated, do not contain the modern source markers needed to separate sources, and do not represent the emissions in areas with the major pollution problems, such as Chinese and Indian cities.

Recent studies show large changes in organic compounds with aging.[71,72] Biomass burning and engine exhaust emit intermediate and semi-volatile organic compounds (I/S-VOCs) that vary with combustion phase and aging. These I/S-VOCs oxidize to low-volatility products that can contribute to BrC and SOA. In the presence of nitrogen oxides (NO_x), biomass burning VOCs such as phenols, cresols, and methoxyphenols can form organonitrate compounds.[73]

Several cost-effective approaches[11,74] can be applied to obtain primary source profiles specific to the area being studied:

- Fugitive dust resuspension: Surface samples can be swept or troweled from different surfaces, sieved, and puffed into a chamber for sampling through size-selective inlets onto filter media amenable to marker analyses.[75] Sethi and Patil[69] demonstrate a simple system for this using Minivol samplers.
- In-plume sampling: This is especially applicable to mobile and area sources. Probes can be inserted into exhaust plumes from tailpipes or chimneys after dilution and cooling in ambient air. When carbon dioxide (CO_2) is measured with the PM samples, fuel-based emission factors can be obtained along with the source profiles.[76,77] Many plumes can be sampled on the same filters to obtain a better representation of the overall mixture.
- Source-dominated environments: Samples from roadsides, tunnels, parking garages, and downwind of wildfires or prescribed burns that overwhelm contributions from other sources provide good representations.[78,79]
- Dilution stack sampling: ISO[80] specifies an approach to diluting industrial stack emissions with cool clean air to allow for condensation of gases on particles. This should replace the hot filter/impinger tests[81] currently applied in stationary source compliance tests.

With regard to the nonlinear summation of species, Assumption 2, it is necessary to measure source profiles, or modify them by some objective method, to account for changes between source and receptor. The conversions of gases to particles and reactions between particles are not inherently linear processes. For the inorganic secondary sulfates and nitrates, the potential end-products are well known, as described by the RM solution. The question then becomes, what are the limiting precursors? There are potential cases in which sulfur dioxide (SO_2) emission reductions can free up NH_3 gas that can then combine with nitric acid (HNO_3) vapor to increase NH_4NO_3. Receptor-oriented equilibrium models have been used to evaluate limiting precursors when total ammonia ($NH_3 + NH_4^+$) and nitrate ($NO_3^- + HNO_3$) along with SO_4^{2-} temperature, and relative humidity measurements are available.[82,83] In NH_3-poor areas, SO_2 emission reductions may have the unintended consequence of increasing NH_4NO_3 contributions.

The increasing importance of SOA contributions presents a challenge for CMB source apportionment, regardless of the solution method. Measureable markers are listed in Tables 1 and 2. These have been identified in smog chamber studies that age model compounds or fresh emissions under ultraviolet (UV) radiation for many hours or days. It is now possible to artificially age profiles more rapidly with a relatively inexpensive and portable Potential Aerosol Mass Flow Reactor (PAMFR).[84] The PAMFR consists of a small chamber illuminated by UV lights into which high ozone (O_3) concentrations are introduced. High oxidant levels are generated that speed up the SOA reactions, leading to a profile that may better represent what is seen at the receptor. A PAMFR can be used as part of the inlet for in-plume, source-dominated, and dilution sampling to simulate changes in the organic compound profile with various degrees of aging. Oxygenated compounds are enriched, while hydrocarbon compounds are depleted in engine exhaust and biomass burning profiles.[85,86] Using such a reactor, Lai *et al.*[87] observed the levoglucosan abundance, a useful biomass burning marker, decrease by 80% after fully reacting. A greater number of such experiments coupled with a larger number of oxygenated organic compounds could be developed into an aerosol evolution model that might be applied to fresh profiles. Some of the SOA end-products contain BrC, potentially interfering with the use of this as a sole biomass burning marker.

Collinearity of source profiles and PMF source factors, Assumption 3, is a major challenge for receptor modeling. Although many abundances may look different, their standard deviations within a source type may overlap. Therefore, one may be substituted for another with little difference in the source contribution estimates. Source types may be separable when they contribute in equal amounts. However, the smaller contributor may be obscured when one contribution greatly exceeds the other; the smaller contributions fall within the uncertainties of the larger one. Collinearity is indicated in the CMB8.2 EV solution by a large positive contribution from one profile offset by a large negative contribution from a similar profile, with large uncertainties for the source contribution estimates. Similarity/uncertainty[88] and influential species[89] metrics in the EV-CMB software quantify the degree of overlap and its effects on inflating the source contribution uncertainty, as well as indicating when the solution is dictated by non-markers. These metrics would be equally useful if incorporated into the PMF software.

Figure 4 compares two profiles from similar vehicles using different fuels to estimate source contributions in India. A liquefied petroleum gas (LPG) profile (not necessarily the one illustrated) was found to fit the ambient data well, so the conclusion was that the switch to LPG as a cleaner fuel was unsuccessful. Sensitivity tests,[90] which interchange similar profiles for selected samples, especially source-dominated ones, would have revealed the collinearity. In this case, the definition of the profile must be expanded to include all of the collinear profiles, which is why terms such as "mobile source" or "engine exhaust" are often used rather than "gasoline exhaust," "diesel exhaust," or "LPG exhaust." Even though the diesel and LPG profiles are too similar to separate with the measured markers, a case could be made based on published emission rates[91] that the diesel exhaust probably dominates contributions represented by this profile. Another alternative would be to re-analyze the source and ambient samples to obtain more specific diesel and LPG markers.

The PMF solution does not solve this collinearity. Many concentrations vary together in time and space for reasons that are not related to their sources. This is evident in Figure 1a, where the morning rush-hour traffic bump is discernable, but the evening rush hour is not, as it also correlates with an increase in home heating as a shallow inversion sets in after sunset. PMF "mobile source" factors often show loadings on elements associated with road dust and brake wear along with the carbonaceous components of engine exhaust, as both are traffic-generated. Source factors labeled "regional" or sometimes "secondary sulfate" or "power plant" may show high carbon and elemental loadings in addition to NH_4^+ and SO_4^{2-}. Sometimes the source factors look similar, as is the case for a recent source apportionment study in Beijing, China,[92] but they are given names that obscure the mixture of sources that they include. In this case, 4% of annual $PM_{2.5}$ was ascribed to a "traffic and waste incineration" factor while 25% of annual $PM_{2.5}$ came from an "industrial pollution" source, even though the sampling location is surrounded by heavily travelled roadways and major industries are far away. The PMF-derived source factors appear similar.

Regarding Assumption 4, there are thousands of individual emitters that can affect a PM sample, many more than the measureable properties (Assumption 5). However, this assumption is complied with by grouping collinear profiles into broader source types, as is necessary to comply with Assumption 3. Examining the presence or absence of the strong source markers, as in the TR solution, is a first step toward determining the potential number of sources. Regionwide emission inventories and micro-inventories around sampling sites are also good starting points, although one of the great values of receptor modeling has been to identify important uninventoried contributors.[93] Examining the ratios of calculated to measured concentrations, normalized by their uncertainties, can reveal deficits owing to missing source types. The PMF solution includes several criteria for selecting the number of sources that can be derived from a given data set,

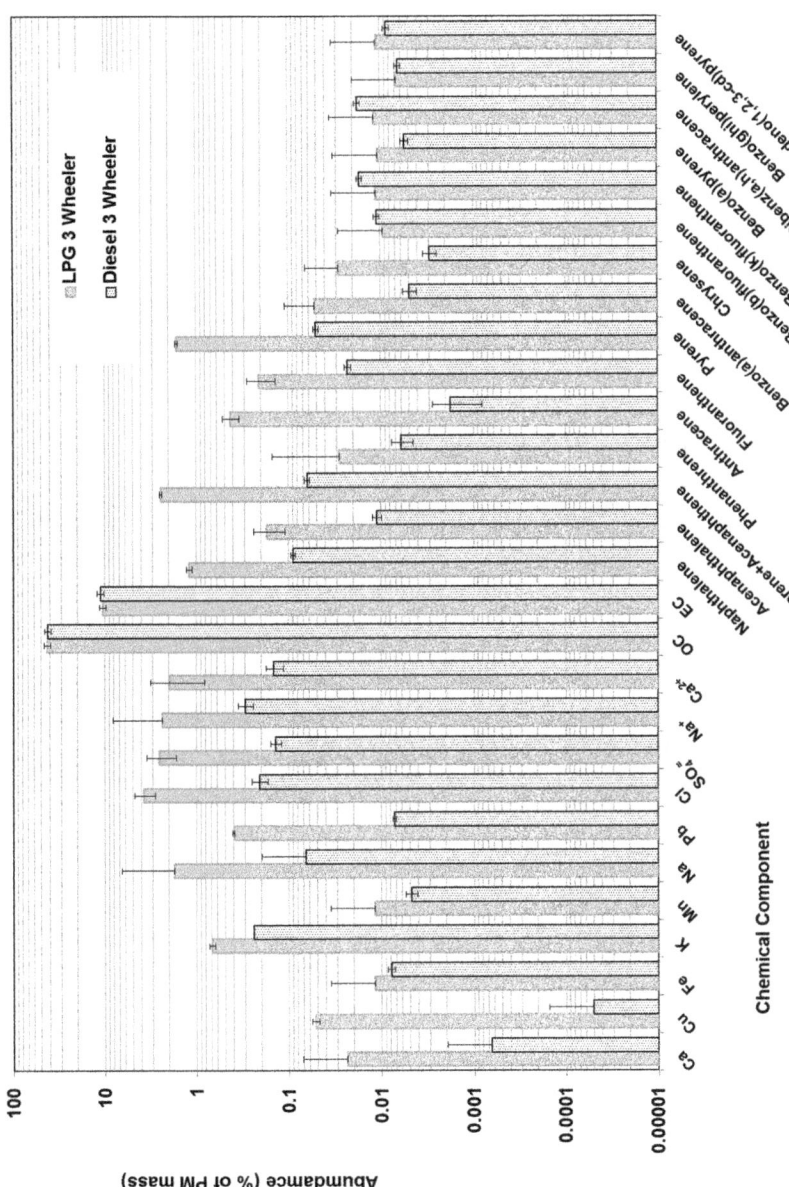

Figure 4 Profile comparisons for emissions from 3-wheeled taxis using liquefied petroleum gas (LPG) and diesel fuel.[68] These were used in a Delhi source apportionment study with the EV-CMB solution[120] that concluded LPG combustion was a major contributor to excessive $PM_{2.5}$ and PM_{10}. Values less than uncertainties were set to half the uncertainty[121] to accommodate the logarithmic scale. Error bars indicate the standard deviation of the average of two tests on the same vehicle. ARAI[68] reports other engine exhaust profiles for gasoline, diesel, LPG, and compressed natural gas that show similarities to these profiles and among themselves.

and these decisions can be guided by results of the TR solution. The sources selected need to make sense, however. For example, even though selenium (Se) is an important marker for coal combustion, this would not be a good indicator in California, which prohibits coal burning and has other Se sources.[94]

Compliance with Assumption 5 is inherent in each of the solution methods, as none will yield a result if the data set is underdetermined. It is evident that elemental analysis alone is insufficient for any of the solutions, and that the major components required for the RM solution are a minimum pre-requisite for any source apportionment study.

Little is known about deviations from Assumption 6, although it is implicit in the least squares minimization of EV and PMF solutions. When the uncertainties are large, as they often are for the source profile abundances, they cannot have a truly normal distribution as abundances cannot be less than zero.

5   More Information from Existing Samples

The availability of elements, ions, and carbon fractions is necessary, but not sufficient, for specific source types relevant to imposing emission reduction measures. Many new *in situ* and laboratory analysis techniques are being reported, but these often require special types of samples and expertise and are only applicable to short-term field projects. It is possible, however, to submit the filters already acquired in compliance and speciation networks to additional chemical analyses, either routinely or applied to archived sample remnants when needed to better understand the causes of excess PM levels.

Figure 5 outlines a potential approach for additional chemical analyses. When properly planned, several of these could be added to existing laboratory operations with little or no added cost, as will soon be the case for IMPROVE multi-wavelength thermal/optical carbon analysis, which will provide additional thermal fractions and light absorption spectra.[46,47] One could envision obtaining similar light absorption spectra as part of the Teflon-membrane filter weighing process, applied in compliance networks, by coupling a transmittance device to the charge neutralization system used to minimize electrostatic attraction. Non-destructive methods, such as ultraviolet-visible light spectroscopy (UV-VIS)[45,95] and Fourier transform infrared spectroscopy (FTIR),[96] can be performed on instruments common to many chemistry laboratories. Organic compound analyses[97] that once required large aliquots of filter samples, hours-long solvent extractions, and extensive spectrum processing can now be done on smaller samples by thermal desorption for non-polar[98] and polar compounds.[99] Although it is possible to apply every method identified in Figure 5 to the available samples, a selection can be made to obtain those markers useful for a specific situation. Table 2 links markers to sources and analysis methods and can be used to guide information gathering efforts.

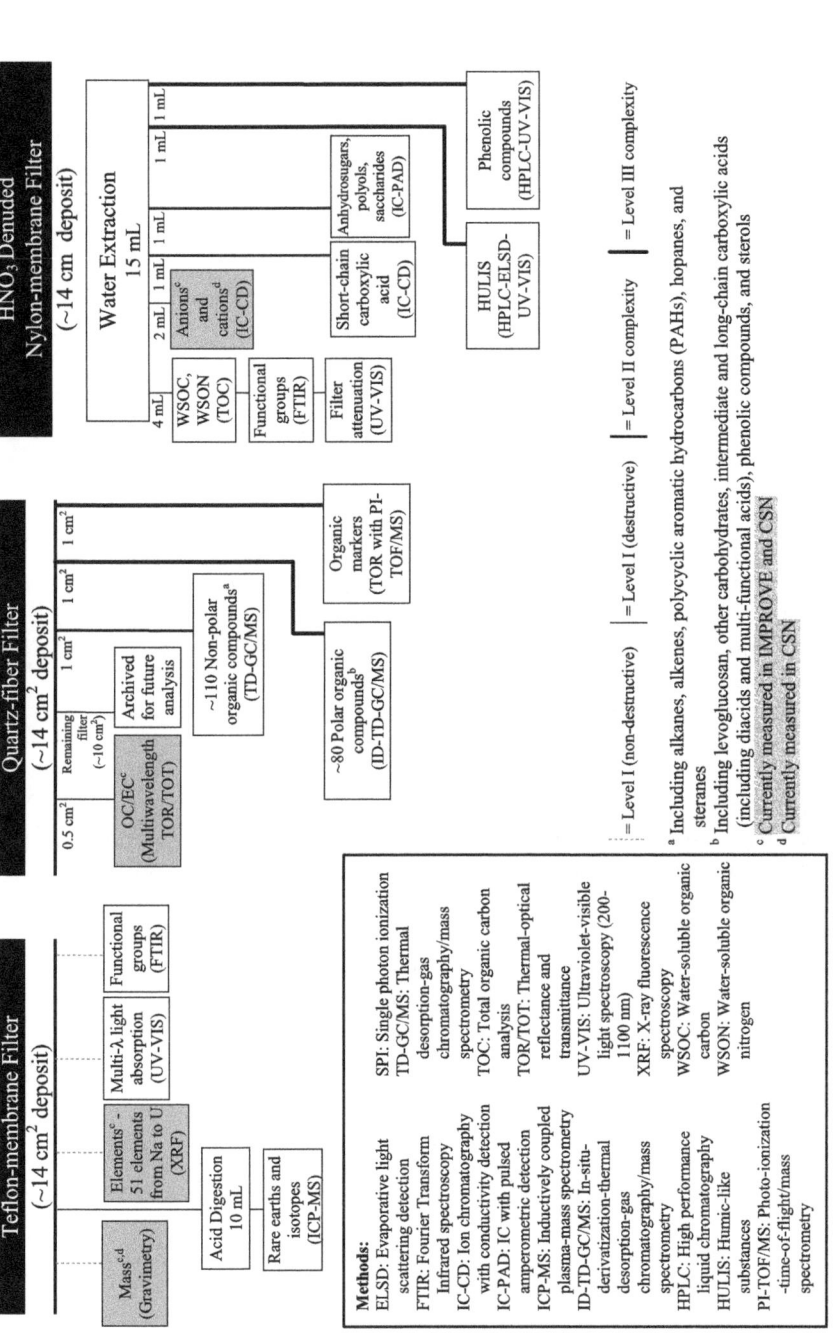

Figure 5 Filter samples are commonly acquired in PM compliance and speciation networks.[29,30] In these networks the Nylon filter is preceded by anodized aluminum denuder tubes[122] that remove gaseous nitric acid so only PM nitrate is collected. Shading indicates the analyses and outputs that are currently obtained from these samples. Additional possible analyses for source markers applicable to sample remnants or solution extracts are classified by level of complexity and cost (Level I = least complex). Table 2 lists the PM properties quantified by each method, associates them with common source types, and provides references to analysis procedures and applications.

An earlier version[4] of Figure 5 included microscopic analysis of particles on polycarbonate substrates, but this has been omitted as such a sample is not commonly acquired in speciation networks. Scanning electron and optical microscopy have been coupled with pattern recognition software to classify particles by size, shape, and elemental composition that can be related to sources.[100,101] Microscopy is applicable to the $PM_{10-2.5}$ fraction for separating different types of minerals[102] and bioaerosols.[103] Despite unsubstantiated claims to the contrary,[104] optical microscopy by itself is insufficient to apportion the $PM_{2.5}$ fraction to sources owing to poor resolution of particles with sizes comparable to the wavelength of light.

6   How to Judge a Source Apportionment Study

This chapter has emphasized more of the limitations than the positive points of source apportionment studies. In reality, other approaches suffer the same limitations. Similar to source profiles measured elsewhere, emission inventories often use emission factor and activity data that have little relationship to the area being studied.[105] Source-oriented air quality models face the same challenges as receptor models when secondary PM contributions begin to equal primary contributions.[106] There is no single measurement system or model that provides an exact answer. However, there is value in taking multiple approaches to source apportionment, recognizing discrepancies, and then formulating ways to resolve them. Receptor models quantified residential wood combustion contributions[107] in the mid-1970s, which were not included in emission inventories, and this led to improved appliances and practices.[108] Receptor models recognized limitations of street sweepers for fugitive dust control,[109] leading to regulations on the particle sizes picked up by modern sweepers.[110] Receptor models inferred meat cooking as a potential contributor,[111] with its addition to the emission inventory and regulations to reduce emissions.[112] There are many other examples not listed here.

Belis *et al.*[113] provide a set of criteria for performing and reporting results of source apportionment studies, and these guidelines are continually being improved. Basic questions to ask of such a study are:

- What is the nature of the study area, potential sources, and interactions between emissions and meteorology?
- What measured source profiles are used as input to the EV solution or to associate PMF source factors with actual sources, and how well do they relate to the study area?
- What is the degree of source mixing within and among the PMF source factors?
- To what extent have sensitivity tests, case studies, and performance measures been evaluated to assure that source types are accurately defined?

Table 2 Summary of PM marker species for different source types.

General source type/marker species[a]	Source subtype	Analysis method[b,c]	Species class	Mnemonics	Molecular formula	Molar mass	Sampling media
1. Primary biogenic emissions							
Palmitoleic acid	Biogenic	ID-TD-GC/MS	Fatty acid	PALMFATC	$C_{16}H_{30}O_2$	254.41	Quartz-fiber
Oleic acid	Biogenic	ID-TD-GC/MS	Fatty acid	OLEIFATC	$C_{18}H_{34}O_2$	282.46	Quartz-fiber
Linoleic acid	Biogenic	ID-TD-GC/MS	Fatty acid	LINOLFATC	$C_{18}H_{32}O_2$	280.45	Quartz-fiber
Erucic acid	Biogenic	ID-TD-GC/MS	Fatty acid	ERUCFATC	$C_{22}H_{42}O_2$	338.58	Quartz-fiber
Threitol	Biogenic	ID-TD-GC/MS	Polyol	THREITC	$C_4H_{10}O_4$	122.12	Quartz-fiber
Erythritol	Biogenic	ID-TD-GC/MS	Polyol	ERYTHOLTC	$C_4H_{10}O_4$	122.12	Quartz-fiber
Xylitol	Biogenic	ID-TD-GC/MS	Polyol	XYLITOLTC	$C_5H_{12}O_5$	152.15	Quartz-fiber
Sorbitol	Biogenic	ID-TD-GC/MS	Polyol	SORBITOLTC	$C_6H_{14}O_6$	182.17	Quartz-fiber
Inositol	Biogenic	ID-TD-GC/MS	Polyol	INOSITOLTC	$C_6H_{12}O_6$	180.16	Quartz-fiber
Arabinose	Biogenic	ID-TD-GC/MS	Monosaccharide	ARABNSETC	$C_5H_{10}O_5$	150.13	Quartz-fiber
Xylose	Biogenic	ID-TD-GC/MS	Monosaccharide	GLXYOSETC	$C_5H_{10}O_5$	150.13	Quartz-fiber
Glucose	Biogenic	ID-TD-GC/MS	Monosaccharide	GLXYOSETC	$C_6H_{12}O_6$	180.16	Quartz-fiber
Galactose	Biogenic	ID-TD-GC/MS	Monosaccharide	GALATC	$C_6H_{12}O_6$	180.16	Quartz-fiber
Fructose	Biogenic	ID-TD-GC/MS	Monosaccharide	FRUCTC	$C_6H_{12}O_6$	180.16	Quartz-fiber

Table 2 Continued

General source type/marker species[a]	Source subtype	Analysis method[b,c]	Species class	Mnemonics	Molecular formula	Molar mass	Sampling media
Sucrose	Biogenic	ID-TD-GC/MS	Disaccharide	SUCRTC	$C_{12}H_{22}O_{11}$	342.30	Quartz-fiber
Trehalose	Biogenic	ID-TD-GC/MS	Disaccharide	TREHLSETC	$C_{12}H_{22}O_{11}$	342.30	Quartz-fiber
Stigmasterol	Biogenic	ID-TD-GC/MS	Sterol	STEROL3TC	$C_{29}H_{48}O$	412.69	Quartz-fiber
β-Sitosterol	Biogenic	ID-TD-GC/MS	Sterol	STEROL4TC	$C_{29}H_{50}O$	414.71	Quartz-fiber
Nonanaldehyde	Biogenic	ID-TD-GC/MS	Aldehyde	NONYDETC	$C_9H_{18}O$	142.24	Quartz-fiber
2. Biogenic secondary organic aerosol (SOA)							
3-Hydroxyglutaric acid (3-HGA)	Biogenic – monoterpene SOA	ID-TD-GC/MS	Carboxylic acid	HGA + E2 + A24:E27	$C_5H_8O_5$	148.11	Quartz-fiber
trans-Norpinic acid	Biogenic – monoterpene SOA	ID-TD-GC/MS	Carboxylic acid	TNORPTC	$C_8H_{12}O_4$	172.18	Quartz-fiber
Pinic acid	Biogenic – monoterpene SOA	ID-TD-GC/MS	Carboxylic acid	PINICTC	$C_9H_{14}O_4$	186.21	Quartz-fiber
cis-Pinonic acid	Biogenic – monoterpene SOA	ID-TD-GC/MS	Carboxylic acid	PINONTC	$C_{10}H_{16}O_3$	184.23	Quartz-fiber
2-Methylglyceric acid	Biogenic – isoprene SOA	ID-TD-GC/MS	Carboxylic acid	MGLYCTC	$C_4H_8O_4$	120.10	Quartz-fiber
Azelaic acid	Biogenic SOA	ID-TD-GC/MS	Carboxylic acid	AZELTC	$C_9H_{16}O_4$	188.22	Quartz-fiber
cis-2-Methyl-1,3,4-trihydroxy-1-butene	Biogenic – isoprene SOA	ID-TD-GC/MS	Alkene triol	ALKTRIOL1TC	$C_5H_{10}O_3$	118.00	Quartz-fiber
3-Methyl-2,3,4-trihydroxy-1-butene	Biogenic – isoprene SOA	ID-TD-GC/MS	Alkene triol	ALKTRIOL2TC	$C_5H_{10}O_3$	118.00	Quartz-fiber

Compound	Source	Method	Type	Code	Formula	MW	Substrate
trans-2-Methyl-1,3,4-trihydroxy-1-butene	Biogenic – isoprene SOA	ID-TD-GC/MS	Alkene triol	ALKTRIOL3TC	$C_5H_{10}O_3$	118.00	Quartz-fiber
2-Methylthreitol	Biogenic – isoprene SOA	ID-TD-GC/MS	Polyol	METHTET1TC	$C_5H_{12}O_4$	136.15	Quartz-fiber
2-Methylerythritol	Biogenic – isoprene SOA	ID-TD-GC/MS	Polyol	METHTET2TC	$C_5H_{12}O_4$	136.15	Quartz-fiber
Pinonaldehyde	Biogenic – monoterpenes SOA	ID-TD-GC/MS	Aldehyde	PINYDETC	$C_{10}H_{16}O_2$	168.23	Quartz-fiber
3. Bioaerosol emissions							
Adonitol	Bioaerosol – bacteria	ID-TD-GC/MS	Polyol	ADONITOLTC	$C_5H_{12}O_5$	152.15	Quartz-fiber
Arabitol	Bioaerosol – fungi	ID-TD-GC/MS	Polyol	ARABITOLTC	$C_5H_{12}O_5$	152.15	Quartz-fiber
Mannitol	Bioaerosol – fungi	ID-TD-GC/MS	Polyol	MANNITOLTC	$C_6H_{14}O_6$	182.17	Quartz-fiber
Ergosterol	Bioaerosol – fungi	ID-TD-GC/MS	Sterol	STEROL2TC	$C_{28}H_{44}O$	396.65	Quartz-fiber
3-Hydroxydecanoic acid	Bioaerosol – GN-bacteria	ID-TD-GC/MS	Hydroxy fatty acid	HOFA1TC	$C_{10}H_{20}O_3$	188.26	Quartz-fiber
3-Hydroxyundecanoic acid	Bioaerosol – GN-bacteria	ID-TD-GC/MS	Hydroxy fatty acid	HOFA2TC	$C_{11}H_{22}O_3$	202.29	Quartz-fiber
3-Hydroxydodecanoic acid	Bioaerosol – GN-bacteria	ID-TD-GC/MS	Hydroxy fatty acid	HOFA3TC	$C_{12}H_{24}O_3$	216.32	Quartz-fiber
3-Hydroxytridecanoic acid	Bioaerosol – GN-bacteria	ID-TD-GC/MS	Hydroxy fatty acid	HOFA4TC	$C_{13}H_{26}O_3$	230.34	Quartz-fiber
3-Hydroxytetradecanoic acid	Bioaerosol – GN-bacteria	ID-TD-GC/MS	Hydroxy fatty acid	HOFA5TC	$C_{14}H_{28}O_3$	244.37	Quartz-fiber
3-Hydroxypentadecanoic acid	Bioaerosol – GN-bacteria	ID-TD-GC/MS	Hydroxy fatty acid	HOFA6TC	$C_{15}H_{30}O_3$	258.40	Quartz-fiber
3-Hydroxyhexadecanoic acid	Bioaerosol – GN-bacteria	ID-TD-GC/MS	Hydroxy fatty acid	HOFA7TC	$C_{16}H_{32}O_3$	272.42	Quartz-fiber

Table 2 Continued

General source type/marker species[a]	Source subtype	Analysis method[b,c]	Species class	Mnemonics	Molecular formula	Molar mass	Sampling media
3-Hydroxyheptadecanoic acid	Bioaerosol – GN-bacteria	ID-TD-GC/MS	Hydroxy fatty acid	HOFA8TC	$C_{17}H_{34}O_3$	286.45	Quartz-fiber
3-Hydroxyoctadecanoic acid	Bioaerosol – GN-bacteria	ID-TD-GC/MS	Hydroxy fatty acid	HOFA9TC	$C_{18}H_{36}O_3$	300.27	Quartz-fiber
4. Primary biomass burning emissions							
Abietic acid	Biomass combustion – softwood	ID-TD-GC/MS	Resin acid	RESA1TC	$C_{20}H_{30}O_2$	302.45	Quartz-fiber
Dehydroabietic acid	Biomass combustion – softwood	ID-TD-GC/MS	Resin acid	RESA2TC	$C_{20}H_{28}O_2$	300.44	Quartz-fiber
7-Oxodehydroabietic acid	Biomass combustion – softwood	ID-TD-GC/MS	Resin acid	RESA3TC	$C_{20}H_{26}O_3$	314.42	Quartz-fiber
Pimaric acid	Biomass combustion – softwood	ID-TD-GC/MS	Resin acid	RESA4TC	$C_{20}H_{30}O_2$	302.45	Quartz-fiber
Isopimaric acid	Biomass combustion – softwood	ID-TD-GC/MS	Resin acid	RESA5TC	$C_{20}H_{30}O_2$	302.45	Quartz-fiber
Sandaracopimaric acid	Biomass combustion – softwood	ID-TD-GC/MS	Resin acid	RESA6TC	$C_{20}H_{30}O_2$	302.46	Quartz-fiber
Vanillic acid	Biomass combustion – softwood	ID-TD-GC/MS	Methoxy phenol	METHOPH1TC	$C_8H_8O_4$	168.15	Quartz-fiber
Homovanillic acid	Biomass combustion – softwood	ID-TD-GC/MS	Methoxy phenol	METHOPH2TC	$C_9H_{10}O_4$	182.17	Quartz-fiber
4-Hydroxybenzoic acid	Biomass combustion – softwood	ID-TD-GC/MS	Aromatic acid	AROMA1TC	$C_7H_6O_3$	138.12	Quartz-fiber
Pinitol	Biomass combustion – softwood	ID-TD-GC/MS	Polyol	PINITOLTC	$C_7H_{14}O_6$	194.18	Quartz-fiber
Acetovanillone	Biomass combustion – softwood	ID-TD-GC/MS	Methoxy phenol	METHOPH3TC	$C_9H_{10}O_3$	166.17	Quartz-fiber

Compound	Source	Method	Class	Abbreviation	Formula	MW	Filter
Retene	Biomass combustion – softwood	TD-GC/MS	PAH	RETENETC	$C_{18}H_{18}$	234.33	Quartz-fiber
syringic acid	Biomass combustion – hard wood	ID-TD-GC/MS	Methoxy phenol	METHOPH4TC	$C_9H_{10}O_5$	198.17	Quartz-fiber
Syringe aldehyde	Biomass combustion – hard wood	ID-TD-GC/MS	Methoxy phenol	METHOPH5TC	$C_9H_{10}O_4$	182.17	Quartz-fiber
Acetosyringone	Biomass combustion – hard wood	ID-TD-GC/MS	Methoxy phenol	METHOPH6TC	$C_{10}H_{12}O_4$	196.20	Quartz-fiber
2,6-Dimethoxyphenol	Biomass combustion – hard wood	ID-TD-GC/MS	Methoxy phenol	METHOPH7TC	$C_8H_{10}O_3$	154.16	Quartz-fiber
Cinnamic acid	Biomass combustion	ID-TD-GC/MS	Methoxy phenol	METHOPH8TC	$C_9H_8O_2$	148.16	Quartz-fiber
4-Hydroxy-3-methoxycinnamic acid	Biomass combustion	ID-TD-GC/MS	Methoxy phenol	METHOPH9TC	$C_{10}H_{10}O_4$	194.18	Quartz-fiber
Sinapinic acid	Biomass combustion	ID-TD-GC/MS	Methoxy phenol	METHOPH10TC	$C_{11}H_{12}O_5$	224.21	Quartz-fiber
Cinnamaldehyde	Biomass combustion	ID-TD-GC/MS	Methoxy phenol	METHOPH11TC	C_9H_8O	132.16	Quartz-fiber
Coniferyl aldehyde	Biomass combustion	ID-TD-GC/MS	Methoxy phenol	METHOPH12TC	$C_{10}H_{10}O_3$	178.19	Quartz-fiber
Levoglucosan	Biomass combustion	IC-PAD	Anhydrosugar	LEVGIC	$C_6H_{10}O_5$	162.14	Quartz-fiber
Mannosan	Biomass combustion	IC-PAD	Anhydrosugar	MANNOIC	$C_6H_{10}O_5$	162.14	Quartz-fiber
Galactosan	Biomass combustion	IC-PAD	Anhydrosugar	GALACTIC	$C_6H_{10}O_5$	162.14	Quartz-fiber
Water-soluble potassium	Biomass combustion	IC-CD	Inorganic ion	KPIC	K^+	39.10	Quartz-fiber/nylon

5. Biomass burning SOA

Compound	Source	Method	Class	Abbreviation	Formula	MW	Filter
Oxalic acid	Biomass combustion SOA	IC-CD	Dicarboxylic acid	OXALATIC	$C_2H_2O_4$	90.03	Quartz-fiber
Malonic acid	Biomass combustion SOA	IC-CD	Dicarboxylic acid	MALONATIC	$C_3H_4O_4$	104.06	Quartz-fiber

Table 2 Continued

General source type/marker species[a]	Source subtype	Analysis method[b,c]	Species class	Mnemonics	Molecular formula	Molar mass	Sampling media
Succinic acid	Biomass combustion SOA	IC-CD	Dicarboxylic acid	SUCCATEIC	$C_4H_6O_4$	118.09	Quartz-fiber
Adipic acid	Biomass combustion SOA	ID-TD-GC/MS	Dicarboxylic acid	ADIPICIC	$C_6H_{10}O_4$	146.14	Quartz-fiber
2-Methyl-3-nitrophenol	Biomass combustion SOA	HPLC-UV-Vis	Nitro-aromatic	NOARO11HC	$C_7H_7NO_3$	153.14	Quartz-fiber
2-Methyl-4-nitrophenol	Biomass combustion SOA	HPLC-UV-Vis	Nitro-aromatic	NOARO12HC	$C_7H_7NO_3$	153.14	Quartz-fiber
2-Methyl-5-nitrophenol	Biomass combustion SOA	HPLC-UV-Vis	Nitro-aromatic	NOARO13HC	$C_7H_7NO_3$	153.14	Quartz-fiber
3-Methyl-2-nitrophenol	Biomass combustion SOA	HPLC-UV-Vis	Nitro-aromatic	NOARO6HC	$C_7H_7NO_3$	153.14	Quartz-fiber
3-Methyl-4-nitrophenol	Biomass combustion SOA	HPLC-UV-Vis	Nitro-aromatic	NOARO3HC	$C_7H_7NO_3$	153.14	Quartz-fiber
4-Methyl-2-nitrophenol	Biomass combustion SOA	HPLC-UV-Vis	Nitro-aromatic	NOARO14HC	$C_7H_7NO_3$	153.14	Quartz-fiber
4-Nitroguaiacol	Biomass combustion SOA	HPLC-UV-Vis	Nitro-aromatic	NOARO15HC	$C_7H_7NO_4$	169.13	Quartz-fiber
6-Nitroguaiacol	Biomass combustion SOA	HPLC-UV-Vis	Nitro-aromatic	NOARO1HC	$C_7H_7NO_4$	169.13	Quartz-fiber
4,6-Dinitroguaiacol	Biomass combustion SOA	HPLC-UV-Vis	Nitro-aromatic	NOARO2HC	$C_7H_6N_2O_6$	214.13	Quartz-fiber
4-Nitrocatechol	Biomass combustion SOA	HPLC-UV-Vis	Nitro-aromatic	NOARO16HC	$C_6H_5NO_4$	155.11	Quartz-fiber
3-Methyl-5-nitrocatechol	Biomass combustion SOA	HPLC-UV-Vis	Nitro-aromatic	NOARO4HC	$C_7H_7NO_4$	169.13	Quartz-fiber
3-Methyl-6-nitrocatechol	Biomass combustion SOA	HPLC-UV-Vis	Nitro-aromatic	NOARO5HC	$C_7H_7NO_4$	169.13	Quartz-fiber
4-Methyl-5-nitrocatechol	Biomass combustion SOA	HPLC-UV-Vis	Nitro-aromatic	NOARO6HC	$C_7H_7NO_4$	169.13	Quartz-fiber

Compound	Source	Method	Class	Code	Formula	MW	Filter
2,3-Dimethyl-4-nitrophenol	Biomass combustion SOA	HPLC-UV-Vis	Nitro-aromatic	NOARO17HC	$C_8H_9NO_3$	167.16	Quartz-fiber
2,5-Dimethyl-4-nitrophenol	Biomass combustion SOA	HPLC-UV-Vis	Nitro-aromatic	NOARO18HC	$C_8H_9NO_3$	167.16	Quartz-fiber
2,6-Dimethyl-4-nitrophenol	Biomass combustion SOA	HPLC-UV-Vis	Nitro-aromatic	NOARO19HC	$C_8H_9NO_3$	167.16	Quartz-fiber
2,4-Dimethyl-6-nitrophenol	Biomass combustion SOA	HPLC-UV-Vis	Nitro-aromatic	NOARO20HC	$C_8H_9NO_3$	167.16	Quartz-fiber
3-5-Dimethyl-4-nitrophenol	Biomass combustion SOA	HPLC-UV-Vis	Nitro-aromatic	NOARO21HC	$C_8H_9NO_3$	167.16	Quartz-fiber
Humic-like substances (HULIS)	Biomass combustion SOA	HPLC-ELSD-UV-Vis	Polyacid	HULISHC			Quartz-fiber

6. Fuel combustion emissions

Compound	Source	Method	Class	Code	Formula	MW	Filter
Acenaphthylene (C12)	Carbonaceous fuel combustion	TD-GC/MS	PAH[d]	ACNAPYTC	$C_{12}H_8$	152.19	Quartz-fiber
Acenaphthene (C12)	Carbonaceous fuel combustion	TD-GC/MS	PAH	ACNAPETC	$C_{12}H_{10}$	154.08	Quartz-fiber
Fluorene (FLU) (C13)	Carbonaceous fuel combustion	TD-GC/MS	PAH	FLUORETC	$C_{13}H_{10}$	166.22	Quartz-fiber
Phenanthrene (PHE) (C14)	Carbonaceous fuel combustion	TD-GC/MS	PAH	PHENANTC	$C_{14}H_{10}$	178.23	Quartz-fiber
Anthracene (ANT) (C14)	Carbonaceous fuel combustion	TD-GC/MS	PAH	ANTHRATC	$C_{14}H_{10}$	178.23	Quartz-fiber
Fluoranthene (FLT) (C16)	Carbonaceous fuel combustion	TD-GC/MS	PAH	FLUORATC	$C_{16}H_{10}$	202.26	Quartz-fiber
Pyrene (PYR) (C16)	Carbonaceous fuel combustion	TD-GC/MS	PAH	PYRENETC	$C_{16}H_{10}$	202.25	Quartz-fiber
Benzo[a]anthracene (BaA) (C18)	Carbonaceous fuel combustion	TD-GC/MS	PAH	BAANTHTC	$C_{18}H_{12}$	228.29	Quartz-fiber
Chrysene (CHR) (C18)	Carbonaceous fuel combustion	TD-GC/MS	PAH	CHRYSNTC	$C_{18}H_{12}$	228.29	Quartz-fiber

Table 2 Continued

General source type/marker species[a]	Source subtype	Analysis method[b,c]	Species class	Mnemonics	Molecular formula	Molar mass	Sampling media
Benzo[b]fluoranthene (C20)	Carbonaceous fuel combustion	TD-GC/MS	PAH	BBFLTC	$C_{20}H_{12}$	252.31	Quartz-fiber
Benzo[j + k]fluoranthene (C20)	Carbonaceous fuel combustion	TD-GC/MS	PAH	BJKFLTC	$C_{20}H_{12}$	252.31	Quartz-fiber
Benzo[a]fluoranthene (C20)	Carbonaceous fuel combustion	TD-GC/MS	PAH	BAFLTC	$C_{20}H_{12}$	252.31	Quartz-fiber
Benzo[e]pyrene (BeP) (C20)	Carbonaceous fuel combustion	TD-GC/MS	PAH	BEPYRNTC	$C_{20}H_{12}$	252.31	Quartz-fiber
Benzo[a]pyrene (BaP) (C20)	Carbonaceous fuel combustion	TD-GC/MS	PAH	BAPYRNTC	$C_{20}H_{12}$	252.31	Quartz-fiber
Perylene (C20)	Carbonaceous fuel combustion	TD-GC/MS	PAH	PERYLETC	$C_{20}H_{12}$	252.31	Quartz-fiber
Indeno[1,2,3-cd]pyrene (IcdP) (C22)	Carbonaceous fuel combustion	TD-GC/MS	PAH	IN123PYRTC	$C_{22}H_{12}$	276.33	Quartz-fiber
Dibenzo[a,h]anthracene (C22)	Carbonaceous fuel combustion	TD-GC/MS	PAH	DBAHANTC	$C_{22}H_{12}$	276.33	Quartz-fiber
Benzo[g,h,i]perylene (BghiP) (C22)	Carbonaceous fuel combustion	TD-GC/MS	PAH	BGHIPETC	$C_{22}H_{12}$	276.33	Quartz-fiber
Coronene (C24)	Carbonaceous fuel combustion	TD-GC/MS	PAH	CORONETC	$C_{24}H_{12}$	300.35	Quartz-fiber
Dibenzo[a,e]pyrene (C24)	Carbonaceous fuel combustion	TD-GC/MS	PAH	DBAEPYRTC	$C_{24}H_{14}$	302.37	Quartz-fiber
Dibenzothiophene (C12)	Carbonaceous fuel combustion	TD-GC/MS	PAH	DBTHIOTC	$C_{12}H_8S$	184.26	Quartz-fiber
9-Fluorenone (C13)	Carbonaceous fuel combustion	TD-GC/MS	PAH	FL9ONETC	$C_{13}H_8O$	180.20	Quartz-fiber
1-Methyl phenanthrene (C15)	Carbonaceous fuel combustion	TD-GC/MS	PAH	M1_PHENANTC	$C_{15}H_{12}$	192.26	Quartz-fiber
2-Methyl phenanthrene (C15)	Carbonaceous fuel combustion	TD-GC/MS	PAH	M2_PHENNTC	$C_{15}H_{12}$	192.26	Quartz-fiber

Compound	Source	Method	Category	Code	Formula	Mass	Filter
9-Methyl anthracene (C15)	Carbonaceous fuel combustion	TD-GC/MS	PAH	M_9ANTTC	$C_{15}H_{12}$	192.26	Quartz-fiber
3,6-Dimethyl phenanthrene (C16)	Carbonaceous fuel combustion	TD-GC/MS	PAH	DMPHENA_36TC	$C_{16}H_{14}$	206.28	Quartz-fiber
Methyl fluoranthene (C17)	Carbonaceous fuel combustion	TD-GC/MS	PAH	M_2FLTC	$C_{17}H_{12}$	216.28	Quartz-fiber
Benzo[g,h,i]fluoranthene (C18)	Carbonaceous fuel combustion	TD-GC/MS	PAH	BGHIFLTC	$C_{18}H_{10}$	226.27	Quartz-fiber
Benzo[c]phenanthrene (C18)	Carbonaceous fuel combustion	TD-GC/MS	PAH	BCPHENANTC	$C_{18}H_{12}$	228.29	Quartz-fiber
Benzo[b]naphtho[1,2-d]thiophene (C16)	Carbonaceous fuel combustion	TD-GC/MS	PAH	BBN12THIOTC	$C_{16}H_{10}S$	234.32	Quartz-fiber
Cyclopenta[c,d]pyrene (C18)	Carbonaceous fuel combustion	TD-GC/MS	PAH	CP_CDPYRTC	$C_{18}H_{10}$	226.27	Quartz-fiber
Benz[a]anthracene-7,12-dione (C18)	Carbonaceous fuel combustion	TD-GC/MS	PAH	BAA7_12TC	$C_{18}H_{10}O_2$	258.27	Quartz-fiber
Methyl chrysene (C19)	Carbonaceous fuel combustion	TD-GC/MS	PAH	CHRYMTC	$C_{19}H_{14}$	242.31	Quartz-fiber
Benzo[b]chrysene (C22)	Carbonaceous fuel combustion	TD-GC/MS	PAH	BBCHRYSNTC	$C_{22}H_{14}$	278.35	Quartz-fiber
Picene (C22)	Carbonaceous fuel combustion	TD-GC/MS	PAH	PICTC	$C_{22}H_{14}$	278.35	Quartz-fiber
Anthanthrene (C22)	Carbonaceous fuel combustion	TD-GC/MS	PAH	ANTHAN	$C_{22}H_{12}$	276.33	Quartz-fiber
Nitrate	Combustion emissions	IC-CD	Inorganic ion	N3IC	NO_3^-	62.00	Quartz-fiber/nylon
Bromine	Combustion emissions	XRF	Element	BRXC	Br	79.90	Teflon
7. Mixed combustion SOA							
2-Nitrophenol	Fossil and biomass combustion SOA	HPLC-UV-Vis	Nitro-aromatic	NOARO7HC	$C_6H_5NO_3$	139.11	Quartz-fiber
3-Nitrophenol	Fossil and biomass combustion SOA	HPLC-UV-Vis	Nitro-aromatic	NOARO8HC	$C_6H_5NO_3$	139.11	Quartz-fiber

Table 2 Continued

General source type/marker species[a]	Source subtype	Analysis method[b,c]	Species class	Mnemonics	Molecular formula	Molar mass	Sampling media
4-Nitrophenol	Fossil and biomass combustion SOA	HPLC-UV-Vis	Nitro-aromatic	NOARO9HC	$C_6H_5NO_3$	139.11	Quartz-fiber
2,4-Dinitrophenol	Fossil and biomass combustion SOA	HPLC-UV-Vis	Nitro-aromatic	NOARO10HC	$C_6H_4N_2O_5$	184.11	Quartz-fiber
Methylhydroquinone	Fossil and biomass combustion SOA	ID-TD-GC/MS	Aromatic	AROM4TC	$C_7H_8O_2$	124.14	Quartz-fiber
o-Phthalic acid	Fossil and biomass combustion SOA	ID-TD-GC/MS	Aromatic	AROMA2TC	$C_8H_6O_4$	166.13	Quartz-fiber
1-Nitro-2-naphthol	Fossil and biomass combustion SOA	HPLC-UV-Vis	Nitro-aromatic	NOARO22HC	$C_{10}H_7NO_3$	189.17	Quartz-fiber
3-(4-Nitrophenoxy)phenol	Fossil and biomass combustion SOA	HPLC-UV-Vis	Nitro-aromatic	NOARO23HC	$C_{12}H_9NO_4$	231.20	Quartz-fiber
4-(4-Nitrophenoxy)phenol	Fossil and biomass combustion SOA	HPLC-UV-Vis	Nitro-aromatic	NOARO24HC	$C_{12}H_9NO_4$	231.20	Quartz-fiber
8. Coal combustion emissions							
Picene	Coal combustion	TD-GC/MS	PAH	PICTC	$C_{22}H_{14}$	278.35	Quartz-fiber
Sulfate	Coal combustion	IC-CD	Inorganic Ion	S4IC	SO_4^{2-}	96.06	Quartz-fiber/nylon
Sulfur	Coal combustion	XRF	Element	SUXC	S	32.06	Teflon
Arsenic	Coal combustion	XRF	Trace Metal	ASXC	As	74.92	Teflon
Selenium	Coal combustion	XRF	Trace Metal	SEXC	Se	78.96	Teflon
Cadmium	Coal combustion	XRF	Trace Metal	CDXC	Cd	112.41	Teflon
Thallium	Coal combustion	XRF	Trace Metal	TLXC	Tl	204.38	Teflon
Mercury	Coal combustion (also mining and biomass combustion)	XRF	Trace Metal	HGXC	Hg	200.59	Teflon

9. Internal combustion engine emissions

Compound	Source	Method	Class	Code	MW	Formula	Filter
18α(H)-22,29,30-Trisnormeophopane (Ts) (C27)	Engine exhaust	TD-GC/MS	Hopane	HOP1TC	370.67	$C_{27}H_{46}$	Quartz-fiber
17α(H)-22,29,30-Trisnorphopane (Tm) (C27)	Engine exhaust	TD-GC/MS	Hopane	HOP2TC	370.67	$C_{27}H_{46}$	Quartz-fiber
17α(H),21ß(H)-30-Norhopane (C29αß)	Engine exhaust	TD-GC/MS	Hopane	HOP3TC	398.72	$C_{29}H_{50}$	Quartz-fiber
22,29,30-Norhopane (Ts) (C29)	Engine exhaust	TD-GC/MS	Hopane	HOP4TC	398.72	$C_{29}H_{50}$	Quartz-fiber
17α(H),21α(H)-30-Norhopane + 17ß(H),21α(H)-30-norhopane (C29αα + 29ßα)	Engine exhaust	TD-GC/MS	Hopane	HOP5TC	398.72	$C_{29}H_{50}$	Quartz-fiber
17α(H),21ß(H)-Hopane (C30αß)	Engine exhaust	TD-GC/MS	Hopane	HOP6TC	412.75	$C_{30}H_{52}$	Quartz-fiber
17α(H),21α(H)-Hopane (C30αα)	Engine exhaust	TD-GC/MS	Hopane	HOP7TC	412.75	$C_{30}H_{52}$	Quartz-fiber
17ß(H),21α(H)-Hopane (C30ßα)	Engine exhaust	TD-GC/MS	Hopane	HOP8TC	412.75	$C_{30}H_{52}$	Quartz-fiber
17α(H),21ß(H)-Homohopane (C31αß 22S)	Engine exhaust	TD-GC/MS	Hopane	HOP9TC	426.78	$C_{31}H_{54}$	Quartz-fiber
17α(H),21ß(H)-(22R)-Homohopane (C31αß 22R)	Engine exhaust	TD-GC/MS	Hopane	HOP10TC	426.78	$C_{31}H_{54}$	Quartz-fiber
17α(H),21ß(H)-(22S)-Bishomohopane (C32αß 22S)	Engine exhaust	TD-GC/MS	Hopane	HOP11TC	440.80	$C_{32}H_{56}$	Quartz-fiber
17α(H),21ß(H)-(22R)-Bishomohopane (C32αß 22R)	Engine exhaust	TD-GC/MS	Hopane	HOP12TC	440.80	$C_{32}H_{56}$	Quartz-fiber
17α(H),21ß(H)-(22RS)-Trishomohopane (C33αß 22S)	Engine exhaust	TD-GC/MS	Hopane	HOP13TC	454.83	$C_{33}H_{58}$	Quartz-fiber

Table 2　Continued

General source type/marker species[a]	Source subtype	Analysis method[b,c]	Species class	Mnemonics	Molecular formula	Molar mass	Sampling media
17α(H),21β(H)-(22RS)-Trishomohopane (C33αβ 22R)	Engine exhaust	TD-GC/MS	Hopane	HOP14TC	$C_{33}H_{58}$	454.83	Quartz-fiber
17α(H),21β(H)-(22S)-Tetrakishomohopane (C34αβ 22S)	Engine exhaust	TD-GC/MS	Hopane	HOP15TC	$C_{34}H_{60}$	468.86	Quartz-fiber
17α(H),21β(H)-(22R)-Tetrakishomohopane (C34αβ 22R)	Engine exhaust	TD-GC/MS	Hopane	HOP16TC	$C_{34}H_{60}$	468.86	Quartz-fiber
17α(H),21β(H)-(22RS)-Pentakishomohopane (C35αβ 22S)	Engine exhaust	TD-GC/MS	Hopane	HOP17TC	$C_{34}H_{60}$	482.88	Quartz-fiber
17α(H),21β(H)-(22RS)-Pentakishomohopane (C35αβ 22R)	Engine exhaust	TD-GC/MS	Hopane	HOP18TC	$C_{34}H_{60}$	482.88	Quartz-fiber
ααα (20S)-Cholestane (C27)	Engine exhaust	TD-GC/MS	Sterane	STER1TC	$C_{27}H_{48}$	372.68	Quartz-fiber
αββ (20R)-Cholestane (C27)	Engine exhaust	TD-GC/MS	Sterane	STER2TC	$C_{27}H_{48}$	372.68	Quartz-fiber
αββ (20S)-Cholestane (C27)	Engine exhaust	TD-GC/MS	Sterane	STER3TC	$C_{27}H_{48}$	372.68	Quartz-fiber
ααα (20R)-Cholestane (C27)	Engine exhaust	TD-GC/MS	Sterane	STER4TC	$C_{27}H_{48}$	372.68	Quartz-fiber
ααα (20S,24S)-24-Methylcholestane (C28)	Engine exhaust	TD-GC/MS	Sterane	STER5TC	$C_{28}H_{50}$	386.71	Quartz-fiber
αββ (20R,24S)-24-Methylcholestane (C28)	Engine exhaust	TD-GC/MS	Sterane	STER6TC	$C_{28}H_{50}$	386.71	Quartz-fiber
αββ (20S,24S)-24-Methylcholestane (C28)	Engine exhaust	TD-GC/MS	Sterane	STER7TC	$C_{28}H_{50}$	386.71	Quartz-fiber
ααα (20R,24R)-24-Methylcholestane (C28)	Engine exhaust	TD-GC/MS	Sterane	STER8TC	$C_{28}H_{50}$	386.71	Quartz-fiber

Compound	Source	Method	Type	Code	Formula	MW	Filter
ααα (20S,24R/S)-24-Ethylcholestane (C29)	Engine exhaust	TD-GC/MS	Sterane	STER9TC	$C_{29}H_{52}$	400.74	Quartz-fiber
αββ (20R,24R)-24-Ethylcholestane (C29)	Engine exhaust	TD-GC/MS	Sterane	STER10TC	$C_{29}H_{52}$	400.74	Quartz-fiber
αββ (20S,24R)-24-Ethylcholestane (C29)	Engine exhaust	TD-GC/MS	Sterane	STER11TC	$C_{29}H_{52}$	400.74	Quartz-fiber
ααα (20R,24R)-24-Ethylcholestane (C29)	Engine exhaust	TD-GC/MS	Sterane	STER12TC	$C_{29}H_{52}$	400.74	Quartz-fiber
Zinc	Engine exhaust (also coal combustion)	XRF	Trace metal	ZNXC	Zn	65.38	Teflon
Barium	Engine exhaust	XRF, ICP-MS	Trace metal	BAXC	Ba	137.33	Teflon

10. Residual-oil combustion emissions

Compound	Source	Method	Type	Code	Formula	MW	Filter
Vanadium	Residual fuel combustion	XRF	Trace metal	VAXC	V	50.94	Teflon
Nickel	Residual fuel combustion	XRF	Trace metal	NIXC	Ni	58.69	Teflon

11. Tobacco smoke emissions

Compound	Source	Method	Type	Code	Formula	MW	Filter
iso-Nonacosane (iso-C29)	Tobacco smoke	TD-GC/MS	iso-Alkane	ISOC29TC	$C_{29}H_{60}$	408.79	Quartz-fiber
anteiso-Nonacosane (anteiso-C29)	Tobacco smoke	TD-GC/MS	anteiso-Alkane	ANISOC29TC	$C_{29}H_{60}$	408.79	Quartz-fiber
iso-Triacontane (iso-C30)	Tobacco smoke	TD-GC/MS	iso-Alkane	ISOC30TC	$C_{30}H_{62}$	422.81	Quartz-fiber
anteiso-Triacontane (anteiso-C30)	Tobacco smoke	TD-GC/MS	anteiso-Alkane	ANISOC30TC	$C_{30}H_{62}$	422.81	Quartz-fiber
iso-Hentriacotane (iso-C31)	Tobacco smoke	TD-GC/MS	iso-Alkane	ISOC31TC	$C_{31}H_{64}$	436.84	Quartz-fiber
anteiso-Hentriacotane (anteiso-C31)	Tobacco smoke	TD-GC/MS	anteiso-Alkane	ANISOC31TC	$C_{31}H_{64}$	436.84	Quartz-fiber
iso-Dotriacontane (iso-C32)	Tobacco smoke	TD-GC/MS	iso-Alkane	ISOC32TC	$C_{32}H_{66}$	450.87	Quartz-fiber

Table 2 Continued

General source type/marker species[a]	Source subtype	Analysis method[b,c]	Species class	Mnemonics	Molecular formula	Molar mass	Sampling media
anteiso-Dotriacontane (anteiso-C32)	Tobacco smoke	TD-GC/MS	anteiso-Alkane	ANISOC32TC	$C_{32}H_{66}$	450.87	Quartz-fiber
iso-Tritriactotane (iso-C33)	Tobacco smoke	TD-GC/MS	iso-Alkane	ISOC33TC	$C_{33}H_{68}$	464.89	Quartz-fiber
anteiso-Tritriactotane (anteiso-C33)	Tobacco smoke	TD-GC/MS	anteiso-Alkane	ANISOC33TC	$C_{33}H_{68}$	464.89	Quartz-fiber
12. Meat cooking emissions							
Palmitic acid	Meat cooking	ID-TD-GC/MS	Fatty acid	PALMTC	$C_{16}H_{32}O_2$	256.42	Quartz-fiber
Stearic acid	Meat cooking	ID-TD-GC/MS	Fatty acid	STEARICTC	$C_{18}H_{36}O_2$	284.48	Quartz-fiber
Cholesterol	Meat cooking	ID-TD-GC/MS	Sterol	STEROL1TC	$C_{27}H_{46}O$	386.65	Quartz-fiber
13. Trash burning emissions							
1,3,5-Triphenylbenzene	Trash burning – plastics	TD-GC/MS	Aromatic	AROM1TC	$C_{24}H_{18}$	306.40	Quartz-fiber
Tris(2,4-tert-butylphenyl)phosphate	Trash burning – plastics	TD-GC/MS	Aromatic	AROM2TC	$C_{42}H_{63}O_4P$	662.89	Quartz-fiber
14. Marine/sea salt sources							
Water-soluble sodium	Marine/sea salt	IC-CD	Inorganic ion	NAIC	Na^+	22.99	Quartz-fiber/nylon
Water-soluble magnesium	Marine/sea salt (also crustal material)	IC-CD	Inorganic ion	MGIC	Mg^{2+}	24.31	Quartz-fiber/nylon
Water-soluble calcium	Marine/sea salt (also crustal material)	IC-CD	Inorganic ion	CAIC	Ca^{2+}	40.08	Quartz-fiber/nylon

Species	Source	Method	Type	Code	Symbol	Atomic weight	Filter
Chloride	Marine/sea salt (also combustion)	IC-CD	Inorganic ion	CLIC	Cl^-	35.45	Quartz-fiber/nylon
Fluoride	Marine/sea salt (also combustion)	IC-CD	Inorganic ion	FLIC	F^-	19.00	Quartz-fiber/nylon
Chlorine	Marine/sea salt (also combustion)	XRF	Element	CLXC	Cl	35.45	Teflon
15. Geological sources							
Aluminum	Crustal material (also engine exhaust)	XRF	Element	ALXC	Al	26.98	Teflon
Silicon	Crustal material	XRF	Element	SIXC	Si	28.09	Teflon
Calcium	Crustal material	XRF	Element	CAXC	Ca	40.08	Teflon
Potassium	Crustal material	XRF	Element	KPXC	K	39.10	Teflon
Titanium	Crustal material	XRF	Trace metal	TIXC	Ti	47.87	Teflon
Iron	Crustal material (also engine exhaust)	XRF	Trace metal	FEXC	Fe	55.85	Teflon
Praseodymium	Mineral	ICP-MS	Trace metal	PRPC	Pr	140.91	Teflon
Neodymium	Mineral	ICP-MS	Trace metal	NDPC	Nd	144.24	Teflon
Gadolinium	Mineral	ICP-MS	Trace metal	GDPC	Gd	157.25	Teflon
Dysprosium	Mineral	ICP-MS	Trace metal	DYPC	Dy	162.50	Teflon
Holmium	Mineral	ICP-MS	Trace metal	HOPC	Ho	164.93	Teflon
Erbium	Mineral	ICP-MS	Trace metal	ERPC	Er	167.26	Teflon
Thulium	Mineral	ICP-MS	Trace metal	TMPC	Tm	168.93	Teflon
Ytterbium	Mineral	ICP-MS	Trace metal	YBPC	Yb	173.04	Teflon
Lutetium	Mineral	ICP-MS	Trace metal	LUPC	Lu	174.97	Teflon
16. Metal processing/smelting/mining sources							
Scandium	Mining	XRF	Trace metal	SCXC	Sc	44.96	Teflon
Chromium	Metal processing (also residual fuel combustion)	XRF	Trace metal	CRXC	Cr	51.99	Teflon

Table 2 Continued

General source type/marker species[a]	Source subtype	Analysis method[b,c]	Species class	Mnemonics	Molecular formula	Molar mass	Sampling media
Manganese	Metal processing	XRF	Trace metal	MNXC	Mn	54.94	Teflon
Cobalt	Metal processing	XRF	Trace metal	COXC	Co	58.93	Teflon
Copper	Smelting (also engine exhaust)	XRF	Trace metal	CUXC	Cu	63.55	Teflon
Rubidium	Mining	XRF	Trace metal	RBXC	Rb	85.47	Teflon
Strontium	Mining	XRF	Trace metal	SRXC	Sr	87.62	Teflon
Yttrium	Mining	XRF	Trace metal	YTXC	Y	88.91	Teflon
Zirconium	Mining	XRF	Trace metal	ZRXC	Zr	91.22	Teflon
Niobium	Mining	XRF	Trace metal	NBXC	Nb	92.91	Teflon
Molybdenum	Mining (also engine exhaust)	XRF	Trace metal	MOXC	Mo	95.94	Teflon
Palladium	Mining	XRF	Trace metal	PDXC	Pd	106.42	Teflon
Silver	Mining	XRF	Trace metal	AGXC	Ag	107.87	Teflon
Indium	Mining	XRF	Trace metal	INXC	In	114.82	Teflon
Tin	Mining	XRF	Trace metal	SNXC	Sn	118.71	Teflon
Antimony	Mining	XRF	Trace metal	SBXC	Sb	121.76	Teflon
Cesium	Mining	XRF, ICP-MS	Trace metal	CSXC	Cs	132.91	Teflon
Lanthanum	Mining (also residual fuel combustion)	XRF, ICP-MS	Trace metal	LAXC	La	138.91	Teflon
Cerium	Mining	XRF, ICP-MS	Trace metal	CEXC	Ce	140.12	Teflon
Samarium	Mining	XRF, ICP-MS	Trace metal	SMXC	Sm	150.36	Teflon
Europium	Mining	XRF, ICP-MS	Trace metal	EUXC	Eu	151.96	Teflon
Terbium	Mining	XRF	Trace metal	TBXC	Tb	158.92	Teflon
Hafnium	Mining	XRF	Trace metal	HFXC	Hf	178.49	Teflon
Tantalum	Mining	XRF	Trace metal	TAXC	Ta	180.95	Teflon
Tungsten	Mining	XRF	Trace metal	WOXC	W	183.84	Teflon

Name	Source	Method	Type	Code	Formula	Mass	Filter
Iridium	Mining	XRF	Trace metal	IRXC	Ir	192.22	Teflon
Gold	Mining	XRF	Trace metal	AUXC	Au	196.97	Teflon
Lead	Smelting (also fuel combustion)	XRF	Trace metal	PBXC	Pb	207.20	Teflon
17. Agricultural sources							
Ammonium	Agriculture	IC-CD	Inorganic ion	N4IC	NH_4^+	18.04	Quartz-fiber/nylon
Phophate	Agriculture	IC-CD	Inorganic ion	P4IC	PO_4^{3-}	94.97	Quartz-fiber/nylon
Phosphorous	Agriculture	XRF	Element	PHXC	P	30.97	Teflon
Glucose	Agriculture (Soil)	ID-TD-GC/MS	Monosaccharide	GLXYOSETC	$C_6H_{12}O_6$	180.16	Quartz-fiber
Trehalose	Agriculture (soil fungi)	ID-TD-GC/MS	Disaccharide	TREHLSETC	$C_{12}H_{22}O_{11}$	342.296	Quartz-fiber
18. Other mixed sources							
OC1[e]	Biogenic and combustion	TOA (TOR/TOT)	Carbon fraction	O1TC	N/A	N/A	Quartz-fiber
OC2[e]	Biogenic and combustion	TOA (TOR/TOT)	Carbon fraction	O2TC	N/A	N/A	Quartz-fiber
OC3[e]	Biogenic and combustion	TOA (TOR/TOT)	Carbon fraction	O3TC	N/A	N/A	Quartz-fiber
OC4[e]	Biogenic and combustion	TOA (TOR/TOT)	Carbon fraction	O4TC	N/A	N/A	Quartz-fiber
OP-R	Biogenic and combustion	TOA (TOR/TOT)	Carbon fraction	OPR	N/A	N/A	Quartz-fiber

Table 2 Continued

General source type/marker species[a]	Source subtype	Analysis method[b,c]	Species class	Mnemonics	Molecular formula	Molar mass	Sampling media
OP-T	Biogenic and combustion	TOA (TOR/TOT)	Carbon fraction	OPT	N/A	N/A	Quartz-fiber
OC (405 nm – R)[e]	Biogenic and combustion	TSA (TOR/TOT)	Carbon fraction	OC405TRC	N/A	N/A	Quartz-fiber
OC (405 nm – T)[e]	Biogenic and combustion	TSA (TOR/TOT)	Carbon fraction	OC405TTC	N/A	N/A	Quartz-fiber
OC (445 nm – R)[e]	Biogenic and combustion	TSA (TOR/TOT)	Carbon fraction	OC445TRC	N/A	N/A	Quartz-fiber
OC (445 nm – T)[e]	Biogenic and combustion	TSA (TOR/TOT)	Carbon fraction	OC445TTC	N/A	N/A	Quartz-fiber
OC (532 nm – R)[e]	Biogenic and combustion	TSA (TOR/TOT)	Carbon fraction	OC532TRC	N/A	N/A	Quartz-fiber
OC (532 nm – T)[e]	Biogenic and combustion	TSA (TOR/TOT)	Carbon fraction	OC532TTC	N/A	N/A	Quartz-fiber
OC (635 nm – R)[e]	Biogenic and combustion	TSA (TOR/TOT)	Carbon fraction	OC635TRC	N/A	N/A	Quartz-fiber
OC (635 nm – T)[e]	Biogenic and combustion	TSA (TOR/TOT)	Carbon fraction	OC635TTC	N/A	N/A	Quartz-fiber
OC (780 nm – R)[e]	Biogenic and combustion	TSA (TOR/TOT)	Carbon fraction	OC780TRC	N/A	N/A	Quartz-fiber

OC (780 nm – T)^e	Biogenic and combustion	TSA (TOR/TOT)	Carbon fraction	OC780TTC	N/A	N/A	Quartz-fiber
OC (808 nm – R)^e	Biogenic and combustion	TSA (TOR/TOT)	Carbon fraction	OC808TRC	N/A	N/A	Quartz-fiber
OC (808 nm – T)^e	Biogenic and combustion	TSA (TOR/TOT)	Carbon fraction	OC808TTC	N/A	N/A	Quartz-fiber
OC (980 nm – R)^e	Biogenic and combustion	TSA (TOR/TOT)	Carbon fraction	OC980TRC	N/A	N/A	Quartz-fiber
OC (980 nm – T)^e	Biogenic and combustion	TSA (TOR/TOT)	Carbon fraction	OC980TTC	N/A	N/A	Quartz-fiber
EC1^e	Carbonaceous mixed combustion	TOA (TOR/TOT)	Carbon fraction	E1TC	N/A	N/A	Quartz-fiber
EC2^e	Carbonaceous mixed combustion	TOA (TOR/TOT)	Carbon fraction	E2TC	N/A	N/A	Quartz-fiber
EC3^e	Carbonaceous mixed combustion	TOA (TOR/TOT)	Carbon fraction	E3TC	N/A	N/A	Quartz-fiber
EC (405 nm – R)^e	Combustion emissions	TSA (TOR/TOT)	Carbon fraction	EC405TRC	N/A	N/A	Quartz-fiber
EC (405 nm – T)^e	Combustion emissions	TSA (TOR/TOT)	Carbon fraction	EC405TTC	N/A	N/A	Quartz-fiber
EC (445 nm – R)^e	Combustion emissions	TSA (TOR/TOT)	Carbon fraction	EC445TRC	N/A	N/A	Quartz-fiber
EC (445 nm – T)^e	Combustion emissions	TSA (TOR/TOT)	Carbon fraction	EC445TTC	N/A	N/A	Quartz-fiber

Table 2 Continued

General source type/marker species[a]	Source subtype	Analysis method[b,c]	Species class	Mnemonics	Molecular formula	Molar mass	Sampling media
EC (532 nm – R)[e]	Combustion emissions	TSA (TOR/TOT)	Carbon fraction	EC532TRC	N/A	N/A	Quartz-fiber
EC (532 nm – T)[e]	Combustion emissions	TSA (TOR/TOT)	Carbon fraction	EC532TTC	N/A	N/A	Quartz-fiber
EC (635 nm – R)[e]	Combustion emissions	TSA (TOR/TOT)	Carbon fraction	EC635TRC	N/A	N/A	Quartz-fiber
EC (635 nm – T)[e]	Combustion emissions	TSA (TOR/TOT)	Carbon fraction	EC635TTC	N/A	N/A	Quartz-fiber
EC (780 nm – R)[e]	Combustion emissions	TSA (TOR/TOT)	Carbon fraction	EC780TRC	N/A	N/A	Quartz-fiber
EC (780 nm – T)[e]	Combustion emissions	TSA (TOR/TOT)	Carbon fraction	EC780TTC	N/A	N/A	Quartz-fiber
EC (808 nm – R)[e]	Combustion emissions	TSA (TOR/TOT)	Carbon fraction	EC808TRC	N/A	N/A	Quartz-fiber
EC (808 nm – T)[e]	Combustion emissions	TSA (TOR/TOT)	Carbon fraction	EC808TTC	N/A	N/A	Quartz-fiber
EC (980 nm – R)[e]	Combustion emissions	TSA (TOR/TOT)	Carbon fraction	EC980TRC	N/A	N/A	Quartz-fiber
EC (980 nm – T)[e]	Combustion emissions	TSA (TOR/TOT)	Carbon fraction	EC980TTC	N/A	N/A	Quartz-fiber

TC	Biogenic and combustion	TOA (TOR/TOT)	Carbon fraction	TCTC	N/A	N/A	Quartz-fiber
$\text{ATN}_{405R}^{\int}$	Light absorbing sources	TSA (TOR/TOT)	Carbon fraction	ATN405R	N/A	N/A	Quartz-fiber
$\text{ATN}_{405T}^{\int}$	Light absorbing sources	TSA (TOR/TOT)	Carbon fraction	ATN405T	N/A	N/A	Quartz-fiber
$\text{ATN}_{445R}^{\int}$	Light absorbing sources	TSA (TOR/TOT)	Carbon fraction	ATN445R	N/A	N/A	Quartz-fiber
$\text{ATN}_{445T}^{\int}$	Light absorbing sources	TSA (TOR/TOT)	Carbon fraction	ATN445T	N/A	N/A	Quartz-fiber
$\text{ATN}_{532R}^{\int}$	Light absorbing sources	TSA (TOR/TOT)	Carbon fraction	ATN532R	N/A	N/A	Quartz-fiber
$\text{ATN}_{532T}^{\int}$	Light absorbing sources	TSA (TOR/TOT)	Carbon fraction	ATN532T	N/A	N/A	Quartz-fiber
$\text{ATN}_{635R}^{\int}$	Light absorbing sources	TSA (TOR/TOT)	Carbon fraction	ATN635R	N/A	N/A	Quartz-fiber
$\text{ATN}_{635T}^{\int}$	Light absorbing sources	TSA (TOR/TOT)	Carbon fraction	ATN635T	N/A	N/A	Quartz-fiber
$\text{ATN}_{780R}^{\int}$	Light absorbing sources	TSA (TOR/TOT)	Carbon fraction	ATN780R	N/A	N/A	Quartz-fiber
$\text{ATN}_{780T}^{\int}$	Light absorbing sources	TSA (TOR/TOT)	Carbon fraction	ATN780T	N/A	N/A	Quartz-fiber

Table 2 Continued

General source type/marker species[a]	Source subtype	Analysis method[b,c]	Species class	Mnemonics	Molecular formula	Molar mass	Sampling media
ATN_{808R}[f]	Light absorbing sources	TSA (TOR/TOT)	Carbon fraction	ATN808R	N/A	N/A	Quartz-fiber
ATN_{808T}[f]	Light absorbing sources	TSA (TOR/TOT)	Carbon fraction	ATN808T	N/A	N/A	Quartz-fiber
ATN_{980R}[f]	Light absorbing sources	TSA (TOR/TOT)	Carbon fraction	ATN980R	N/A	N/A	Quartz-fiber
ATN_{980T}[f]	Light absorbing sources	TSA (TOR/TOT)	Carbon fraction	ATN980T	N/A	N/A	Quartz-fiber
WSON	Mixed sources	TOC/TN	Organic nitrogen	WSONHC	N/A	N/A	Quartz-fiber/nylon
Nitrite	Mixed sources	IC-CD	Inorganic ion	N2IC	NO_2^-	46.01	Quartz-fiber/nylon
Gallium	Mixed sources	XRF	Trace metal	GAXC	Ga	69.72	Teflon
Pentadecane (n-C15)	Mixed sources	TD-GC/MS	n-Alkane	PENTADTC	$C_{15}H_{32}$	212.41	Quartz-fiber
Hexadecane (n-C16)	Mixed sources	TD-GC/MS	n-Alkane	HEXADTC	$C_{16}H_{34}$	226.44	Quartz-fiber
Heptadecane (n-C17)	Mixed sources	TD-GC/MS	n-Alkane	HEPTADTC	$C_{17}H_{36}$	240.47	Quartz-fiber
Octadecane (n-C18)	Mixed sources	TD-GC/MS	n-Alkane	OCTADTC	$C_{18}H_{38}$	254.49	Quartz-fiber
Nonadecane (n-C19)	Mixed sources	TD-GC/MS	n-Alkane	NONADTC	$C_{19}H_{40}$	268.52	Quartz-fiber

Eicosane (n-C20)	Mixed sources	TD-GC/MS	n-Alkane	EICOSATC	$C_{20}H_{42}$	282.55	Quartz-fiber
Heneicosane (n-C21)	Mixed sources	TD-GC/MS	n-Alkane	HENEICTC	$C_{21}H_{44}$	296.57	Quartz-fiber
Docosane (n-C22)	Mixed sources	TD-GC/MS	n-Alkane	DOCOSATC	$C_{22}H_{46}$	310.60	Quartz-fiber
Tricosane (n-C23)	Mixed sources	TD-GC/MS	n-Alkane	TRICOSATC	$C_{23}H_{48}$	324.63	Quartz-fiber
Tetracosane (n-C24)	Mixed sources	TD-GC/MS	n-Alkane	TETCOSTC	$C_{24}H_{50}$	338.65	Quartz-fiber
Pentacosane (n-C25)	Mixed sources	TD-GC/MS	n-Alkane	PENCOSTC	$C_{25}H_{52}$	352.68	Quartz-fiber
Hexacosane (n-C26)	Mixed sources	TD-GC/MS	n-Alkane	HEXCOSTC	$C_{26}H_{54}$	366.71	Quartz-fiber
Heptacosane (n-C27)	Mixed sources	TD-GC/MS	n-Alkane	HEPCOSTC	$C_{27}H_{56}$	380.73	Quartz-fiber
Octacosane (n-C28)	Mixed sources	TD-GC/MS	n-Alkane	OCTCOSTC	$C_{28}H_{58}$	394.76	Quartz-fiber
Nonacosane (n-C29)	Mixed sources	TD-GC/MS	n-Alkane	NONCOSTC	$C_{29}H_{60}$	408.79	Quartz-fiber
Triacontane (n-C30)	Mixed sources	TD-GC/MS	n-Alkane	TRICONTTC	$C_{30}H_{62}$	422.81	Quartz-fiber
Hentriacotane (n-C31)	Mixed sources	TD-GC/MS	n-Alkane	HTRICONTTC	$C_{31}H_{64}$	436.84	Quartz-fiber
Dotriacontane (n-C32)	Mixed sources	TD-GC/MS	n-Alkane	DTRICONTTC	$C_{32}H_{66}$	450.87	Quartz-fiber
Tritriactotane (n-C33)	Mixed sources	TD-GC/MS	n-Alkane	TTRICONTTC	$C_{33}H_{68}$	464.89	Quartz-fiber
Tetratriactoane (n-C34)	Mixed sources	TD-GC/MS	n-Alkane	TETRICNTTC	$C_{34}H_{70}$	478.92	Quartz-fiber
Pentatriacontane (n-C35)	Mixed sources	TD-GC/MS	n-Alkane	PTRICONTTC	$C_{35}H_{72}$	492.95	Quartz-fiber
Hexatriacontane (n-C36)	Mixed sources	TD-GC/MS	n-Alkane	HXTRICNTTC	$C_{36}H_{74}$	506.97	Quartz-fiber

Table 2 Continued

General source type/marker species[a]	Source subtype	Analysis method[b,c]	Species class	Mnemonics	Molecular formula	Molar mass	Sampling media
Heptatriacontane (n-C37)	Mixed sources	TD-GC/MS	n-Alkane	HPTRICNTTC	$C_{37}H_{76}$	521.00	Quartz-fiber
Octatriacontane (n-C38)	Mixed sources	TD-GC/MS	n-Alkane	OTRICONTTC	$C_{38}H_{78}$	535.02	Quartz-fiber
Nonatriacontane (n-C39)	Mixed sources	TD-GC/MS	n-Alkane	NTRICONTTC	$C_{39}H_{80}$	549.05	Quartz-fiber
Tetracontane (n-C40)	Mixed sources	TD-GC/MS	n-Alkane	TETRCONTTC	$C_{40}H_{82}$	563.08	Quartz-fiber
2-Methylnonadecane (C20)	Mixed sources	TD-GC/MS	alkyl-Alkane	NONAD2MTC	$C_{20}H_{42}$	282.55	Quartz-fiber
3-Methylnonadecane (C20)	Mixed sources	TD-GC/MS	alkyl-Alkane	NONAD3MTC	$C_{20}H_{42}$	282.55	Quartz-fiber
Pristane (C19)	Mixed sources	TD-GC/MS	Alkane	PRISTTC	$C_{19}H_{40}$	268.52	Quartz-fiber
Phytane (C20)	Mixed sources	TD-GC/MS	Alkane	PHYTANTC	$C_{20}H_{42}$	282.55	Quartz-fiber
Squalane (C30)	Mixed sources	TD-GC/MS	Alkane	SQUALTC	$C_{30}H_{62}$	422.81	Quartz-fiber
Octylcyclohexane (C14)	Mixed sources	TD-GC/MS	Cycloalkane	OCYCYHXTC	$C_{14}H_{28}$	196.37	Quartz-fiber
Decylcyclohexane (C16)	Mixed sources	TD-GC/MS	Cycloalkane	DECYHXTC	$C_{16}H_{32}$	224.42	Quartz-fiber
Tridecylcyclohexane (C19)	Mixed sources	TD-GC/MS	Cycloalkane	DEC3YHXTC	$C_{19}H_{38}$	266.50	Quartz-fiber
Heptadecylcyclohexane (C23)	Mixed sources	TD-GC/MS	Cycloalkane	N_HDCYHXTC	$C_{23}H_{46}$	322.61	Quartz-fiber
Nonadecylcyclohexane (C25)	Mixed sources	TD-GC/MS	Cycloalkane	DEC9YHXTC	$C_{25}H_{50}$	350.66	Quartz-fiber
1-Octadecene (C18)	Mixed sources	TD-GC/MS	Alkene	OCTDECENTC	$C_{18}H_{36}$	252.48	Quartz-fiber

[a]WSOC = water-soluble organic carbon; WSON = water-soluble organic nitrogen.

[b]Analysis Methods are abbreviated as follows:

FTIR = Fourier Transform Infrared Spectroscopy.

IC-CD = Ion chromatography-conductivity detection.

IC-PAD = Ion chromatography-pulsed amperometric detection.

ICP-MS = Inductively coupled plasma-mass spectrometry.

ID-TD-GC/MS = *In situ* derivatization-thermal desorption-gas chromatography/mass spectrometry.

HPLC-ELSD-UV/VIS = High performance liquid chromatography-ultraviolet-visible light spectroscopy.

TD-GC/MS = Thermal desorption-gas chromatography/mass spectrometry.

TOA (TOR/TOT) = Thermal optical analysis (thermal-optical reflectance/thermal-optical transmittance).

TOC = Total organic carbon.

TON/TN = Total organic nitrogen/total nitrogen.

TSA (TOR/TOT) = Thermal spectral analysis (thermal-optical reflectance/thermal-optical transmittance).

XRF = X-ray fluorescence spectroscopy.

[c]Analysis Method References are as follows:

FTIR

C. Coury and A. M. Dillner, ATR-FTIR characterization of organic functional groups and inorganic ions in ambient aerosols at a rural site, *Atmos. Environ.*, 2009 **43**, 940–948.

S. Takahama, A. Johnson, L. M. Russell, Quantification of Carboxylic and Carbonyl Functional Groups in Organic Aerosol Infrared Absorbance Spectra, *Aerosol Sci. Technol*, 2013, **47**, 310–325.

IC-CD

T. Lee, S. M. Kreidenwels and J. L. Collett, Aerosol ion characteristics during the Big Bend Regional Aerosol and Visibility Observational Study, 2004, *J. Air Waste Manage. Assoc.*, **54**, 585–592.

J. C. Chow and J. G. Watson, Ion chromatography in elemental analysis of airborne particles, in Elemental Analysis of Airborne Particles, ed. S. Landsberger and M. Creatchman, Gordon and Breach Science, 1999, Amsterdam, vol. 1, pp. 97–137.

IC-PAD

G. Engling, C. M. Carrico, S. M. Kreidenweis, J. L. Collett Jr., D. E. Day, W. C. Malm, E. Lincoln, W. Min Hao, Y. Iinuma and H. Herrmann, Determination of levoglucosan in biomass combustion aerosol by high-performance anion-exchange chromatography with pulsed amperometric detection, *Atmos. Environ.*, 2006, **40**, Suppl. 2, 299–311.

Table 2 Continued

Y. Iinuma, G. Engling, H. Puxbaum and H. Herrmann, A highly resolved anion-exchange chromatographic method for determination of saccharidic tracers for biomass combustion and primary bio-particles in atmospheric aerosol, 2009, *Atmos. Environ.*, **43**, 1367–1371.

Z.-S. Zhang, G. Engling, C.-Y. Chan, Y.-H. Yang, M. Lin, S. Shi, J. He, Y.-D. Li and X.-M. Wang, Determination of isoprene-derived secondary organic aerosol tracers (2-methyltetrols) by HPAEC-PAD: Results from size-resolved aerosols in a tropical rainforest, *Atmos. Environ.*, 2013, **70**, 468–476.

ICP-MS

N. Clements, J. Eav, M. J. Xie, M. P. Hannigan, S. L. Miller, W. Navidi, J. L. Peel, J. J. Schauer, M. M. Shafer and J. B. Milford, Concentrations and source insights for trace elements in fine and coarse particulate matter, *Atmos. Environ.*, 2014, **89**, 373–381.

ID-TD-GC/MS

J. Orasche, J. Schnelle-Kreis, G. Abbaszade and R. Zimmermann, Technical Note: In-situ derivatization thermal desorption GC-TOFMS for direct analysis of particle-bound non-polar and polar organic species, *Atmos. Chem. Phys.*, 2011, **11**, 8977–8993.

E. Grandesso and P. P. Ballesta, Derivatization and analysis of levoglucosan and PAHs in ambient air particulate matter by moderate temperature thermal desorption coupled with GC/MS, *Anal. Methods*, 2014, **6**, 6900–6908.

HPLC-ELSD

C. Emmenegger, A. Reinhardt, C. Hueglin, R. Zenobi and M. Kalberer, Evaporative light scattering: A novel detection method for the quantitative analysis of humic-like substances in aerosols, *Environ. Sci. Technol.*, 2007, **41**, 2473–2478.

P. Lin, G. Engling and J. Z. Yu, Humic-like substances in fresh emissions of rice straw burning and in ambient aerosols in the Pearl River Delta Region, China, *Atmos. Chem. Phys.*, 2010, **10**, 6487–6500.

HPLC-UV/VIS

Y. Iinuma, O. Boge, R. Grafe and H. Herrmann, Methyl-Nitrocatechols: Atmospheric Tracer Compounds for Biomass Burning Secondary Organic Aerosols, *Environ. Sci. Technol.*, 2010 **44**, 8453–8459.

Z. Kitanovski, I. Grgíc, F. Yasmeen, M. Claeys and A. Cusak, Development of a liquid chromatographic method based on ultraviolet-visible and electrospray ionization mass spectrometric detection for the identification of nitrocatechols and related tracers in biomass burning atmospheric organic aerosol, *Rapid Commun. Mass Spectrom.*, 2012, **26**, 793–804.

TD-GC/MS

S. S. H. Ho and J. Z. Yu, In-injection port thermal desorption and subsequent gas chromatography-mass spectrometric analysis of polycyclic aromatic hydrocarbons and n-alkanes in atmospheric aerosol samples, *J. Chromatogr. A*, 2004, **1059**, 121–129.

J. Schnelle-Kreis, J. Orasche, G. Abbaszade, K. Schafer, D. P. Harlos, A. D. A. Hansen and R. Zimmermann, Application of direct thermal desorption gas chromatography time-of-flight mass spectrometry for determination of nonpolar organics in low-volume samples from ambient particulate matter and personal samplers, *Anal. Bioanal. Chem.*, 2011, **401**, 3083–3094.

M. D. Hays and R. J. Lavrich, Developments in direct thermal extraction gas chromatography-mass spectrometry of fine aerosols, *TrAC, Trends Anal. Chem.*, 2007, **26**, 88–102.

TOA

J. C. Chow, J. G. Watson, L.-W. A. Chen, M.-C. O. Chang, N. F. Robinson, D. L. Trimble and S. D.Kohl, The IMPROVE_A temperature protocol for thermal/optical carbon analysis: Maintaining consistency with a long-term database, *J. Air Waste Manage. Soc.*, 2007, **57**, 1014–1023.

TOC

H. Yang, Q. F. Li and J. Z. Yu, Comparison of two methods for the determination of water-soluble organic carbon in atmospheric particles, *Atmos. Environ.*, 2003, **37**, 865–870.

X. Zhang, Z. Liu, A. Hecobian, M. Zheng, N. H. Frank, S. Edgerton and R. J. Weber, Spatial and seasonal variations of fine particle water-soluble organic carbon (WSOC) over the southeastern United States: implications for secondary organic aerosol formation, *Atmos. Chem. Phys.*, 2012, **12**, 6593–6607.

Z. Y. Du, K. B. He, Y. Cheng, F. K. Duan, Y. L. Ma, J. M. Liu, X. L. Zhang, M. Zheng and R. Weber, A yearlong study of water-soluble organic carbon in Beijing I: Sources and its primary vs. secondary nature, *Atmosp. Environ.*, 2014, **92**, 514–521.

TON/TN

Q. Zhang, C. Anastasio and M. Jimenez-Cruz, Water-soluble organic nitrogen in atmospheric fine particles (PM2.5) from northern California, *J. Geophys. Res.Atmos.*, 2002 107.

K. Violaki, J. Sciare, J. Williams, A. R. Baker, M. Martino and N. Mihalopoulos, Atmospheric water-soluble organic nitrogen (WSON) over marine environments: a global perspective, *Biogeosciences*, 2015, **12**, 3131–3140.

K. F. Ho, S. S. H. Ho, R. J. Huang, S. X. Liu, J. J. Cao, T. Zhang, H. C. Chuang, C. S. Chan, D. Hu and L. W. Tian, Characteristics of water-soluble organic nitrogen in fine particulate matter in the continental area of China, *Atmos. Environ.*, 2015, **106**, 252–261.

TSA

L.-W. A. Chen, J. C. Chow, X. L. Wang, J. A. Robles, B. J. Sumlin, D. H. Lowenthal, R. Zimmermann, J. G. Watson, Multi-wavelength optical measurement to enhance thermal/optical analysis for carbonaceous aerosol, *Atmos. Meas. Tech*, 2015, **8**, 451–461.

J. C. Chow, L.-W. A. Chen, X. L. Wang, B. J. Sumlin, S. B. Gronstal, L.-W. A. Chen, D. L. Trimble, S. D. Kohl, S. R. Mayorga, G. Riggio, P. R. Hurbain, R. Zimmermann and J. G. Watson, Optical calibration and equivalence of a multiwavelength thermal/optical carbon analyzer, *Aerosol Air Qual. Res.*, 2015, **15**, 1145–1159.

UV-VIS

R. M. B. O. Duarte, C. A. Pio and A. C. Duarte, Spectroscopic study of the water-soluble organic matter isolated from atmospheric aerosols collected under different atmospheric conditions, *Anal. Chim. Acta*, 2005, **530**, 7–14.

XRF

J. G. Watson, J. C. Chow and C. A. Frazier, X-ray fluorescence analysis of ambient air samples, in Elemental Analysis of Airborne Particles, ed. S. Landsberger and M. Creatchman, Gordon and Breach Science, Amsterdam, The Netherlands, 1999, vol. 1, pp. 67–96.

Table 2 Continued

[a]PAH = Polycyclic aromatic hydrocarbons; diagnostic PAH ratios are as follows:

BaP/(BaP + BeP)—Carbonaceous Fuel Combustion—Indicator of particle aging, *e.g.*, fresh exhaust emissions typically contain similar amounts of BeP and BaP [ratio = 0.5], while BaP is more easily decomposed by light and oxidants than BeP.

BaP/(BaP + CHR)—Carbonaceous Fuel Combustion—0.5: diesel emissions; and 0.7–0.9: gasoline emissions.

BaA/(BaA + CHR)—Carbonaceous Fuel Combustion—0.16: rural samples; 0.33: urban samples with catalyst-equipped vehicles; 0.50: industrial and heavy-truck emissions; 0.38–0.65: diesel emissions; 0.4–0.5: gasoline emissions; >0.35 coal or biomass combustion; and <0.2: liquid fossil fuel sources.

PHE/(PHE + ANT)—Carbonaceous Fuel Combustion—Identifies the importance of petrogenic hydrocarbons in relation to emissions from biomass burning. >0.7: lubricant oils and fossil fuels; 0.5: gasoline emissions; 0.65–0.75: diesel emissions; and 0.8: coal combustion.

ANT/(ANT + PHE)—Carbonaceous Fuel Combustion—higher ratios: stronger influence from petroleum combustion; <0.1: liquid fossil fuel sources; and >0.1: combustion sources.

FLU/(FLU + PYR)—Carbonaceous Fuel Combustion—higher ratios (>0.5): diesel emissions; lower ratios (<0.5): gasoline emission; >0.5: coal and wood combustion; <0.4: unburned petrogenic sources; and 0.4–0.5: liquid fossil fuel combustion.

FLT/(FLT + PYR)—Carbonaceous Fuel Combustion—0.35–0.70: diesel emissions; 0.4–0.6: gasoline emissions; 0.5–0.6: biomass combustion; and 0.6: coal combustion.

IcdP/(IcdP + BghiP)—Carbonaceous Fuel Combustion—0.2: gasoline emissions; 0.35–0.70: diesel emissions; 0.4–0.6: wood combustion; >0.5: coal or wood combustion; <0.2: unburned petrogenic sources; and 0.2–0.4: liquid fossil fuel combustion.

[b]OC1–OC4 are organic carbon (OC) evolved from a 0.5 cm² circular filter punch in a pure helium (He) [>99.999%] atmosphere at 140, 280, 480, and 580 °C, respectively, and EC1–EC3 are elemental carbon (EC) evolved from the filter punch in a 98% He/2% O₂ atmosphere at 580, 740, and 840 °C, respectively, following the IMPROVE_A protocol,[123] for optical reflectance (R) and transmittance (T). Pyrolyzed OC (OP) is defined as the carbon evolving between the introduction of oxygen (O₂) in the helium (He) atmosphere and the return of reflectance (R) or transmittance (T) to its initial value for R and T as OPR and OPT (the OC/EC split). OC equals OC1 + OC2 + OC3 + OC4 + OP and EC equals EC1 + EC2 + EC3 – OP. In a multiwavelength system, OC and EC are represented as a function of wavelength (*e.g.*, OC [405 nm, R] and OC [405 nm, T] are OC at 405 nm by reflectance and transmittance, respectively). OC and EC are listed for each of the seven wavelengths (λ) for either R or T, where OC$_\lambda$ = OC1 + OC2 + OC3 + OC4 + OP$_\lambda$ and EC$_\lambda$ = EC1 + EC2 + EC3 – OP$_\lambda$.

[c]ATNR = ln(R$_{final}$/R$_{initial}$); ATNT = ln(T$_{final}$/T$_{initial}$). ATNR and ATNT are listed for each of the seven wavelengths (*e.g.*, ATN$_{405R}$ and ATN$_{405T}$ for 405 nm reflectance and transmittance, respectively).

Acknowledgements

This work benefited from support by the US National Science Foundation (CHE 1214163 and FAIN 1464501), Washington, D.C.; the US National Park Service IMPROVE Carbon Analysis Contract (C2350000894), Washington, D.C.; and the Wood Buffalo Environmental Association, Alberta, Canada.

References

1. G. M. Hidy and S. K. Friedlander, in *Proceedings of the Second International Clean Air Congress*, ed. H. M. Englund and W. T. Beery, Academic Press, New York, 1971, pp. 391–404.
2. J. C. Chow, D. H. Lowenthal, L.-W. A. Chen, X. L. Wang and J. G. Watson, *Air Qual., Atmos. Health*, 2015, **8**, 243–263.
3. J. J. Cao, J. C. Chow, S. C. Lee and J. G. Watson, *Aerosol Air Qual. Res.*, 2013, **13**, 1197–1211.
4. J. G. Watson and J. C. Chow, in *Introduction to Environmental Forensics*, ed. B. L. Murphy and R. D. Morrison, Elsevier, Amsterdam, The Netherlands, 3rd edn, 2015, ch. 20, pp. 677–706.
5. I. Cheng, X. Xu and L. Zhang, *Atmos. Chem. Phys.*, 2015, **15**, 7877–7895.
6. C. A. Belis, F. Karagulian, B. R. Larsen and P. K. Hopke, *Atmos. Environ.*, 2014, **85**, 275–276.
7. C. A. Belis, F. Karagulian, B. R. Larsen and P. K. Hopke, *Atmos. Environ.*, 2013, **69**, 94–108.
8. P. Pant and R. M. Harrison, *Atmos. Environ.*, 2012, **49**, 1–12.
9. T. M. Johnson, S. K. Guttikunda, G. J. Wells, P. Artaxo, T. C. Bond, A. G. Russell, J. G. Watson and J. West, *Tools for improving air quality management: A review of top-down source apportionment techniques and their application in developing countries*, Report 339/11, Washington DC, 2011, http://www.esmap.org/esmap/sites/esmap.org/files/7607-Source%20Web(Small).pdf.
10. J. G. Watson, L.-W. A. Chen, J. C. Chow, D. H. Lowenthal and P. Doraiswamy, *J. Air Waste Manage. Assoc.*, 2008, **58**, 265–288.
11. J. G. Watson, T. Zhu, J. C. Chow, J. P. Engelbrecht, E. M. Fujita and W. E. Wilson, *Chemosphere*, 2002, **49**, 1093–1136.
12. M. Viana, T. A. J. Kuhlbusch, X. Querol, A. Alastuey, R. M. Harrison, P. K. Hopke, W. Winiwarter, A. Vallius, S. Szidat, A. S. H. Prevot, C. Hueglin, H. Bloemen, P. Wahlin, R. Vecchi, A. I. Miranda, A. Kasper-Giebl, W. Maenhaut and R. Hitzenbergerq, *J. Aerosol Sci.*, 2008, **39**, 827–849.
13. A. Reff, S. I. Eberly and P. V. Bhave, *J. Air Waste Manage. Assoc.*, 2007, **57**, 146–154.
14. J. R. Brook, E. Vega and J. G. Watson, in *Particulate Matter Science for Policy Makers – A NARSTO Assessment, Part 1.*, ed. J. M. Hales and G. M. Hidy, Cambridge University Press, London, UK, 2004, pp. 235–281.

15. P. K. Hopke, *J. Chemom.*, 2003, **17**, 255–265.
16. R. C. Henry, *Chemom. Intell. Lab. Syst.*, 2002, **60**, 43–48.
17. P. K. Hopke, *Receptor Modeling for Air Quality Management*, Elsevier Press, Amsterdam, The Netherlands, 1991.
18. P. K. Hopke, *Receptor Modeling in Environmental Chemistry*, John Wiley & Sons, Inc., New York, 1985.
19. M. L. Pitchford, W. C. Malm, B. A. Schichtel, N. K. Kumar, D. H. Lowenthal and J. L. Hand, *J. Air Waste Manage. Assoc.*, 2007, **57**, 1326–1336.
20. USEPA, *Receptor modeling*, Research Triangle Park, NC, 2016, http://www.epa.gov/scram001/receptorindex.htm.
21. I. Linkov, D. Loney, S. Cormier, F. K. Satterstrom and T. Bridges, *Sci. Total Environ.*, 2009, **407**, 5199–5205.
22. USEPA, *Guidance on the use of models and other analyses for demonstrating attainment of air quality goals for ozone, $PM_{2.5}$, and regional haze*, Report EPA -454/B-07-002, Research Triangle Park, NC, 2007, http://www.epa.gov/ttn/scram/guidance/guide/final-03-pm-rh-guidance.pdf.
23. J. G. Watson, J. A. Cooper and J. J. Huntzicker, *Atmos. Environ.*, 1984, **18**, 1347–1355.
24. W. T. Davis, *Air Pollution Engineering Manual*, Wiley-Interscience, New York, NY, 2000.
25. P. Paatero, *Chemom. Intell. Lab. Syst.*, 1997, **37**, 23–35.
26. P. Paatero, *J. Comput. Graph. Stat.*, 1999, **8**, 854–888.
27. P. Paatero and U. Tapper, *Environmetrics*, 1994, **5**, 111–126.
28. N. M. Donahue, I. K. Ortega, W. Chuang, I. Riipinen, F. Riccobono, S. Schobesberger, J. Dommen, U. Baltensperger, M. Kulmala, D. R. Worsnop and H. Vehkamaki, *Faraday Discuss.*, 2013, **165**, 91–104.
29. IMPROVE, *Interagency Monitoring of PRotected Visual Environments*, Ft. Collins, CO, 2016, http://vista.cira.colostate.edu/IMPROVE.
30. USEPA, *Chemical speciation*, Research Triangle Park, NC, 2016, http://www.epa.gov/ttn/amtic/speciepg.html.
31. E. Dabek-Zlotorzynska, T. F. Dann, P. K. Martinelango, V. Celo, J. R. Brook, D. Mathieu, L. Y. Ding and C. C. Austin, *Atmos. Environ.*, 2011, **45**, 673–686.
32. D. Salameh, A. Detournay, J. Pey, N. Perez, F. Liguori, D. Saraga, M. C. Bove, P. Brotto, F. Cassola, D. Massabo, A. Latella, S. Pillon, G. Formenton, S. Patti, A. Armengaud, D. Piga, J. L. Jaffrezo, J. Bartzis, E. Tolis, P. Prati, X. Querol, H. Wortham and N. Marchand, *Atmos. Res.*, 2015, **155**, 102–117.
33. X. Y. Zhang, Y. Q. Wang, T. Niu, X. C. Zhang, S. L. Gong, Y. M. Zhang and J. Y. Sun, *Atmos. Chem. Phys.*, 2012, **12**, 779–799.
34. P. K. K. Louie, J. C. Chow, L.-W. A. Chen, J. G. Watson, G. Leung and D. Sin, *Sci. Total Environ.*, 2005, **338**, 267–281.
35. P. K. K. Louie, J. G. Watson, J. C. Chow, L.-W. A. Chen, D. W. M. Sin and A. K. H. Lau, *Atmos. Environ.*, 2005, **39**, 1695–1710.
36. S. Tiwari, A. K. Srivastava and A. K. Singh, *Atmos. Environ.*, 2013, **77**, 738–747.

37. J. C. Chow, J. P. Engelbrecht, J. G. Watson, W. E. Wilson, N. H. Frank and T. Zhu, *Chemosphere*, 2002, **49**, 961–978.
38. A. Charron, R. M. Harrison and P. G. Quincey, *Atmos. Environ.*, 2007, **41**, 1960–1975.
39. J. G. Watson, J. C. Chow and T. G. Pace, in *Air Pollution Engineering Manual*, ed. W. T. Davis, John Wiley & Sons, Inc., New York, 2nd edn, 2000, ch. 4, pp. 117–135.
40. L. M. Zhou, P. K. Hopke, P. Paatero, J. M. Ondov, J. P. Pancras, N. J. Pekney and C. I. Davidson, *Atmos. Environ.*, 2004, **38**, 4909–4920.
41. E. Pere-Trepat, E. Kim, P. Paatero and P. K. Hopke, *Atmos. Environ.*, 2007, **41**, 5921–5933.
42. J. G. Watson and J. C. Chow, *J. Air Waste Manage. Assoc.*, 2001, **51**, 1522–1528.
43. M. O. Andreae and A. Gelencser, *Atmos. Chem. Phys.*, 2006, **6**, 3131–3148.
44. L. Drinovec, G. Mocnik, P. Zotter, A. S. H. Prevot, C. Ruckstuhl, E. Coz, M. Rupakheti, J. Sciare, T. Muller, A. Wiedensohler and A. D. A. Hansen, *Atmos. Meas. Tech.*, 2015, **8**, 1965–1979.
45. R. Vecchi, V. Bernardoni, C. Paganelli and G. Valli, *J. Aerosol Sci.*, 2014, **70**, 15–25.
46. J. C. Chow, X. L. Wang, B. J. Sumlin, S. B. Gronstal, L.-W. A. Chen, D. L. Trimble, S. D. Kohl, S. R. Mayorgal, G. M. Riggio, P. R. Hurbain, M. Johnson, R. Zimmermann and J. G. Watson, *Aerosol Air Qual. Res.*, 2015, **15**, 1145–1159.
47. L.-W. A. Chen, J. C. Chow, X. L. Wang, J. A. Robles, B. J. Sumlin, D. H. Lowenthal, R. Zimmermann and J. G. Watson, *Atmos. Meas. Tech.*, 2015, **8**, 451–461.
48. D. Liu, J. Allan, B. Corris, M. Flynn, E. Andrews, J. Ogren, K. Beswick, K. Bower, R. Burgess, T. Choularton, J. Dorsey, W. Morgan, P. I. Williams and H. Coe, *Atmos. Chem. Phys.*, 2011, **11**, 1603–1619.
49. J. Sandradewi, A. S. H. Prevot, E. Weingartner, R. Schmidhauser, M. Gysel and U. Baltensperger, *Atmos. Environ.*, 2008, **42**, 101–112.
50. J. Sandradewi, A. S. H. Prevot, S. Szidat, N. Perron, M. R. Alfarra, V. A. Lanz, E. Weingartner and U. Baltensperger, *Environ. Sci. Technol.*, 2008, **42**, 3316–3323.
51. USEPA., *Chemical Mass Balance (CMB) model*, Research Triangle Park, NC, 2004, http://www.epa.gov/scram001/receptor_cmb.htm.
52. U.S. EPA, *Positive Matrix Factorization Model for Environmental Data Analyses*, U.S. Environmental Protection Agency, Research Triangle Park, NC, 2016, https://www.epa.gov/air-research/positive-matrix-factorization-model-environmental-data-analyses.
53. M. R. Heal, *Anal. Bioanal. Chem.*, 2014, **406**, 81–98.
54. L. R. Crilley, W. J. Bloss, J. Yin, D. C. S. Beddows, R. M. Harrison, J. D. Allan, D. E. Young, M. Flynn, P. Williams, P. Zotter, A. S. H. Prevot, M. R. Heal, J. F. Barlow, C. H. Halios, J. D. Lee, S. Szidat and C. Mohr, *Atmos. Chem. Phys.*, 2015, **15**, 3149–3171.

55. M. M. Shafer, B. M. Toner, J. T. Oyerdier, J. J. Schauer, S. C. Fakra, S. H. Hu, J. D. Herner and A. Ayala, _Environ. Sci. Technol._, 2012, **46**, 189–195.
56. A. Katsoyiannis, A. J. Sweetman and K. C. Jones, _Environ. Sci. Technol._, 2011, **45**, 8897–8906.
57. H. Simon, P. V. Bhave, J. L. Swall, N. H. Frank and W. C. Malm, _Atmos. Chem. Phys._, 2011, **11**, 2933–2949.
58. T. Zeng and Y. H. Wang, _Atmos. Environ._, 2011, **45**, 578–586.
59. D. M. Murphy, J. C. Chow, E. M. Leibensperger, W. C. Malm, M. L. Pitchford, B. A. Schichtel, J. G. Watson and W. H. White, _Atmos. Chem. Phys._, 2011, **11**, 4679–4686.
60. L.-W. A. Chen, J. C. Chow, J. G. Watson and B. A. Schichtel, _Atmos. Meas. Tech_, 2012, **5**, 2329–2338.
61. J. G. Watson, J. C. Chow, L. C. Pritchett, J. E. Houck and R. A. Ragazzi, _Sci. Total Environ._, 1990, **93**, 183–190.
62. B. J. Turpin and H. J. Lim, _Aerosol Sci. Technol._, 2001, **35**, 602–610.
63. J. J. Schauer, W. F. Rogge, M. A. Mazurek, L. M. Hildemann, G. R. Cass and B. R. T. Simoneit, _Atmos. Environ._, 1996, **30**, 3837–3855.
64. J. G. Watson, J. C. Chow and T. G. Pace, in _Receptor Modeling for Air Quality Management_, ed. P. K. Hopke, Elsevier Press, New York, NY, 1991, vol. 7, pp. 83–116.
65. T. Pachauri, V. Singla, A. Satsangi, A. Lakhani and K. M. Kumari, _Environ. Sci. Pollut. Res._, 2013, **20**, 5737–5752.
66. CARB, _Speciation profiles used in ARB modeling_, Sacramento, CA, 2016, http://arb.ca.gov/ei/speciate/speciate.htm.
67. USEPA, _SPECIATE Version 4.4_, Research Triangle Park, NC, 2014, http://www.epa.gov/ttn/chief/software/speciate/index.html.
68. ARAI, _Source profiling for vehicular emissions_, Report ARAI/VSP-III/SP/RD/08-09/60, Pune, India, 2009, http://www.cpcb.nic.in/Source_Profile_Vehicles.pdf.
69. V. Sethi and R. S. Patil, _Development of Air Pollution Source Profiles – Stationary Sources Volume 1_, Mumbai, India, 2008, http://www.cpcb.nic.in/Source_Emission_%20Profiles_NVS_Volume%20One.pdf.
70. V. Sethi and R. S. Patil, _Development of Air Pollution Source Profiles - Stationary Sources Volume 2_, Mumbai, India, 2008, http://www.cpcb.nic.in/Source_Emission_Profiles_NVS_Volume%20Two.pdf.
71. N. M. Donahue, S. A. Epstein, S. N. Pandis and A. L. Robinson, _Atmos. Chem. Phys._, 2011, **11**, 3303–3318.
72. N. M. Donahue, J. H. Kroll, S. N. Pandis and A. L. Robinson, _Atmos. Chem. Phys._, 2012, **12**, 615–634.
73. L. D. Yee, K. E. Kautzman, C. L. Loza, K. A. Schilling, M. M. Coggon, P. S. Chhabra, M. N. Chan, A. W. H. Chan, S. P. Hersey, J. D. Crounse, P. O. Wennberg, R. C. Flagan and J. H. Seinfeld, _Atmos. Chem. Phys._, 2013, **13**, 8019–8043.
74. J. G. Watson, J. C. Chow, X. L. Wang, S. D. Kohl, L.-W. A. Chen and V. R. Etyemezian, in _Alberta Oil Sands: Energy, Industry, and the_

Environment, ed. K. E. Percy, Elsevier Press, Amsterdam, The Netherlands, 2012, ch. 7, vol. 11, pp. 145–170.

75. J. C. Chow, J. G. Watson, J. E. Houck, L. C. Pritchett, C. F. Rogers, C. A. Frazier, R. T. Egami and B. M. Ball, *Atmos. Environ.*, 1994, **28**, 3463–3481.
76. B. C. Singer and R. A. Harley, *J. Air Waste Manage. Assoc.*, 1996, **46**, 581–593.
77. X. L. Wang, J. G. Watson, J. C. Chow, S. Gronstal and S. D. Kohl, *Aerosol Air Qual. Res.*, 2012, **12**, 145–160.
78. J. K. Gietl, R. Lawrence, A. J. Thorpe and R. M. Harrison, *Atmos. Environ.*, 2010, **44**, 141–146.
79. W. R. Pierson and W. W. Brachaczek, *Aerosol Sci. Technol.*, 1983, **2**, 1–40.
80. ISO, *ISO 25597:2013: Stationary source emissions – Test method for determining PM$_{2.5}$ and PM$_{10}$ mass in stack gases using cyclone samplers and sample dilution*, Geneva, Switzerland, 2013, http://www.iso.org/iso/home/store/catalogue_tc/catalogue_detail.htm?csnumber=43029.
81. USEPA, *Method 5. Particulate matter (PM), Determination of particulate matter emissions from stationary sources*, Research Triangle Park, NC, 2000, http://www.epa.gov/ttn/emc/promgate/m-05.pdf.
82. J. J. West, A. S. Ansari and S. N. Pandis, *J. Air Waste Manage. Assoc.*, 1999, **49**, 1415–1424.
83. S. L. Clegg and A. S. Wexler, *Extended AIM Aerosol Thermodynamics Model*, Norwich, UK, 2012, http://www.aim.env.uea.ac.uk/aim/aim.php.
84. E. Kang, M. J. Root, D. W. Toohey and W. H. Brune, *Atmos. Chem. Phys.*, 2007, **7**, 5727–5744.
85. D. S. Tkacik, A. T. Lambe, S. Jathar, X. Li, A. A. Presto, Y. L. Zhao, D. Blake, S. Meinardi, J. T. Jayne, P. L. Croteau and A. L. Robinson, *Environ. Sci. Technol.*, 2014, **48**, 11235–11242.
86. A. M. Ortega, D. A. Day, M. J. Cubison, W. H. Brune, D. Bon, J. A. de Gouw and J. L. Jimenez, *Atmos. Chem. Phys.*, 2013, **13**, 11551–11571.
87. C. Y. Lai, Y. C. Liu, J. Z. Ma, Q. X. Ma and H. He, *Atmos. Environ.*, 2014, **91**, 32–39.
88. R. C. Henry, in *Proceedings, Receptor Models Applied to Contemporary Air Pollution Problems*, ed. P. K. Hopke and S. L. Dattner, Air Pollution Control Association, Pittsburgh, PA, 1982, pp. 141–162.
89. B. M. Kim and R. C. Henry, *J. Air Waste Manage. Assoc.*, 1999, **49**, 1449–1455.
90. J. G. Watson, *Protocol for applying and validating the CMB model for PM$_{2.5}$ and VOC*, Report EPA-451/R-04-001, Research Triangle Park, NC, 2004, www.epa.gov/scram001/models/receptor/CMB_Protocol.pdf.
91. Z. Ning, M. Wubulihairen and F. H. Yang, *Atmos. Environ.*, 2012, **61**, 265–274.
92. R. Zhang, J. Jing, J. Tao, S. C. Hsu, G. Wang, J. J. Cao, C. S. L. Lee, L. Zhu, Z. Chen, Y. Zhao and Z. Shen, *Atmos. Chem. Phys.*, 2013, **13**, 7053–7074.

93. T. G. Pace, *An empirical approach for relating annual TSP concentrations to particulate microinventory emissions data and monitor siting characteristics*, Report EPA-450/4-79-012, Research Triangle Park, NC, 1979.

94. A. De Santiago, A. F. Longo, E. D. Ingall, J. M. Diaz, L. E. King, B. Lai, R. J. Weber, A. G. Russell and M. Oakes, *Environ. Sci. Technol.*, 2014, **48**, 8988–8994.

95. T. W. Kirchstetter, T. Novakov and P. V. Hobbs, *J. Geophys. Res. Atmos.*, 2004, **109**, D21208.

96. A. M. Dillner and S. Takahama, *Atmos. Meas. Tech.*, 2015, **8**, 1097–1109.

97. M. A. Mazurek, G. R. Cass and B. R. T. Simoneit, *Aerosol Sci. Technol.*, 1989, **10**, 408–420.

98. S. S. H. Ho and J. Z. Yu, *Environ. Sci. Technol.*, 2004, **38**, 862–870.

99. J. Orasche, J. Schnelle-Kreis, G. Abbaszade and R. Zimmermann, *Atmos. Chem. Phys.*, 2011, **11**, 8977–8993.

100. G. S. Casuccio, P. B. Janocko, R. J. Lee, J. F. Kelly, S. L. Dattner and J. S. Mgebroff, *JAPCA*, 1983, **33**, 937–943.

101. P. K. Hopke and X. H. Song, *Anal. Chim. Acta*, 1997, **348**, 375–388.

102. B. Gajdzik, R. Wieszala and T. Wieczorek, *Metalurgija*, 2012, **51**, 101–104.

103. J. C. Chow, X. F. Yang, X. L. Wang, S. D. Kohl and J. G. Watson, *Aerosol Air Qual. Res.*, 2015, **15**, 1433–1447.

104. A. W. Gertler, D. Moshe and Y. Rudich, *Sci. Total Environ.*, 2014, **488**, 458–462.

105. J. D. Mobley, M. Deslauriers and L. Rojas-Brachos, *Improving emission inventories for effective air-quality management across North America – A NARSTO assessment*, Report NARSTO-05-001, Pasco, WA, 2005.

106. J. Kukkonen, T. Olsson, D. M. Schultz, A. Baklanov, T. Klein, A. I. Miranda, A. Monteiro, M. Hirtl, V. Tarvainen, M. Boy, V. H. Peuch, A. Poupkou, I. Kioutsioukis, S. Finardi, M. Sofiev, R. Sokhi, K. E. J. Lehtinen, K. Karatzas, R. S. Jose, M. Astitha, G. Kallos, M. Schaap, E. Reimer, H. Jakobs and K. Eben, *Atmos. Chem. Phys.*, 2012, **12**, 1–87.

107. J. G. Watson, Ph.D. Dissertation, Oregon Graduate Center, 1979, http://digitalcommons.ohsu.edu/etd/317/.

108. P. G. Burnet, R. C. McCrillis and S. J. Morgan, in *Transactions, PM$_{10}$: Implementation of Standards*, ed. C. V. Mathai and D. H. Stonefield, Air Pollution Control Association, Pittsburgh, PA, 1988, pp. 664–672.

109. J. C. Chow, J. G. Watson, R. T. Egami, C. A. Frazier, Z. Lu, A. Goodrich and A. Bird, *J. Air Waste Manage. Assoc.*, 1990, **40**, 1134–1142.

110. SCAQMD, *Rule 1186.1. Less-polluting sweepers*, Diamond Bar, CA, 2005, http://www.arb.ca.gov/DRDB/SC/CURHTML/R1186-1.PDF.

111. W. F. Rogge, L. M. Hildemann, M. A. Mazurek, G. R. Cass and B. R. T. Simoneit, *Environ. Sci. Technol*, 1991, **25**, 1112–1125.

112. SCAQMD, *Rule 1138: Control of emissions from restaurant operations*, South Coast Air Quality Management District, Diamond Bar, CA, 1997,

http://www.aqmd.gov/docs/default-source/rule-book/reg-xi/rule-1138. pdf?sfvrsn=4.

113. C. A. Belis, B. R. Larsen, F. Amato, I. E. Haddad, O. Favez, R. M. Harrison, P. K. Hopke, S. Nava, P. Paatero, A. Prevot, U. Quass, R. Vecchi and M. Viana, *European Guide on Air Pollution Source Apportionment with Receptor Models*, Ispra, Italy, 2014, http://source-apportionment.jrc.ec.europa.eu/Docu/EU_guide_on_SA.pdf.

114. L.-W. A. Chen, J. G. Watson, J. C. Chow, M. C. Green, D. Inouye and K. Dick, *Atmos. Chem. Phys.*, 2012, **12**, 10051–10064.

115. M. C. Green, J. C. Chow, J. G. Watson, K. Dick and D. Inouye, *J. Appl. Meteorol. Climatol.*, 2015, **54**, 1191–1201.

116. J. G. Watson, J. C. Chow, L.-W. A. Chen, S. D. Kohl and G. S. Casuccio, *Atmos. Res.*, 2012, **106**, 181–189.

117. A. D. A. Hansen and G. Mocnik, in *Proceedings, Leapfrogging Opportunities for Air Quality Improvement*, ed. J. C. Chow, J. G. Watson and J. J. Cao, Air & Waste Management Association, Pittsburgh, PA, 2010, pp. 984–989.

118. X. L. Wang, J. C. Chow, S. D. Kohl, K. E. Percy, A. H. Legge and J. G. Watson, *J. Air Waste Manage. Assoc.*, 2015, **65**, 1421–1433.

119. X. L. Wang, J. G. Watson, J. C. Chow, S. D. Kohl, L.-W. A. Chen, D. A. Sodeman, A. H. Legge and K. E. Percy, in *Alberta Oil Sands: Energy, Industry, and the Environment*, ed. K. E. Percy, Elsevier Press, Amsterdam, The Netherlands, 2012, ch. 8, vol. 11, pp. 171–192.

120. NEERI, *Air quality monitoring, emission inventory & source apportionment studies for Delhi*, Nagpur, India, 2010, http://www.cpcb.nic.in/Delhi.pdf.

121. E. J. Kushner, *Atmos. Environ.*, 1976, **10**, 975–979.

122. J. C. Chow, J. G. Watson, J. L. Bowen, C. A. Frazier, A. W. Gertler, K. K. Fung, D. Landis and L. L. Ashbaugh, in *Sampling and Analysis of Airborne Pollutants*, ed. E. D. Winegar and L. H. Keith, Lewis Publishers, Ann Arbor, MI, 1993, ch. 13, pp. 209–228.

123. J. C. Chow, J. G. Watson, L.-W. A. Chen, M.-C. O. Chang, N. F. Robinson, D. L. Trimble and S. D. Kohl, *J. Air Waste Manage. Assoc.*, 2007, **57**, 1014–1023.

Case Studies of Source Apportionment from North America

PHILIP K. HOPKE

ABSTRACT

An important aspect of air quality planning is the identification of air pollution sources and their importance in contributing to the observed ambient conservations. Since the 1960s, there have been efforts to use the measured ambient concentrations and what is known about the nature of source emissions. The methods have been formalized into a set of techniques termed receptor models and they have been extensively applied to a variety of air quality problems. This chapter outlines the history of the application of source apportionment tools. A number of studies are highlighted that have been important in the development or adoption of source apportionment into air quality strategy development. For example, an early application of the chemical mass balance model in Portland, OR, led to improvements in their deterministic dispersion model and enabled it to more accurately reflect the source/receptor relationships in this city. Positive matrix factorization developed in the early 1990s has now become the most widely used receptor model and provides a flexible approach to apportion pollution sources using only the ambient data. Such applications include conventional composition data, volatile organic compounds, particle size distribution data, and high time resolved data from systems like aerosol mass spectrometers or rotating drum impactors. PMF can now also incorporate external information like known source profiles. It is possible to develop conceptual models that

Issues in Environmental Science and Technology No. 42
Airborne Particulate Matter: Sources, Atmospheric Processes and Health
Edited by R.E. Hester, R.M. Harrison and X. Querol
Published by the Royal Society of Chemistry, www.rsc.org

align with the nature of the data such as composition as a function of particle size and composition or composition as a function of location and time across a large-scale monitoring network. Illustrative examples of this variety of applications are presented.

1 Introduction

As part of the development of effective and efficient air quality management strategies, it is important to be able to identify the major sources of air pollution and quantitatively apportion the concentrations of the pollutant(s) of interest to those sources. One major approach to source apportion are a set of data analysis tools termed *receptor modeling*[1,2] that have evolved over the past 30 years into being an accepted part of air quality planning processes. These approaches are based on the assumption that different source types emit different patterns of chemical species such that the measured concentrations of those species in the atmosphere at any given location (the receptor site) would be a linearly additive combination of the input from these independent sources.

In addition to the primary emissions that come directly from their sources, there are atmospheric processes that modify the composition of those emissions, such as chemical reactions or size dependent dry deposition, cause them to change phase (gas to particle conversion) or condense onto the surface of existing particles. In particular, fine particles less than 2.5 μm in aerodynamic diameter ($PM_{2.5}$) consist mostly of secondary materials like sulfate, nitrate, ammonium, and secondary organic species. Fully representing the atmospheric processes that convert gaseous precursors to ambient PM in receptor models is difficult. Thus, there are difficulties in directly identifying the sources of secondary particulate matter.

The fundamental principle of receptor modeling is that mass conservation can be assumed and a mass balance analysis can be used to identify and apportion sources of contaminants in the atmosphere. The approach to obtaining a data set for receptor modeling is to determine a large number of chemical constituents such as chemical species concentrations in a number of samples. Alternatively, methods like automated electron microscopy or Aerosol Time-of-Flight Mass Spectrometry can be used to characterize the composition and size of particles for a large number of particles. In either case, a mass balance equation can be written to account for all *m* chemical species in the *n* samples as contributions from *p* independent sources.

$$x_{ij} = \sum_{k=1}^{p} g_{ij} f_{kj} \tag{1}$$

where x_{ij} is the *j*th chemical species concentration measured in the *i*th sample, f_{kj} is the concentration of the *j*th species in material from the *k*th source, and g_{ik} is the airborne contribution of material from the *k*th source contributing to the *i*th sample. This basic conceptual model can then be fit to the various kinds of available data.

There exist a set of natural physical constraints on the system that must be considered in developing any model for identifying and apportioning the sources of airborne particle mass.[3] The fundamental, natural physical constraints that must be obeyed are:

1. The original data must be reproduced by the model; the model must explain the observations.
2. The predicted source compositions must be non-negative; a source cannot have a negative percentage of an element.
3. The predicted source contributions to the aerosol must all be non-negative; a source cannot emit negative mass.
4. The sum of the predicted elemental mass contributions for each source must be less than or equal to total measured mass for each element; the whole is greater than or equal to the sum of its parts.

There are then methods to solve eqn (1) depending on what *a priori* information is available. The details of the various models are described elsewhere in this volume.[4] In this chapter, the development of source apportionment methods will be presented as well as a number of examples of the application of these methods to specific air quality problems.

2   Historic Development

Source apportionment dates back to the 1960s. Colucci and Begeman[5] were the first to report the apportionment of pollutants (polycyclic aromatic hydrocarbons, PAHs) to a specific source type (automobile emissions) based on the concentrations of the co-emitted carbon monoxide (CO) and lead.

Blifford and Meeker[6] used a principal component analysis with several types of axis rotations to examine particle composition data collected by the National Air Sampling Network (NASN) during 1957–1961 in 30 US cities. They were generally not able to extract much interpretable information from their data. Since there are a very wide variety of particle sources among these 30 cities and only 13 elements were measured, it is not surprising that they were not able to provide much specificity to their factors. Prinz and Stratmann[7] examined both the aromatic hydrocarbon content of the air in 12 West German cities and data on the air quality of Detroit using factor analysis methods. In both cases, they found solutions that yielded readily interpretable results.

The concept of an atmospheric mass balance model was suggested independently by Miller *et al.*[8] and by Winchester and Nifong.[9] In these initial models, specific elements were associated with particular source types to develop a mass balance for airborne particles. Subsequently, more chemical species than sources were used in a least-squares fit to provide estimates of the mass contributions of the sources.[10]

There were a number of these early application of the mass balance analysis including Gent, Belgium,[11] Heidelberg, Germany,[12] and Chicago, Illinois.[13] Several major research efforts have subsequently resulted in substantially better source data. The source emission studies led to much improved resolution of the particle sources in Washington, DC.[14,15] In the first of these studies, Kowalczyk *et al.*[14] introduced weighted least-squares regression to fit six sources with eight elements for ten ambient samples. Subsequently, Kowalczyk *et al.*[15] examined 130 samples using 7 sources with 28 elements included in the fit. They obtained an excellent fit of the ambient concentration data and a quite good understanding of the major sources of airborne particles in the Washington, D.C. area.

Mayrsohn and Crabtree[16] presented the use of an iterative least-squares approach to apportion six sources of airborne hydrocarbon compounds in the Los Angeles basin. The sources were automotive exhaust, volatilization of gasoline and release of gasoline vapor, commercial natural gas, geological natural gas, and liquified petroleum gas. They performed the least-squares fit to the hydrocarbon compound concentrations using gas chromatography to determine the concentrations of eight compounds. Their ordinary least-squares source reconciliation algorithm recognized that not all sources may contribute to every sample, and, if negative contributions were obtained, a different configuration of sources was employed with certain qualifying assumptions. Each possible configuration with positive coefficients was considered and the one with the lowest standard error was chosen as the optimum solution. On average, automotive exhaust was the source of almost 50% of observed hydrocarbons. Gasoline and its vapor contributed up to 30% by weight and the balance resulted from commercial and geological natural gas. Thus, automobiles and other highway related sources were responsible for the majority of these hydrocarbons. A similar study utilizing this mass balance approach for resolving hydrocarbon sources has been made by Nelson *et al.*[17] in Sydney, Australia. Thus, it is possible to identify the impact of emission sources on gaseous as well as particulate pollutants. CMB analysis has also been used to determine the sources of benzene in rural New York State.[18]

In 1979, Watson[19] and Dunker[20] independently suggested a mathematical formalism called effective variance weighting that included the uncertainty in the measurement of the source composition profiles as well as the uncertainties in the ambient concentrations. As part of this analysis, a method was also developed to permit the calculation of the uncertainties in the mass contributions. Effective-variance least squares has been incorporated into the standard personal computer software developed by the US EPA for receptor modeling. The most extensive use of effective-variance fitting has been made by Watson and colleagues in their work on data from Portland, OR.[21] Since that study, a number of other applications of this approach have been made in a wide variety of locations and extensive libraries of compositional profiles of emission sources have been developed to be used in the mass balance models. These models are described in detail by Watson *et al.*[22]

Factor analysis was reintroduced in the mid-1970s by Hopke *et al.*[23] and Gaarenstroom *et al.*[24] in their analyses of particle composition data from Boston, MA and Tucson, AZ, respectively. A problem that exists with these forms of factor analysis is that they do not permit quantitative source apportionment of particle mass or of specific elemental concentrations. In an effort to find alternative methods that would provide information on source contributions when only the ambient particulate analytical results are available, Hopke *et al.*[25] introduced target transformation factor analysis by demonstrating the source resolution of collected and physically fractionated street dust. It was applied in a number of applications[26–29] and reviewed by Hopke.[30]

Henry and coworkers[31–34] developed alternative methods based on eigenvector methods. The initial model was SAFER and it has evolved into Unmix.[35] It is based on an eigenvector decomposition of the data as the basis for finding the edges.[34] Unmix has been applied to a number of air quality data sets including airborne PM,[36–39] particle number size distribution,[40] and semivolatile compounds associated with airborne PM.[41–46]

Positive Matrix Factorization (PMF) was developed by Paatero[47,48] and has become the most widely used source resolution method following its release by the US EPA.[49] The conceptual framework of this method is to utilize an explicit least-squares formulation of the mass balance problem presented in eqn (1). The eigenvector approaches used in TTFA and Unmix represent an implicit unweighted least-squares fit to the data.[50] Since environmental data do not possess uniform uncertainties, the regression problem should not be solved without proper weighting, but an eigenvector approach only permits scaling by a row or a column. Data points cannot be individually weighted.[47] PMF solves the receptor modeling problem by minimizing a weighted objective function.

PMF has been implemented in two algorithms. Initially, PMF2 was developed to provide a rapid solution to the model presented in eqn (1).[51] PMF2 was initially applied to data sets of major ion compositions of daily precipitation samples collected over a number of sites in Finland[52] and samples of bulk precipitation[53] in which they are able to obtain considerable information on the sources of these ions. Polissar *et al.*[54] applied the PMF2 program to Arctic data from seven National Park Service sites in Alaska as a method to resolve the major source contributions more quantitatively. Polissar *et al.*[55] reanalyzed the Alaska data and proposed an approach to uncertainty estimation that has now been widely used in PMF applications. It should be noted that the rules of thumb provided to estimate the uncertainties were derived by extensive testing to find an approach that provided useful results. It has no statistical basis and other approaches to error estimation may provide superior results.

In the late 1990s, a need was identified to be able to solve more complex problems because not all mass balance problems are represented by the bilinear model in eqn (1). For example, data obtained from a series of samples collected with a cascade impactor as a function of time are a 3-way data tensor (size, composition, time). Information is lost if the data are combined to produce a matrix for analysis. Thus, a more flexible solver was needed and Paatero[56] developed a multilinear engine that can solve any problem that can be expressed as a sum of product terms. It allows the development of a variety of conceptual models to fit the data set and has been incorporated into the version of PMF being distributed by the US Environmental Protection Agency.

3 Applications

3.1 Chemical Mass Balance

3.1.1 Portland Air Quality Study (PAQS). One of the earliest studies conducted specifically with the chemical mass balance model as the basis of the study design was the Portland Air Quality Study conducted in Portland, OR in the late 1970s.[57,58] Because of the violations of ambient air quality standards in the Portland area, economic development was being inhibited and better tools for air quality planning were needed. However, there were problems with the Eulerian grid model predictions where the measured TSP values were significantly larger than those predicted by the model (Figure 1(left)).

Sampling of major source types was performed and the samples were analyzed to provide locally relevant source profiles. Ambient samples were collected at 4 sites across the city. The CMB model was applied to identify the various TSP sources. Based on comparisons between the CMB results and the dispersion model results, problems with the modeling could be identified. These issues included the assumption of constant emissions from residual oil combustion over the entire year. The dispersion model was modified to reflect the actual operating schedules for the various meteorological regimes that had been identified for Portland. One of the point source stack heights was improperly estimated relative to the related receptor site. The road dust emission factor was underpredicted based on the EPA generalized paved road emission factor. An error in the estimated average motor vehicle tailpipe emissions at the 4 sites can be seen in Figure 2 (left). After appropriate corrections, Figure 2 (right) shows that there is much better agreement between the dispersion and CMB results. Making all of the changes in the dispersion model identified by the CMB modeling resulted in the results shown in Figure 1 (right). It can be seen that the model now provides substantially improved estimates of the TSP.

This work resulted in significant changes in the emissions inventory for the major sources as reflected in Figure 3. The changes resulting from the study increased the total estimated emissions by a factor of 2.66, added a significant impact from wood combustion, and shifted the major focus from

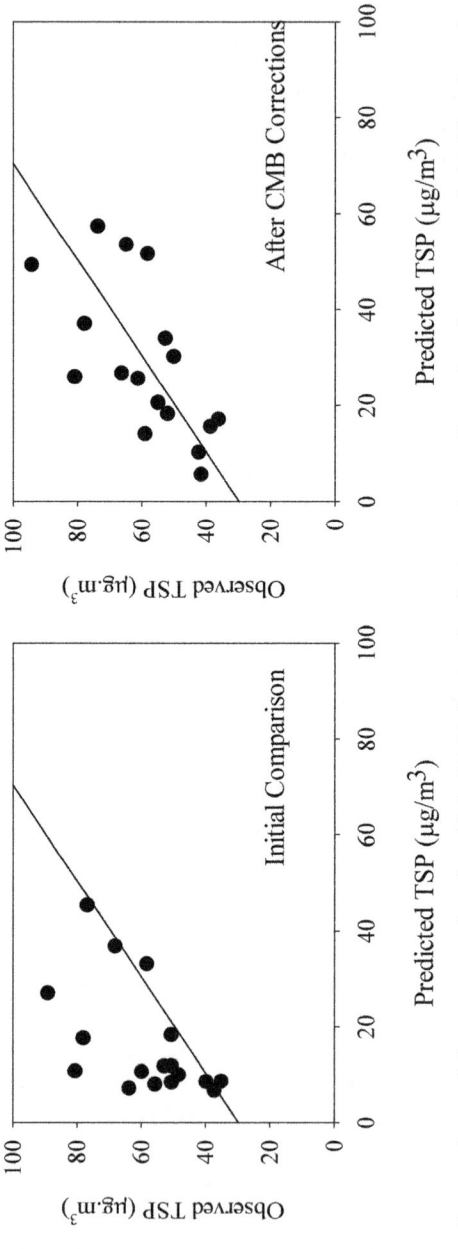

Figure 1 Comparison of the measured and modeled TSP in Portland, OR redrawn from data in Core *et al.* (Left) Before the emissions inventory adjustments. (Right) after the emissions inventory adjustments. The line is the 1:1 line. Data for plots taken from Core *et al.*[58]

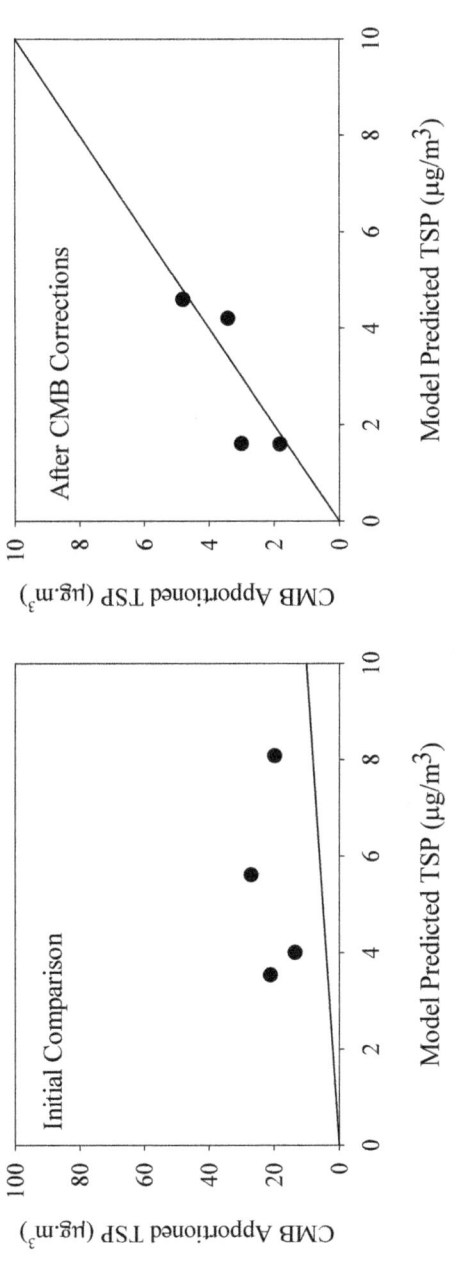

Figure 2 Comparison of the CMB and dispersion model results for motor vehicle exhaust redrawn from data in Core et al. (Left) Before the emissions inventory adjustments. (Right) after the emissions inventory adjustments. Data for plots taken from Core et al.[58]

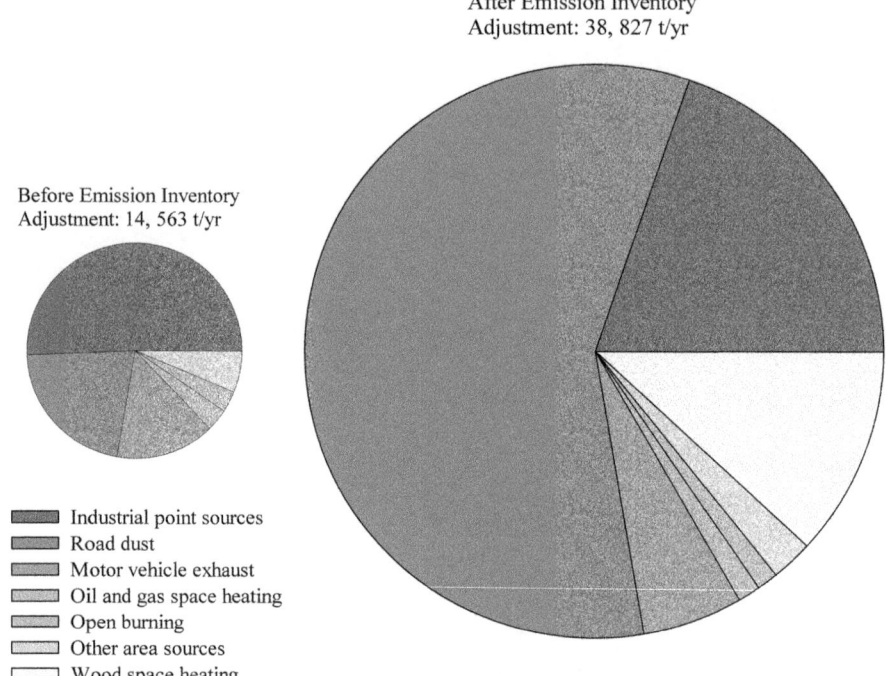

Figure 3   Comparison of the emissions inventory for Portland, OR before (left) and after (right) the CMB adjustments were made.
Data for plots taken from Core *et al.*[58]

industrial point sources to road dust as the single largest source of airborne particulate matter. Thus, the application of the CMB modeling provided a much more accurate tool for future air quality decision making.

3.1.2   Primary Organic Carbon.   Based on the initial work by Simoneit and coworkers[59-62] on organic species in the rural ambient aerosol, Cass and coworkers[63-81] developed sampling and analysis methods for the urban carbonaceous fine particle aerosol that have now been used extensively for apportioning primary organic carbon (OC) contributions to ambient $PM_{2.5}$ using specific organic tracers known as *molecular markers*. Table 1 lists the compounds typically measured in these studies. Kleindienst *et al.*[82] has identified additional species (Table 2) that can serve as markers of secondary organic aerosol (SOA).

Schauer *et al.*[83] then utilized these measured molecular marker profiles to apportion the primary OC in $PM_{2.5}$ samples collected at 4 air quality monitoring sites in Los Angeles basin of Southern California. They were able to apportion approximately 85% of the primary OC to up to 9 particle source types. Major contributors to the fine particle mass concentrations were

Table 1 Commonly measured molecular marker species.

n-Alkanes	Aliphatic dicarboxylic acids	Hopanes and steranes
n-Pentacosane	Propanedioic acid (malonic acid)	18α(*H*)22,29,30-Trisnorneohopane
n-Hexacosane	Butanedioic acid (succinic acid)	17α(*H*)-22,29,30-Trisnorhopane
n-Heptacosane	Methylsuccinic acid	17α(*H*),21β(*H*)-29-Norhopane
n-Octacosane	Pentanedioic acid (glutaric acid)	18α(*H*)-29-Norneohopane
n-Nonacosane	Hydroxybutanedioic acid (malic)	17α(*H*),21β(*H*)-Hopane
n-Triacontane	Hexanedioic acid (adipic acid)	22S,17α(*H*),21β(*H*)-30-Homohopane
n-Hentriacontane	Octanedioic acid (suberic acid)	22R,17α(*H*),21β(*H*)-30-Homohopane
n-Dotriacontane	Nonanedioic acid (azelaic acid)	22S,17α(*H*),21β(*H*)-30-Bishomohopane

n-Alkanes
- *n*-Pentacosane
- *n*-Hexacosane
- *n*-Heptacosane
- *n*-Octacosane
- *n*-Nonacosane
- *n*-Triacontane
- *n*-Hentriacontane
- *n*-Dotriacontane

n-Alkanoic acids
- *n*-Nonanoic acid
- *n*-Decanoic acid
- *n*-Undecanoic acid
- *n*-Dodecanoic acid
- *n*-Tridecanoic acid
- *n*-Tetradecanoic acid
- *n*-Pentadecanoic acid
- *n*-Hexadecenoic acid
- *n*-Heptadecanoic acid
- *n*-Octadecanoic acid
- *n*-Octadedenoic acid
- *n*-Nonadecanoic acid
- *n*-Eicosanoic acid
- *n*-Heneicosanoic acid
- *n*-Docosanoic acid
- *n*-Tricosanoic acid
- *n*-Tetracosanoic acid
- *n*-Pentacosanoic acid
- *n*-Hexacosanoic acid
- *n*-Heptacosanoic acid
- *n*-Octacosanoic acid
- *n*-Nonacosanoic acid
- *n*-Triacontanoic acid

Resin & aromatic acids
- Abietic acid
- *cis*-Pinonic Acid
- Vanillic acid

Aliphatic dicarboxylic acids
- Propanedioic acid (malonic acid)
- Butanedioic acid (succinic acid)
- Methylsuccinic acid
- Pentanedioic acid (glutaric acid)
- Hydroxybutanedioic acid (malic)
- Hexanedioic acid (adipic acid)
- Octanedioic acid (suberic acid)
- Nonanedioic acid (azelaic acid)

Aromatic polycarboxylic acids
- 1,2-Benzenedicarboxylic acid
- 1,3-Benzenedicarboxylic acid

Sterols
- Cholesterol
- β-Sitosterol
- Stigmasterol

n-Alkanols
- 1-Hexacosanol
- 1-Tetracosanol
- 1-Octacosanol
- 1-Triacontanol

Sugars
- Levoglucosan
- Sucrose
- Mannose
- 1,6-Anhydro-β-D-Mannopyranose
- 1,6-Anhydro-β-D-Galactopyranose
- Glucose
- D(+)-Xylose
- D(+)-Maltose
- D(+) Trehalose (Mycose)

Aromatic ketones
- 1,4-Naphthoquinone
- Phenanthrenequinone
- 1,2-Naphthoquinone
- Anthraquinone

Hopanes and steranes
- 18α(*H*)22,29,30-Trisnorneohopane
- 17α(*H*)-22,29,30-Trisnorhopane
- 17α(*H*),21β(*H*)-29-Norhopane
- 18α(*H*)-29-Norneohopane
- 17α(*H*),21β(*H*)-Hopane
- 22S,17α(*H*),21β(*H*)-30-Homohopane
- 22R,17α(*H*),21β(*H*)-30-Homohopane
- 22S,17α(*H*),21β(*H*)-30-Bishomohopane
- 22R,17α(*H*),21β(*H*)-30-Bishomohopane
- 20R,5α(*H*),14β(*H*),17β(*H*)-Cholestane
- 20S,5α(*H*),14β(*H*),17β(*H*)-Cholestane
- 20R,5α(*H*),14α(*H*),17α(*H*)-Cholestane
- αββ,20*R*,24*S*-Methylcholestane
- αββ,20*R*,24*R*-Ethylcholestane
- ααα,20*R*,24*R*-Ethylcholestane

Polycyclic aromatic hydrocarbons
- Phenanthrene
- Anthracene
- 4-*H*-cyclopenta[*def*]phenanthrene
- Fluoranthene
- Pyrene
- Benzo[*a*]anthracene
- Chrysene
- Triphenylene
- 1-Methylnaphthalene
- 2,6-Dimethylnaphthalene
- 2-Methylanthracene
- 1-Methylpyrene
- 3-Methylchrysene
- Retene
- Benzo[*b*]fluoranthene
- Benzo[*k*]fluoranthene
- Benzo[*e*]pyrene
- Benzo[*a*]pyrene
- Indeno[1,2,3-*cd*]fluoranthene
- Indeno[1,2,3-*cd*]pyrene
- Dibenz[*a,h*]+[*a,c*]anthracene
- Benzo[*ghi*]perylene
- Coronene

Table 2 Molecular marker compounds for secondary organic aerosol.

Organic compound name	Compound MW	Precursor Hydrocarbon
2-Methylglyceric acid	134	Isoprene
2-Methylthreitol	136	Isoprene
2-Methylerythritol	136	Isoprene
3-Isopropylpentanedioic acid	174	α-Pinene
3-Acetylpentanedioic acid	174	α-Pinene
2-Hydroxy-4-isopropyladipic acid	204	α-Pinene
3-Acetyl hexanedioic acid	188	α-Pinene
3-Hydroxyglutaric acid	148	α-Pinene
2-Hydroxy-4,4-dimethylglutaric acid	176	α-Pinene
3-(2-Hydroxy-ethyl)-2,2-dimethylcyclobutane-carboxylic acid	172	α-Pinene
Pinic acid	186	α-Pinene
Pinonic acid	184	α-Pinene
2,3-Dihydroxy-4-oxopentanoic acid	148	Toluene
b-Caryophyllinic acid	254	*b*-Caryophyllen

identified as diesel engine exhaust, paved road dust, gasoline-powered vehicle exhaust, plus emissions from food cooking and wood smoke. Smaller impacts were estimated for tire dust, plant fragments, natural gas combustion aerosol, and cigarette smoke. The remainder of the fine PM mass was secondary sulfate, nitrate, and organics so that essentially all of the annual mean mass can be apportioned, albeit the origins of the secondary species cannot be assigned. Figure 4 shows the annual average distributions of the identified sources for the 4 measurement sites.

3.1.3 Expanded Chemical Mass Balance Analysis. Marmur et al.[84] introduced an extended CMB approach using the Lipschitz global optimizer (LGO) program (CMB-LGO). The approach incorporates ratios of gaseous pollutant concentrations (CO, SO$_2$, and NO$_x$) to PM$_{2.5}$ values as constraints to assist in separating source profiles that would be unresolvable owing to their collinearity. These species are not included within the suite of chemical species in the source profiles that are fit to the ambient data, but are used to constrain the solutions to reproduce the known gas/particle ratios for each source type. The technique was applied to 25 months of daily PM$_{2.5}$ measurements (total mass and composition) made at the Atlanta Jefferson Street SEARCH site as described by Marmur et al.[84] An ordinary weighted least-squares fit was made using a Lipschitz (continuous) global optimizer (LGO) rather than the effective variance least squares approach that is used in the conventional CMB approach. The CMB-LGO method forces constraints on the calculated concentrations of SO$_2$, CO, and NO$_x$.

A comparison of the average apportionments is shown in Figure 5. The three solutions are characterized by high correlation coefficients for the fit obtained (0.97–0.99), good mass closure (91–93%), and calculated to

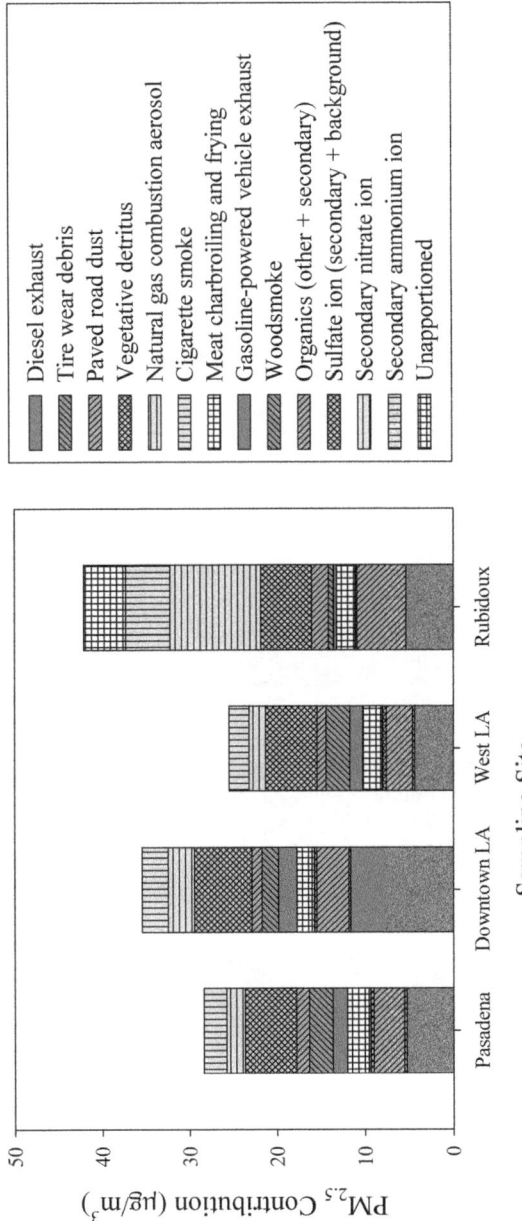

Figure 4 Average contributions of sources to PM$_{2.5}$ samples collected at multiple sites in Los Angeles and analyzed using molecular marker CMB. Figure draw from data presented in Schauer *et al.*[83]

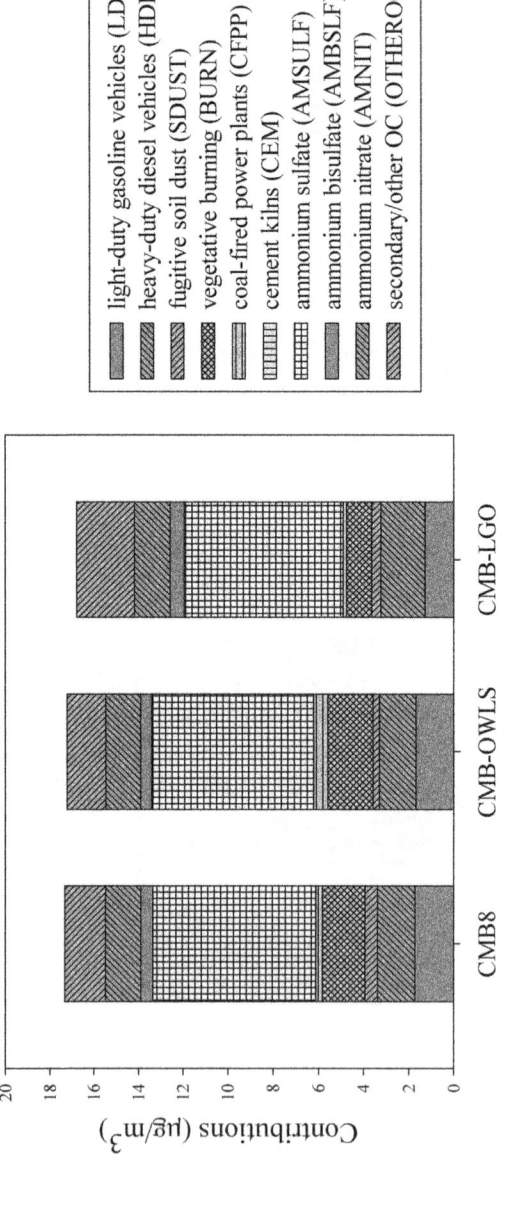

Figure 5   Comparison of source apportionments using different CMB algorithms. Results taken from Marmur *et al.*[84]

observed ratios nearing one for the major $PM_{2.5}$ components. For the EV solution, the average chi-square value, and most individual values, lay within an acceptable range (<4). The chi-square values based on the OWLS and LGO solutions are not comparable to that of the EV solution because of the different weighting used in the EV approach. The chi-square based on the LGO solution is significantly higher than that in the OWLS solution, and reflects the "penalty" of bounding acceptable solutions based on the gas-phase species. However, the correlation coefficient is higher, and the overall and trace-metal-based normalized mean-square-error (NMSE) values are lower for the LGO solution compared to the OWLS solution. More detailed analyses of these results are provided in Marmur *et al.*[84] This approach appears to hold considerable promise, but faces the difficulty of not being as readily available as the EPA supplied CMB software.

3.2 Unmix

Unmix has been applied to a number of air quality data sets including airborne PM,[36–39] particle number size distribution,[40] and VOCs.[41–46] Lewis *et al.*[37] applied Unmix to a data set of $PM_{2.5}$ compositions analyzed in samples collected in Phoenix, AR between 1995 and 1997. This data set was also used as one of the test sets for the receptor modeling intercomparison conducted in 2003 under the auspices of the EPA PM and Health Centers.[85] Comparisons of the analyses of these data were provided by Hopke *et al.*[86] Unmix uses a singular value decomposition and the edges are then identified in the reduced dimensional space.[34]

Five sources were identified: diesel exhaust, vegetative burning, secondary aerosol, gasoline vehicle emissions, and crustal/soil. Two sets of results are presented. In the initial set, the secondary aerosol profile included a significant amount of organic carbon that they claim to be a positive artifact and modify the data set to make it disappear from the secondary aerosol factor in their second solution. Given what is now known about the catalysis of secondary organic aerosol formation by acidic surfaces[87] as well as the fact that the sulfate particles represented the bulk of the available surface area onto which primary and secondary organic aerosol could condense, it would appear that their initial solution was more likely to be the correct one. It is quite reasonable for there to be secondary organic aerosol (SOA) associated with secondary inorganic particles and most factor analysis-based source apportionments show OC associated with secondary sulfate and secondary nitrate particles since there is generally not going to be sufficient SOA produced that they can nucleate on their own to form particles only containing OC.

3.3 Positive Matrix Factorization

Positive Matrix Factorization (PMF) has been used as a source apportionment tool in very many airborne PM composition studies,[88–101] particularly during the mid-2000s when the results were employed in many of the 2008 state implementation plans for dealing with areas out of compliance with the $PM_{2.5}$ ambient air quality standard.

PMF has also been applied to VOC data,[102] to Aerosol Mass Spectrometry (AMS) data,[103,104] and to particle number size distributions.[40,105–108] With the distribution of a version of PMF by the US EPA,[49] it is now being routinely applied to many air pollution data sets and has become the most widely used receptor model[109] with hundreds of applications world-wide.

3.3.1 Application to St. Louis Supersite Data. St. Louis, MO has had chronic problems meeting the US national air quality standards for particulate air pollution dating back to the 1970s. There have therefore been multiple studies in St. Louis to examine the concentrations and compositions within and across the city beginning with the Regional Air Pollution Study of the mid-1970s.[27–29] It is one of the few locations in the United States where there were still large point sources. These sources were a steel mill, a copper processing plant, a primary zinc smelter and a primary lead smelter. It was chosen to be one of the US Environmental Protection Agency's "Supersites" with high intensity sampling and analysis operating at a site in East St. Louis, IL from 2001 to 2003. Lee *et al.*[97] presented the analysis of the daily 24-hour PM samples collected at this site and analyzed for a suite of elements, ions, and organic and elemental carbon (OC/EC). A total of 709 samples and 33 chemical species were used in the PMF analysis.

Ten factors were identified as shown in Figure 6. They were identified as secondary sulfate, carbon-rich secondary sulfate, gasoline exhaust, secondary nitrate, steel production, airborne soil, diesel emissions/railroad, zinc smelting, lead smelting, and copper processing. Lee *et al.*[97] discuss these assignments in detail. The average apportionment of the PM mass to these sources over the two years of measurements is presented in Table 3.

To further assess the assignments of the profiles to sources, conditional probability function (CPF) analyses were performed.[90] CPF is a powerful approach for identifying the location of local point sources whose emissions significantly impact the receptor site. To minimize the effect of atmospheric dilution, daily fractional mass contribution from each source relative to the total of all sources was used rather than the absolute source contributions. The same daily fractional contribution was assigned to each hour of a given day to match the hourly wind data. These results are shown in Figure 7. It can be seen that the CPF plots for each of the 4 major point sources points clearly to that source. There are interesting differences between the gasoline and diesel plots. The gasoline CPF points largely to the south where there is a large residential area in which there is considerable car traffic with limited heavy duty diesel trucks. The diesel figure points more toward the north and northwest where there is a nexus of major interstate highways on which there would be significant heavy duty diesel traffic.

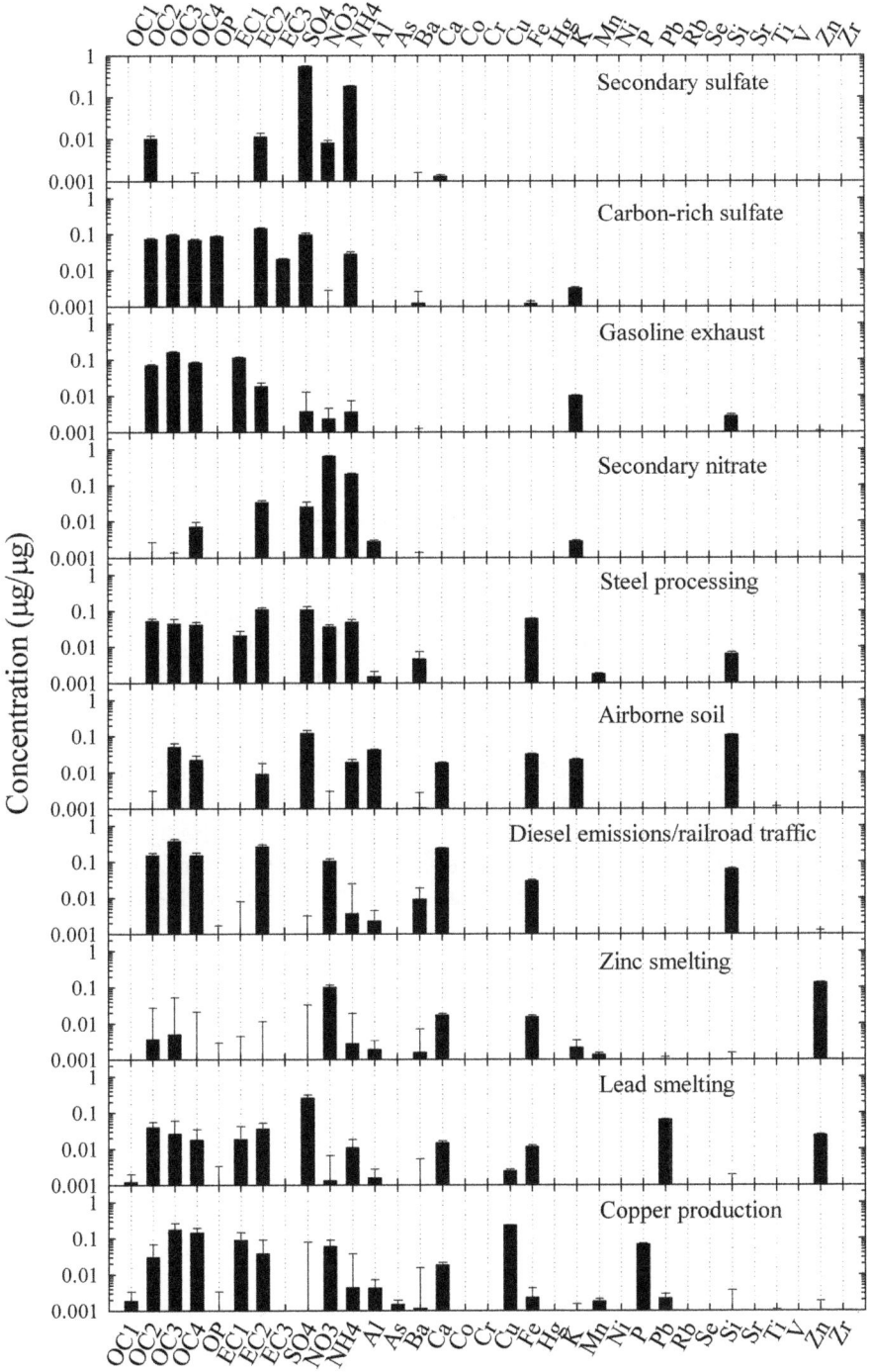

Figure 6 Source profiles derived from the 24-hour PM$_{2.5}$ compositional data collected at the St. Louis–Midwest Supersite.
Figure taken from Lee *et al.*[97]

Table 3 Average source contributions from PMF to measured $PM_{2.5}$ mass concentration at the East St. Louis–Midwest Supersite.

	Average source contribution (standard error), %
Secondary sulfate	32.6 (1.1)
Carbon-rich sulfate	19.6 (0.4)
Gasoline exhaust	16.4 (0.6)
Secondary nitrate	15.3 (0.5)
Steel processing	6.8 (0.2)
Airborne soil	4.2 (0.3)
Diesel emissions/railroad traffic	2.1 (0.1)
Zinc smelting	1.3 (0.1)
Lead smelting	1.3 (0.1)
Copper production	0.5 (0.04)

CPF is not useful for more distant sources, such as those of sulfate and nitrate. Given that it typically takes time for the gas-to-particle conversion processes, the sources of the gaseous precursors of sulfate and nitrate are typically distant from the receptor site. An easy to use method is the potential source contribution function (PSCF).[110,111] Figure 8 shows the PSCF analysis for the secondary sulfate factor and highlights high source probability areas in the Ohio River and Tennessee River valleys where there are many coal-fired power plants.

PMF has also been applied to the apportionment of organic carbon at the St. Louis Supersite by Jaeckels *et al.* [2007]. One hundred and twenty five particulate matter samples that had been collected over the 2 year sampling period at the St. Louis–Midwest Supersite were analyzed for 24 hour average organic carbon (OC), elemental carbon (EC), and particle-phase organic compound (molecular markers) concentrations. They found an 8-factor solution resolved 2 point sources, two winter combustion factors, a biomass-burning factor, a mobile source factor, a secondary organic aerosol factor, and a resuspended soil factor. The point sources have characteristics that suggest they are not any of the 4 point sources described above. There are large chemical production facilities in the area such as one in Sauget, IL. Lee *et al.*[97] had not resolved a biomass burning factor and several of the other sources like natural gas combustion could not be identified with only the elements, ions, and OC/EC concentrations. Thus, this analysis found additional sources that could not be resolved without the availability of the molecular marker species.

3.3.2 Source Apportionment in Rochester, NY. Rochester is a small city (city population $\approx 200\,000$; area population $\approx 1\,100\,000$) along the southern shore of Lake Ontario. There is an EPA Chemical Speciation Network (CSN) site operated by the New York State Department of Environmental Conservation (NYSDEC) at which $PM_{2.5}$ samples are collected and analyzed

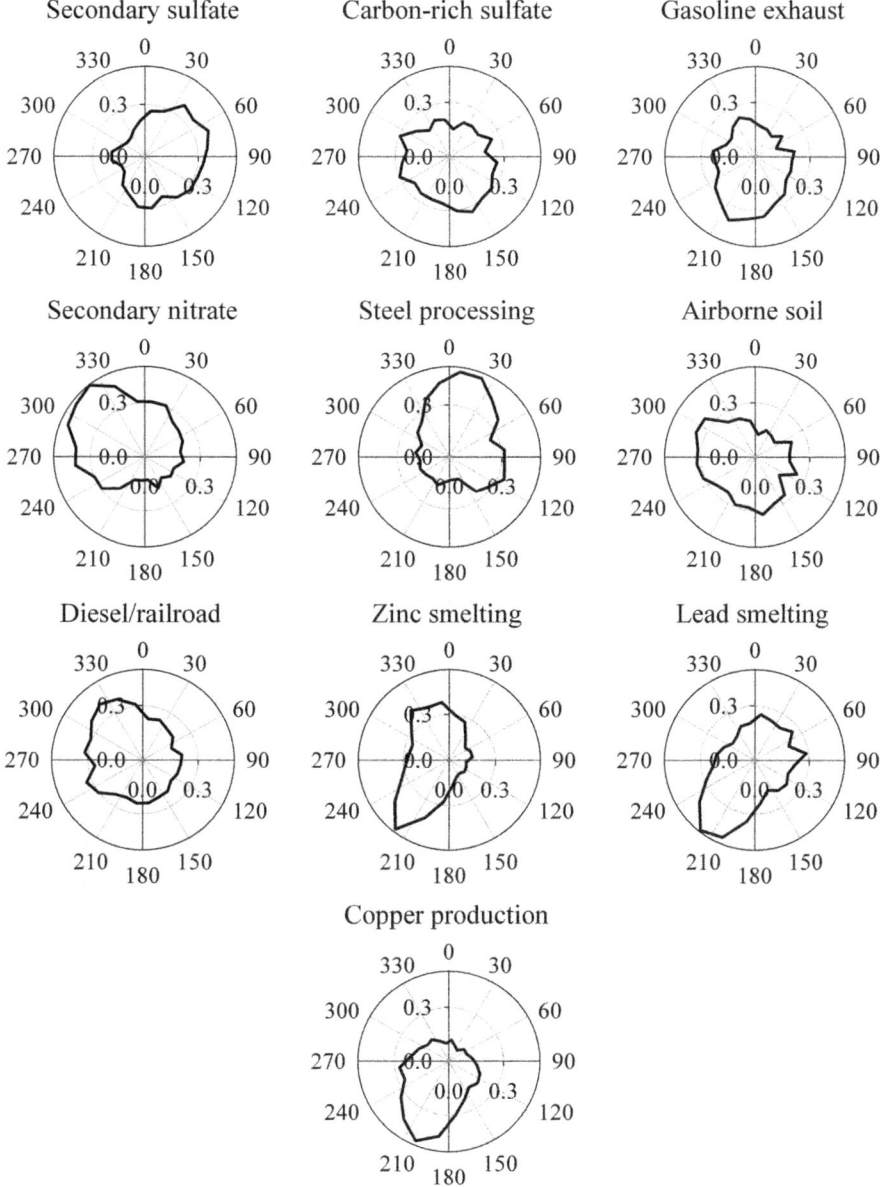

Figure 7 CPF results for the PMF resolved source contributions to the PM$_{2.5}$ mass concentrations measured at the St. Louis–Midwest Supersite. Figure taken from Lee *et al.*[97]

on an every third day basis. In 2008, a model AE-22 aethalometer (Magee Scientific) was added to the other measurements. During the period from October 2009 to October 2010, a Tisch Environmental TE1000 PUF high volume sampler with a PM$_{2.5}$ inlet operating at 220 LPM was used to

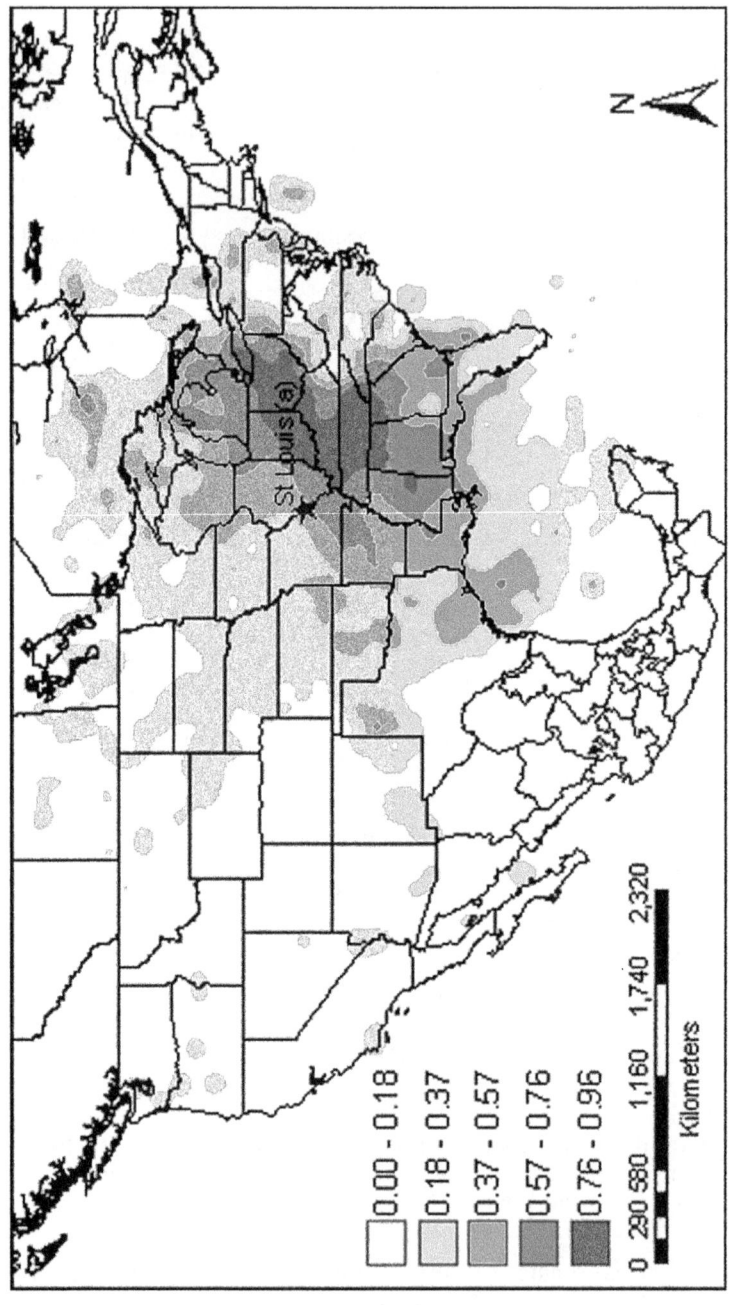

Figure 8 PSCF plot for secondary sulfate measured in St. Louis. Figure adapted from Lee and Hopke.[96]

collect 24 hour samples on the same schedule as the CSN network sampling. These samples were then analyzed for molecular markers so that a more complete set of chemical species was available for analysis.

From the analysis of these results, Wang *et al.*[112] showed that the difference between black carbon (BC) measured at 880 nm and UVBC, black carbon measured at 370 nm, called Delta-C (Delta-C = UVBC – BC) was highly correlated with levoglucosan during the winter months (December to February) with an r^2 of 0.89. Thus, as was suggested by Allen *et al.*,[113] Delta-C can be a useful indicator of woodsmoke.

Wang *et al.*[100] added Delta-C to the usual suite of speciation network data collected January 2007 to December 2010 to perform a PMF apportionment of the $PM_{2.5}$ measured at the NYSDEC site. The addition of Delta-C provided a clear separation of woodsmoke from traffic (particularly diesel). Wang *et al.* report that the annual average $PM_{2.5}$ concentrations apportioned to diesel emissions in 2007, 2008, 2009, and 2010 were 1.34 $\mu g\,m^{-3}$, 1.25 $\mu g\,m^{-3}$, 1.13 $\mu g\,m^{-3}$, and 0.97 $\mu g\,m^{-3}$, respectively, suggesting an impact of the US EPA 2007 Heavy-Duty Highway Rule on the decrease of BC and $PM_{2.5}$ concentrations over this period.

When the molecular markers were added to the data for the more restricted time period of 2009 to 2010,[101] there was again a clear separation of a woodsmoke source with most of the explained variation of Delta-C and levoglucosan in this factor. This work showed that during the winter of 2009–2010 about 30% of the $PM_{2.5}$ at the NYSDEC site is woodsmoke. Two secondary organic aerosol sources were identified with one related to isoprene oxidation and the other arising from other precursor compounds particularly biogenic terpenes.

A comparison of the results of the two analyses is presented in Table 4. It can be seen that the woodsmoke contribution over the 2009–2010 year was comparable to that resolved over the longer period where molecular marker data were not available. The molecular marker results for motor vehicles are lower than only the speciation network species results because these organic species measurements were made late in the study period when traffic emissions had decreased.

In Rochester, particle size number distribution measurements have been ongoing since the end of 2001. These data along with $PM_{2.5}$ concentrations

Table 4 Comparison of source apportionments in Rochester, NY.

Wang *et al.*[101]	($\mu g\ m^{-3}$)	Wang *et al.*[100]	($\mu g\ m^{-3}$)
Soil	1.02	Soil	0.45
Wood combustion	0.72	Wood combustion	0.84
Diesel emissions	0.62	Diesel emissions	1.45
Gasoline vehicles	0.63	Gasoline vehicles	1.73
Secondary nitrate	0.50	Secondary nitrate	1.00
Secondary sulfate	3.62	Secondary sulfate	3.72
Isoprene SOA	0.55	Road salt	0.12
Other SOA	0.29		

measured with a TEOM, and gaseous pollutant data have been subjected to PMF analysis.[107,108] The measurement site was moved in 2004 from downtown Rochester to the current NYSDEC site on the eastern side of the city. The data were divided into 3 seasons, summer, winter, and transition (spring and fall) and the data from each site and each season were analyzed separately. Kasumba *et al.*[108] identified ten sources at both sites and these include traffic, nucleation, residential/commercial heating, industrial emissions, secondary nitrate, ozone-rich secondary aerosol, secondary sulfate, regionally transported aerosol, and a mixed source of nucleation and traffic.

The important resolved sources were traffic, nucleation, mixed source (traffic and nucleation), and industrial emissions, collectively contributing more than 80% of the apportioned concentrations for each individual season considered in the study. Nucleation accounted for the smallest particles, while traffic was mostly accountable for particles ranging from 20 to 40 nm. Both sampling sites were in close proximity to significant traffic. A source that was thought to be a mixture of traffic and nucleation accounted for particles between 30 and 40 nm. It was not possible to separate this factor into two independent factors. An industrial emissions factor was mainly accountable for particles ranging from 40 to 60 nm, and they were attributed to the coal-fired power plants and the coal-fired cogeneration plant in Rochester based on the CPF analysis of their directionality. Distant sources were found to be mostly accountable for the secondary aerosol or accumulation mode particles probably owing to the long distances between the sources and the sampling sites. These results are generally consistent with the analyses of the particle composition data although these results show the effect of sources on small particle sizes that contribute little mass to the $PM_{2.5}$ concentrations.

4   Advanced Model Applications

4.1   Constrained Models

Adding constraints to the least square fitting process can reduce the extent of rotational ambiguity. With PMF being applied through the use of the multilinear engine (Paatero, 1999), it is possible to build constraints into the model as was done in two studies where there was an effort to separate multiple sources of similar composition.[114,115] Amato *et al.*[114] applied the multilinear engine to data from an urban background site in Barcelona (Spain) to quantify the contribution of road dust resuspension to PM_{10} and $PM_{2.5}$ concentrations. A recent emission profile of local resuspended road dust had been previously obtained.[116] This *a priori* information was introduced into the model as auxiliary terms in the object function to be minimized by the implementation of so-called "pulling equations."[117]

Amato and Hopke[118] have applied constraints to combine the analysis of the Supersite data previously analyzed by Lee *et al.*[96] as well as the two EPA speciation sites in the St. Louis[97] area into a single analysis. Amato and

Hopke used fixed source profiles for some of the sources in the analysis to independently analyze the compositional data collected at each of the three locations. To obtain good target profiles for the major point sources, additional high time-resolution particle composition data collected as part of the St. Louis–Midwest Supersite study[119] were analyzed. Applying PMF to these data permitted extraction of profiles for the copper products plant, the lead smelter, the zinc smelter and the steel mill. Average tailpipe emissions profiles were taken from Schauer *et al.*[120] These profiles were taken as targets and introduced in ME continuation runs with the aim of extend the number of sources found. The results were substantially improved over the original results as shown in Table 5. The constrained solutions provided more realistic results in which the source impacts of all of the major point sources were assessed at each site.

Constraints have shown to be of sufficient value that they have now been incorporated into the US EPA version of PMF in version 5.0.14.[49] Using them requires a multiple step analysis in which an initial solution is obtained and then constraints applied to the continuation run. Details of how to perform such analyses are provided in the user's manual.[121] Sofowote *et al.*[122] described how to construct constraints using available data. They analyzed data from 5 sites across Ontario assuming that all of the sites would be affected by the same distant sources *via* long-range transport. Eight factors were found to be common across these sites. These factors had profiles that varied greatly from one site to the other, suggesting that the PMF solutions were impacted by some rotational ambiguity. The features in the EPA PMF V5 were used to impose mathematical constraints that guided the factor solutions. These constraints reduce the rotational space. In situations where major emissions sources are known and located in the neighborhood of receptors, or emissions inventories and literature source profiles exist, it is easy to use these profiles to force the factor solutions to conform to the expected signatures. In this case, reported source profiles were neither available nor applicable owing to the large spatial span of potential sources and receptor sites. This work described how such constraints can be generated and used in these complex situations. The constrained solution was then applied to provide the source apportionments for the $PM_{2.5}$ at each of the 5 sites across Ontario.[123] Thus, there are now tools available where *a priori* information can be used in EPA PMF to fit known sources and then be able to derive the remaining source profiles to appropriately fit the data.

4.2   Multiple Sample Type Data

In many panel studies of the effects of airborne particles on health, measurements are made in multiple environments. For example, Hopke *et al.*[124] report on the analysis of elderly subjects living in a single multi-family residence. Measurements were made at a central outdoor site, an unoccupied room in the building and using personal samplers on specific

Table 5 Comparison of results after incorporation of known profiles taken from Amato and Hopke.[118]

	Blair St. (BS) μg m⁻³ (%)		Arnold St. (AS) μg m⁻³ (%)		St. Louis Supersite (SS) μg m⁻³ (%)	
	ME-2	PMF2	ME-2	PMF2	ME-2	PMF2
Sulfate	6.2 (38)	6.5 (40)	6.1 (39)	5.5 (36)	6.6 (37)	5.8 (33)
Nitrate	2.8 (17)	3.2 (20)	2.0 (13)	2.0 (13)	2.4 (13)	2.7 (15)
Zinc smelting	1.4 (9)	0.5 (3)	0.6 (4)		1.4 (8)	0.2 (1.3)
Copper products	0.1 (1)	0.6 (3)	0.03 (0.2)		0.2 (1)	0.1 (0.5)
Diesel	0.7 (5)		0.4 (3)	0.8 (5)	0.7 (4)	0.4 (2)
Lead smelting	0.4 (3)		0.4 (3)	0.5 (3)	0.6 (3)	0.2 (1)
Gasoline	2.6 (16)		2.9 (19)	3.2 (21)	1.3 (7)	2.9 (16)
Ca-rich			1.9 (12)	1.8 (12)		
Soil	1.6 (10)	2.5 (15)	0.4 (3)	0.5 (3)	0.5 (3)	0.8 (4)
Biomass			0.5 (3)	0.5 (3)		
Metal (Fe, Cu, Zn)				0.4 (2)		
				0.4 (3)		
Steel processing	0.3 (2)	0.05 (0. 3)			0.6 (3)	1.2 (7)
Motor vehicles (total)		2.8 (17)				
C-rich sulfate					3.4 (19)	3.5 (20)

individuals. Thus, different sources will affect different sample types. Only "external" sources of ambient particles will affect the outdoor samples. However, ambient particles will penetrate into indoor air and add to the exposure observed in the indoor and personal samples. Indoor sources such as cooking and the use of personal care products will not affect the outdoor samples. The expanded receptor model for this study can be expressed as:

$$x_{ijdt} = \sum_{p=1}^{N} g_{ipdt} f_{jp} + \sum_{p=N+1}^{N+H} g_{ipdt} f_{jp} \quad (t=1/2: \text{personal/indoor}) \qquad (2)$$

$$x_{jdt} = \sum_{p=1}^{N} g_{pdt} f_{jp} \quad (t=3: \text{outdoor}) \qquad (3)$$

where i is the individual (subject or participant) index, j is the species index, d is the sampling date index, t is the type index, N is the number of external sources, H is the number of internal sources. x_{ijdt} denotes the concentration of species j in the sample of type t collected by subject i on date d, g_{ipdt} denotes the contribution of source p to the sample of type t collected by subject i on date d, and f_{jp} denotes the relative concentration of species j in source p.

This model has been used to analyze data for cardiac patients in the Raleigh–Chapel Hill area of North Carolina[125] and of data for asthmatic children attending a special school for moderate to severe asthmatics in Denver.[126] In the case of the Denver study, four external sources and three internal sources were resolved from the PM$_{2.5}$ data for the three different environments. Secondary nitrate and motor vehicle emissions were the two largest external sources in this study. Cooking was the largest internal source. A significant influence of indoor tobacco smoking on daily personal exposures to particles was observed for those houses in which smokers reside and the environmental tobacco smoke contribution correlated with urinary cotinine levels in these urban schoolchildren. The influence of the high traffic flow outside the school on the indoor air quality was also observed.

4.3   Time Synchronization Model

One of the major developments in atmospheric monitoring over the past 15 years has been the deployment of more real-time and near real-time instruments. However, these instruments collect data at different frequencies ranging from a few minutes to a few hours. Higher frequency data have the advantage that transient events can be observed that can often provide edge points that would otherwise be averaged out of a longer interval sample.[127] Thus, it is not desirable to average the higher frequency data to the longer time interval instrument data in the suite of data. There is no way to split the longer integration time data down to the shorter time intervals so it is necessary to have models that permit each set of data to be included within its own measurement frequency. Such models have been applied to several of the sets of data from the US EPA's Supersite program. Zhou *et al.*[128] analyzed data from Pittsburgh, PA while Ogulei *et al.*[129] used the same model for data from Baltimore, MD. The model has been examined further using simulated data[130] and found that the model performed well. It has also been applied to provide VOC and $PM_{2.5}$ source apportionments that then was used to apportion risk in the exposed populations.[131]

4.4   Spatially Distributed Data

Paatero *et al.*[132] examined a spatial data set of $PM_{2.5}$ mass concentrations measured every third day at over 300 locations in the eastern United States during 2000. The basic PMF model was enhanced by modeling the dependence of $PM_{2.5}$ concentrations on temperature, humidity, pressure, ozone concentrations, and wind velocity vectors. The model comprises 12 general factors, augmented by 5 urban-only factors intended to represent excess concentration present in urban locations only. The flux density maps showed the major transport patterns of $PM_{2.5}$. For example, they show the increase in particle mass as the air moves from the regions of the gaseous precursor (SO_2) and is converted in sulfate. Recognition of this combination of transport and transformation is necessary in order that control procedures can be targeted to significant causes of high $PM_{2.5}$ concentrations.

A different spatial model was developed by Chueinta *et al.*[133] for the analysis of the spatial patterns and possible sources affecting haze and its visual effects in the southwestern United States. The data are from the Measurement of Haze and Visual Effects (MOHAVE) project that were collected during the late winter and midsummer of 1992 at the monitoring sites in four states (*i.e.*, California, Arizona, Nevada and Utah). The resulting three-way data array was analyzed by a four product-term model. This study makes a direct effort to include wind patterns as a component in the model in order to obtain the information of the spatial patterns of source contributions. The solution is computed using the conjugate gradient algorithm with applied non-negativity constraints. For the winter data set, reasonable solutions contained six sources and six wind patterns. The analysis of summer data required seven sources and seven wind patterns.

4.5   Mixed Way Data

Airborne particulate samples have been collected at Alert, Nanuvut, Canada on a weekly basis since July 1980. Details of the sampling and chemical analyses were given by Barrie and coworkers.[134,135] Major ions were analyzed by ion chromatography (IC) and trace elements data were obtained by instrumental neutron activation analysis (INAA) and by inductively coupled plasma emission spectroscopy (ICP). Data for 24 chemical species measured in samples obtained between September 1980 and August 1991 were analyzed by Xie *et al.*[136,137] in two different studies. In the first study,[136] the data were analyzed using the basic model presented in eqn (1) and as a tridimensional system as defined by:

$$\mathbf{X} = \mathbf{A} \otimes \mathbf{B} \otimes \mathbf{C}$$

$$x_{ijk} = \sum_{h=1}^{p} a_{ih} b_{jn} c_{kh} + e_{ijk} \tag{4}$$

where $\mathbf{X}$ is an $n \times m \times q$ three-way data array, $\mathbf{A}$, $\mathbf{B}$, $\mathbf{C}$ are the resulting 2-way factor matrices for each of the three modes and $\mathbf{E}$ is the un-modeled part of $\mathbf{X}$. Based on the same non-negativity constraints and minimization criterion, PMF can solve the trilinear model [Paatero, 1997b].

The data were fit to both the bilinear [eqn (1)] and trilinear [eqn (4)] models. There were problems in both cases in fitting the data. Several species like Br and I were well fit in the trilinear model but poorly fit in the bilinear model. Alternatively, elements like Na, Al, and other crustal elements were very poorly fit in the trilinear model.[136] Thus, a new mixed bilinear/trilinear model was developed[137] as defined by:

$$x_{ijk} = \sum_{q=1}^{Q} t_{ikq} g_{jq} + \sum_{p=1}^{P} a_{ip} b_{jp} c_{kp} + e_{ijk} \tag{5}$$

The first term represents a customary 2-way PMF model with score and loading matrices T and G, while the second term is a P factor 3-way trilinear model with the A, B, and C matrices being the three modes, respectively.

For the Alert data, factor T contains time variation over the entire 11 years, and G provides concentration profiles of the corresponding sources from 2-way modeling. A and C are factors corresponding to time variation during a year and across 11 years, respectively, and B is the source concentration profile matrix in the 3-way part of the model. The application of ME to the Alert data has been described by Xie *et al.*[137] A five PMF factor solution was found to fit the data and was the easiest to physically interpret for both the 2-way and 3-way analyses.

A biogenic factor was identified that was characterized by the presence of methane sulfonic acid (MSA), which is the product of the oxidation of sulfur-containing compounds like dimethylsulfide (DMS) or dimethyldisulfide (DMDS) emitted by biogenic activity in the surface layer of the ocean. The ratio of MSA to $SO_4^=$ in this profile is 0.34 ± 0.04. Isotopic studies by Li and Barrie[138] have shown that this ratio should be approximately 0.31 for biogenic sulfur aerosol in the Arctic region. This factor, therefore, can reasonably be attributed as being a biogenic component of the Arctic aerosol. Further corroboration of the origin of this factor can be obtained by examining the temporal variation. They found two peaks in the seasonal variation of the factor; a large one around April/May and another, smaller peak about August. Li *et al.*[139] attributed the spring peak to the sea surface temperature in the North Atlantic Ocean west of the coast of continental Europe, and the summer peak to the ocean region further north in the Atlantic Ocean off the coast of Norway and in the northwestern north Pacific Ocean. Hopke *et al.*[140] identified possible source locations for MSA in these Alert samples and these findings are consistent with those of Li *et al.*[139]

The year-to-year variation of the biogenic factor showed a strong rise during the late 1980s and a sharp drop from 1990 to 1991. Because biological activity should be related to average temperature, the scores for the biogenic factor obtained from both analyses are plotted in Figure 9 against the Northern Hemisphere temperature anomaly as obtained from Jones *et al.*[141] The anomaly is the average temperature for the year minus the average temperature for a reference period of 1950 to 1979 and corrected for El Nino. There is a good correlation between these quantities with a squared correlation coefficient (r^2) of 0.78 for the trilinear analysis and 0.59 for the multiway fit. Both results show the same slope so that these two analyses are extracting the same feature from the data. It was only possible to identify this factor because of the availability of methane sulfonic acid data.

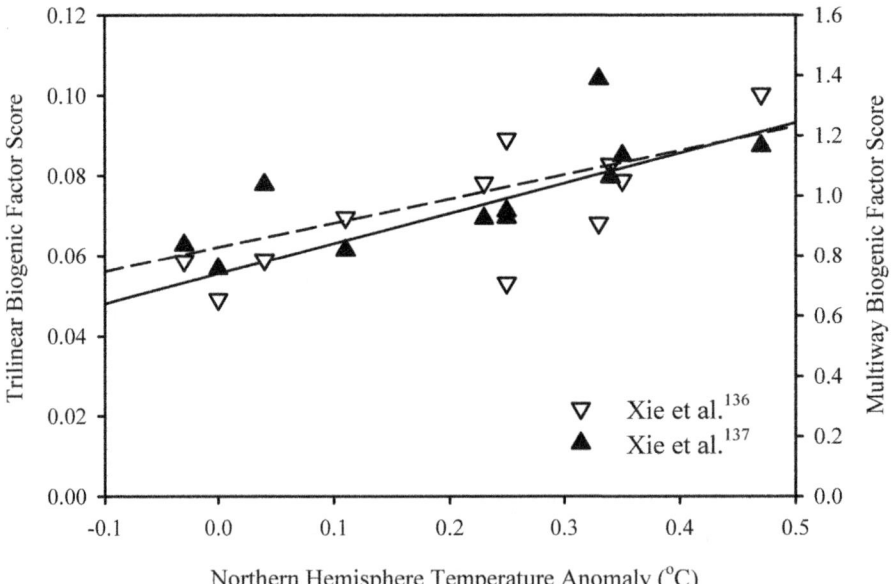

Northern Hemisphere Temperature Anomaly (°C)

Figure 9 Plot of the factor scores in arbitrary units for the year-by-year biogenic sulfur factor from the ME1 factor analysis against the Northern Hemisphere Temperature Anomaly for the time period from 1981 to 1991 relative to a reference period from 1950 to 1979 and corrected for El Nino. Data for figure taken from Xie *et al.*[136,137] Solid line is the regression line for the trilinear analysis[136] while the dotted line is the regression line for the multiway analysis.[137]

The strength of this relationship points to the strong possibility that there is a real interaction between the changing hemispheric temperature and the extent of biological sulfur production as predicted by Shaw[142] and Charlson *et al.*[143] These results suggest further examination of compositional data from remote sites to determine if this signal is observed elsewhere. Its strength and worldwide importance can then be fully assessed.

4.6 Size–Composition–Time Data

There are a number of devices that can separate particles by size such that samples can be collected that represent a relatively limited particle size range. The most common of these systems is a cascade impactor in which particles are sequentially separated and collected for analysis. Most of these systems are manually operated so there is considerable effort involved in collecting a series of samples. However, there have been several systems developed for collecting a time-series of time- and size-resolved samples that can then be analyzed. One of these systems is the rotating DRUM impactor sampler[144] that collects the particles on Mylar films placed on a rotating drum under the nozzle that determines the aerodynamic behavior of the particles. The resulting samples can be analyzed using synchrotron XRF[145] to provide the 3-way data set.

Different sources have different size-composition profiles in their emissions.[146] Thus, a source profile for size segregated data is a matrix of composition as a function of size and therefore, a special model is required to properly account for the processes by which the particles are formed and emitted into the atmosphere. The main equation of the model is as follows:

$$\vec{X} = A \otimes \vec{B} + \vec{E} \tag{6}$$

where $\vec{X}(I,J,K)$ is the three-way array of observed data, $\otimes$ represents a Kroneker product[147,148] of the source profile array $\vec{B}(I,J,K)$ with the contribution matrix, $A(I,P)$, P is the number of factors, and $\vec{E}(I,J,K)$ is the three-way array of residuals.

This model has been applied to several data sets including three-stage DRUM impactor data from Detroit, MI with the samples collected between February and April 2002[149] and eight-stage DRUM impactor data from the Washington-Dulles International Airport.[150] For the Detroit data,[149] nine factors were identified: road salt, industrial (Fe + Zn), cloud processed sulfate, two types of metal works, road dust, local sulfate source, sulfur with dust, and homogeneously formed sulfate. Road salt had high concentrations of Na and Cl. Mixed industrial emissions are characterized by Fe and Zn. The cloud processed sulfate had a high concentration of S in the intermediate size mode. The first metal works was represented by Fe in all three size modes and by Zn, Ti, Cu, and Mn. The second included a high concentration of small size particle sulfur with intermediate size Fe, Zn, Al, Si, and Ca. Road dust contained Na, Al, Si, S, K, and Fe in the large size mode. The local

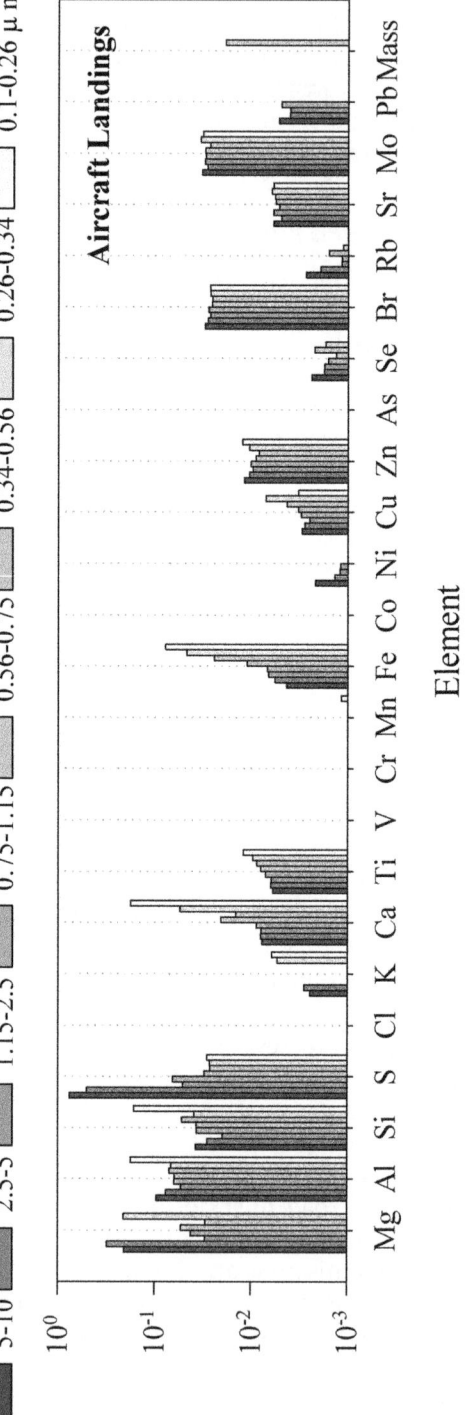

Figure 10 Source profile for aircraft landing extracted from the rotating drum impactor data collected at Dulles International Airport. Concentrations are in arbitrary units.
Figure derived from results presented by Li *et al.*[150]

and homogeneous sulfate factors show high concentrations of S in the smallest size mode, but different time series behavior in their contributions. Sulfur with dust is characterized by S and a mix of Na, Mg, Al, Si, K, Ca, Ti, and Fe from the medium and large size modes. The analysis utilized light absorption measurements at 4 wavelengths, 350, 450, 550, and 650 nm, to provide limited information on the carbonaceous components in the samples.

At Dulles International Airport, five major emission sources: soil, road salt, aircraft landings, transported secondary sulfate, and local sulfate/construction were identified.[150] Aircraft landing was notable for it had not previously been identified as a significant source of $PM_{2.5}$. Its pattern (Figure 10) showed particles containing sulfur, zinc, bromine, zirconium and molybdenum. This factor is assigned to particles that are emitted during landings. The sulfur and zinc come from tire wear. These elements are key constituents in tires. Often a visible puff of smoke is observed at touchdown. There is considerable frictional heat produced at this instant and particles are generated across the particle size range. Both zirconium and molybdenum are used in high temperature greases as might be used to lubricate bearings that would undergo significant heat stress. The energy deposited in the bearings can be expected to liberate particles from the lubricants. The CPF analysis for this source pointed strongly to the runways supporting the assignment of this factor to aircraft landings as an important local source of particulate matter.

5 Summary

Receptor modeling has evolved substantially from the early efforts of 40 years ago. With increasing computational capabilities, improving sampling, analysis and continuous monitoring technologies and better data analysis methods, it is now possible to extract substantially greater information from a wide variety of data characterizing air pollution. Although initially these techniques were not readily accepted as inputs into the air quality planning process, they have now become routine with the US Environmental Protection Agency providing significant resources to collect particle composition data in urban and rural areas across the United States. They have also made the effort to make the major data analysis tools (CMB, Unmix, PMF) readily available to the global atmospheric community where they are being widely applied to a variety of air quality problems.

Although the basic approaches are now well established, there are still active development efforts such that complex models can be developed and applied to further utilize the richness of the available data. It is important to be able to develop conceptual models that fit the nature of the available data rather than to fit all of the data into a single data analysis framework. For example, multiway data can be analyzed to examine the full suite of correlations across the dimensions in the data rather than unfolding it into simple matrices.

There is also a need for more complete error assessments. CMB currently does not account for any source profile variability. Unmix does not properly reflect that some sources (traffic, for example) may never get sufficiently low so as to provide the true edge in the system and PMF error methods currently only estimate the uncertainties in the profiles. Errors in the mass contributions are important since they really represent the policy-relevant information in the solutions.

Source apportionments are now being incorporated into health effects modeling and eventually such efforts may permit the identification of source types that are of greater relevance to the induction of specific adverse health effects.

References

1. P. K. Hopke, *Receptor Modeling in Environmental Chemistry*, John Wiley & Sons, Inc., New York, 1985.
2. *Receptor Modeling for Air Quality Management*, ed. P. K. Hopke, Elsevier Science Publishers, Amsterdam, 1991.
3. R. C. Henry, Multivariate Receptor Models, in *Receptor Modeling for Air Quality Management*, ed. P. K. Hopke, Elsevier, Amsterdam, 1991, pp. 117–147.
4. Watson and Chow Chapter this volume.
5. J. M. Colucci and C. R. Begeman, The Automotive Contribution to Air-Borne Polynuclear Aromatic Hydrocarbons in Detroit, *J. Air Pollut. Control Assoc.*, 1965, **15**, 113–122.
6. I. H. Blifford and G. O. Meeker, A Factor Analysis Model of Large Scale Pollution, *Atmos. Environ.*, 1967, **1**, 147–157.
7. B. Prinz and H. Stratmann, The Possible Use of Factor Analysis in Investigating Air Quality, *Staub – Reinhalt. Luft*, 1968, **28**, 33–39.
8. M. S. Miller, S. K. Friedlander and G. M. Hidy, A Chemical Element Balance for the Pasadena Aerosol, *J. Colloid Interface Sci.*, 1972, **39**, 65–176.
9. J. W. Winchester and G. D. Nifong, Water Pollution in Lake Michigan by Trace Elements from Pollution Aerosol Fallout, *Water, Air, Soil Pollut.*, 1971, **1**, 50–64.
10. S. K. Friedlander, Chemical Element Balances and Identification of Air Pollution Sources, *Environ. Sci. Technol.*, 1973, 7, 235–240.
11. R. Heindryckx and R. Dams, Continental, Marine, and Anthropogenic Contributions to the Inorganic Composition of the Aerosol of an Industrial Zone, *J. Radioanal. Chem.*, 1974, **19**, 339–349.
12. J. Bogen, Trace Elements in Atmospheric Aerosol in the Heidelberg Area Measured by Instrumental Neutron Activation Analysis, *Atmos. Environ.*, 1973, 7, 1117–1125.
13. D. F. Gatz, Relative Contributions of Different Sources of Urban Aerosols: Application of a New Estimation Method to Multiple Sites in Chicago, *Atmos. Environ.*, 1975, **9**, 1–18.

14. G. S. Kowalczyk, C. E. Choquette and G. E. Gordon, Chemical Element Balances and Identification of Air Pollution Sources in Washington, D.C., *Atmos. Environ.*, 1978, **12**, 1143–1153.
15. G. S. Kowalczyk, G. E. Gordon and S. W. Rheingrover, Identification of Atmospheric Particulate Sources in Washington, D.C. Using Chemical Element Balances, *Environ. Sci. Technol.*, 1982, **16**, 79–90.
16. H. Mayrsohn and J. H. Crabtree, Source Reconciliation of Atmospheric Hydrocarbons, *Atmos. Environ.*, 1976, **10**, 137–143.
17. P. F. Nelson, S. M. Quigley and M. Y. Smith, Sources of Atmospheric Hydrocarbons in Sydney: A Quantitative Determination Using a Source Reconciliation Technique, *Atmos. Environ.*, 1983, **17**, 439–449.
18. R. Li, S. Kalenge, P. K. Hopke, R. Lebouf, A. Rossner and A. Benedict, Source Apportionment of Benzene Downwind of a Major Point Source, *Atmos. Pollut. Res.*, 2011, **2**, 138–143.
19. J. G. Watson, Chemical Element Balance Receptor Model Methodology for Assessing the Source of Fine and Total Suspended Particulate Matter in Portland, Oregon, Ph.D. Thesis, Oregon Graduate Center, Beaverton, OR, 1979.
20. A. M. Dunker, 1979, A Method for Analyzing Data on the Elemental Composition of Aerosols, General Motors Research Laboratories Report MR-3074 ENV-67, Warren, MI.
21. J. G. Watson, J. A. Cooper and J. J. Huntzicker, The Effective Variance Weighting for Least Squares Calculations Applied to the Mass Balance Receptor Model, *Atmos. Environ.*, 1984, **18**, 1347–1355.
22. J. G. Watson, J. C. Chow and T. G. Pace, Chemical Mass Balance, in *Receptor Modeling for Air Quality Management*, ed. P. K. Hopke, Elsevier Science Publishers, Amsterdam, 1991, pp. 83–116.
23. P. K. Hopke, E. S. Gladney, G. E. Gordon, W. H. Zoller and A. G. Jones, The Use of Multivariate Analysis to Identify Sources of Selected Elements in the Boston Urban Aerosol, *Atmos. Environ.*, 1976, **10**, 1015–1025.
24. P. D. Gaarenstroom, S. P. Perone and J. P. Moyers, Application of Pattern Recognition and Factor Analysis for Characterization of Atmospheric Particulate Composition in Southwest Desert Atmosphere, *Environ. Sci. Technol.*, 1977, **11**, 795–800.
25. P. K. Hopke, R. E. Lamb and D. F. S. Natusch, Multielemental characterization of urban street dust, *Environ. Sci. Technol.*, 1980, **14**, 164–172.
26. D. J. Alpert and P. K. Hopke, A quantitative determination of sources in the Boston urban aerosol, *Atmos. Environ.*, 1980, **14**, 1137–1146.
27. D. J. Alpert and P. K. Hopke, A determination of the sources of airborne particles collected during the regional air pollution study, *Atmos. Environ.*, 1981, **15**, 675–687.
28. C. K. Liu, B. A. Roscoe, K. G. Severin and P. K. Hopke, The application of factor analysis to source apportionment of aerosol mass, *Am. Ind. Hyg. Assoc. J.*, 1982, **43**, 314–318.

29. K. G. Severin, B. A. Roscoe and P. K. Hopke, The use of factor analysis in source determination of particulate emissions, *Part. Sci. Technol.*, 1983, **1**, 189–196.

30. P. K. Hopke, Target transformation factor analysis as an aerosol mass apportionment method: a review and sensitivity study, *Atmos. Environ.*, 1988, **22**, 1777–1792.

31. R. C. Henry and B.-M. Kim, Extension of Self-Modeling Curve Resolution Mixtures of More Than Three Components. Part 1. Finding the Basic Feasible Region, *Chemom. Intell. Lab. Syst.*, 1989, **8**, 205–216.

32. B.-M. Kim and R. C. Henry, Extension of self-modeling curve resolution to mixtures of more than three components: Part II. Finding the complete solution, *Chemom. Intell. Lab. Syst.*, 1999, **49**, 67–77.

33. B.-M. Kim and R. C. Henry, Application of the SAFER model to Los Angeles PM10 data, *Atmos. Environ.*, 2000, **34**, 1747–1759.

34. R. C. Henry, Multivariate receptor modeling by N-dimensional edge detection, *Chemom. Intell. Lab. Syst.*, 2003, **65**, 179–189.

35. U.S. Environmental Protection Agency, 2004. Receptor Modeling, CMB, http://www.epa.gov/ttn/scram/receptor_cmb.htm.

36. L.-W. A. Chen, B. G. Doddridge, R. R. Dickerson, J. C. Chow and R. C. Henry, Origins of fine aerosol mass in the Baltimore–Washington corridor: implications from observation, factor analysis, and ensemble air parcel back trajectories, *Atmos. Environ.*, 2002, **36**, 4541–4554.

37. C. W. Lewis, G. A. Norris, T. L. Conner and R. C. Henry, Source apportionment of Phoenix PM2.5 aerosol with the Unmix receptor model, *J. Air Waste Manage. Assoc.*, 2003, **53**, 325–338.

38. S. H. Hu, R. McDonald, D. Martuzevicius, P. Biswas, S. A. Grinshpun, A. Kelley, T. Reponen, J. Lockey and G. LeMasters, UNMIX modeling of ambient PM2.5 near an interstate highway in Cincinnati, OH, USA, *Atmos. Environ.*, 2006, **40**, S378–S395.

39. C. Li, T. X. Wen, Z. Q. Li, R. R. Dickerson, Y. J. Yang, Y. A. Zhao, Y. S. Wang and S. C. Tsay, Concentrations and origins of atmospheric lead and other trace species at a rural site in northern China, *J. Geophys. Sci.-Atmos.*, 2010, D00K23, DOI: 10.1029/2009JD013639.

40. E. Kim, P. K. Hopke, T. V. Larson and D. S. Covert, Analysis of ambient particle size distributions using Unmix and positive matrix factorization, *Environ. Sci. Technol.*, 2004, **38**, 202–209.

41. S. L. Miller, M. J. Anderson, E. P. Daly and J. B. Milford, Source apportionment of exposures to volatile organic compounds. I. Evaluation of receptor models using simulated exposure data, *Atmos. Environ.*, 2002, **36**, 3629–3641.

42. H. Hellén, H. Hakola and T. Laurila, Determination of source contributions of NMHCs in Helsinki (60°N, 25°E) using chemical mass balance and the Unmix multivariate receptor models, *Atmos. Environ.*, 2003, **37**, 1413–1424.

43. S. Mukerjee, G. A. Norris, L. A. Smith, C. A. Noble, L. M. Neas, A. H. Ozkaynak and M. Gonzales, Receptor model comparisons and wind direction analyses of volatile organic compounds and sub-micrometer particles in an arid binational, urban air shed, *Environ. Sci. Technol.*, 2008, **38**, 2317–2327.
44. Y. Song, W. Dai, M. Shao, Y. Liu, S. H. Lu, W. Kuster and P. Goldan, Comparison of receptor models for source apportionment of volatile organic compounds in Beijing, China, *Environ. Pollut.*, 2008, **156**, 174–183.
45. M. A. Khairy and R. Lohmann, Source apportionment and risk assessment of polycyclic aromatic hydrocarbons in the atmospheric environment of Alexandria, Egypt, *Chemosphere*, 2013, **91**, 895–903.
46. J. Patokoski, T. M. Ruuskanen, H. Hellen, R. Taipale, T. Gronholm, M. K. Kajos, T. Petaja, H. Hakola, M. Kulmala and J. Rinne, Winter to spring transition and diurnal variation of VOCs in Finland at an urban background site and a rural site, *Boreal Environ. Res.*, 2014, **19**, 79–103.
47. P. Paatero and U. Tapper, Analysis of different modes of factor analysis as least squares fit problems, *Chemom. Intell. Lab. Syst.*, 1993, **18**, 183–194.
48. P. Paatero and U. Tapper, Positive Matrix Factorization: a non-negative factor model with optimal utilization of error estimates of data values, *Environmetrics*, 1994, **5**, 111–126.
49. U.S. Environmental Protection Agency, 2014. Receptor Modeling, Positive Matrix Factorization, http://www.epa.gov/heasd/research/pmf.html.
50. E. R. Malinowski, *Factor Analysis in Chemistry*, Wiley, New York, 2nd edn, 1991.
51. P. Paatero, Least squares formulation of robust, non-negative factor analysis, *Chemom. Intell. Lab. Syst.*, 1997, **37**, 23–35.
52. S. Juntto and P. Paatero, Analysis of daily precipitation data by positive matrix factorization, *Environmetrics*, 1994, **5**, 127–144.
53. P. Anttila, P. Paatero, U. Tapper and O. Jaarvinen, Application of positive matrix factorization to source apportionment: results of a study of bulk deposition chemistry in Finland, *Atmos. Environ.*, 1995, **29**, 1705–1718.
54. A. V. Polissar, P. K. Hopke, W. C. Malm and J. F. Sisler, The Ratio of Aerosol Optical Absorption Coefficients to Sulfur Concentrations, as an Indicator of Smoke from Forest Fires when Sampling in Polar Regions, *Atmos. Environ.*, 1996, **30**, 1147–1157.
55. A. V. Polissar, P. K. Hopke, P. Paatero, W. C. Malm and J. F. Sisler, Atmospheric aerosol over Alaska 2. Elemental composition and sources, *J. Geophys. Res.*, 1998b, **103**, 19045–19057.
56. P. Paatero, The multilinear engine-a table-driven least squares program for solving multilinear problems, including the n-way parallel factor analysis model, *J. Comput. Graph. Stat.*, 1999, **8**, 854–888.

57. J. E. Core, P. L. Hanrahan and J. A. Cooper, Air Particulate Control Strategy Development: A New Approach Using Chemical Mass Balance, in *Atmospheric Aerosol Source/Air Quality Relationships*, ed. E. Macias and P. Hopke, ACS Symposium Series 167, 1981, pp. 107–123.

58. J. E. Core, J. A. Cooper, P. L. Hanrahan and W. H. Cox, Particulate Dispersion Model Evaluation: A New Approach Using Receptor Models, *J. Air Pollut. Contr. Assoc.*, 1982, **32**, 1142–1147.

59. B. R. T. Simoneit, M. A. Mazurek and T. A. Cahill, Contamination of the Lake Tahoe air basin by high molecular weight petroleum residues, *J. Air Pollut. Control Assoc.*, 1980, **30**, 387–390.

60. B. R. T. Simoneit and M. A. Mazurek, Air pollution: the organic constituents, *Crit. Rev. Environ. Control*, 1981, **11**, 219–276.

61. B. R. T. Simoneit and M. A. Mazurek, Organic matter of the troposphere. II. natural background of biogenic lipid matter in aerosols over the rural western United States, *Atmos. Environ.*, 1982, **16**, 2139–2159.

62. M. A. Mazurek and B. R. T. Simoneit, Characterization of biogenic and petroleum-derived organic matter in aerosols over remote, rural and urban areas, in *Identification and Analysis of Organic Pollutants in Air*, ed. L. H. Keith, Ann Arbor Science/Butterworth, Boston, MA, 1984, pp. 353–370.

63. L. M. Hildemann, G. R. Cass and G. R. Markowski, A dilution stack sampler for collection of organic aerosol emissions: Design, characterization and field tests, *Aerosol Sci. Technol.*, 1989, **10**, 193–204.

64. L. M. Hildemann, G. R. Cass, M. A. Mazurek and B. R. T. Simoneit, Mathematical modeling of urban organic aerosol: Properties measured by high-resolution gas chromatography, *Environ. Sci. Technol.*, 1993, **27**, 2045–2055.

65. L. M. Hildemann, M. A. Mazurek, G. R. Cass and B. R. T. Simoneit, Seasonal trends in Los Angeles ambient organic aerosol observed by high-resolution gas chromatography, *Aerosol Sci. Technol.*, 1994, **20**(303–317), 1994.

66. L. M. Hildemann, W. F. Rogge, G. R. Cass, M. A. Mazurek and B. R. T. Simoneit, Contribution of primary aerosol emissions from vegetation- derived sources to fine particle concentrations in Los Angeles, *J. Geophys. Res.*, 1996, **101**, 19541–19549.

67. M. A. Mazurek, B. R. T. Simoneit, G. R. Cass and H. A. Gray, Quantitative high-resolution gas chromatography and high-resolution gas chromatography/mass spectrometry analysis of carbonaceous fine aerosol particles, *Int. J. Anal. Chem.*, 1987, **29**, 119–139.

68. M. A. Mazurek, G. R. Cass and B. R. T. Simoneit, Interpretation of high-resolution gas chromatography and high-resolution gas chromatography/mass spectrometry data acquired from atmospheric organic aerosol samples, *Aerosol Sci. Technol.*, 1989, **10**, 408–420.

69. M. A. Mazurek, G. R. Cass and B. R. T. Simoneit, Biological input to visibility-reducing aerosol particles in the remote arid southwestern

United States, *Environ. Sci. Technol.*, 1991, **25**, 684–694; M. A. Mazurek, M. Mason-Jones, H. Mason-Jones, L. G. Salmon, G. R. Cass, K. A. Hallock and M. Leach, Visibility-reducing organic aerosols in the vicinity of Grand Canyon National Park: 1. Properties observed by high resolution gas chromatography, *J. Geophys. Res.*, 1997, **102**, 3779–3793.

70. W. F. Rogge, L. M. Hildemann, M. A. Mazurek, G. R. Cass and B. R. T. Simoneit, Sources of fine organic aerosol: 1. Charbroilers and meat cooking operations, *Environ. Sci. Technol.*, 1991, **25**, 1112–1125.

71. W. F. Rogge, L. M. Hildemann, M. A. Mazurek, G. R. Cass and B. R. T. Simoneit, Sources of fine organic aerosol: 2. Noncatalyst and catalyst-equipped automobiles and heavy-duty diesel trucks, *Environ. Sci. Technol.*, 1993, **27**, 636–651.

72. W. F. Rogge, L. M. Hildemann, M. A. Mazurek, G. R. Cass and B. R. T. Simoneit, Sources of fine organic aerosol: 3. Road dust, tire debris, and organometallic brake lining dust - roads as sources and sinks, *Environ. Sci. Technol.*, 1993, **27**, 1892–1904.

73. W. F. Rogge, L. M. Hildemann, M. A. Mazurek, G. R. Cass and B. R. T. Simoneit, Sources of fine organic aerosol: 4. Particulate abrasion products from leaf surfaces of urban plants, *Environ. Sci. Technol.*, 1993, **27**, 2700–2711.

74. W. F. Rogge, L. M Hildemann, M. A. Mazurek, G. R. Cass and B. R. T. Simoneit, Sources of Fine Organic Aerosol. 5. Natural Gas Home Appliances, *Environ. Sci. Technol.*, 1993, **27**, 2736–2744.

75. W. F. Rogge, L. M. Hildemann, M. A. Mazurek and G. R. Cass, Sources of fine organic aerosol: 6. Cigarette smoke in the urban atmosphere, *Environ. Sci. Technol.*, 1994, **28**, 1375–1388.

76. W. F. Rogge, L. M. Hildemann, M. A. Mazurek, G. R. Cass and B. R. T. Simoneit, Sources of fine organic aerosol: 7: Hot asphalt roofing tar pot fumes, *Environ. Sci. & Technol.*, 1997, **32**, 2726–2730.

77. W. F. Rogge, L. M. Hildemann, M. A. Mazurek, G. R. Cass and B. R. T. Simoneit, Sources of fine organic aerosol: 8. Boilers burning No. 2 distillate fuel oil, *Environ. Sci. Technol.*, 1997, **32**, 2731–2737.

78. W. F. Rogge, L. M. Hildemann, M. A. Mazurek, G. R. Cass and B. R. T. Simoneit, Sources of fine organic aerosol. 9. Pine, oak and synthetic log combustion in residential fireplaces, *Environ. Sci. Technol.*, 1997c, **32**, 13–22.

79. B. R. T. Simoneit, W. F. Rogge, M. A. Mazurek, L. J. Standley, L. M. Hildemann and G. R. Cass, Lignin pyrolysis products, lignans and resin acids as specific tracers of plant classes in emissions from biomass combustion, *Environ. Sci. Technol.*, 1993, **27**, 2533–2541.

80. P. M. Fine, G. R. Cass and B. R. T. Simoneit, Chemical Characterization of Fine Particle Emissions from Fireplace Combustion of Woods Grown in the Northeastern United States, *Environ. Sci. Technol.*, 2001, **35**, 2665–2675.

81. P. M. Fine, G. R. Cass and B. R. T. Simoneit, Chemical Characterization of Fine Particle Emissions from the Wood Stove Combustion of Prevalent United States Tree Species, *Environ. Eng. Sci.*, 2004, **21**, 705–721.
82. T. E. Kleindienst, M. Jaoui, M. Lewandowski, J. H. Offenberg, C. W. Lewis, P. V. Bhave and E. O. Edney, Estimates of the contributions of biogenic and anthropogenic hydrocarbons to secondary organic aerosol at a southeastern US location, *Atmos. Environ.*, 2007, **41**, 8288–8300.
83. J. J. Schauer, W. F. Rogge, L. M. Hildemann, M. A. Mazurek, G. R. Cass and B. R. T. Simoneit, Source apportionment of airborne particulate matter using organic compounds as tracers, *Atmos. Environ.*, 1996, **30**, 3837–3855.
84. A. Marmur, A. Unal, J. A. Mulholland and A. G. Russell, Optimization based source apportionment of $PM_{2.5}$ incorporating gas-to-particle ratios, *Environ. Sci. Technol.*, 2005, **39**, 3245–3254.
85. G. D. Thurston, K. Ito, T. Mar, W. F. Christensen, D. J. Eatough, R. C. Henry, E. Kim, F. Laden, R. Lall, T. V. Larson, H. Liu, L. Neas, J. Pinto, M. Stölzel, H. Suh and P. K. Hopke, Workgroup Report: Workshop on Source Apportionment of Particulate Matter Health Effects–Intercomparison of Results and Implications, *Environ. Health Perspect.*, 2005, **113**, 1768–1774.
86. P. K. Hopke, K. Ito, T. Mar, W. Christensen, D. J. Eatough, R. C. Henry, E. Kim, F. Laden, T. V. Larson, H. Liu, L. Neas, J. Pinto, M. Stölzel, H. Suh, P. Paatero and G. D. Thurston, PM Source Apportionment and Health Effects: Intercomparison of Source Apportionment Results, *J. Expo. Sci. Environ. Epidemiol.*, 2005, **16**, 275–286.
87. C. George, M. Ammann, B. D'Anna, D. J. Donaldson and S. A. Nizkorodov, Heterogeneous, *Photochem. Atmos., Chem. Rev*, 2015, **115**, 4218–4258.
88. N. N. Maykut, J. Lewtas, E. Kim and T. V. Larson, Source Apportionment of PM2.5 at an Urban IMPROVE Site in Seattle, Washington, *Environ. Sci. Technol.*, 2003, **37**, 5135–5142.
89. W. Liu, P. K. Hopke and R. A. VanCuren, Origins of Fine Aerosol Mass in the Western United States Using Positive Matrix Factorization, *J. Geophys. Res. Atmos.*, 2003, **108**(D23), 4716, DOI: 10.1029/2003JD003678.
90. E. Kim, P. K. Hopke and E. Edgerton, Source Identification of Atlanta Aerosol by Positive Matrix Factorization, *J. Air Waste Mange. Assoc.*, 2003, **53**, 731–739.
91. E. Kim and P. K. Hopke, Source apportionment of fine particles at Washington, DC utilizing temperature resolved carbon fractions, *J. Air Waste Mange. Assoc.*, 2004, **54**, 773–785.
92. E. Kim, P. K. Hopke and E. S. Edgerton, Improving source identification of Atlanta aerosol using temperature resolved carbon fractions in Positive Matrix Factorization, *Atmos. Environ.*, 2004, **38**, 3349–3362.

93. W. Zhao and P. K. Hopke, Source Apportionment for Ambient Particles in the San Gorgonio Wilderness, *Atmos. Environ.*, 2004, **38**, 5901–5910.

94. W. Zhao and P. K. Hopke, Source Investigation for Ambient PM$_{2.5}$ in Indianapolis, IN, *Aerosol Sci. Technol.*, 2006, **40**, 898–909.

95. N. J. Pekney, C. I. Davidson, A. L. Robinson, L. Zhou, P. K. Hopke and D. Eatough, Major Source Categories for PM$_{2.5}$ in Pittsburgh using PMF and UNMIX, *Aerosol Sci. Technol.*, 2006, **40**, 910–924.

96. J.-H. Lee and P. K. Hopke, Apportioning sources of PM2.5 in St. Louis, MO using speciation trends network data, *Atmos. Environ.*, 2006, **40**(Suppl. 2), S360–S377.

97. J.-H. Lee, P. K. Hopke and J. R. Turner, Source Identification of Airborne PM-2.5 at the St. Louis – Midwest Supersite, *J. Geophys. Res.*, 2006, **111**, D10S10, DOI: 10.1029/2005JD006329.

98. I.-J. Hwang and P. K. Hopke, Comparison of Source Apportionments of Fine Particulate Matter at Two San Jose Speciation Trends Network Sites, *J. Air Waste Manage. Assoc.*, 2006, **56**, 1287–1300.

99. I.-J. Hwang and P. K. Hopke, Estimation of source apportionment and potential source locations of PM$_{2.5}$ at a west coastal IMPROVE site, *Atmos. Environ.*, 2007, **41**, 506–518.

100. Y. Wang, P. K. Hopke, O. V. Rattigan, D. C. Chalupa and M. J. Utell, Multi-year black carbon measurement and source apportionment using Delta-C in Rochester, NY, *J. Air Waste Manage. Assoc.*, 2012, **62**, 880–887.

101. Y. Wang, P. K. Hopke, X. Xia, D. C. Chalupa, Y. Zhang and M. J. Utell, Source Apportionment Using Positive Matrix Factorization on Daily Measurements of Organic and Inorganic Speciated PM2.5, *Atmos. Environ.*, 2012, **55**, 525–532.

102. E. Kim, S. G. Brown, H. R. Hafner and P. K. Hopke, Characterization of Non-Methane Volatile Organic Compounds Sources in Houston during 2001 using Positive Matrix Factorization, *Atmos. Environ.*, 2005, **39**, 5934–5946.

103. V. A. Lanz, A. S. H. Prevôt, M. R. Alfarra, S. Weimer, C. Mohr, P. F. Decarlo, M. F. D. Gianini, C. Hueglin, J. Schneider, O. Favez, B. D'Anna, C. George and U. Baltensperger, Characterization of aerosol chemical composition with aerosol mass spectrometry in Central Europe: an overview, *Atmos. Chem. Phys.*, 2010, **10**, 10453–10471.

104. I. M. Ulbrich, M. R. Canagaratna, Q. Zhang, D. R. Worsnop and J. L. Jimenez, Interpretation of organic components from Positive Matrix Factorization of aerosol mass spectrometric data, *Atmos. Chem. Phys.*, 2009, **9**, 2891–2918.

105. L. Zhou, P. K. Hopke, P. Paatero, J. M. Ondov, J. P. Pancras, N. J. Penney and C. I. Davidson, Advanced factor analysis for multiple time resolution aerosol composition data, *Atmos. Environ.*, 2004, **38**, 4909–4920.

106. L. Zhou, E. Kim, P. K. Hopke, C. Stanier and S. N. Pandis, Mining Airborne Particulate Size Distribution Data by Positive Matrix Factorization (PMF), *J. Geophys. Res.*, 2005, **110**, D07S19.

107. D. Ogulei, P. K. Hopke, D. Chalupa and M. Utell, Modeling Source Contributions to Submicron Particle Number Concentrations Measured in Rochester, NY, *Aerosol Sci. Technol.*, 2007, **41**, 179–201.

108. J. Kasumba, P. K. Hopke, D. Chalupa and M. J. Utell, Comparison of Sources of Submicron Particle Number Concentrations Measured at Two Sites in Rochester, NY, M, *Sci. Total Environ.*, 2009, **407**, 5071–5084.

109. C. A. Belis, F. Karagulian, B. R. Larsen and P. K. Hopke, Critical review and meta-analysis of ambient particulate matter source apportionment using receptor models in Europe, *Atmos. Environ.*, 2013, **69**, 94–108.

110. L. L. Ashbaugh, W. C. Malm and W. Z. Sadeh, A residence time probability analysis of sulfur concentrations at Grand Canyon National Park, *Atmos. Environ.*, 1985, **19**, 1263–1270.

111. Y. Zeng and P. K. Hopke, A study on the sources of acid precipitation in Ontario, Canada, *Atmos. Environ.*, 1989, **23**, 1499–1509.

112. Y. Wang, P. K. Hopke, O. V. Rattigan and X. Xia, Characterization of Residential Wood Combustion Particles Using the Two-Wavelength Aethalometer, *Environ. Sci. Technol.*, 2011, **45**, 7387–7393.

113. G. A. Allen, P. Babich and R. L. Poirot, 2004, Evaluation of a New Approach for Real-Time Assessment of Woodsmoke PM, in Proceedings of the Regional and Global Perspectives on Haze: Causes, Consequences, and Controversies, Air and Waste Management Association Visibility Specialty Conference, Asheville, NC, paper #16.

114. F. Amato, M. Pandolfi, A. Escrig, X. Querol, A. Alastuey, J. Pey, N. Perez and P. K. Hopke, Quantifying road dust resuspension in urban environment by multilinear engine: a comparison with PMF2, *Atmos. Environ.*, 2009, **43**, 2770–2780.

115. A. Escrig Vidal, E. Monfort, I. Celades, X. Querol, F. Amato, M. C. Minguillón and P. K. Hopke, Application of optimally scaled target factor analysis for assessing source contribution of ambient PM10, *J. Air Waste Manage. Assoc.*, 2009, **59**, 1296–1307.

116. F. Amato, M. Pandolfi, M. Viana, X. Querol, A. Alastuey and T. Moreno, Spatial and chemical patterns of PM10 in road dust deposited in urban environment, *Atmos. Environ.*, 2009, **43**, 1650–1659.

117. P. Paatero and P. K. Hopke, Rotational tools for factor analytic models, *J. Chemom.*, 2009, **23**, 91–100.

118. F. Amato and P. K. Hopke, Source apportionment of the ambient $PM_{2.5}$ across St. Louis using constrained positive matrix factorization, *Atmos. Environ.*, 2012, **46**, 329–337.

119. J. M. Ondov, J. P. Pancras, N. Poor, J. R. Turner, M. N. S. Yu, E. Lipsky, E. Weitkamp and A. Robinson, 2003, PM Emission Rates from Highly Time-Resolved Ambient Concentration Measurements, 2003 Annual Meeting of the American Association for Aerosol Research (October 2003, Anaheim, CA).

120. J. J. Schauer, G. C. Lough, M. M. Shafer, W. F. Christensen, M. F. Arndt, J. T. DeMinter and J.-S. Park, 2006, Characterization of metals emitted from motor vehicles, Health Effects Institute, MA.

121. U.S. Environmental Protection Agency, 2014, EPA Positive Matrix Factorization (PMF) 5.0 Fundamentals and User Guide, available at http://www.epa.gov/heasd/documents/EPA%20PMF%205.0%20User%20Guide.pdf.

122. U. M. Sofowote, Y. Su, E. Dabek-Zlotorzynska, A. K. Rastogi, J. Brook and P. K. Hopke, Constraining the Factor Analytical Solutions Obtained from Multiple-Year Receptor Modeling of Ambient $PM_{2.5}$ Data from Five Speciation Sites in Ontario, Canada, *Atmos. Environ.*, 2015, **108**, 151–157.

123. U. M. Sofowote, Y. Su, E. Dabek-Zlotorzynska, A. K. Rastogi, J. Brook and P. K. Hopke, Sources and Temporal Variations of Constrained PMF Factors Obtained from Multiple-Year Receptor Modeling of Ambient PM2.5 Data from Five Speciation Sites in Ontario, Canada, *Atmos. Environ.*, 2015, **108**, 140–150.

124. P. K. Hopke, Z. Ramadan, P. Paatero, G. Norris, M. Landis, R. Williams and C. W. Lewis, Receptor Modeling of Ambient and Personal Exposure Samples: 1998 Baltimore Particulate Matter Epidemiology-Exposure Study, *Atmos. Environ.*, 2003, **37**, 3289–3302.

125. W. Zhao, P. K. Hopke, G. Norris, R. Williams and P. Paatero, Source Apportionment and Analysis on Ambient and Personal Exposure Samples with a Combined Receptor Model and an Adaptive Blank Estimation Strategy, *Atmos. Environ.*, 2006, **40**, 3788–3801.

126. W. Zhao, P. K. Hopke, E. W. Gelfand and N. Rabinovitch, Use of an Expanded Receptor Model for Personal Exposure Analysis in Schoolchildren with Asthma, *Atmos. Environ.*, 2007, **41**, 4084–4096.

127. P. J. Lioy, M. P. Zelenka, M. D. Cheng, N. M. Reiss and W. E. Wilson, The effect of sampling duration of the ability to resolve source types using factor analysis, *Atmos. Environ.*, 1989, **23**, 239–254.

128. L. Zhou, P. K. Hopke, P. Paatero, J. M. Ondov, J. P. Pancras, N. J. Penney and C. I. Davidson, Advanced factor analysis for multiple time resolution aerosol composition data, *Atmos. Environ.*, 2004, **38**, 4909–4920.

129. D. Ogulei, P. K. Hopke, P. Paatero, S.-S. Park and J. M. Ondov, Receptor modeling for highly-time resolved species: The Baltimore supersite, *Atmos. Environ.*, 2005, **39**, 3751–3762.

130. H.-T. Liao, C.-P. Kuo, P. K. Hopke and C.-F. Wu, Evaluation of a modified receptor model for solving multiple time resolution equations: a simulation study, *Aerosol Air Qual. Res.*, 2013, **13**, 1253–1262.

131. H.-T. Liao, C. C.-K. Chou, J. C. Chow, J. G. Watson, P. K. Hopke and C.-F. Wu, Source and risk apportionment of selected VOCs and PM2.5 species using partially constrained receptor models with multiple time resolution data, *Environ. Pollut.*, 2015, **205**, 121–130.

132. P. A. Paatero, P. K. Hopke, J. Hoppenstock and S. Eberly, Advanced Factor Analysis of Spatial Distributions of PM2.5 in the Eastern U.S, *Environ. Sci. Technol.*, 2003, **37**, 2460–2476.

133. W. Chueinta, P. K. Hopke and P. Paatero, Multilinear Model for Spatial Pattern Analysis of the Measurement of Haze and Visual Effects Project, *Environ. Sci. Technol.*, 2004, **38**, 544–554.

134. L. A. Barrie and R. M. Hoff, Five years of air chemistry observations in the Canadian Arctic, *Atmos. Environ.*, 1985, **19**, 1995–2010.

135. L. A. Barrie and M. J. Barrie, Chemical Components of Lower Tropospheric Aerosols in the High Arctic: Six Years of Observations, *J. Atmos. Chem.*, 1990, **11**, 211–226.

136. Y.-L. Xie, P. K. Hopke, P. Paatero, L. A. Barrie and S.-M. Li, Identification of Source Nature and Seasonal Variations of Arctic Aerosol by Positive Matrix Factorization, *J. Atmos. Sci.*, 1999, **56**, 249–260.

137. Y.-L. Xie, P. K. Hopke, P. Paatero, L. A. Barrie and S.-M. Li, Identification of Source Nature and Seasonal Variations of Arctic Aerosol by the Multilinear Engine, *Atmos. Environ.*, 1999b, **33**, 2549–2562.

138. S. M. Li and L. A. Barrie, Biogenic Sulfur in the Arctic Troposphere 1, Contributions to Total Sulfate, *J. Geophys. Res.*, 1993, **98**, 20613–20622.

139. S. M. Li, L. A. Barrie and A. Sirois, Biogenic sulphur aerosol in the Arctic troposphere: 2. Trends and seasonal variations, *J. Geophys. Res.*, 1993, **98**, 20623–20631.

140. P. K. Hopke, L. A. Barrie, S.-M. Li, C. Li, M.-D. Cheng and Y. Xie, Possible Sources and Preferred Pathways for Biogenic and Non-Seasalt Sulfur for the High Arctic, *J. Geophys. Res.*, 1995, **100**, 16595–16603.

141. P. D. Jones, T. M. L. Wigley and P. B. Wright, 1997, Global and Hemispheric Annual Temperature Variations Between 1854 and 1991, Data set no. NDP022R2 available from CDIAC, ORNL.

142. G. E. Shaw, Bio-Controlled Thermostasis Involving the Sulfur Cycle, *Clim. Change*, 1983, **5**, 297–303.

143. R. J. Charlson, J. E. Lovelock, M. O. Andreae and S. G. Warren, Oceanic Phytoplankton, Atmospheric Sulphur, Cloud Albedo, and Climate, *Nature*, 1987, **326**, 655–661.

144. O. G. Raabe, D. A. Braaten, R. L. Axelbaum, S. V. Teague and T. A. Cahill, Calibration studies of the DRUM impactor, *J. Aerosol Sci.*, 1988, **19**, 183–195.

145. A. Knochel, Basic principles of XRF with synchrotron radiation, *2nd International Workshop on XRF and PIXE Applications in Life Science, Capri, Italy*, World Scientific Publishing Co, Singapore, 1989.

146. J. A. Dodd, J. M. Ondov, G. Tuncel, T. G. Dzubay and R. K. Stevens, Multimodal size spectra of submicrometer particles bearing various elements in rural air, *Environ. Sci. Technol.*, 1991, **25**, 890–903.

147. D. S. Burdick, An introduction to tensor products with applications to multiway data analysis, *Chemom. Intell. Lab. Syst.*, 1995, **28**, 229–237.

148. H. A. L. Kiers, Towards a standardized notation and terminology in multiway analysis, *J. Chemom.*, 2000, **14**, 105–122.
149. E. Pere-Trepat, P. K. Hopke and P. Paatero, Source Apportionment of Time and Size Resolved Ambient Particulate Matter Measured with a Rotating DRUM Impactor, *Atmos. Environ.*, 2007, **41**, 5921–5933.
150. N. Li, P. K. Hopke, P. Kumar, S. S. Cliff, Y. Zhao and C. Navasca, Source apportionment of time- and size-resolved ambient particulate matter, *Chemom. Intell. Lab. Syst.*, 2013, **129**, 15–20.

Case Studies of Source Apportionment and Suggested Measures at Southern European Cities

F. AMATO,* F. LUCARELLI, S. NAVA, G. CALZOLAI, A. KARANASIOU,
C. COLOMBI, V. L. GIANELLE, C. ALVES, D. CUSTÓDIO, K. ELEFTHERIADIS,
E. DIAPOULI, C. RECHE, A. ALASTUEY, M. C. MINGUILLÓN, M. SEVERI,
S. BECAGLI, T. NUNES, M. CERQUEIRA, C. PIO, M. MANOUSAKAS,
T. MAGGOS, S. VRATOLIS, R. M. HARRISON AND X. QUEROL

ABSTRACT

This chapter reports the results of the PM_{10} and $PM_{2.5}$ source apportionment at 3 urban background sites (Barcelona, Florence and Milan, BCN-UB, FI-UB, MLN-UB) 1 suburban background site (Athens, ATH-SUB) and 1 traffic site (Porto, POR-TR). Road traffic (sum of vehicle exhaust, non-exhaust and traffic-related secondary nitrate) is the most important source of PM_{10} (23–38% at all sites) and $PM_{2.5}$ (22–39%, except for ATH-SUB and BCN-UB). The second most important source of PM_{10} (20–26%) is secondary sulphate/OC at BCN-UB, FI-UB and ATH-SUB, while it represents 10–14% in MLN-UB and POR-TR. The relative importance of this source is higher in $PM_{2.5}$ (19–37% at SUB-UB sites). Biomass burning contributions vary widely from 14–24% of PM_{10} in POR-TR, MLN-UB and FI-UB, 7% in ATH-SUB to <2% in BCN-UB. In $PM_{2.5}$, BB is the second most important source in MLN-UB (21%) and in POR-TR (18%). This large variability is due to the degree of penetration of biomass for residential heating. Other significant sources are local dust, industries (metallurgy), remaining

---

*Corresponding author.

Issues in Environmental Science and Technology No. 42
Airborne Particulate Matter: Sources, Atmospheric Processes and Health
Edited by R.E. Hester, R.M. Harrison and X. Querol
© The Royal Society of Chemistry 2016
Published by the Royal Society of Chemistry, www.rsc.org

secondary nitrate (from industries, shipping and power generation), sea salt and Saharan dust. The same analysis is performed for exceedances days. Based on the above, a priority list of measures to improve PM levels is proposed for each city.

1 Introduction

Atmospheric Particulate Matter (PM) concentrations can vary widely across Europe owing to different climatic conditions and local features, such as anthropogenic source types, emission rates and dispersion patterns. Moreover, natural contributions, such as biogenic aerosols, forest fires, sea salt and Saharan dust intrusions, impact differently from one region to another.[1–4,18]

Urban PM_{10} concentrations show significant variability across Europe as reported from research studies[5–7] and by routine monitoring networks.[8] For $PM_{2.5}$, the spatial variability across Southern Europe is less known because in most air quality zones it is not as widely measured as PM_{10}. As a consequence, there is limited information on the geographical variability of the coarse fraction ($PM_{10-2.5}$), which is often linked to local sources and whose evidence of health concern is increasing.[9]

Comparability of data is also hampered by the fact that most research studies analyzed PM data from different periods or with different sampling calendars. An example of this is given by the multi-city studies aimed at investigating the short- and long-term health effects of exposure to PM_x mass concentrations, NO_x and SO_2.[10–13]

Moreover, the comparison of bulk PM concentrations only, without the necessary chemical characterization of collected samples and source apportionment analysis, does not allow for an in-depth investigation of sources, limiting the scope for air quality management purposes.

For instance, PM_{10} and $PM_{2.5}$ mass and particle number were continuously measured for 18 months in urban background locations across Europe to determine the spatial and temporal variability of particulate matter in Helsinki, Athens, Amsterdam and Birmingham, but no information on PM composition and sources was provided.[7] Another study[5] compared the PM_{10} and $PM_{2.5}$ levels and chemistry of seven selected EU regions with at least one year of data coverage at regional, urban background and kerbside sites. However, available datasets were not simultaneous, within 1998 to 2002. Kukkonen *et al.*[6] analyzed in detail four selected episodes involving substantially high concentrations of PM_{10} that occurred in Oslo, Helsinki, London and Milan but in different periods. More recently, the ESCAPE project[10] investigated the health effects of long-term exposure to ambient air pollution across Europe. $PM_{2.5}$, PM_{10} and particle composition were compared at 20 sites across 2008–2011, but measurements were done 3 times for 14 days in different seasons without covering the full year period.

The AIRUSE LIFE project created the first harmonized dataset of Southern European cities for PM_{10} and $PM_{2.5}$ levels and composition, following the same sampling protocol and 1 year (2013) calendar in Barcelona (Spain), Porto (Portugal), Florence and Milan (Italy), and Athens (Greece). The goal is to characterize the similarities and heterogeneities in PM sources and contributions across the Mediterranean region. Originally AIRUSE focused on 4 cities, but Milan was added to the list given the high demographic and scientific interest of this big city and the offer of ARPA Lombardia to collaborate in the project.

Once the main sources of PM_{10} and $PM_{2.5}$ are identified, the strategic goal of the AIRUSE project is to test and develop specific measures to improve air quality in Southern Europe, targeted to meet air quality standards and to approach as closely as possible the WHO guidelines.

2 Methods

2.1 PM Sampling and Measurements

PM measurements were carried out from January 2013 to February 2014, simultaneously at five urban stations in Barcelona, Porto, Florence, Milan and Athens. PM_{10} and $PM_{2.5}$ samples were collected simultaneously over 24 hours, every third day, on quartz microfiber and/or Teflon filters. Furthermore, in order to evaluate the chemical fingerprint of Saharan dust, additional PM_{10} and $PM_{2.5}$ samplings were performed at each city under selected Saharan dust intrusions after forecasting the occurrence of this phenomenon. The forecast was based on the interpretation of: (i) air mass back trajectories calculated with the HYSPLIT4 model from NOAA; and (ii) predictions of dust concentrations by the SKIRON model, University of Athens (http://forecast.uoa.gr.html) and Barcelona Supercomputing Center (NMMB/BSC-Dust forecasts) prediction models.

For shorter periods, aerosol was also sampled with the Streaker sampler, which allows the collection of the coarse and fine aerosol (*i.e.* $PM_{2.5-10}$ and $PM_{2.5}$, respectively) with 1 hour time resolution.[14] The details of each monitoring site and instrumentation used are described below:

- **BCN-UB: Barcelona (Spain).** PM_{10} and $PM_{2.5}$ were collected by means of sequential DIGITEL DH1080 high volume samplers on quartz fiber filters at the Palau Reial station ($41°23'14''$N, $2°6'56''$E). This is an urban background (UB) site located within the University Campus (South West part of the city) and part of the local air quality network. The nearest trafficked road (Diagonal Avenue, 90 000 vehicles per day) is located 200 m away.
- **POR-TR: Porto (Portugal).** The urban traffic (TR) station is located in Praça Francisco Sá Carneiro ($41°09'46.10''$ N, $8°35'26.95''$ W) and is part of the National Air Quality Network, QualAr. It is located in the eastern side of Porto city, next to the Fernão de Magalhães Avenue and at 600 meters from the Inner Circular Motorway.
- **FI-UB: Florence (Italy).** The urban site Bassi is an air quality UB monitoring station ($43°47'8.33''$N, $11°17'13.19''$E) of the Environmental

Protection Agency of Tuscany. PM_{10} and $PM_{2.5}$ samples were collected by means of two low volume sequential samplers.

- **ATH-SUB: Athens (Greece).** The Demokritos sub urban background station is located in NCSR "Demokritos" campus ($37°.99'50''°N$, $23°.81'60''°E$), at the North East corner of the Greater Athens Metropolitan Area and at an altitude of 270 m a.s.l. The suburban site is away from direct emission sources in a vegetated area (pine). PM_{10} and $PM_{2.5}$ samples were collected on Teflon filters by means of low volume samplers and on quartz microfiber filters by means of high volume samplers.
- **MLN-UB: Milan (Italy).** The MI-Pascal urban background station is part of the ARPA Lombardia Air Quality Network, and it is one of the Italian Supersites (2008/50/CE). It is located in the eastern side of Milan, the University area called "Città Studi" ($45°28'44''$ N, $9°14'07''$ E), in a playground about 130 m from the road traffic.

In addition to these fixed long-term sampling campaigns, shorter sampling campaigns were performed at traffic sites in Barcelona (BCN-TR) and Athens (ATH-TR).

- **BCN-TR: Barcelona (Spain).** The traffic site was installed during April–May 2013 in Valencia Road (typically 11 000 vehicles per day, $41°24'11.08''N$, $2°10'37.60''E$). Valencia Road is a 19 m wide five-lane road, with the outer right lane exclusive for buses and taxis and one additional parking lane on the left side. The orientation ($45°–225°$) of the road and the 7-story buildings may generate a street canyon effect during the early afternoon hours. $PM_{2.5}$ samples from BCN-TR were only used for the determination of the PM mass concentration and were not processed for chemical analysis, which was done with PM_{10}.
- **ATH-TR: Athens (Greece).** The Aristotelous Monitoring station of the National Air Quality Network was selected as a typical traffic impacted site. The station is located at the 1st floor open balcony of the Ministry of Health facing a busy crossroad ($37.99°N$, $23.72°E$, at an altitude of 64 m a.s.l.). The urban traffic site is located at the Athens commercial center and it is mainly affected by vehicle emissions.

2.2 Sample Treatment and Analysis

Before sampling, quartz microfiber filters were dried for 5 h and conditioned for 48 h at 20 °C and 50% relative humidity and weighed by means of a microbalance (1 μg sensitivity). After sampling, filters were brought back to the laboratory to be weighed at the same temperature and relative humidity as the first weighing. Then the filters were destined for several analytical determinations:[15]

- Major elements and trace elements were determined:
  - In Teflon filters by Particle Induced X-Ray Emission (PIXE)[16] or by Inductively Coupled Plasma Mass Spectrometry (ICP-MS) and

Inductively Coupled Plasma Atomic Emission Spectroscopy (ICP-AES)[17] after acid digestion (5 ml HF, 2.5 ml HNO_3, 2.5 ml $HClO_4$).

- ○ In quartz filters by ICP-MS and ICP-AES after acid digestion (5 ml HF, 2.5 ml HNO_3, 2.5 ml $HClO_4$).[17]
- ○ In Teflon and MCE filters by XRF (X-ray Fluorescence).
- Water soluble ions by IC (Ion Chromatography).
- On quartz filters organic carbon (OC) and elemental carbon (EC) by thermal/optical analysis with the EUSAAR2 temperature program by means of Sunset analyzers. Filters from POR-TR were analyzed in a thermo-optical transmission system described in detail elsewhere.[19,20]
- On the PM_{10} quartz microfiber filters, Carbonate Carbon (CC), by means of the procedure described by Pio *et al.*[21]
- On the $PM_{2.5}$ quartz filters levoglucosan, by means of Ion Chromatography.

As concerns the samples with hourly resolution, collected with the Streaker sampler, they were analyzed by PIXE at the LABEC laboratory of Florence (Italy) allowing for the determination of the elemental concentrations (atomic number $Z > 10$) with hourly time resolution.

A summary of the performed analyses is reported in Tables 1 and 2.

2.3 Source Apportionment

Source apportionment studies of atmospheric particulate matter are often performed by means of receptor models that are based on the mass conservation principle:

$$x_{ij} = \sum_{k=1}^{p} g_{ik} f_{jk} \quad i = 1, 2, \ldots, m \quad j = 1, 2 \ldots n \tag{1}$$

Table 1 Summary of the number and type of samples collected for each city and analytical technique during the one-year measurements.

			BCN-UB	FI-UB	MLN-UB	POR-TR	ATH-SUB
Daily	PM_{10}	**Mass**	**122**	**226**	**379**	**123**	**197**
		Elements	125	226	241	123	197
		Ions	122	226	337	123	197
		ECOC	122	226	348	123	197
		CC	122	226	89	123	197
		Levoglucosan			324		
	$PM_{2.5}$	**Mass**	**126**	**243**	**378**	**126**	**243**
		Elements	126	243	361	126	243
		Ions	126	243	374	126	243
		ECOC	126	243	370	126	243
		Levoglucosan	126	243	356	126	243
Hourly	$PM_{2.5-10}$	Elements		640		800	480
	$PM_{2.5}$	Elements		640		800	480

Table 2 Summary of the number and type of samples collected for each city and analytical technique during the short time campaigns focusing on traffic sites.

Traffic site campaigns			BCN-TR	ATH-TR
Daily	PM_{10}	**Mass**	**59**	**64**
		Elements	59	64
		Ions	59	64
		ECOC	31	78
	$PM_{2.5}$	**Mass**	**59**	**64**
		Elements		64
		Ions		64
		ECOC		64
		Levoglucosan		14
Hourly	$PM_{2.5-10}$	Elements	210	480
	$PM_{2.5}$	Elements	210	480

where x_{ij} is the concentration of the species *j in the ith sample*, g_{ik} is the contribution of the source k^{in} *the ith sample* and f_{jk} is the concentration of the species *j* in source *k*. When both g_{ik} and f_{jk} are unknown, factor analysis (FA) techniques such as Principal Components Analysis (PCA)[22,23] and Positive Matrix Factorization (PMF)[24] are used for solving eqn (1). In this study, the US EPA PMF v5 (http://www.epa.gov/heasd/research/pmf.html) was applied to the five datasets obtained at BCN-UB, FI-UB, ATH-SUB, MLN-UB and POR-TR. Since PMF is a weighted least-squares method, individual estimates of the uncertainty in each data value are needed. The uncertainty estimates were based on the approaches by Polissar *et al.*[25] and Amato *et al.*[26] Species with S/N < 0.2 are generally defined as bad variables and removed from the analysis, and species with 0.2 < S/N < 2 are generally defined as weak variables and down weighted by a factor of 3.[27] Nevertheless, since S/N is very sensitive to sporadic values much higher than the level of noise, the percentage of data above the detection limit was used as a complementary criterion.

In those cases where the PMF model was not able to resolve a Saharan dust source, the Saharan dust contribution was estimated according to the methodology proposed by SEC.[28]

Once the source contributions are obtained for hourly data, these were plotted in polar coordinates as a function of wind direction and speed by means of the PolarPlot function available in the OpenAir software.[29,30]

3 Results

3.1 PM Levels and Seasonality

The observed mean PM_x levels during the study period (January 2013 to February 2014) were as follows:

- PM_{10} in the urban background (UB) reached 19–22 µg m^{-3} in FI and BCN, and 39 µg m^{-3} in MLN. In the suburban background (SUB) site in

ATH, levels reached 20 $\mu g\ m^{-3}$, whereas at the traffic (TR) sites of POR, BCN and ATH concentrations reached 34 $\mu g\ m^{-3}$.

- For $PM_{2.5}$, levels reached 13–15 $\mu g\ m^{-3}$ at the UB sites of FI and BCN, 31 $\mu g\ m^{-3}$ in MLN, 11 $\mu g\ m^{-3}$ in SUB site of Athens, 20 $\mu g\ m^{-3}$ in TR site of ATH and 26 $\mu g\ m^{-3}$ in POR-TR and BCN-TR.

All sites meet EU legal requirements (2008/50/EC) for the annual limit value of PM_{10} (40 $\mu g\ m^{-3}$). However, the legal requirement of not exceeding 50 $\mu g\ m^{-3}$ for the percentile 90.4 of the annual values (daily limit value) was exceeded at MLN-UB (72 $\mu g\ m^{-3}$), POR-TR (52 $\mu g\ m^{-3}$) and ATH-TR (52 $\mu g\ m^{-3}$). The annual EU target value of $PM_{2.5}$ (25 $\mu g\ m^{-3}$, 2008/50/EC) was exceeded at POR-TR (26 $\mu g\ m^{-3}$) and MLN-UB (31 $\mu g\ m^{-3}$), as well as in the short measurement campaign of BCN-TR (26 $\mu g\ m^{-3}$).

For the UB sites, mean levels are similar in the cities with the exception of MLN with relatively higher levels, owing to intense local and regional PM source contributions and to the specific meteorology of the Po Valley, with frequent and intensive atmospheric thermal inversions that induce regional accumulation of pollutants.

For the TR sites very similar mean PM_{10} levels were recorded at ATH, BCN and POR ($\mu g\ m^{-3}$), and in the first two cities these exceeded by around 55% the PM_{10} levels measured at SUB and UB sites, respectively. This highlights the relevance of traffic emissions for PM_{10} levels at the study cities.

For $PM_{2.5}$, similar and relatively lower values (as compared with PM_{10}) were recorded at UB and SUB sites (11–15 $\mu g\ m^{-3}$), with the exception of MLN (31 $\mu g\ m^{-3}$). In the case of the TR sites, mean levels reached 26 and 20 $\mu g\ m^{-3}$ for BCN and ATH, respectively, whereas for POR these reached 27 $\mu g\ m^{-3}$. Since levels of PM_{10} were very similar in the three TR sites (34 $\mu g\ m^{-3}$ in all cases), it is expected to be additional non-traffic related source contributions that account for the POR and BCN $PM_{2.5}$ increase with respect to ATH-TR.

UB and SUB $PM_{2.5}/PM_{10}$ ratios were close to 0.7 in the case of FI and BCN, 0.5 for ATH and 0.8 for MLN. At TR sites these ratios increased up to 0.8 in BCN and POR and 0.6 in ATH. Although at TR sites levels of resuspension of road dust are expected to be higher than at UB sites (and consequently the $PM_{2.5}/PM_{10}$ are lower), the proximity to the exhaust emissions (dominated by fine PM) causes a shift of the PM load to the fine fraction.

When considering the stricter WHO guidelines, all cities exceeded both the PM_{10} and $PM_{2.5}$ annual mean thresholds, with the exception of PM_{10} in FI. It is worth noting that the sampling year was particularly rainy in most of the study sites, including Florence. The daily $PM_{2.5}$ WHO threshold was not attained only in POR.

Seasonal trends for PM_{10} at the 5 AIRUSE cities were very different. ATH-SUB was characterized by higher spring–summer and lower autumn–winter PM_{10} levels, probably owing to higher African dust influence in the warm seasons. A similar but much smoother trend was observed for BCN-UB. For POR-TR, the highest levels were recorded in summer and the lowest in spring. At MLN-UB, lower levels were recorded in spring and summer, when PM_{10} levels reached similar levels to BCN, FI or ATH; however in autumn and winter levels at MLN were higher by a factor of more than 2 with respect to the other seasons and most of the other cities. This is owing to the aforementioned atmospheric stagnation features of the Po Valley that favored the accumulation of intensive winter PM emissions. At FI-UB a similar trend was observed but with a less pronounced winter increase than that described for MLN. In fact, Florence is also located in a closed basin (the Arno river valley), which is characterized by stagnant conditions during the cold season. $PM_{2.5}$ levels followed similar seasonal patterns to those described above for PM_{10} at each city.

3.2 PM Chemical Characterization

EC traces the influence of emissions from traffic, mainly from diesel vehicles;[31] it is a carrier of highly health relevant organic species.[32] Mean levels of EC reached 1.1 µg m^{-3} in PM_{10} at the UB sites of BCN and FI and 1.9 µg m^{-3} at the UB site of MLN (Figure 1). At the SUB site of ATH, EC levels were lower (0.4 µg m^{-3} in PM_{10}, Figure 1) owing to the higher distance from traffic ways, but also owing to the lower proportion of diesel vehicles in the fleet of ATH compared to the other 3 cities.

Until recently (2012), use of diesel for private cars was not allowed in this city. As expected, levels of EC at traffic sites were higher than at UB sites by a factor ranging from 3.6 to 4.9. EC levels recorded for PM_{10} are very close to those of $PM_{2.5}$ ($PM_{2.5}/PM_{10}$ ratios for EC levels from 0.9 to 1.0) pointing to the very fine size of the EC-bearing diesel soot particles. The results revealed thus that EC (mostly from diesel soot) levels are increased by a factor from 4 to 5 in the TR sites with respect to the UB and SUB sites. Organic matter (OM) was calculated by multiplying OC by a factor accounting for non-C atoms, which varies according to location and existing sources. This factor ranged within 1.2–1.8. Mean levels of organic matter (OM, Figure 1) reached 5.0 and 5.5 µg m^{-3} in PM_{10} at the UB and SUB sites of BCN and ATH, but increased up to 9.8 and 15.7 µg m^{-3} at the UB sites of FI and MLN. At traffic sites, levels of OM also increased with respect to the UB sites up to values of 7.9 to 13.6 µg m^{-3} at BCN, POR and ATH. OM also has a fine grain size since $PM_{2.5}/PM_{10}$ ratios for this component ranged from 0.7 to 1.0. OM/EC ratios for PM_{10} ranged from 1.6 to 4.1 at the TR sites, reaching values of 4.4 at the UB site of BCN, 8.2 and 8.7 at the UB sites of MLN and FI and 13 at the SUB site of ATH. This ratio indicates the proximity to the emission sources but also the possible influence of biomass burning. Thus primary diesel soot is characterized by an OM/EC ratio close to 1.0, whereas a higher ratio indicates a major relative abundance of secondary OM, owing to a longer distance from emission hotspots (from TR to UB to SUB) and/or a higher contribution from biomass burning to PM_x levels, since OM/EC from biomass burning is much higher.[33] In addition, biogenic emissions can increase the OM/EC ratio. Levels of OM + EC accounted for around 27–30%

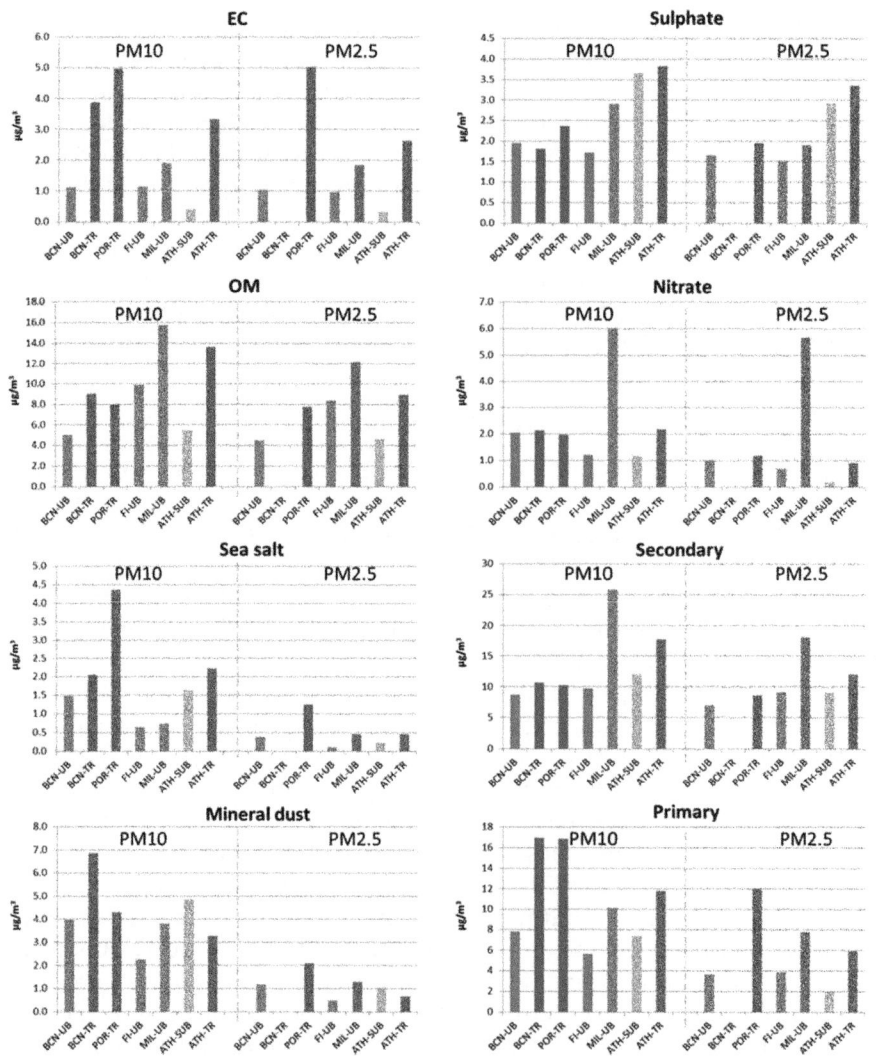

Figure 1   Levels of PM components in PM_{10} and $PM_{2.5}$ in the five AIRUSE cities.

of the PM_{10} load at the UB and SUB sites, with the exception of FI and MLN (45–58%). At the TR sites this contribution increased with respect to the respective UB sites by a factor of 40 to 100%. In $PM_{2.5}$ the OM + EC load increased as compared with PM_{10} and reached from 37 to 69% of the $PM_{2.5}$ mass. This is mainly owing to the decrease of mineral dust and sea salt in $PM_{2.5}$ when compared with PM_{10} (see following sections). The higher levels of OM recorded at MLN, FI, and partially at POR, are probably the result of a high contribution to PM_{10} levels from biomass burning (BB). This is also supported by the high OC/EC ratios. Biomass burning PM emissions are characterized by high OM loads but also by a high OC/EC ratio when compared with traffic emissions. In the case of MLN, the frequent atmospheric stagnation episodes also favor the formation of secondary OM, as is also described later for nitrate. To a lesser extent this also applies for FI, where stagnant conditions are common during winter.

Sea salt was calculated based on the main composition of sea spray (ss) as the sum: $ssNa + Cl + ssMg + ssK + ssCa + ssSO_4^{2-}$. Mean levels of sea salt (Figure 1) in PM_{10} reached 0.6–0.7 $\mu g\ m^{-3}$ at the inland Italian cities (FI and MLN, 2–3% of the PM_{10} load) and 1.5–1.6 $\mu g\ m^{-3}$ (7–8% of the PM_{10} load) at the Mediterranean coastal sites. However, at the Atlantic site (POR) much higher levels were recorded (4.4 $\mu g\ m^{-3}$ in PM_{10}, 13% of the PM_{10} load at the TR site). As expected, owing to the coarse mass size distribution of sea salt, levels were reduced by a 71–86% in $PM_{2.5}$ with respect to PM_{10}, with the exception of MLN, where levels of sea salt were only reduced by around 38% in $PM_{2.5}$.

Mineral dust was calculated based on average crust composition as: $Al_2O_3 + SiO_2 + (1.42 \times Fe) + (1.94 \times Ti) + (1.2 \times nssK) + nssMg + nssCa + CO_3^{2-}$ (nss corresponds to non-sea spray fraction). Mineral dust at UB sites reached around 2.2 $\mu g\ m^{-3}$ as annual mean in FI-UB and 3.8–4.8 $\mu g\ m^{-3}$ at the UB and SUB sites of BCN, MLN and ATH (Figure 1). These levels account for 18 and 25% of the PM_{10} mass at BCN and ATH and 12 and 10% at FI and MLN, respectively. At the TR sites levels of mineral dust increased by around 10 to 70% with respect to UB sites as a consequence of the road dust emissions. At TR sites, mineral dust accounts for 10–20% of the PM_{10}. As also expected from the coarse mode of occurrence of mineral dust, levels of this PM component were much lower in $PM_{2.5}$, down to 20–49% of the PM_{10} levels. The contribution of dust reached from 4 to 10% of $PM_{2.5}$ for all sites.

There is a clear trend to markedly increase levels of sulfate from FI-BCN to POR-MLN to ATH (1.7–1.9 to 2.4–2.9 to 3.7 $\mu g\ m^{-3}$ in PM_{10} and 1.5–1.7 to 1.9–2.0 to 2.9 $\mu g\ m^{-3}$ in $PM_{2.5}$, Figure 1). This is probably owing to the influence of the use of coal and petroleum coke/fuel oil for power generation in or/and around the high sulfate regions of this study, but it may also be caused by the influence of SO_2 emissions from petrochemical plants. In the case of POR, when the site is under the influence of NW winds, one of the major sources of sulfate could be the Porto Refinery, which began operating in 1970. It is a crude oil industrial processing plant that has an annual installed capacity of 4.5 million tons and produces a wide range of products, including fuels, lubricants, aromatics (BTX) for the petrochemical industry, industrial solvents and petroleum waxes. Emissions from shipping in the harbor may represent another possible source. As expected from the fine mode of occurrence of ammonium sulfate $((NH_4)_2SO_4$ or $NH_4HSO_4)$, 76 to 85% of sulfate in PM_{10} is present in $PM_{2.5}$.

Levels of nitrate in PM_{10} show a less marked spatial variability, with the exception of MLN, with mean annual levels reaching 1.2 (FI-ATH) to 2.0 (BCN-POR) $\mu g\ m^{-3}$ and 6.0 $\mu g\ m^{-3}$ in MLN. This marked difference between MLN and the rest of the other AIRUSE regions is mainly owing to the specific meteorological and emission patterns of the Po Valley. In this case, it coincides with a large urban and industrial agglomeration (with the associated road traffic), with the consequent elevated atmospheric emissions and a peculiar meteorology favoring frequent and marked thermal inversions that cause the accumulation of pollutants and the formation of high levels of ammonium nitrate (NH_4NO_3) from the high anthropogenic NO_x and NH_3 emissions. In particular, high NH_3 levels, emitted from agricultural and animal husbandry activities, can be transported from the southern part of the Po Valley to the urbanized northern part, inducing ammonium nitrate formation. On the other hand, the relatively coarse mode of nitrate in all cases is noticeable, since levels in $PM_{2.5}$ are generally 50–60% lower than in PM_{10}, with the exception of MLN by 40% and the ATH SUB area by 85%. These large fractions of coarse nitrate are probably owing to the high temperatures and dry conditions reached in summer in the study regions. At ambient temperature exceeding 25 °C, NH_4NO_3 may be decomposed in gaseous HNO_3 and NH_3, and in turn a fraction of this HNO_3 may react with NaCl or $CaCO_3$ to give coarse $NaNO_3$ or $Ca(NO_3)_2$. This likely accounts for the large differences observed for nitrate in the coarse and fine size fractions at MLN and ATH, with 60% of nitrate residing in the fine aerosol at the first site and only 15% at the latter. The hypothesis of a temperature-driven behaviour of nitrate is reinforced by the seasonal trends of the ratios between fine ($PM_{2.5}$) nitrate and PM_{10} nitrate: most of the nitrate, *i.e.* 60–80% of it, is in $PM_{2.5}$ during fall and winter, while the percentage drastically decreases during spring–summer, down to roughly 15%. Seasonal trends are less and less pronounced when passing from cities with continental climates, such as FI (and MLN), to coastal ones with mild climates, such as POR and BCN, and it is not visible in ATH, with a warmer climate.

The contribution of sulfate, nitrate and ammonium (secondary inorganic aerosols, SIA) to the PM_{10} load reaches 15–21% of PM_{10} in BCN-UB, BCN-TR, POR-TR, ATH-TR and FI-UB, but 30–31% at both MLN-UB and ATH-SUB sites. In spite of the prevailing fine mode of these PM components, the contribution of SIA to the $PM_{2.5}$ load remains similar to that of PM_{10}, with 13–23% and 31–37% for the above groups of sites. The latter two cities (MLN and ATH) are both characterized by similarly high SIA loads, but owing to different causes, whereas at MLN SIA is dominated by ammonium nitrate, whilst at ATH ammonium sulfate prevails.

The secondary fraction of PM was obtained as:

$$\text{Secondary aerosols: } NO_3^- + SO_4^{2-} + NH_4^+ + (OC - (EC \times a)) \times b$$

where a expresses the ratio of primary OC/EC ratio (averaged among existing sources, and varying within 0.7–2.2)) and b accounts for the non-C atoms in secondary aerosol mass (varying within 1.6–2.1). Table 3 shows that for PM_{10}, at the UB and SUB sites the secondary fraction dominated the PM_{10} mass by contributing from 53% (BCN) to 71% (MLN), whereas at traffic sites the primary contribution prevailed (60–63%). For the calculation of the secondary and primary fractions, all values are normalized by the sum of

Table 3 Annual average concentration (%) of primary and secondary aerosol groups to PM_{10} and $PM_{2.5}$.

	Secondary organics	Secondary inorganics	POM + EC	Mineral dust	Sea salt
PM_{10}					
BCN-UB	25	28	14	24	9
FI-UB	36	20	27	13	4
MLN-UB	35	36	17	10	2
ATH-SUB	32	30	4	25	9
BCN-TR	23	16	29	25	7
ATH-TR	37	23	21	11	8
POR-TR	18	19	31	16	16
$PM_{2.5}$					
BCN-UB	34	32	20	11	3
FI-UB	49	21	25	4	1
MLN-UB	32	38	23	5	2
ATH-SUB	46	36	6	10	2
ATH-TR	38	29	27	4	2
POR-TR	23	19	42	10	6

reconstructed PM mass. This high load of secondary PM mass is very important to be taken into account in designing air quality plans, since these components are formed into the atmosphere from organic and inorganic gaseous precursors. The organic and inorganic secondary contributions to PM_{10} are very well balanced (close to 50–50% in most cases), but not in the case of PM_{10} in FI-UB, where 65% of the secondary PM_{10} load is from organic aerosols (Table 3), probably indicating a relative contribution from biomass burning compared with the other sources. The contributions of secondary PM to $PM_{2.5}$ increase when compared with PM_{10} owing to the lower contributions of dust and sea salt to the fine fraction. This secondary contribution then ranged from 66 (BCN) to 82% (ATH) in the UB and SUB sites and from 42 to 67% at the traffic sites (Table 3).

Levels of levoglucosan and K are usually considered as tracers of biomass burning.[34–36] In fact, levoglucosan is a sugar emitted into the atmosphere exclusively by breaking cellulose chains during biomass burning. Levoglucosan levels varied by one order of magnitude: 22 (BCN), 53 (ATH), 287 (FI), 303 (MLN) and 407 (POR) ng m^{-3} in $PM_{2.5}$ (Figure 2), reflecting a very clear difference of the impact of biomass burning on air quality across southern European cities. The values obtained should be considered as minimum levels because it is well known that levoglucosan may be degraded in high oxidizing environments.[37] This impact is mostly owing to differences in using biomass burning for domestic purposes, but peak events were also detected in summer in POR and MLN as a consequence of the impact of the emissions of forest fires and/or agricultural fires. The decreasing impact of biomass burning on $PM_{2.5}$ levels when passing from MLN to FI to POR to ATH and to BCN is clearly demonstrated

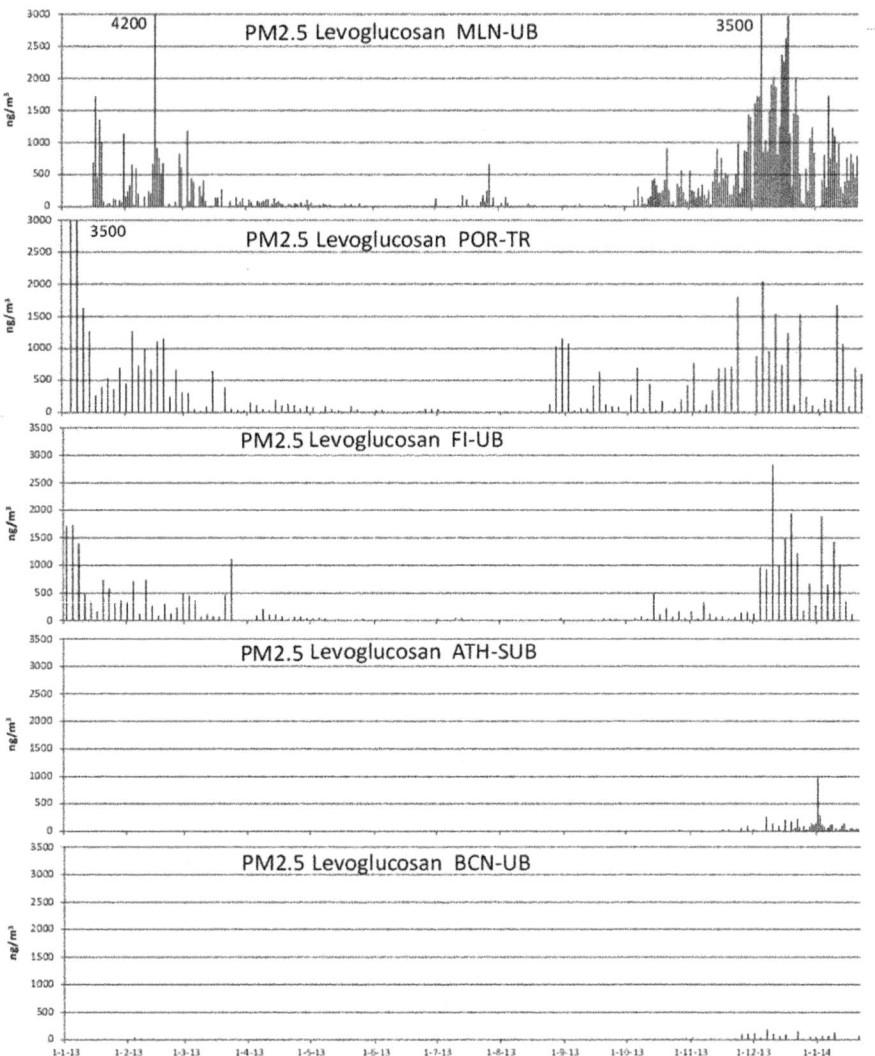

Figure 2   Daily levels of levoglucosan in PM$_{2.5}$ from the five AIRUSE cities.

by the time series of daily levels of levoglucosan (Figure 2). On the other hand, levels of K in the fine fraction (PM$_{2.5}$) may also increase as a consequence of the influence of inorganic biomass burning ash contributions to the PM load. In this case, the differences are also evident but less marked than for levoglucosan: 89 (BCN), 121 (ATH), 178 (FI), 327 (POR) and 348 (MLN) ng m^{-3} in PM$_{2.5}$. This lower differentiation is owing to the fact that K may be partially supplied by mineral dust (clay minerals and feldspars) and probably by specific industrial sources. In fact, K in the TR sites exhibits higher levels owing to road dust contributions.[26]

The trace elements were divided into 5 groups:

- *Metallurgy and heavy industry tracers*: Levels of Ti, Mn, Sr, Cu, Zn, Sn, Ni, Ba and Pb are higher in MLN, and in some cases in POR (Zn, Ba, Pb, Cd), when compared with the other sites.
- *Coal combustion tracers:* Levels of As and Se are relatively higher in POR and ATH, pointing to a possible local, regional or long range transport influence of emissions from coal combustion sources on air quality. Northern Greece has a number of large coal fired power plants that may partly account for these concentration levels. Transport of emissions from coal fired power plants from the Balkans, Eastern Europe and Turkey may also contribute. In the case of POR, the coal power plant located 15 km southeast of Porto was deactivated in December 2004 and there was no coal combustion activity at the area during the campaign. The closest coal fired power plant is located in Northwestern Spain, at around 300 km from Porto. Is it possible that other industrial sources of As and Se (ceramic, glass, and cement production, among others) may also contribute to increased levels of these elements.
- *Non-exhaust vehicle emission tracers*: Levels of Cu, Sb, Ba and Sn are also relatively high in the traffic sites (BCN-TR, POR-TR and ATH-TR) when compared with UB sites.
- *Shipping emission tracers*: Levels of V are 4 to 2 times higher in the cities with a harbor (POR, BCN and ATH) pointing to the fact that, in spite of the possible use of fuel oil or petroleum coke for power generation or industrial processes, shipping emissions are the main sources of these heavy oil combustion tracers in the AIRUSE cities.
- *Rare earth elements (REEs)* are usually occurring in mineral dust and the ratio Ce/La in the Earth's crust is close to 2. The marked decrease of this ratio may be used as a tracer of the influence of emission from La-based oil cracking in petrochemical plants. In this study most of the cities have a ratio Ce/La close to 2.0 with the exception of FI and ATH (1.0 and 1.3), indicating the possible influence of this type of emissions at both sites, although other sources for La cannot be discounted.

3.3 PM Mass Closure

Concerning the composition and origin of the main components of PM_{10}, important differences were observed among the cities. Thus in FI-UB, MLN-UB, POR-TR and ATH-TR, OM + EC was the main PM_{10} constituent (Table 4), accounting for 58, 45, 37 and 50% of the PM_{10} load and 69, 45, 48 and 59% of the $PM_{2.5}$. However, at BCN-UB and ATH-SUB these PM components were still dominant but made up only 27 and 30% of PM_{10} and 37 and 45% of $PM_{2.5}$ (Table 4).

The PM fraction accounted for by secondary inorganic aerosols (sulfate, nitrate, and ammonium) also showed significant variations with a load of 18–21% of PM_{10} for FI-UB and BCN-UB, 30% for ATH-SUB and 31% in

Table 4 PM$_{10}$ and PM$_{2.5}$ mass closure: annual average concentration (%) of aerosol groups.

	OM	EC	Sulfate	Nitrate	Ammonium	Sea salt	Soil dust	Unaccounted
PM$_{10}$								
BCN-UB	22	5	9	9	3	6	18	28
FI-UB	51	6	9	6	3	3	12	10
MLN-UB	38	5	7	14	8	2	9	17
ATH-SUB	28	2	19	6	5	8	25	7
BCN-TR	28	12	6	7	2	6	22	17
ATH-TR	40	10	11	6	3	7	10	13
POR-TR	23	15	7	6	2	13	12	23
PM$_{2.5}$								
BCN-UB	30	7	11	6	5	2	8	31
FI-UB	63	7	11	5	4	1	4	5
MLN-UB	40	6	6	19	7	2	4	16
ATH-SUB	41	3	26	2	8	2	9	9
ATH-TR	46	14	17	5	5	2	3	8
POR-TR	29	20	7	5	2	5	8	25

MLN-UB, and 20–23% of PM$_{2.5}$ for FI-UB and BCN-UB, 31% in MLN-UB and 37% for ATH-SUB (Table 4). For traffic sites, the SIA percentage was reduced owing to the increase of contributions from OM + EC (mostly from exhaust emissions) and mineral dust (mostly from road dust), and made up from 13 to 15% of PM$_{10}$ and PM$_{2.5}$, with the exception of ATH-TR, with 20 and 27% of PM$_{10}$ and PM$_{2.5}$, owing to the high load.

Mineral dust was one of the main PM$_{10}$ components at all sites ranging from 10 (MLN-UB) to 25% (ATH-SUB) of the PM$_{10}$ load. In PM$_{2.5}$, as expected from the coarser mode of occurrence of mineral dust, the contribution was reduced down to 4 (FI) to 10% (ATH) in all cases (Table 4).

The sea salt contribution reached the maximum at the Atlantic site (Table 4), 13% of PM$_{10}$ and 5% of PM$_{2.5}$ in POR, followed by the other two Mediterranean coastal sites, BCN and ATH (7 and 8% of PM$_{10}$ and 3 and 2% of PM$_{2.5}$); the minimum sea salt contributions were recorded in the inland Italian cities, MLN and FI (2–3% of PM$_{10}$ and 1% of PM$_{2.5}$).

3.4 PM Source Apportionment

3.4.1 Barcelona (BCN-UB)

3.4.1.1 Source Apportionment
For BCN-UB (January 2013–January 2014), the best PMF solution was found from pooling PM$_{10}$ and PM$_{2.5}$ samples in a single input matrix for PMF, comprising 234 samples, 30 strong species (EC, OC, NH$_4^+$, Cl$^-$, NO$_3^-$, Al, Ca, Fe, K, Mg, Na, S, Li, Ti, V, Cr, Mn, Ni, Cu, Zn, Ga, As, Rb, Sr, Cd, Sn, Sb, La, Ce and Pb), 2 weak species (Se and Ba) and setting PM as total variable with 400% uncertainty.[38] The distribution of residuals, G-space plots, Fpeak values and Q values were explored for solutions with numbers

of factors varying between 6 and 10. The most reliable solution was found with 8 factors/sources, adding 7% extra uncertainty and a minimum (base run) Q robust value of 7054, found over 30 runs (seed number 33), which exceeded the theoretical Q value (5650) by 24%. Species concentrations were reconstructed within 79–101%. Several constraints were added to the base run solution. Such constraints were of both physical and chemical nature and aimed at reducing the rotational ambiguity of the PMF problem, driving the solution towards *a priori* information based on mass conservation principles or partial knowledge of emission sources.[26,39] More specifically, in the case of Barcelona, the following constraints were introduced:

1. Pulling the difference of source contributions between PM_{10} and $PM_{2.5}$ to zero, only for those days and sources where the $PM_{2.5}$ contribution was higher than the PM_{10} contribution in the base run solution (96 cases). The % of dQ was set at 5% for each constraint and the converged results used totally 2.9% dQ.
2. Pulling the chemical profile (24 species) of the non-exhaust source towards the experimental profile of road dust obtained in Barcelona.[26] The % dQ was set at 5% for each elemental constraint and the converged results used totally 13.2% dQ.
3. Pulling the ratios Cl^-/Na, S/Na, K/Na, Ca/Na, Mg/Na of the sea salt profile to the literature values of 1.8, 0.084, 0.037, 0.038, 0.119 respectively. The % dQ was set at 5% for each ratio and the converged results used totally 3.2% dQ.

The 8 sources identified in Barcelona were: vehicle exhaust (VEX, Table 5), vehicle non-exhaust (NEX, Table 6), secondary nitrate (SNI, Table 7), mineral (MIN, Table 8), secondary sulfate and organics (SSO, Table 9), industrial (IND, Table 10), heavy oil combustion (HOC, Table 11) and aged sea salt (SEA, Table 12), similarly to those found by previous studies.[5,17,26]

The vehicle exhaust (VEX) comprises organic particles (42% of OC explained variation) from motor exhaust, as well as EC (51%) and trace metals from brake wear (Cu, Sb, Sn, Ba, Cr) (Tables 13–17). The ratio OC/EC of 1.8 reveals a significant proportion of secondary organic aerosols[40] from oxidation of primary volatile organic compounds (VOCs) both from fuel combustion (including exhaust emissions) and probably from enhanced anthropogenic transformation of biogenic VOCs by NO_x.[3] Another secondary product, in this case mostly from motor exhaust emissions, is ammonium nitrate, which was identified as a separate factor. Average source contributions from VEX were 3.3 µg m^{-3} (14%) to PM_{10} levels and 3.0 µg m^{-3} (21%) to $PM_{2.5}$ levels with an average $PM_{2.5}$/PM_{10} ratio of 0.91 owing to the very fine mode of motor exhaust particles (Table 18). However, source contributions are highly season-dependent with maxima in winter owing to the coupled effect of the lower temperature (favoring condensation of VOCs), shallower height of the planetary boundary mixing layer (PBL) and the lower sea breeze ventilation (Figure 13).

Table 5 PMF factor profile for the vehicle exhaust source at the five cities ($\mu g/\mu g$).

	Vehicle exhaust				
	BCN-UB	MLN-UB	POR-TR	FI-UB	ATH-SUB
EC	0.20080	0.09595	0.36687	0.15494	0.05252
OC	0.35957	0.25281	0.29652	0.56761	0.86342
Levo	n.i.[a]	0.00000	0.00277	0.00000	n.i.[a]
Si	n.i.[a]	0.00000	0.00181	0.00000	0.00046
Al	0.00000	0.00000	0.00113	0.00023	0.00000
Ca	0.00003	0.00367	0.00037	0.00000	0.01070
Fe	0.01509	0.00402	0.00000	0.00000	0.00361
K	0.00995	0.00000	0.01166	0.01152	0.00699
Mg	0.00152	0.00000	0.00000	0.00000	0.00000
Na	0.00713	0.00000	0.00325	0.00166	0.00000
Cl	0.00477	0.00010	0.00000	0.00000	0.00028
Ti	0.00001	0.00004	0.00006	0.00004	0.00025
V	0.00000	0.00000	0.00003	0.00000	0.00000
Cr	0.00016	0.00017	0.00000	0.00009	n.i.[a]
Mn	0.00026	0.00102	0.00008	0.00015	0.00010
Ni	0.00001	0.00017	0.00004	0.00004	0.00000
Cu	0.00069	0.00114	0.00022	0.00006	0.00018
Zn	0.00000	0.02055	0.00000	0.00090	0.00000
As	0.00002	n.i.[a]	0.00004	0.00003	0.00000
Se	0.00002	n.i.[a]	n.i.[a]	0.00004	n.i.[a]
Rb	0.00001	0.00000	0.00004	0.00000	n.i.[a]
Sr	0.00000	n.i.[a]	n.i.[a]	0.00007	0.00000
Cd	0.00000	n.i.[a]	0.00000	0.00001	0.00000
Sn	0.00020	n.i.[a]	0.00000	n.i.[a]	n.i.[a]
Sb	0.00010	n.i.[a]	0.00000	n.i.[a]	0.00002
Ba	0.00051	n.i.[a]	0.00000	0.00013	n.i.[a]
Pb	0.00004	0.00183	0.00006	0.00037	0.00004
Br	n.i.[a]	0.00136	0.00027	0.00053	0.00015
S	0.01751	0.00000	0.01380	0.00000	0.00000
NH$_4$	0.00000	0.00000	0.00000	0.00000	0.00000
NO$_3$	0.00000	0.00000	0.00266	0.00000	0.00000

[a]n.i.: Not included in PMF.

The vehicle non-exhaust source (NEX) was the object of several auxiliary equations aimed at pulling the original (base-run) factor, already traced by Cu, Mn, Fe, Sn and Sb (Table 6), towards the well-known emission profile of road dust in Barcelona, as experimentally obtained by a previous study.[26] This approach was proved to be effective in separating the mineral component owing to traffic emissions (road dust) from other mineral sources. However, a contribution from direct brake and tire wear cannot be disregarded in this factor. As the target source profile, the NEX factor is traced (Tables 13–17) by both the aforementioned brake wear metals/metalloids (22–31%) and crustal species from road surface wear (Ca, Al, Ti, Rb and Sr, among others; 20–37%). NEX emissions increased levels of PM$_{10}$ by 2.6 $\mu g\ m^{-3}$ (11%) and PM$_{2.5}$ levels by only 0.2 $\mu g\ m^{-3}$ (2%) with a very low PM$_{2.5}$/PM$_{10}$ ratio (0.1, Table 18). Conversely to the MIN source, NEX contributions were rather constant across the year with slightly higher values in

Table 6 PMF factor profile for the vehicle non-exhaust source at the five cities ($\mu g/\mu g$).

| | Vehicle non-exhaust | | | | |
	BCN-UB	MLN-UB	POR-TR	FI-UB	ATH-SUB
EC	0.02398	0.16829	0.28159	0.15340	0.04245
OC	0.05503	0.04432	0.00000	0.28355	0.00000
Si	n.i.[a]	0.03124	0.03034	0.02815	0.00273
Al	0.02999	0.00196	0.01212	0.00752	0.00000
Ca	0.10099	0.00452	0.01776	0.03353	0.08371
Fe	0.07329	0.06699	0.11834	0.13092	0.05966
K	0.01100	0.00000	0.00489	0.00329	0.00524
Mg	0.01000	0.00000	0.00271	0.00413	0.00219
Na	0.00372	0.00000	0.01371	0.00610	0.00000
Cl	0.00589	0.00000	0.00000	0.00166	0.00200
Ti	0.00250	0.00104	0.00113	0.00079	0.00000
V	0.00000	0.00000	0.00000	0.00000	0.00000
Cr	0.00029	0.00016	0.00070	0.00052	n.i.[a]
Mn	0.00067	0.00037	0.00114	0.00120	0.00072
Ni	0.00006	0.00000	0.00000	0.00007	0.00066
Cu	0.00322	0.00129	0.00542	0.00866	0.00230
Zn	0.00150	0.00079	0.00627	0.00363	0.00491
As	0.00001	n.i.[a]	0.00007	0.00003	0.00000
Rb	0.00005	0.00001	0.00006	0.00035	n.i.[a]
Sr	0.00021	n.i.[a]	n.i.[a]	0.00000	0.00000
Cd	0.00000	n.i.[a]	0.00004	0.00001	0.00001
Sn	0.00060	n.i.[a]	0.00196	n.i.[a]	n.i.[a]
Sb	0.00040	n.i.[a]	0.00131	n.i.[a]	0.00020
Ba	0.00136	n.i.[a]	0.00456	0.00207	n.i.[a]
La	0.00002	n.i.[a]	0.00001	n.i.[a]	n.i.[a]
Ce	0.00003	n.i.[a]	0.00004	n.i.[a]	n.i.[a]
Pb	0.00023	0.00109	0.00149	0.00000	0.00014
S	0.00549	0.00014	0.00000	0.02127	0.08882
NH_4	0.00245	0.00000	0.00377	0.00000	0.08243
NO_3	0.00308	0.00000	0.11532	0.00000	0.02014

[a]n.i.: Not included in PMF.

winter, as already found by Amato *et al.* (2009), and probably owing to low seasonal variation of the emission potential (Figure 13).

The secondary nitrate (SNI) factor explained about 35% of NH_4^+ and most of the variance of NO_3^- (Tables 7 and 13–17) as oxidation products of local gaseous NO_x emissions (road traffic and industrial plants). Based on the NO_x emission inventory of the Special Environmental Protection Area of Barcelona (BCN-SEPA), 50% of SNI can be apportioned to road traffic[26] while in other cities a large proportion may derive from biomass burning activities, which are not found in Barcelona. As expected, in winter ammonium nitrate formation is favored by low temperature and higher humidity while in summer it becomes volatilized more quickly owing to higher temperature. On the annual basis, average SNI contributions were 3.2 and 2.0 $\mu g \ m^{-3}$ in PM_{10} and $PM_{2.5}$, respectively (ratio $PM_{2.5}/PM_{10}$ of 0.63, Table 18), including a proportion of semivolatile organic aerosols, which easily condense on the high surface area of ammonium nitrate particles.

Table 7 PMF factor profile for the secondary nitrate factor at the five cities ($\mu g/\mu g$).

	Secondary nitrate				
	BCN-UB	MLN-UB	POR-TR	FI-UB	ATH-SUB
EC	0.01461	0.04473	0.00000	0.01477	0.01995
OC	0.08609	0.13081	0.09753	0.17814	0.00000
Si	n.i.[a]	0.00000	0.04286	0.00041	0.00401
Al	0.00000	0.00067	0.01749	0.00013	0.00263
Ca	0.00508	0.00183	0.01668	0.00000	0.01044
Fe	0.00521	0.00138	0.04182	0.00556	0.00436
K	0.00690	0.00331	0.00959	0.00986	0.01448
Mg	0.00012	0.00065	0.00654	0.00000	0.01536
Na	0.00200	0.00301	0.02173	0.00000	0.10777
Cl	0.00251	0.00000	0.00000	0.00345	0.00000
Ti	0.00000	0.00008	0.00104	0.00006	0.00000
V	0.00006	0.00000	0.00012	0.00000	0.00000
Cr	0.00006	0.00000	0.00019	0.00000	n.i.[a]
Mn	0.00012	0.00011	0.00040	0.00009	0.00007
Ni	0.00004	0.00000	0.00003	0.00002	0.00002
Cu	0.00033	0.00056	0.00147	0.00040	0.00024
Zn	0.00060	0.00081	0.00242	0.00087	0.00000
As	0.00001	n.i.[a]	0.00000	0.00001	0.00001
Se	0.00001	n.i.[a]	n.i.[a]	0.00002	n.i.[a]
Rb	0.00001	0.00000	0.00000	0.00003	n.i.[a]
Sr	0.00000	n.i.[a]	n.i.[a]	0.00000	0.00000
Cd	0.00000	n.i.[a]	0.00000	0.00002	0.00000
Sn	0.00013	n.i.[a]	0.00000	n.i.[a]	n.i.[a]
Sb	0.00008	n.i.[a]	0.00006	n.i.[a]	0.00000
Ba	0.00044	n.i.[a]	0.00030	0.00003	n.i.[a]
Pb	0.00010	0.00012	0.00000	0.00016	0.00000
Br	n.i.[a]	0.00020	0.00023	0.00003	0.00057
S	0.00000	0.00806	0.01574	0.00000	0.02933
NH_4	0.09265	0.16130	0.04615	0.06216	0.00000
NO_3	0.40996	0.50058	0.38860	0.32248	0.27792

[a]n.i.: Not included in PMF.

The total contribution from traffic can be estimated as the sum VEX + NEX + (0.5×SNI), which results in 7.4 and 4.2 $\mu g\ m^{-3}$ in PM_{10} (33%) and $PM_{2.5}$ (28%), respectively (Table 19). This indicates a clear relative decrease of road traffic emissions with respect to other sources in Barcelona, when compared to the 2003–2007 period, when mean contributions of 46 and 51% for PM_{10} and $PM_{2.5}$, respectively, were estimated in a nearby site.[26] The effect of lowering PM emissions in EURO5 vehicles, the reduction of the urban traffic flow in the city from 2007 to 2012 (around 9% according the city council, Ajuntament de Barcelona, 2014) and the measures taken by the public transport company (TMB) may account for this decrease.

The mineral factor (MIN) was identified by the typical crustal species, such as Al, Ca, Li, Ti, Rb, Sr (all with explained variation above 40%) Mg, Ga, La and Ce (above 30%) (Tables 8 and 13–17). This source is also responsible for significant concentrations of K, Fe and Na owing to their presence in clay minerals and feldspars. The MIN source is interpreted as a

Table 8 PMF factor profile for the mineral dust source at the five cities ($\mu g/\mu g$).

	Mineral dust				
	BCN-UB	MLN-UB	POR-TR	FI-UB	ATH-SUB
EC	0.00251	0.01855	0.00000	0.00000	0.00000
OC	0.07267	0.33746	0.04797	0.12831	0.02244
Levo	n.i.[a]	0.00000	0.00000	0.00243	n.i.[a]
Si	n.i.[a]	0.14882	0.09721	0.11381	0.17143
Al	0.04804	0.07275	0.04584	0.03008	0.06870
Ca	0.10446	0.07190	0.02486	0.18842	0.08326
Fe	0.03922	0.03117	0.03822	0.03968	0.04652
K	0.01832	0.05176	0.02340	0.01430	0.01833
Mg	0.01821	0.00228	0.00682	0.01096	0.01774
Na	0.02666	0.00000	0.00715	0.00160	0.00000
Cl	0.00315	0.00574	0.00000	0.00000	0.00000
Ti	0.00275	0.00531	0.00253	0.00260	0.00411
V	0.00021	0.00028	0.00002	0.00000	0.00011
Cr	0.00009	0.00000	0.00015	0.00011	n.i.[a]
Mn	0.00079	0.00137	0.00075	0.00116	0.00081
Ni	0.00003	0.00000	0.00002	0.00008	0.00004
Cu	0.00032	0.00199	0.00056	0.00107	0.00004
Zn	0.00201	0.00655	0.00130	0.00078	0.00017
Li	0.00004	n.i.[a]	0.00008	n.i.[a]	n.i.[a]
Ga	0.00001	n.i.[a]	n.i.[a]	n.i.[a]	n.i.[a]
As	0.00001	n.i.[a]	0.00002	0.00002	0.00001
Se	0.00001	n.i.[a]		0.00003	n.i.[a]
Rb	0.00007	0.00014	0.00017	0.00017	n.i.[a]
Sr	0.00031	n.i.[a]		0.00040	0.00023
Cd	0.00000	n.i.[a]	0.00000	0.00001	0.00000
Sn	0.00002	n.i.[a]	0.00000	n.i.[a]	n.i.[a]
Sb	0.00000	n.i.[a]	0.00005	n.i.[a]	0.00000
Ba	0.00067	n.i.[a]	0.00085	0.00036	n.i.[a]
La	0.00003	n.i.[a]	0.00003	n.i.[a]	n.i.[a]
Ce	0.00005	n.i.[a]	0.00006	n.i.[a]	n.i.[a]
Pb	0.00004	0.00090	0.00000	0.00001	0.00001
Br	n.i.[a]	0.00072	0.00008	0.00000	0.00000
S	0.00130	0.05329	0.00779	0.01715	0.01874
NH_4	0.00000	0.07157	0.00000	0.00000	0.00000
NO_3	0.08134	0.01881	0.00000	0.04638	0.01273

[a]n.i.: Not included in PMF.

mixture of several sources, including soil resuspension, urban works, regional mineral dust and Saharan dust. However, the Saharan dust contribution (SAH) could be separated from the non-Saharan (or Local DUST, LDU). The overall composition of MIN reveals a significant enrichment in Ca when compared to the average crust composition,[41,42] indicating an anthropogenic component, such as particles from concrete or limestone. The overall MIN contribution was 2.8 $\mu g\ m^{-3}$ (12%) to PM_{10} levels (0.3 $\mu g\ m^{-3}$ from SAH) and 1.0 $\mu g\ m^{-3}$ (7%) to $PM_{2.5}$ levels (Table 18). The average $PM_{2.5}/PM_{10}$ ratio of 0.36 confirms the expected coarse size distribution. Contributions from LDU are generally higher in warmer months, although high concentrations were observed in November and December 2013 (Figure 14).

Table 9 PMF factor profile for the secondary sulfate and organics source at the five cities (μg/μg).

	Secondary sulfate and organics				
	BCN-UB	MLN-UB	POR-TR	FI-UB	ATH-SUB
EC	0.04856	0.04015	0.00000	0.00000	0.00000
OC	0.15195	0.20151	0.15848	0.19929	0.21131
Levo	n.i.[a]	0.00000	0.00000	0.00000	n.i.[a]
Si	n.i.[a]	0.00497	0.00331	0.00668	0.00168
Al	0.00185	0.00157	0.00183	0.00166	0.00000
Ca	0.00000	0.00000	0.00000	0.00635	0.00000
Fe	0.00210	0.00000	0.00000	0.00207	0.00075
K	0.00233	0.00000	0.00568	0.00620	0.00838
Mg	0.00128	0.00000	0.00152	0.00088	0.00029
Na	0.00582	0.00000	0.00947	0.00266	0.00580
Cl	0.00000	0.00000	0.01012	0.00000	0.00000
Ti	0.00026	0.00014	0.00017	0.00012	0.00022
V	0.00012	0.00019	0.00057	0.00000	0.00003
Cr	0.00002	0.00014	0.00000	0.00003	n.i.[a]
Mn	0.00001	0.00006	0.00002	0.00015	0.00009
Ni	0.00003	0.00042	0.00033	0.00003	0.00006
Cu	0.00003	0.00070	0.00002	0.00000	0.00001
Zn	0.00000	0.00000	0.00000	0.00063	0.00061
Li	0.00000	n.i.[a]	0.00001	n.i.[a]	n.i.[a]
Ga	0.00000	n.i.[a]	n.i.[a]	n.i.[a]	n.i.[a]
As	0.00001	n.i.[a]	0.00002	0.00002	0.00000
Se	0.00001	n.i.[a]	n.i.[a]	0.00004	n.i.[a]
Rb	0.00001	0.00002	0.00003	0.00002	n.i.[a]
Sr	0.00002	n.i.[a]	n.i.[a]	0.00004	0.00000
Cd	0.00000	n.i.[a]	0.00001	0.00001	0.00000
Sn	0.00000	n.i.[a]	0.00008	n.i.[a]	n.i.[a]
Sb	0.00002	n.i.[a]	0.00005	n.i.[a]	0.00001
Ba	0.00000	n.i.[a]	0.00009	0.00001	n.i.[a]
La	0.00000	n.i.[a]	0.00001	n.i.[a]	n.i.[a]
Ce	0.00001	n.i.[a]	0.00001	n.i.[a]	n.i.[a]
Pb	0.00014	0.00000	0.00028	0.00012	n.i.[a]
Br	n.i.[a]	0.00028	0.00035	0.00016	0.00037
S	0.08035	0.05608	0.14403	0.14130	0.20360
NH$_4$	0.05419	0.08556	0.09567	0.09629	0.20809
NO$_3$	0.00000	0.01904	0.01446	0.00000	0.00000

[a]n.i.: Not included in PMF.

The factor secondary sulfate and organics (SSO), traced by SO_4^{2-} and NH_4^+ (Tables 9 and 13–17), is the result of the formation of secondary sulfate in the atmosphere from the photochemical oxidation of locally emitted gaseous sulfur oxides (in Barcelona mostly from shipping and with lower relevance from industrial activities in this region) and from long-range transport. Condensation of VOCs is suggested by the relatively high content of OC (15%). Source contributions from SSO were generally higher in summer than in winter owing to the higher photochemical activity during warmer months. Annual average contributions of SSO in Barcelona were 5.9 (26%) and 5.6 (39%) μg m^{-3} in PM$_{10}$ and PM$_{2.5}$, respectively (ratio PM$_{2.5}$/PM$_{10}$ of 0.95) (Table 18).

Table 10 PMF factor profile for the industrial emission sources at three cities ($\mu g/\mu g$).

	Industrial emissions		
	BCN-UB	MLN-UB	POR-TR
EC	0.00000	0.04467	0.40920
OC	0.09362	0.68874	0.41120
Levo	n.i.[a]	0.00000	0.00450
Si	n.i.[a]	0.00000	0.00467
Al	0.00000	0.00265	0.00000
Ca	0.00414	0.06681	0.00420
Fe	0.02026	0.03155	0.01842
K	0.00635	0.00000	0.01004
Mg	0.00033	0.00049	0.00055
Na	0.00000	0.00000	0.01922
Cl	0.02607	0.03190	0.00000
Ti	0.00022	0.00080	0.00027
V	0.00000	0.00008	0.00006
Cr	0.00032	0.00038	0.00022
Mn	0.00182	0.00112	0.00173
Ni	0.00007	0.00134	0.00021
Cu	0.00195	0.00809	0.00112
Zn	0.02705	0.00000	0.05371
Li	0.00001	n.i.[a]	0.00001
Ga	0.00001	n.i.[a]	n.i.[a]
As	0.00006	n.i.[a]	0.00006
Se	0.00000	n.i.[a]	n.i.[a]
Rb	0.00001	0.00011	0.00004
Sr	0.00002	n.i.[a]	n.i.[a]
Cd	0.00005	n.i.[a]	0.00008
Sn	0.00026	n.i.[a]	0.00011
Sb	0.00018	n.i.[a]	0.00011
Ba	0.00080	n.i.[a]	0.00000
La	0.00000	n.i.[a]	0.00001
Ce	0.00001	n.i.[a]	0.00002
Pb	0.00295	0.00046	0.00359
Br	n.i.[a]	0.00040	0.00039
S	0.00327	0.00000	0.02713
NH_4	0.00000	0.00000	0.00098
NO_3	0.00000	0.39971	0.00000

[a]n.i.: Not included in PMF.

The industrial factor (IND) explains most of the variance of Zn, Cd and Pb, and in minor proportion As, Mn and Cu, with OC being the major constituent (Tables 10 and 13–17). This source has been identified in prior studies at the urban background of Barcelona and allocated to the Llobregat industrial basin (SW of Barcelona) with higher impact on the SW part of the city and during night-time owing to the land breeze transport.[43,44] Mean IND concentrations during the AIRUSE campaign were 1.7 $\mu g\ m^{-3}$ (8%) and 1.3 $\mu g\ m^{-3}$ (9%) in PM_{10} and $PM_{2.5}$, respectively. No discernable seasonality is observed (Figure 14).

Similarly to road traffic, industries are responsible for a significant part (30%) of NO_x emissions, therefore the total contribution from IND should be

Table 11 PMF factor profile for the heavy oil combustion source at three cities ($\mu g/\mu g$).

	Heavy oil combustion		
	BCN-UB	FI-UB	ATH-SUB
EC	0.06805	0.09634	0.10097
OC	0.01066	0.07908	0.28524
Levo	n.i.[a]	0.00000	n.i.[a]
Si	n.i.[a]	0.00925	0.00000
Al	0.00000	0.00144	0.00095
Ca	0.00000	0.00748	0.00000
Fe	0.00979	0.00867	0.00581
K	0.00000	0.00507	0.00514
Mg	0.00000	0.00000	0.00033
Na	0.00810	0.00305	0.00000
Cl	0.00000	0.00000	0.00000
Ti	0.00019	0.00034	0.00038
V	0.00352	0.00134	0.00297
Cr	0.00025	0.00006	n.i.[a]
Mn	0.00051	0.00000	0.00000
Ni	0.00114	0.00054	0.00072
Cu	0.00085	0.00074	0.00044
Zn	0.00949	0.00000	0.00000
Li	0.00000	n.i.[a]	n.i.[a]
Ga	0.00001	n.i.[a]	n.i.[a]
As	0.00004	0.00000	0.00011
Se	0.00004	0.00017	n.i.[a]
Rb	0.00000	0.00000	n.i.[a]
Sr	0.00005	0.00000	0.00063
Cd	0.00001	0.00001	0.00002
Sn	0.00098	n.i.[a]	n.i.[a]
Sb	0.00008	n.i.[a]	0.00019
Ba	0.00113	0.00026	n.i.[a]
La	0.00002	n.i.[a]	n.i.[a]
Ce	0.00001	n.i.[a]	n.i.[a]
Pb	0.00047	0.00060	0.00096
Br	n.i.[a]	0.00040	0.00041
S	0.10361	0.01709	0.03096
NH_4	0.00000	0.00000	0.04984
NO_3	0.00000	0.00000	0.00000

[a]n.i.: Not included in PMF.

calculated as $IND + (0.3 \times SNI)$, which results in 2.7 and 1.9 $\mu g\ m^{-3}$ in PM_{10} and $PM_{2.5}$, respectively (Table 19).

The heavy oil combustion (HOC) factor is characterized by high concentrations of EC and S, and traced by V and Ni, reflecting the influence of residual oil combustion processes mostly from shipping and in a minor proportion from industries in this region (Tables 11 and 13–17). HOC particles are rather fine (0.84 $PM_{2.5}/PM_{10}$ ratio) and contributed more in summer than in winter, similarly to SSO (Figure 14). On annual average, HOC increased PM_{10} and $PM_{2.5}$ concentrations by 0.9 (4%) and 0.7 (5%) $\mu g\ m^{-3}$, respectively (Table 18).

The aged sea salt factor (SEA), traced by Na and Cl^- (Table 12), is interpreted as sea salt particles including aged sodium nitrate, thus this is the

Table 12 PMF factor profile for the aged and fresh sea salt sources at the five cities (µg/µg).

	Sea salt					
	Aged sea salt	Aged sea salt	Fresh sea salt	Aged sea salt	Fresh sea salt	Fresh sea salt
	BCN-UB	MLN-UB	POR-TR	FI-UB	FI-UB	ATH-SUB
EC	0.00000	0.00000	0.00000	0.00000	0.00000	0.00000
OC	0.00000	0.00000	0.00000	0.23166	0.00000	0.00000
Levo	n.i.[a]	0.00000	0.00020	0.00000	0.00986	n.i.[a]
Si	n.i.[a]	0.00000	0.00000	0.00000	0.00510	0.00159
Al	0.00000	0.00876	0.00028	0.00016	0.00115	0.00100
Ca	0.00733	0.01337	0.00961	0.00997	0.01952	0.03709
Fe	0.01193	0.04138	0.00478	0.02629	0.00008	0.00160
K	0.00715	0.00000	0.00792	0.00915	0.00800	0.00944
Mg	0.02302	0.00000	0.03224	0.03061	0.02125	0.04459
Na	0.19346	0.19324	0.22281	0.19476	0.14507	0.26806
Cl	0.18936	0.31477	0.36186	0.00000	0.28633	0.55010
Ti	0.00000	0.00078	0.00002	0.00000	0.00019	0.00017
V	0.00000	0.00000	0.00009	0.00000	0.00000	0.00000
Cr	0.00009	0.00000	0.00003	0.00018	0.00001	
Mn	0.00015	0.00117	0.00003	0.00022	0.00005	0.00000
Ni	0.00003	0.00000	0.00005	0.00007	0.00002	0.00005
Cu	0.00078	0.00607	0.00028	0.00147	0.00005	0.00000
Zn	0.00026	0.00000	0.00054	0.00109	0.00000	0.00000
Li	0.00000	n.i.[a]	0.00001	n.i.[a]	n.i.[a]	n.i.[a]
As	0.00001	n.i.[a]	0.00001	0.00002	0.00000	0.00000
Se	0.00003	n.i.[a]	n.i.[a]	0.00006	0.00000	n.i.[a]
Rb	0.00000	0.00000	0.00000	0.00007	0.00000	n.i.[a]
Sr	0.00015	n.i.[a]	n.i.[a]	0.00011	0.00017	0.00022
Cd	0.00000	n.i.[a]	0.00000	0.00002	0.00000	0.00000
Sn	0.00015	n.i.[a]	0.00002	n.i.[a]	n.i.[a]	n.i.[a]
Sb	0.00011	n.i.[a]	0.00004	n.i.[a]	n.i.[a]	0.00002
Ba	0.00036	n.i.[a]	0.00010	0.00033	0.00000	n.i.[a]
Pb	0.00006	0.00855	0.00002	0.00021	0.00000	0.00000
Br	n.i.[a]	0.00497	0.00068	0.00084	0.00031	0.00102
S	0.01631	0.00000	0.02278	0.06868	0.00000	0.03312
NH$_4$	0.00000	0.00000	0.00000	0.00000	0.00000	0.00000
NO$_3$	0.19140	0.06430	0.01644	0.37613	0.00000	0.01127

[a]n.i.: Not included in PMF.

contribution of a mixture of natural and anthropogenic emissions. SEA contributed on average 2.5 (11%) and 0.4 (3%) µg m^{-3}, respectively, to PM$_{10}$ and PM$_{2.5}$ samples, with a coarse PM$_{2.5}$/PM$_{10}$ ratio (0.16), without significant seasonal variation (Table 18).

Finally, average source contributions to PM$_{10}$ and PM$_{2.5}$ during the days in which the PM$_{10}$ concentration is above 40 µg m^{-3} are reported in Table 20 (no days with PM$_{10}$ above 50 µg m^{-3} were recorded during the 126 sampling days). Traffic (as sum of VEX, NEX and about 50% of SNI) and secondary sulfate are the sources that give the most relevant contributions, with 56 and 20% of PM$_{10}$ and 54 and 26% of PM$_{2.5}$, respectively.

3.4.1.2   Validation with Hourly Elemental Data

In Barcelona, the AIRUSE Streaker campaign was carried out at one traffic site (Valencia avenue, used also as control site for the dust suppressants campaign in Barcelona). Six and three PMF factors were identified in the fine and in the coarse fraction, respectively (Figures 3 and 4): Sulfate and oil

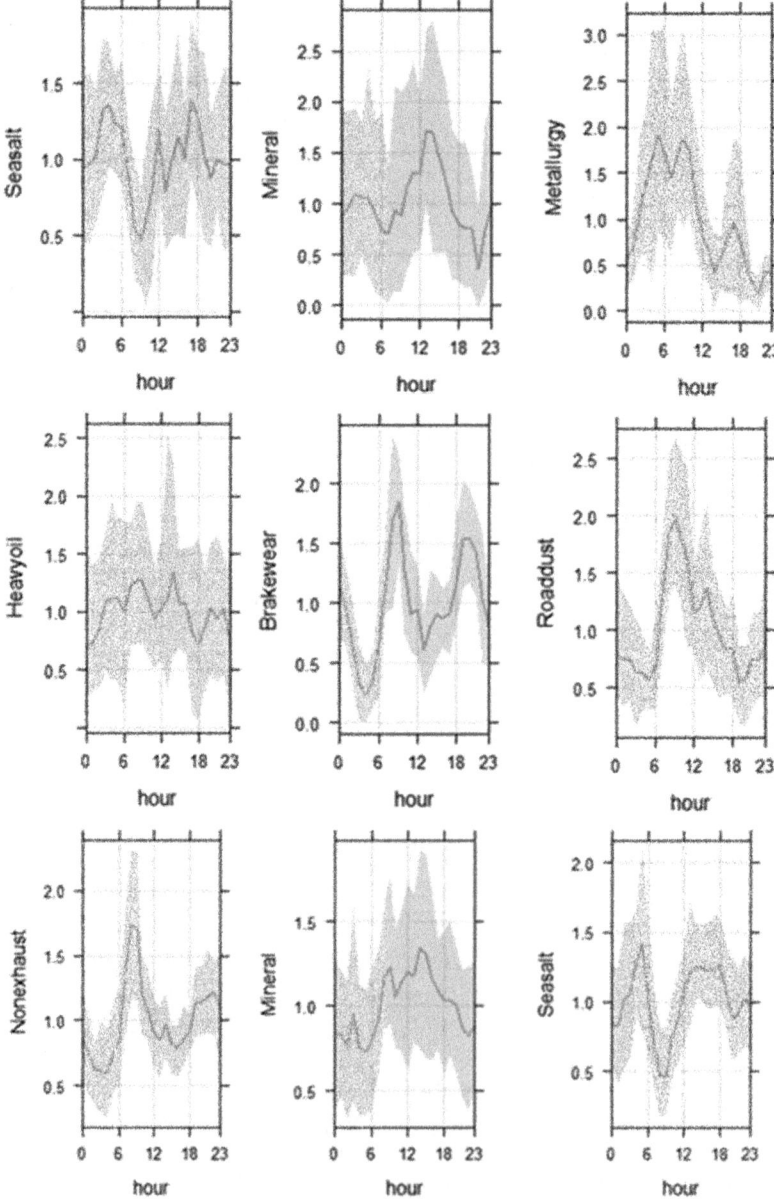

Figure 3   Daily cycle of sources impact on an hourly basis in BCN-TR in the fine (six top) and coarse fraction (three at the bottom) (arbitrary units).

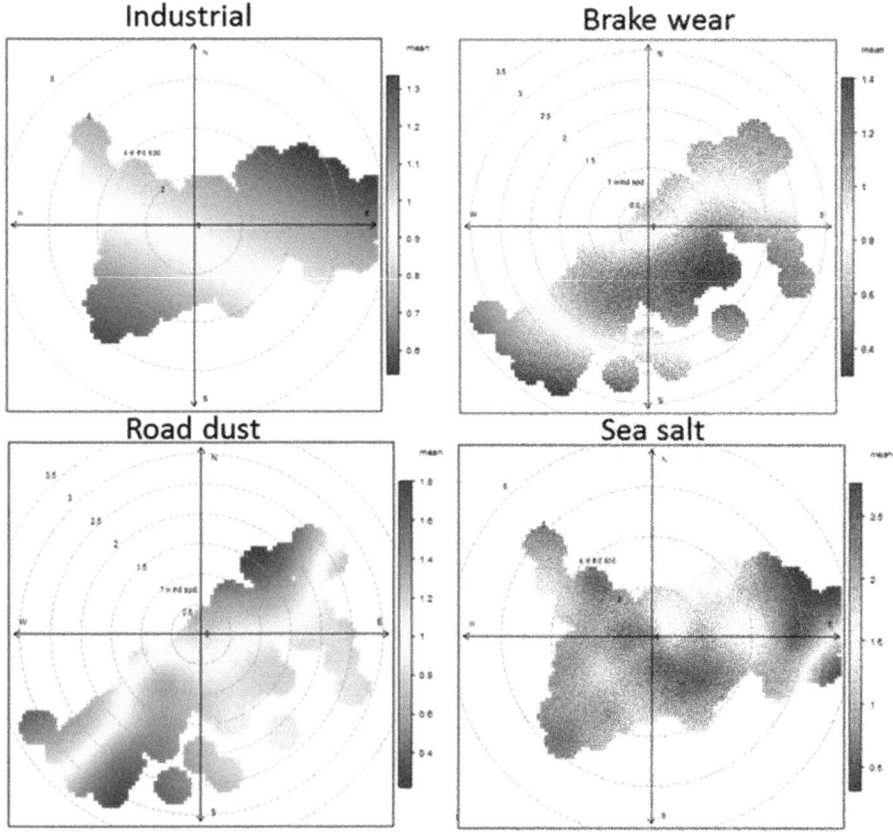

Figure 4 Polar plots of hourly source contributions (a.u.) in BCN-TR.

combustion, brake wear, mineral dust, road dust, sea salt and metallurgy in the fine fraction, mineral dust, vehicle non-exhaust and sea salt in the coarse one.

The sulfate and oil combustion source corresponds to a mix of the above SSO and HOC factors. This factor does not exhibit a clear diurnal variation, indicating a regional rather than a local source. The hourly $PM_{2.5}$ samples allowed separation between brake wear and road dust (Figure 5) instead of the obtained NEX source with daily samples, the first showing two peaks during traffic rush hours while the latter only showing a morning peak. The polar plots are clearly oriented with the direction of the street canyon (Figure 6). The vehicle exhaust source (VEX) cannot be identified because, as reported in the methodology, EC and OC cannot be detected by PIXE, the only technique applicable to the Streaker sample analysis. The industrial factor in the fine fraction shows maxima during the first hours of the day, when the sea breeze is not yet developed and the air mass may come from a land-breeze (Figure 5). The polar pot clearly points to the SW, where the metallurgy facilities are located (Figure 6). The Sea salt source was identified both in the coarse and

Table 13 Explained variation (%) of species in PM_{10} and $PM_{2.5}$ (BCN-UB).

FI-UB	Heavy oil combustion	Vehicle non-exhaust	Mineral	Fresh sea salt	Aged sea salt	Vehicle exhaust	Secondary nitrate	Secondary sulfate and organics	Saharan dust	Biomass burning
EC	8	15	0	0	0	36	3	0	0	30
OC	1	5	3	0	2	26	7	15	0	36
Levo	0	0	1	1	0	0	0	0	1	84
Na	1	3	1	29	56	2	0	5	2	0
Mg	0	7	22	14	28	0	0	6	20	0
Al	1	7	35	0	0	1	0	6	44	1
Si	3	9	45	1	0	0	0	8	34	1
S	2	3	3	0	5	0	0	76	2	6
Cl	0	1	0	89	0	0	5	0	0	4
K	2	1	8	1	2	13	9	11	7	38
Ca	2	10	69	2	2	0	0	7	6	0
Ti	4	10	41	1	0	1	2	6	33	0
Mn	0	28	33	0	3	9	4	13	7	1
Fe	3	52	19	0	6	0	4	3	12	0
Ni	40	6	8	1	3	8	4	11	1	12
Cu	5	65	10	0	6	1	6	0	3	0
Zn	0	25	7	0	4	15	12	17	0	12
As	0	7	7	0	3	19	4	21	1	11
Se	26	0	7	0	6	17	7	26	0	3
Br	9	0	0	3	12	33	1	16	0	19
Rb	0	31	18	0	3	0	5	7	12	13
Sr	0	0	35	5	4	13	0	12	11	2
NH$_4$	0	0	0	0	0	0	24	72	0	0
NO$_3$	0	0	6	0	22	0	66	0	0	2
V	97	0	0	0	0	0	0	0	3	0
Ba	5	45	10	0	4	7	1	1	6	3
Pb	11	0	0	0	2	19	7	10	0	20
Cr	3	36	9	0	7	14	0	9	3	8
Cd	4	3	5	0	4	8	18	13	0	16

Table 14 Explained variation (%) of species in PM_{10} and $PM_{2.5}$ (FI-UB).

BCN-UB	Heavy oil combustion	Vehicle non-exhaust	Mineral	Industrial	Aged sea salt	Vehicle exhaust	Secondary nitrate	Secondary sulfate and organics
EC	5	3	0	0	0	58	3	26
OC	0	3	5	5	0	42	8	33
NH4	0	0	0	0	0	0	34	44
Cl	0	2	1	10	70	4	2	0
NO3	0	0	10	0	18	0	68	0
Al	0	29	62	0	0	0	0	7
Ca	0	36	51	2	3	0	3	0
Fe	2	32	23	9	5	14	4	4
K	0	11	25	7	8	22	13	10
Mg	0	14	35	1	35	5	0	7
Na	1	1	12	0	65	5	1	8
S	13	1	0	1	4	8	0	70
Li	3	20	46	7	4	4	3	11
Ti	1	33	48	3	0	0	0	14
V	69	0	10	0	0	0	4	17
Cr	9	18	8	22	6	23	7	5
Mn	5	12	20	36	3	11	4	1
Ni	58	5	3	7	3	2	7	10
Cu	5	33	4	22	8	16	6	1
Zn	13	4	7	71	1	0	3	0
Ga	8	17	35	22	0	1	0	14
As	9	4	5	29	2	17	8	22
Se	13	4	7	1	14	19	5	19
Rb	0	23	41	5	0	7	5	18
Sr	3	22	43	3	16	7	1	9
Cd	3	1	0	57	1	0	6	19
Sn	22	24	1	11	6	17	9	0
Sb	3	30	0	14	8	17	11	5
Ba	11	22	15	14	6	19	13	0
La	10	16	35	4	1	13	0	18
Ce	3	17	32	5	2	21	0	17
Pb	6	5	1	67	1	2	4	12

Table 15 Explained variation (%) of species in PM_{10} and $PM_{2.5}$ (ATH-SUB).

ATH-SUB	Heavy oil combustion	Secondary sulfate and organics	Vehicle non-exhaust	Biomass burning	Vehicle exhaust	Mineral	Secondary nitrate	Sea salt
EC	22	0	13	15	27	0	11	0
OC	8	27	0	5	55	2	0	0
Na	0	6	0	0	0	0	52	35
Mg	0	1	2	0	0	44	24	19
Al	0	0	0	1	0	85	2	0
Si	0	1	1	1	0	86	1	0
Cl	0	0	1	0	0	0	0	87
K	2	17	3	12	7	31	15	3
Ca	0	0	20	0	4	53	4	4
Ti	2	5	0	1	3	77	0	1
Mn	0	7	18	5	4	53	3	0
Fe	2	1	25	1	2	53	3	0
Ni	27	10	36	1	0	6	2	1
Cu	8	1	58	0	8	3	10	0
Zn	0	16	40	26	0	4	0	0
Br	7	31	0	12	6	0	24	12
NH_4	4	79	10	4	0	0	0	0
NO_3^-	0	0	3	0	0	6	76	0
SO_4^{2-}	2	70	9	0	0	5	5	2
V	85	4	0	0	0	12	0	0
As	19	0	0	69	0	4	3	0
Sr	27	0	0	12	0	38	0	6
Cd	12	9	8	50	0	0	0	0
Sb	22	8	34	13	7	1	5	1
Pb	22	0	5	67	2	1	0	0

Table 16 Explained variation (%) of species in PM$_{10}$ and PM$_{2.5}$ (POR-TR).

POR-TR	Biomass burning	Secondary nitrate	Heavy oil and secondary sulfate	Mineral	Sea salt	Industrial	Vehicle non-exhaust	Vehicle exhaust
EC	13	0	0	0	0	11	12	60
OC	29	4	9	4	0	8	0	38
Levo	84	0	0	0	0	2	0	6
Na	0	5	3	4	78	3	3	3
Mg	0	8	3	19	59	0	3	0
Al	1	12	2	68	0	0	7	3
Si	1	14	2	70	0	1	9	2
S	3	4	58	5	9	4	0	13
Cl	9	0	2	0	82	0	0	0
K	17	6	5	31	7	3	3	24
Ca	0	15	0	50	13	2	15	1
Ti	1	12	3	64	0	2	12	2
V	2	8	56	4	9	2	0	7
Cr	1	13	0	23	3	9	45	1
Mn	0	8	0	32	1	19	21	5
Fe	0	16	0	32	3	4	41	0
Ni	5	3	42	4	6	10	0	11
Cu	1	14	0	12	4	6	47	7
Zn	0	6	0	7	2	70	14	0
Br	11	6	13	4	25	5	0	24
NH$_4$	13	16	50	0	0	0	1	0
NO$_3$	14	58	3	0	4	0	16	1
Li	1	0	7	62	5	3	7	16
As	11	0	6	9	2	9	16	41
Rb	14	0	5	51	1	3	7	17
Cd	12	0	8	0	0	37	28	14
Sn	8	0	4	0	1	2	63	0
Sb	6	2	3	5	2	3	53	0
Ba	0	4	2	28	2	0	61	0
La	0	1	10	54	10	3	9	16
Ce	1	3	4	48	3	3	13	20
Pb	11	0	7	0	0	34	23	3

F. Amato et al.

Table 17 Explained variation (%) of species in PM$_{10}$ and PM$_{2.5}$ (MLN-UB).

MLN-UB	Vehicle exhaust	Vehicle non exhaust	Secondary nitrate	Heavy oil and secondary sulfate	Mineral dust	Industrial	Aged sea salt	Biomass burning
EC	18	12	12	7	0	21	0	24
OC	5	17	11	9	5	14	4	13
Levo	0	7	0	0	0	0	0	88
Al	0	12	2	4	49	4	0	1
Si	1	27	0	6	60	0	0	0
S	0	0	3	67	0	24	0	0
Cl	0	0	0	0	0	20	67	0
K	1	0	7	1	23	9	0	44
Ca	1	23	3	0	44	4	2	0
Ti	2	25	3	3	53	3	1	2
V	0	0	0	79	1	32	0	0
Cr	11	31	0	11	2	15	0	5
Mn	14	24	7	2	17	3	4	4
Fe	3	50	2	2	8	4	1	4
Ni	8	21	0	23	8	23	0	3
Cu	5	42	7	0	4	21	0	0
Zn	46	12	8	3	4	0	5	2
Br	21	0	11	16	0	16	18	5
Rb	0	7	3	4	31	12	0	31
Ba	2	16	0	8	14	25	4	9
Pb	22	9	8	0	4	13	17	11
Sn	4	5	2	19	15	32	8	17
Na$^+$	0	12	19	0	0	0	35	0
NH$_4$$^+$	0	0	50	23	1	0	5	3
Mg$^+$	2	22	3	0	19	2	0	2
NO$_3$$^-$	2	1	60	1	1	0	5	7

Table 18 Annual average PMF source contributions at the five cities (µg m^{-3}).

	Vehicle exhaust	Vehicle non-exhaust	Secondary nitrate	Secondary sulfate and organics	Heavy oil combustion	Local dust	Saharan dust	Industrial	Aged sea salt	Fresh sea salt	Biomass burning	Unaccounted
µg m^{-3}												
PM$_{10}$												
BCN-UB	3.3	2.6	3.2	5.9	0.9	2.5	0.3	1.7	2.5		3	0.2
FI-UB	2.5	1.9	2.2	3.9	0.9	2.3	0.7		1	0.8	1.4	1.6
ATH-SUB	2.1	1.8	3.2	4	0.8	2.1	3			1	4.2	
POR-TR	7.9	2.9	3.2	3.4		6.3		1.2		5.5		
MLN-UB	2.8	3.4	10	5.6		2.6		3.7	1		7.8	2.1
PM$_{2.5}$												
BCN-UB	3	0.2	2	5.6	0.7	1		1.3	0.4	0.1	2.9	0.7
FI-UB	2.5	0.3	1.9	4	0.8	0.3	0.2		0.2	0.1	1.2	0.7
ATH-SUB	1.7	0.6	0.7	3.8	0.8	0.6	0.7				4.4	1.3
POR-TR	8.1	1.3	1.4	3.3		3.8		1.3		1.1		0.9
MLN-UB	1.8	2.6	8.5	5.6		1.5		1.4	0.4		5.3	2.6
%												
PM$_{10}$												
BCN-UB	14	11	14	26	4	11	1	8	11		15	1
FI-UB	13	10	11	20	5	12	4		5	4	7	8
ATH-SUB	10	8	15	19	4	10	14			5	12	
POR-TR	23	8	9	10		18		4		16		
MLN-UB	7	9	26	19		7		9	3		20	5
PM$_{2.5}$												
BCN-UB	20	1	13	37	5	7		9	3	1	21	5
FI-UB	18	2	14	29	6	2	1		1	1	11	5
ATH-SUB	15	5	6	33	7	5	6				17	11
POR-TR	32	5	5	13		15		5		4		4
MLN-UB	6	9	28	19		5		5	1		18	9

Table 19 Annual average PMF source contributions at the five cities (μg m^{-3}) per source category.

	Traffic	Non-traffic nitrate	Secondary sulfate and organics	Heavy oil combustion	Local dust	Saharan dust	Industrial	Sea salt	Biomass burning	Unaccounted
μg m^{-3}										
PM$_{10}$										
BCN-UB	7.4	1.8	5.9	0.9	2.5	0.3	2.5	1.5	3.1	0.2
FI-UB	6.1	1.2	4	0.9	2.3	0.7		0.6	1.4	1.6
ATH-SUB	4.8	1.5	4.2	0.8	2.1	3		1.6		
POR-TR	12.7	0.8	3.5		6.3		1.2	4.4	4.7	
MLN-UB	13	2	5.5		2.6		3.7	0.7	9.1	2.1
PM$_{2.5}$										
BCN-UB	4.2	0.5	5.6	0.7	1	0.2	1.8	0.4	2.9	0.7
FI-UB	4.3	0.5	4.1	0.8	0.3			0.1	1.2	0.7
ATH-SUB	2.5	0.3	3.8	0.8	0.6	0.7		0.2		1.3
POR-TR	10.3	0.3	3.4		4		1.3	1.3	4.7	0.9
MLN-UB	10.1	1.7	5.6		1.5		1.4	0.5	6.4	2.7
%										
PM$_{10}$										
BCN-UB	32	8	26	4	11	1	11	7	16	2
FI-UB	31	6	21	5	12	4		3	7	8
ATH-SUB	23	7	20	4	10	14		7		
POR-TR	38	2	10		19		4	13	14	
MLN-UB	34	5	14		7		9	2	24	5
PM$_{2.5}$										
BCN-UB	28	3	37	5	7	1	12	3	21	5
FI-UB	31	4	29	6	2			1	11	5
ATH-SUB	22	3	34	7	4	6		2		11
POR-TR	39	1	13		15		5	5	18	4
MLN-UB	34	6	19		5		5	1	21	9

Table 20 Average contribution (%) of PM_{10} and $PM_{2.5}$ sources during high pollution days ($PM_{10} > 50$ µg m^{-3} and > 40 µg m^{-3} in BCN-UB).

	BCN-UB		FI-UB		ATH-SUB		POR-TR		MLN-UB	
	PM_{10}	$PM_{2.5}$	PM_{10}	$PM_{2.5}$	PM_{10}	$PM_{2.5}$	PM_{10}	$PM_{2.5}$	PM_{10}	$PM_{2.5}$
Aged sea salt	2	1	<1	<1			3	<1	2	1
Saharan dust	<1	<1	<1	<1	52	45	<1	<1	<1	<1
Local dust	4	2	<1	<1	1	2	27	22	3	2
Sec. sulfate and organics	19	22	6	6	2	5	5[a]	2[a]	9[a]	11[a]
Vehicle non-exhaust	14	2	9	1	3	1	6	3	14	8
Vehicle exhaust	13	20	5	5	4	9	25	30	5	5
Vehicle nitrate	18	20	29	26	2	1	5	2	22	26
Heavy oil combustion	4	6	3	3	3	10				
Industrial	17	18			<1	<1	2	1	4	3
Non-traffic nitrate	9	9	7	6	1	3	2	1	6	8
Fresh sea salt			<1	<1	7	1				
Biomass burning			30	33	1	2	25	33	35	26
Unaccounted			8	20	24	21	5			10

[a]Includes heavy oil combustión.

in the fine fractions (Figure 5), in both cases showing a rather constant value across the 24 hours, although a minimum is observed during 7–9 am. The polar plots clearly point to the coast line (Figure 6).

3.4.1.3 Suggested Measures to Improve Air Quality

In the Special Environmental Protection Area of Barcelona (BCN-SEPA), the responsibility for improving air quality belongs to the Regional Government of Catalonia. The exceedances of PM_{10} (and NO_2) limit values during the last decade have led to the development of two Air Quality Plans (2007–2010 and 2011–2015). Concerning the city of Barcelona, a clear PM_{10} downward trend has been observed (2013 was the first year without PM_{10} exceedances) during the last decade (also present but less marked for NO_2), revealing a general improvement in air quality, probably as result of multiple effects, such as the European, national, regional and local actions, the economic downturn and the favorable meteorology. In spite of the aforementioned decrease of PM concentrations, more effort has to be made in order to attain the WHO guidelines, and to prevent future exceedances under the economic recovery and less favorable meteorological scenarios.

According to our AIRUSE results, road traffic is clearly the main source of PM_{10} and the second most important for $PM_{2.5}$ (only smaller than secondary sulfate, which has an important regional contribution). It is therefore evident that future efforts from air quality plans for the city of Barcelona must focus on the road traffic sector (Table 21), other local sources being of less importance, such as metallurgy, heavy oil combustion and urban works, or

Table 21 Measures and priority degree proposed by AIRUSE for Barcelona based on PM speciation and source apportionment obtained in 2013 at BCN-UB.

Sector	Measure	Competence	Description	Priority
Road traffic 33% of PM$_{10}$: Exhaust 14% Non-exhaust 11% Nitrate 8% 28% of PM$_{2.5}$: Exhaust 20% Non-exhaust 1% Nitrate 7%	Restricted entry to the city centre	Local	Limit the access of vehicles to the city centre of municipalities with >100 000 inhabitants. Barcelona has one of the highest car densities (per km^2) in Europe.	***
	Lowering number of cars in the urban areas	Local	Implementation of measures that attribute a direct fee to a car entering the city, such as road and congestion charges. Restrict vehicle circulation to neighborhoods in specific areas of the city.	***
	Low emission zones (LEZ)	National and local	Adopting a national labelling scheme to enforce local emission-related traffic restrictions. Already applied in 220 EU cities. In Berlin a 58% reduction of exhaust particle emissions was achieved.	***
	BUS-HVO lanes	Local	This measure will reduce private vehicles entering the city from the province. We recommend to convert one existing lane to BUS-VAO.	***
	Park and ride	Local	Creation of 4–5 new strategic parking lots at main transport interfaces (train and metro stations) at the outskirts of the city in order to promote the combined use of car and public transport. High charges for parking in the city centre for non-residents and low charges for eco-cars.	***

Measure	Scope	Description	
Street cleaning	Local	Tandem use of sweeping and, more importantly, water washing. It is effective (7–10% on a daily basis) to reduce road dust emissions and must be performed before morning traffic rush hour (5–6 h am) and mostly during dry periods. It is especially recommended after African dust episodes or after >10 days without rain.	***
Pedestrians lanes and cycling paths	Local	Increase the city surface reserved for pedestrian and cyclists. Improvement in traffic planning for a shift in modal split from motor traffic towards public transport, cycling and pedestrian traffic.	***
Promoting low-carbon and low-NO$_x$ vehicles and diesel retrofit	National and local	• Reduce highway toll and parking fees for low emission vehicles (for NO$_2$ and PM). Incentive for the installation of particle filters in heavy duty commercial vehicles. • Development of electric vehicles. • Infrastructure. Barcelona has 30% of motorbikes in the fleet; the use of electric bikes may have a large influence in decreasing emissions.	***
Improving public transportation	Local	• Continuous update and improvement for an economic, environmentally friendly and faster public transportation (metro, train, and tram) • Barcelona busses are already new and clean. Spatial coverage and price need to be improved. • Intercity railway connection should be improved in the metropolitan area.	***
Renewal of car/taxi/motorcycle fleet.	National	• Subsidies and bonuses that either support the retrofitting of a vehicle, or the scrapping of it. • Increase the share of hybrid (now 11%) natural gas (now 6%) and LPG (now 6%) taxis. • Promoting the replacement of 2-stroke motorcycles by 4-strokes. • Low-emission machinery (*e.g*, diesel particle filters and electric machinery) at urban works.	***

Table 21 Continued

Sector	Measure	Competence	Description	Priority
	Improving public fleet	Local	The bus fleet of Barcelona already comprises 39% natural gas vehicles and 43% of SCRT-equipped diesel. This example should be followed by the villages of the metropolitan area, mostly for the buses entering Barcelona.	**
	Reducing pricing of public transport	National and local	• Income tax return by accumulating public transportation bonds to make public transport more economically attractive. • Free tickets for public transport during anticyclonic pollution episodes.	**
	Reducing road transportation for goods	National and local	Creating a train-harbor interface for transport of goods, which is currently carried by trucks.	**
	Vehicle and road maintenance	National and local	• Increase the frequency of inspection programmes of public vehicles to ensure that in-use engines continue to have functional controls and proper maintenance. • Maintaining roads in good repair to reduce the contribution of PM from road surface wear.	**
	Car-sharing	Local	Reduce single-occupancy car journeys by promoting car sharing, linking car sharing to public transport services and cycling.	**
	Awareness	National	Integration of mobility awareness into school curricula.	***

Source	Scope	Measure	Rating
Other sources **Shipping/heavy oil combustion** 4% of PM$_{10}$ 5% of PM$_{2.5}$	National and local	Stricter legislation for harbor	Use of plug-in shore power; permission to dock in Barcelona harbour only given to vessels with engines operating with low sulfur fuel; best available technologies applied to marine ship flue gas must be required; alternative at-dock technologies (like channeling exhaust through barge-mounted control devices).	**
Industrial 11% of PM$_{10}$ 12% of PM$_{2.5}$	Local	Industrial facilities	Impose high standards for fuels and BATs (channelled and fugitive emissions) increase inspections to facilities.	**
Precursors of secondary particles	Local	Reduce NH$_3$ emissions	Application of the "guidelines" for regional containment of ammonia resulting from agricultural, livestock and waste practices.	***
Urban works/ mineral 11% of PM$_{10}$ 6% of PM$_{2.5}$	Local	Water spraying	Apply water during construction operations (earthmoving, demolition, grading).	***
	Local	Storage and handling of dusty materials	Cover bulk material during storage; covering the cargo beds of haul trucks with a tarp or other suitable closure to minimise wind-blown dust emissions and spillage; green curtains in areas with high diffusive dust emissions; reduction of vehicle speed and eco-driving test for workers.	***
	Local	Unpaved areas	Apply water and/or nano-polymers or pave unpaved parking lots or roads; set additional control requirements for unpaved roads (*e.g.* vegetating, adopting speed limits).	**

Table 21 Continued

Sector	Measure	Competence	Description	Priority
Biomass burning	Biomass combustion use	Local	Communication campaign through the media and dissemination of "best practices". Central theme: the improper use of firewood in small household systems involves a number of negative effects on the environment and health. The practical suggestions relate to the choice of the type of plant and wood, the need for proper installation and maintenance and monitoring of the adequacy of combustion.	***
<3% of PM$_{2.5}$ (measures to prevent future impact)	Biomass combustion regulations	National and local	Energy and emission classification for biomass burning appliances. Rules for new systems installation and for proper maintenance. Census of domestic plants using wood for heating (including existing ones) in order to contain emissions. In 2015, following the introduction of the classes of emission, regulate the use of equipment characterised by high emissions.	***
	Reduction of low efficiency wood burning for residential heating	Local	Information material and training to discourage citizens from this inefficient form of energy are needed.	***
	Open biomass burning	Local	Regulatory for open burning activities with particular reference to agriculture, forestry and construction. In general, it is forbidden but the regional government allows them with some restrictions and conditions.	***
Inter-sectorial	Environmental education and awareness raising	Local	Communication campaigns through the media and dissemination of "best practices" should be promoted in order to raise awareness of the population on the opportunity of the previous measures.	***

negligible, such as power generation and biomass combustion, the latter however being important in some villages of the Protection Area 2 of Barcelona owing to domestic heating. The high penetration of natural gas in the domestic and residential heating of Barcelona accounts for a relatively low contribution of this source to PM levels. Furthermore, the 2007–2010 AQ Plan banned the use of fuels for power generation other than natural gas in and around the city of Barcelona. Since there was an oil combustion source detected in our study, it should be attributed near exclusively to shipping emissions. Finally, the large effort made by the public transport company that has achieved a bus fleet made by 39% of natural gas buses, 43% buses retrofitted with SCRT, and 8% hybrid buses is noteworthy. Based on the PM speciation and source apportionment results and on the measures already implemented, AIRUSE has therefore selected two main categories of mitigation measures: priority (three stars in Table 21 for road traffic, mainly private traffic) and ancillary (two stars for other local sources, such as metallurgy, heavy oil combustion and urban works). Moreover, AIRUSE encourages horizontal (or inter-sectorial) measures that will have a medium-long-term beneficial effect on air quality.

3.4.2 Florence (FI-UB)

3.4.2.1 Source Apportionment

For the city of Florence (January 2013–January 2014), the best solution was found pooling PM_{10} and $PM_{2.5}$ samples in a single input matrix for PMF, comprising 232 samples, 25 strong species (EC, OC, Levoglucosan, Na, Mg, Al, Si, S, Cl, K, Ca, Ti, Mn, Fe, Ni, Cu, Zn, As, Se, Br, Rb, Sr, NH_4, NO_3 and V), 4 weak species (Ba, Pb, Cr and Cd) and setting PM as total variable with 400% uncertainty.[40] The distribution of residuals, G-space plots, Fpeak values and Q values were explored for solutions with number of factors varying between 6 and 11. The most reliable solution was found with 10 factors/sources, with a minimum (base run) Q robust value of 3878, found over 30 runs (seed number 25), which resulted 13% lower than the theoretical Q value (4460). Species concentrations were reconstructed within 80–100%, with few exceptions (As 72%, Pb 68% and Cd 69%).

Some constraints were added to the base run solution:

1. Pulling down the EC and OC contributions in the Saharan Dust source profile and the NO_3 contribution in the Sea Salt profile. The % dQ was set at 1% for each elemental constraint and the converged results used totally 0.9% dQ.
2. Pulling down the Saharan Dust source contributions during a period (1-22/07/13) when the advection of desert dust can be excluded on the basis of all the used transport models (Skiron, NMMB and Hysplit). The % dQ was set at 0.5% for each day and the converged results used totally 1.4% dQ.
3. Pulling the difference of source contributions between PM_{10} and $PM_{2.5}$ to zero, only for those days and sources where the $PM_{2.5}$ contribution

was higher than the PM_{10} contribution in the base run solution (55 days). The % of dQ was set at 0.5% for each constraint and the converged results used totally 7.0% dQ.

The 10 sources identified in Florence were: aged sea salt (SEA, Table 12), Saharan dust (SAH, Table 22), secondary sulfate and organics (SSO, Table 9), vehicle non-exhaust (NEX, Table 6), biomass burning (BB, Table 23), secondary nitrate (SNI, Table 7), vehicle exhaust (VEX, Table 5), heavy oil combustion (HOC, Table 11), local dust (LDU, Table 8) and fresh sea salt (FSS, Table 12).

The VEX is mainly composed of organic particles (it accounts for 26% of OC) from motor exhaust, as well as EC (36%) (Tables 5 and 13–17). The ratio OC/EC of 3.7 reveals a significant proportion of secondary organic aerosols as in BCN-UB. Another secondary product, in this case mostly from motor

Table 22   PMF factor profile for the biomass burning sources at four cities ($\mu g/\mu g$).

	Biomass burning			
	MLN-UB	POR-TR	FI-UB	ATH-SUB
EC	0.08192	0.14343	0.10723	0.04372
OC	0.37959	0.41872	0.65430	0.11529
Levo	0.04312	0.07195	0.08322	n.i.[a]
Si	0.00000	0.00170	0.00072	0.00701
Al	0.00254	0.00060	0.00028	0.00290
Ca	0.00000	0.00000	0.00000	0.00000
Fe	0.00513	0.00069	0.00003	0.00153
K	0.04254	0.01523	0.02940	0.01736
Mg	0.00000	0.00000	0.00000	0.00000
Na	0.00000	0.00000	0.00000	0.00000
Cl	0.01592	0.03067	0.00161	0.00114
Ti	0.00017	0.00005	0.00000	0.00019
V	0.00000	0.00002	0.00000	0.00000
Cr	0.00000	0.00001	0.00004	n.i.[a]
Mn	0.00008	0.00000	0.00001	0.00018
Ni	0.00000	0.00003	0.00005	0.00001
Cu	0.00004	0.00006	0.00000	0.00000
Zn	0.00000	0.00000	0.00061	0.00286
As	n.i.[a]	0.00002	0.00001	0.00024
Se	n.i.[a]	n.i.[a]	0.00001	n.i.[a]
Rb	0.00008	0.00005	0.00005	n.i.[a]
Sr	n.i.[a]	n.i.[a]	0.00001	0.00017
Cd	n.i.[a]	0.00001	0.00001	0.00005
Sn	n.i.[a]	0.00011	n.i.[a]	n.i.[a]
Sb	n.i.[a]	0.00007	n.i.[a]	0.00007
Ba	n.i.[a]	0.00000	0.00004	n.i.[a]
Pb	0.00059	0.00034	0.00033	0.00181
Br	0.00011	0.00023	0.00026	0.00043
S	0.00274	0.00564	0.01447	0.00000
NH_4	0.00000	0.01997	0.00000	0.03196
NO_3	0.00000	0.04935	0.00755	0.00000

[a]n.i.: Not included in PMF.

Table 23 PMF factor profile for the Saharan dust source at FI-UB (μg/μg).

	Saharan dust FI-UB
EC	0.00000
OC	0.00000
Levo	0.00471
Si	0.25522
Al	0.11004
Ca	0.04530
Fe	0.07380
K	0.03775
Mg	0.02803
Na	0.01018
Cl	0.00000
Ti	0.00607
V	0.00009
Cr	0.00009
Mn	0.00068
Ni	0.00004
Cu	0.00107
Zn	0.00000
As	0.00001
Se	0.00000
Rb	0.00034
Sr	0.00036
Cd	0.00000
Ba	0.00063
Pb	0.00000
Br	0.00000
S	0.03541
NH_4	0.00000
NO_3	0.00013

exhaust emissions, is ammonium nitrate, which was found as a separate factor (SNI). The average source contribution from VEX was 2.5 μg m^{-3} both in PM$_{2.5}$ and PM$_{10}$ indicating that this source is completely in the fine mode. This contribution corresponds to 13% of PM$_{10}$ and 18% of PM$_{2.5}$ mass concentrations (Table 18).

The NEX source is traced by Cu, Fe, Ba and Mn (Tables 6 and 13–17); it is mainly composed of OC, EC and Fe, but it also receives significant mass contributions by crustal species like Ca, Si and Al. This source represents the road dust that is resuspended by traffic, but it also includes a contribution from direct brake and tire wear. The average source contribution from NEX is 1.9 μg m^{-3} (10%) in PM$_{10}$ and 0.3 μg m^{-3} (2%) in PM$_{2.5}$ (Table 18), with a very low PM$_{2.5}$/PM$_{10}$ ratio (0.15), which is reasonable for a dust resuspension/abrasion source. Higher values are observed in winter (Figure 13).

The SNI factor explained about 24% of NH_4^+ and 66% of NO_3^- (Tables 7 and 13–17) as the oxidation product of local gaseous NO$_x$ emissions. On the basis of the emission inventory, in Florence $\sim$80% of NO$_x$ is owing to road traffic and $\sim 10\%$ to domestic heating, with only $\sim 2\%$ owing to stoves and chimneys (http://servizi2.regione.toscana.it/aria/); it is, however, highly suspected that this datum underestimates the contribution of domestic heating by BB to NO_x. On the annual basis, average contributions were 2.2 µg m^{-3} and 1.9 µg m^{-3} in PM_{10} (11%) and $PM_{2.5}$ (14%), respectively (ratio $PM_{2.5}/PM_{10}$ of 0.9) (Table 18). This source shows a strong seasonality, with higher values during the cold period (when most of the exceedances of the daily PM_{10} limit value occur) and very low contributions in summertime. This behaviour may be owing not only to higher emissions, but also to lower mixing layer heights.[45]

The total contribution from traffic can be therefore estimated as the sum $VEX + NEX + 0.8 \times SNI$, which results in 6.1 and 4.3 µg m^{-3} in PM_{10} (31%) and $PM_{2.5}$ (31%), respectively (Table 19).

The BB profile is characterized by levoglucosan (84%), EC (30%), OC (36%), K, and, to a lesser extent, by S, Cl, Zn, Br, Pb and nitrate (Tables 13–17 and 23). During bark and waste wood combustion Zn, Br and Pb may also become relevant.[46,47] The OC/EC ratio in this profile (6.1) is within the ranges reported in literature for this source.[48] The average source contribution from BB is 3.0 µg m^{-3} (16%) in PM_{10} and 2.9 µg m^{-3} (21%) in $PM_{2.5}$ (Table 18), with a very high $PM_{2.5}/PM_{10}$ ratio (0.95), which is reasonable for a combustion source. As expected, the contributions of this source are highly season-dependent with maxima in winter when, as already observed, most of the exceedances of the limits occur (Figure 14).

The SSO factor, which accounts for 76% of S and 72% of NH_4^+ (Tables 9 and 13–17), is the result of the formation of secondary sulfate in the atmosphere from the photochemical oxidation of gaseous sulfur oxides (which in Florence are mostly of regional origin). On the basis of the emission inventory, in Florence $\sim 50\%$ of SO_2 (http://servizi2.regione.toscana.it/aria/) is owing to the power production sector. It is worth noting that in Tuscany there is the most important geothermal power plant of Italy and a fraction of sulfate may be produced by H_2S transformations in the atmosphere. Condensation of VOCs is suggested by the high content of OC. The annual average contribution is about 4 µg m^{-3} both in PM_{10} and $PM_{2.5}$, indicating that this source is completely in the fine mode. This amount corresponds to 20% of PM_{10} and 29% of $PM_{2.5}$ mass concentrations (Table 18). Source contributions were generally higher in summer than in winter (Figure 14) owing to the enhanced photochemical activity and the higher air circulation during warmer months, which favor the formation of secondary sulfates and their distribution over the regional area.

The HOC factor is traced by V and Ni and is characterized by high concentrations of EC, OC and S (Tables 11 and 13–17). This source reflects the influence of residual oil combustion processes, mostly from activities that are located outside the city, like energy production, refinery, industrial plants and shipping. HOC particles are rather fine (0.85 as $PM_{2.5}/PM_{10}$ ratio) and, similarly to SSO, they contributed more in summer than in winter, owing to the higher air circulation, which favors their transport and distribution on the regional area. On annual average, HOC increased PM_{10} and $PM_{2.5}$ concentrations by 0.9 (5%) and 0.8 (6%) µg m^{-3}, respectively (Table 18).

PMF identified two mineral dust sources: SAH and LDU, both traced by the typical crustal species, such as Al, Si, K, Ca, Ti, Fe, Rb and Sr (Tables 8 and 22). The first one is characterized by a source profile that is very close to the average crust composition, with enrichment factors (EF), calculated with respect to Al using the average continental crust composition reported by Mason (1966) and Rahn (1976), that are all close to one: 0.99 (Mg), 0.83 (Si), 1.08 (K), 0.92 (Ca), 1.0 (Ti) and 1.1 (Fe). Conversely. the LDU source profile is contaminated by OC and nitrate, it is highly enriched in Ca (EF 14.0) and Fe (2.1) and also the EFs of Mg, Si, K and Ti deviate from unity by 30–60%. While LDU contributes all year long, the SAH time trend is characterized by concentration peaks during days classified as Saharan intrusions (on the basis of both models and satellite observations) and it is close to zero in all other days (Figure 14). As expected, for both these sources the average $PM_{2.5}/PM_{10}$ ratio is quite low: 0.24 for SAH and 0.12 for LDU (the lower ratio for local dust with respect to Saharan dust seems reasonably explained by the particle size fractionation experienced during the transport).

PMF identified two sea salt sources: fresh sea salt (FSS) and aged sea salt (SEA). The first one is traced by typical marine elements like Na, Cl and, to a lesser extent, Mg and Br (Table 12), with inter-elemental ratios (Cl/Na: 1.9, Mg/Na: 0.15) in accordance with those reported in the literature for seawater. The time pattern is characterized by short episodic peaks, occurring when air masses are directly transported by strong winds to Florence from the seaside (mainly from the Tyrrhenian Sea). The SEA source is also characterized by typical sea salt elements like Na, Mg and Br, but completely depleted in Cl and enriched in sulfate, nitrate and OC (Table 12). It is well known that different heterogeneous reaction between airborne sea-salt particles and gaseous pollutants, like nitric and sulfuric acid, may lead to Cl volatilization and nitrates and sulfate formation.[49,50] Thus, this SEA source is a mixture of anthropogenic and natural source contributions. As expected, for both sources, the $PM_{2.5}/PM_{10}$ ratio is quite low: 0.18 for SEA and 0.06 for FSS. The overall sea salt contribution (SEA + FSS) was 1.8 (9%) and 0.2 (2%) µg m^{-3}, respectively for PM_{10} and $PM_{2.5}$ (Table 19).

Finally, average source contributions to PM_{10} and $PM_{2.5}$ during the days in which the PM_{10} concentration is above 50 µg m^{-3} are reported in Table 20. Traffic (42.7 and 32.4% of PM_{10} and $PM_{2.5}$, respectively, as the sum of VEX, NEX and about 80% of secondary nitrate) and biomass burning (29.7 and 31.8% of PM_{10} and $PM_{2.5}$) are the sources that give the most relevant contribution.

3.4.2.2 Validation of PMF Results From Daily Data With Hourly Data

Five and four factors were identified in the fine and in the coarse fraction, respectively: traffic, aged sea salt, local dust in both fractions, secondary

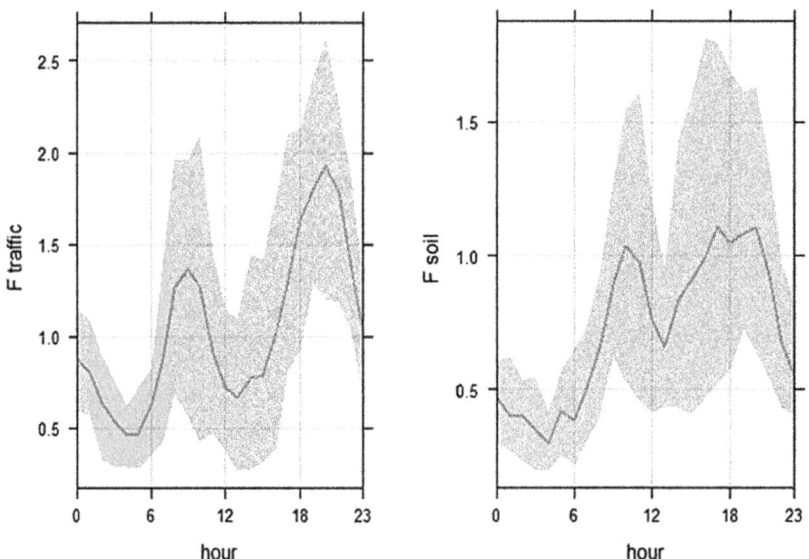

Figure 5   Daily cycle of traffic and local dust sources contributions in Florence (FI-UB) in the fine fraction (arbitrary units).

sulfate and biomass burning in the fine fraction, and fresh sea salt in the coarse fraction.

The traffic source is mainly traced by Fe, Cu and Mn, both in the fine and in the coarse fraction. It corresponds to the non-exhaust source (NEX) and represents the road dust that is re-suspended by traffic and the contribution from direct brake and tire wear. This is confirmed by its time trend with peaks during traffic rush hours (Figure 7) and by the source polar plot, which shows that it is a local source (Figure 8). The biomass burning source is characterized by K, and, to a lesser extent, by S, Cl, Zn, Pb. The time trend of this source supports the identification of this source as biomass burning for domestic heating: it is characterized by a periodic pattern with peaks starting in the evening and lasting several hours. The absence of the evening-night peak on some days is explained by the meteorological conditions. The polar plot shows that it is a local source, also with some influence from the hills surrounding Florence on the north/east side of the town, where many houses are equipped with chimneys.

The secondary sulfate source does not exhibit a clear diurnal variation, indicating a regional rather than a local source, as confirmed by the polar plot. The smooth hourly time pattern is typical of secondary aerosols. The absence of the secondary nitrates source is owing to the fact that Streaker samples cannot be analyzed by IC to obtain nitrate concentration (the sulfates source is on the contrary identified as traced by elemental S).

As for the daily samples, the PMF identified two sea salt sources in the coarse fraction: fresh sea salt (FSS) and aged sea salt (SEA). The first one's

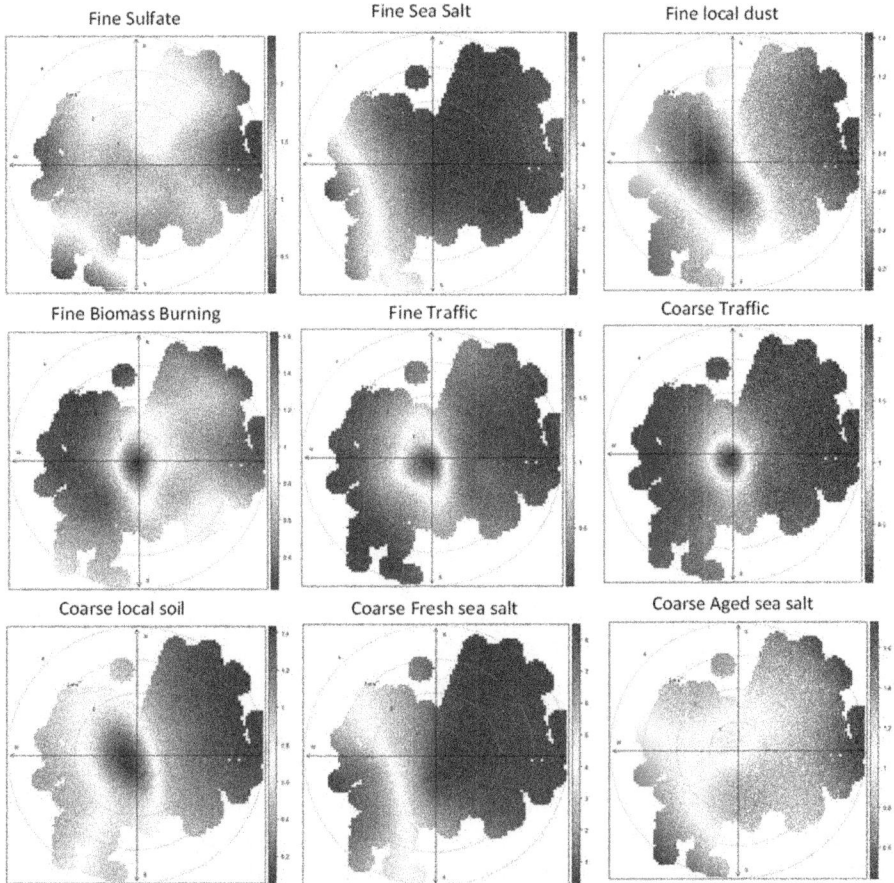

Figure 6 Polar plots of hourly source contributions (a.u.) in Florence (FI-UB).

time pattern is characterized by episodic peaks, occurring when air masses are directly transported by winds to Florence from the Tyrrhenian Sea (see the polar plot in Figure 14). The SEA source again shows its provenance from the Tyrrhenian Sea. The source identified in the fine fraction is intermediate between the two.

The local dust time trend normally shows peaks during the day when anthropogenic activities are maxima. No Saharan dust intrusions were present during the weeks when the Streaker sampler was used. The polar plots show a contribution in the fine and the coarse fraction both local and from the center of the town where most of the construction works take place.

The La/Ce ratio may be used as a tracer of the influence of emissions from La-based oil cracking in petrochemical plants. In FI the average value of 1.0 indicates the possible influence of this type of emissions; this is supported by the polar plot showing high La/Ce values for western winds (Figure 8), where a petrochemical plant is located, about 90 km away, in the province of Livorno.

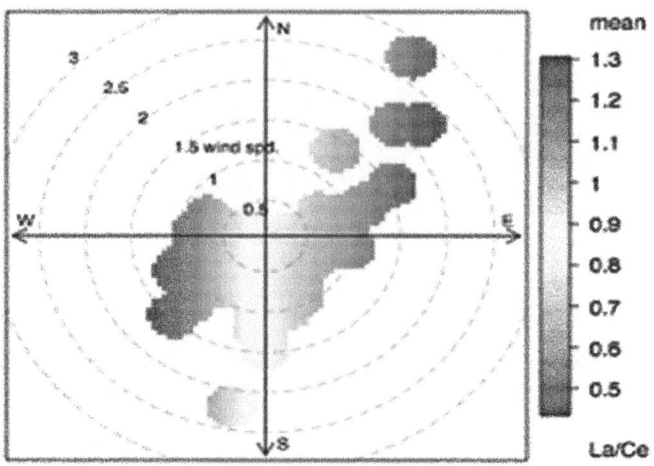

Figure 7   Polar plots of La/Ce in Florence (FI-UB).

3.4.2.3   Suggested Measures to Improve Air Quality

During the study, the AIRUSE urban background site of Florence (FI-UB) met the requirements of the Directive 2008/50/EC for PM_{10} and $PM_{2.5}$. The study year, 2013, was characterized by a meteorology that favored dispersion and cleansing of atmospheric pollution. According to the national legislation, in Italy, the regional agencies for environmental protection (ARPA in each regions) are responsible for the Air Quality Monitoring; the competence of regional/local measures to ensure the attainment of the air quality legal limits is the responsibility of the Regional Administration, with the technical support of their ARPA, and they have to take any necessary measure to ensure the attainment of the air quality legal limits in national territory. The exceedances of PM_{10} and NO_2 limit values during the last decade have led to the development of the Regional Air Quality Plan 2008–2010. Concerning the city of Florence, a clear PM_{10} downward trend has been observed revealing a general improvement in air quality, probably as the result of multiple influences, such as the European, national, regional and local actions, the economic downturn and the favorable meteorology. In spite of the aforementioned decrease of PM concentrations, more effort has to be made in order to attain the WHO guidelines and to prevent future exceedances under the economic recovery and less favorable meteorological scenarios.

Road traffic is clearly the main local source of PM_{10} and $PM_{2.5}$ in Florence, but during the most polluted days biomass burning also plays a relevant role (in recent years the use of biomass burning for domestic heating has continuously increased). It is therefore evident that mitigation measures must focus on the transport sector and on domestic heating systems using biomass (Table 24). Another important contribution is owing to non-traffic secondary particles (which also have a regional origin). Other sources are less important (such as urban works and heavy oil combustion). Based on the PM speciation and source apportionment results and on the measures

Table 24 Measures and priority degree proposed by AIRUSE for Florence based on PM speciation and source apportionment obtained in 2013 at FI-UB.

Sector	Measure	Competence	Description	Priority
Road traffic	Improving car and public transport fleet	National and local	Facilitate the replacement of vehicles <Euro 3 diesel with Euro 6 adopting restrictions on their movement and/or subsidies. Prohibition on movement for <Euro 2 diesel. Promotion of the LPG, hybrid engine and electric vehicles. Incentive to the installation of particle filters in heavy duty commercial vehicles. Promoting the replacement of 2-stroke motorcycles by 4-strokes. Upgrading of vehicles for public transportation with the introduction of high environmental standards.	***
31% of PM₁₀: Exhaust 13% Non-exhaust 10% Nitrate 8%	Low emission zones (LEZ)	Local	Compulsory establishment of LEZ in all municipalities with >15 000 inhabitants.	***
31% of PM₂.₅: Exhaust 18%	Improving public transportation	National and local	Continuous update and improvement for an economic, environmentally friendly and faster public transportation (train, bus and tram). Reduction of congestion delays by using bus lanes.	***
Non-exhaust 2%		Local	Increase the number of routes and frequency of service, in order to reduce waiting and exchange times.	***
Nitrate 11%		Local	Regular inspection programmes for public vehicles for improved maintenance.	***
		Local	Make public transport more appealing (reduced crowding, better seats, cleaner vehicles, on-board internet access and other amenities). More efficient user information through signs, websites and mobile telephone applications that provide information on transit routes, fares and real-time vehicle arrival predictions.	*
	Random tests on vehicle emissions	Local	Random tests on vehicles (mainly diesel trucks) to ensure that smoke emissions are within acceptable limits.	***

Table 24 Continued

Sector	Measure	Competence	Description	Priority
	Mobility card	Local	Improve the public transport system. Incentives to alternative mobility with a prepaid "Mobility Card" for public transport.	***
	Lowering number of cars in the urban areas	Local	Implementation of measures that attribute a direct fee to a car entering the city, such as road and congestion charges. Restrict vehicle circulation to neighbourhoods in specific areas of the city.	***
	Parking	Local	New strategic parking lots at main transport interfaces (train, bus and tram stations) at the outskirts of the city in order to promote the combined use of car and public transport. High charges for parking in the city center for nonresidents and cheap charges for eco-cars	***
	Cycling paths and pedestrian lanes	Local	Facilitating alternative mobility by the construction of new cycling paths and the increase of the city surface reserved for pedestrians. Develop cycle parking and changing facilities; develop cycle sharing programmes; improved connectivity to public transport.	***
	Car-sharing	Local	Reduce single-occupancy car journeys by promoting car sharing with a proper connection to public transport services and cycling.	**
	Lowering traffic speed	Local	Reduction of traffic speed on specific roads to reduce congestion and emissions from stop-and-go traffic.	**
	Awareness	National	Integration of mobility awareness into school curricula.	***
Biomass burning 16% of PM_{10} 21% of $PM_{2.5}$	Biomass burning regulations	Local	Rules for new systems installation and for proper maintenance. Mandatory certification of residential combustion equipment. Census of domestic heating systems using wood for heating. Regulation for the use of equipment characterised by high emissions. Regulatory program to ban or restrict some wood burning devices in new homes, create financial incentives for old stove replacement, obligation to substitute old stoves upon re-sale of a house with energy efficient and emission controlled devices.	***

(30% of PM$_{10}$ and 32% of PM$_{2.5}$ during high pollution episodes)	Local	Updating of the regulations regarding the installation of biomass burning plants for energy production. The use of biomass for energy production must be steered towards producing heat, favouring the use of heat in the district heating network. Prohibition of the use of chimneys and wood stoves (when they are not the only heating system) during stagnant meteorological periods.	***
Biomass burning use	Local	Communication campaign through the media (web, radio, TV, newspapers) and dissemination of "best practices". Central theme: the improper use of firewood in small household systems involves a number of negative effects on the environment and health (including indoor air pollution) that, otherwise, may be at least partially contained by the correct use of biomass leading to a reduction in fuel consumption, lower emissions and increased safety in the home. The practical suggestions relate to the choice of the type of plant and wood, the need for proper installation and maintenance and monitoring of the adequacy of combustion. Educational programmes (what to burn, how to burn, choosing the right appliance).	***
Open biomass burning	Local	Because of the generalised practice of agricultural biomass burning, regulations for open burning activities with particular reference to agriculture and forestry are necessary. Development of a system for collecting firewood from to feed small controlled biomass heating plants.	***
Reduction of low efficiency wood burning for residential heating	Local	Information material and training to discourage citizens from the use of low efficiency wood are needed.	***
Secondary particles Reduction of secondary sulfates precursors (SOx, NH$_3$ and H$_2$S)	Local	Promote the conversion of oil burning power plants to gas power plants. Control of H$_2$S emissions from geothermal powered plants and obligation to install AMIS plants.	***

Table 24 Continued

Sector	Measure	Competence	Description	Priority
27% of PM$_{10}$	Reduce NH$_3$ emissions	Local	Application of "guidelines" for regional containment of ammonia resulting from agricultural, livestock and waste practices.	***
31% of PM$_{2.5}$	NO$_x$ reduction	Local	Introduction of a centralized system in buildings with more than 5 units. Favour the extensive use of valves for heat metering in each residential building.	***
Local dust/urban works	Water spraying	Local	Apply water during construction, operations (earthmoving, demolition, grading).	***
12% of PM$_{10}$	Storage and handling of dusty materials	Local	Cover bulk material during storage; covering the cargo beds of haul trucks with a suitable closure to minimise wind-blown dust emissions and spillage; green curtains in areas with high diffusive dust emissions; reduction of vehicle speed	***
1% of PM$_{2.5}$	Unpaved areas	Local	Apply water, nano-polymers, gravel, or pave unpaved parking lots or roads; set additional control requirements for unpaved roads (*e.g.* vegetating, adopting speed limits).	**
Inter-sectorial	Environmental education and awareness raising	Local	Communication campaigns through the media and dissemination of "best practices" should be promoted in order to sensitise population on the opportunity of the previous measures. Integration of mobility awareness into school curricula.	***

already implemented, AIRUSE has therefore selected three main categories of mitigation measures: priority (three stars in Table 24, mainly for road traffic and biomass burning), medium-priority (two stars) and ancillary (one star). Moreover, AIRUSE encourages horizontal (or inter-sectorial) measures that will have a medium-long-term beneficial effect on air quality.

3.4.3 Athens (ATH-SUB)

3.4.3.1 Source Apportionment

The best solution for Athens (ATH-UB) was achieved using the combined dataset of available PM_{10} and $PM_{2.5}$ data. The results of the combined dataset were compared and found to be very consistent with the results obtained using PM_{10} and $PM_{2.5}$ datasets separately. After the examination of the residuals, G-space plots, Fpeak values and Q values with number of factors varying between 5 and 10, a solution of 8 factors was found to be the optimum. The matrix comprised 256 samples with 22 strong species (EC, OC, Na, Mg, Al, Si, Cl, K, Ca, Ti, Mn, Fe, Ni, Cu, Zn, Br, V, As, Pb, NH_4^+, NO_3^-, SO_4^{2-}) and 3 weak species (Sr, Cd, Sb). PM mass was set as total variable with 400% uncertainty assigned to it, while 5% extra modeling uncertainty was added to the individual species. Q robust was found to be similar to the theoretical Q (Q robust $= 4301$ *versus* Q theoretical $= 4216$, difference of 2%).

For the cases with daily $PM_{2.5}$ contribution significantly higher than PM_{10} contribution, $PM_{2.5}$ contribution was either set to zero (2 cases) or pulled down maximally (4 cases), depending on the respective PM_{10} contribution. The % of dQ was left at the default option of 0.5%.

The 8 factors identified for both PM_{10} and $PM_{2.5}$ in Athens were attributed to the following sources: heavy oil combustion (HOC, Table 11), vehicle exhaust (VEX, Table 5), secondary nitrate (SNI, Table 7), mineral (MIN, Table 8), vehicle non-exhaust (NEX, Table 6), biomass burning (BB, Table 23), secondary sulfate and organics (SSO, Table 9) and fresh sea salt (SEA, Table 12).

OC was pulled up in the heavy oil combustion factor. This constraint resulted in biomass burning factor losing all its OC. For this reason, finally two constraints on the source profiles were applied simultaneously: (i) OC was pulled up in the heavy oil combustion factor and (ii) OC was set to have the original value of the unconstrained solution in the biomass burning factor. The % of dQ was left at the default option of 0.5%.

SNI explains the highest percentage of the nitrates, while it also includes major sea salt components such as Na and Mg (Tables 7 and 13–17). Cl is absent from this factor owing to Cl depletion, a phenomenon known to occur when urban pollutants such as nitric acid remaining in the gaseous phase in poor NH_3 environments interact with sea-salt and other available species (minerals).[51,52] This factor is therefore associated with the mixing of urban gaseous emissions (from traffic, fossil fuel combustion or biomass burning) with marine aerosol and is representative of relatively fresh local urban emissions. The secondary nitrate factor contributes more in the colder months of the year. This factor contributes 15% of PM_{10} (3.2 µg m^{-3}) mass and 6% of $PM_{2.5}$ mass (0.7 µg m^{-3}, Table 18). The higher contribution to coarse particle fraction ($PM_{10}/PM_{2.5}$ ratio equal to 0.22) is owing to the availability of sea salt and other basic species (such as K and Ca) in this fraction.

NEX, as expected, is a mixture of elements from various origins. This factor is traced by metals that originate from brake, tire and vehicle body wear (Cu, Zn, Cd, Sb, Fe) 9–67% and from the wear of the road surface (Ca) 23% (Tables 6 and 13–17). This source contributes 8% of PM_{10} mass (1.8 µg m^{-3}) and 5% of $PM_{2.5}$ mass (0.6 µg m^{-3}, Table 18). It has no clear seasonal pattern probably owing to low seasonal variation of the emission potential. Particles from this source are rather coarse ($PM_{2.5}/PM_{10}$ ratio of 0.31).

The VEX factor contains organic aerosol (56% of OC explained variation) from motor vehicle exhaust, as well as EC (31%) and trace metals from brake wear (Cu, Sb, Cd) (Tables 5 and 13–17). Average source contributions from VEX were 2.1 µg m^{-3} (10%) to PM_{10} levels and 1.7 µg m^{-3} (15%) to $PM_{2.5}$ levels with an average $PM_{2.5}/PM_{10}$ ratio of 0.82 (Table 18). The high $PM_{2.5}/PM_{10}$ is to be expected because primary particles that originate from this source are fine. The contribution from this source for both PM fractions shows no significant seasonal trend (Figure 13).

The HOC factor is traced by the high variance of EC, V and Ni (Tables 11 and 13–17). Heavy oil combustion in the area may originate from residential heating, the petrochemical industries at the west of the Athens Metropolitan area and shipping emissions from the Piraeus harbor and the close Aegean shipping routes.[53] On annual average, HOC contributes 4% of PM_{10} and 7% of $PM_{2.5}$ mass (0.8 µg m^{-3} for both size fractions, Table 18). $PM_{2.5}/PM_{10}$ ratio for this source is 0.94 as expected for aerosols originating from combustion processes. No clear seasonal trend is observed for HOC probably because of the variety of the sources that may contribute to this factor (residential heating and shipping), and may have higher contribution in different seasons, however large contributions during the winter months (December–February) are observed while contributions decreased significantly in early spring (Figure 14).

The fresh sea salt factor (SEA) is characterized by naturally occurring seawater bearing Na, Cl and Mg (Tables 12–17). This source has an impact almost exclusively on the PM_{10} fraction. It increases PM_{10} mass by 5% (1.0 µg m^{-3}) and $PM_{2.5}$ by 1% (0.1 µg m^{-3}) per year (Table 18). This source has no clear seasonal trend (Figure 14) and offers a number of high intensity incidents, probably when the conditions are favorable for direct transfer of marine aerosol from the sea to the sampling location.

The factor secondary sulfate and organics (SSO) explains the highest percentage of SO_4^{2-} and NH_4^+ (Tables 9 and 13–17), while it also contains OC. These components are the result of secondary aerosol formation in the atmosphere from the photochemical oxidation of emitted gaseous sulfur oxides and VOCs, the former combined with NH_3. These processes may have variable time scales depending on the kinetics of reactive species and meteorological conditions. Aerosol formation may be owing to local emissions converted shortly after emission or be the result of formation during regional and long range transport of precursor species released by distant anthropogenic and natural sources. Source contributions were generally higher in summer than in winter owing to the higher photochemical activity during warm months. Annual average contributions were 19% (4.0 µg m^{-3}) and 33% (3.8 µg m^{-3}) for PM$_{10}$ and PM$_{2.5}$, respectively (PM$_{2.5}$/PM$_{10}$ ratio of 0.96, Table 18). It is known from previous studies that these aerosol components display relatively high and homogeneous concentrations across the greater area of Greece and it is more the product of long range transport than of local processes.[52,54] In the case of SO$_2$, there is more evidence that it is transferred to the Mediterranean Region where it is transformed to sulfate owing to the high photochemical activity in the area.

The mineral factor (MIN) contains elements with crustal origin, all with a share of mass higher than 34% (Mg, Al, Si, K, Ca, Ti, Mn, Fe, Sr) (Tables 8 and 13–17). The MIN source is interpreted as a mixture of several sources, including soil resuspension, urban works, regional mineral dust and Saharan dust. However, ongoing work within the AIRUSE project can provide us with means to quantify Saharan dust contribution (SAH) dust and separate it from the local dust (LDU). The overall MIN contribution was 5.1 µg m^{-3} (24%) to PM$_{10}$ levels (3.0 µg m^{-3} from SAH) and 1.2 µg m^{-3} (11%) to PM$_{2.5}$ levels (0.7 µg m^{-3} from SAH, Table 18). The average PM$_{2.5}$/PM$_{10}$ ratio of 0.23 confirms the expected coarse size of the particles coming from this source. Contributions from MIN are higher in the warm season (spring and summer months), while the peak contributions correspond to Saharan dust events (Figure 14).

The biomass burning factor (BB) is identified by EC, OC and K with variations of 6–18% in the factor (Tables 13–17 and 23). The factor contains tracers of waste burning, such as As, Cd, Sb and Pb with variations ranging from 12 to 72%. This source has a significant contribution to both PM$_{10}$ (1.4 µg m^{-3}, 7%) and PM$_{2.5}$ (1.2 µg m^{-3}, 11%, Table 18) mass. Owing to the economic crisis and the increased diesel oil prices (which was the most common means of residential heating in Greece), citizens of Athens have turned to alternative ways of heating, such as wood burning. In many cases, treated wood or even combustible waste are used as fuel. This source has a clear seasonal pattern contributing to PM mass exclusively in the colder months of the year, with the exception of some events taking place during August, which could be attributed to some uncontrolled waste burning incidents or forest fires. Finally, average source contributions to PM$_{10}$ and PM$_{2.5}$ during the days in which the PM$_{10}$ concentration is above 50 µg m^{-3} are reported in Table 20. The mineral dust is the source that gives the most relevant contribution, therefore in Athens Saharan dust intrusions play a relevant role in air quality.

3.4.3.2 Validation with Hourly Elemental Data
Seven and five factors were identified in the fine and in the coarse fractions, respectively: traffic, aged sea salt, mineral dust, mineral dust (Ca enriched)

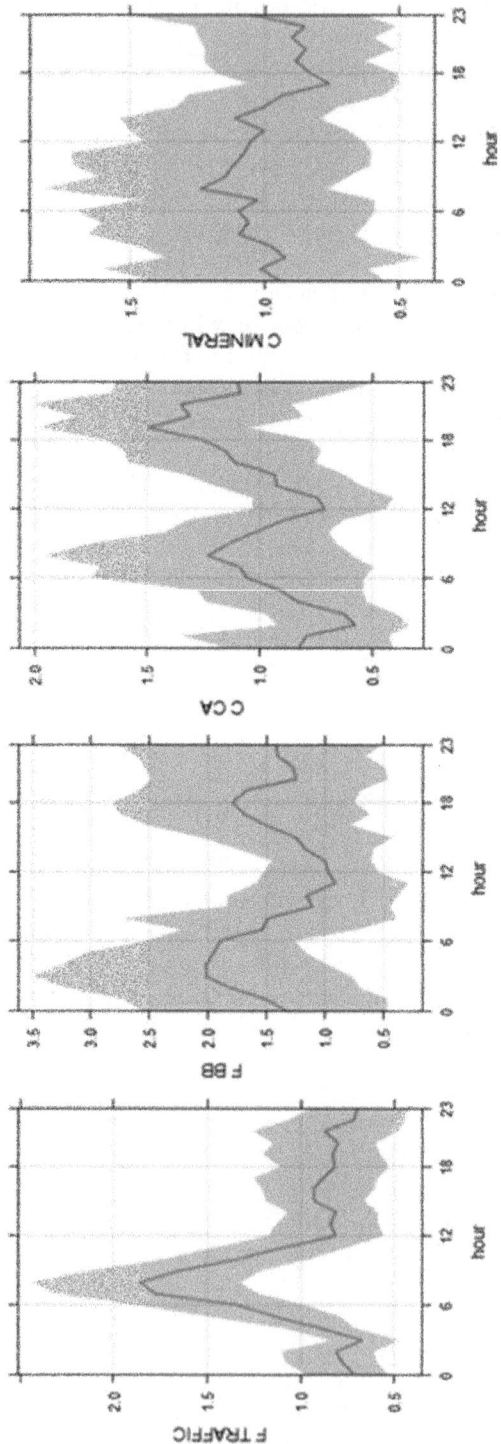

Figure 8   Daily cycle of traffic, biomass burning (fine fraction), mineral dust Ca enriched and mineral dust (coarse fraction) sources contributions in (ATH-SUB) in the fine fraction (arbitrary units).

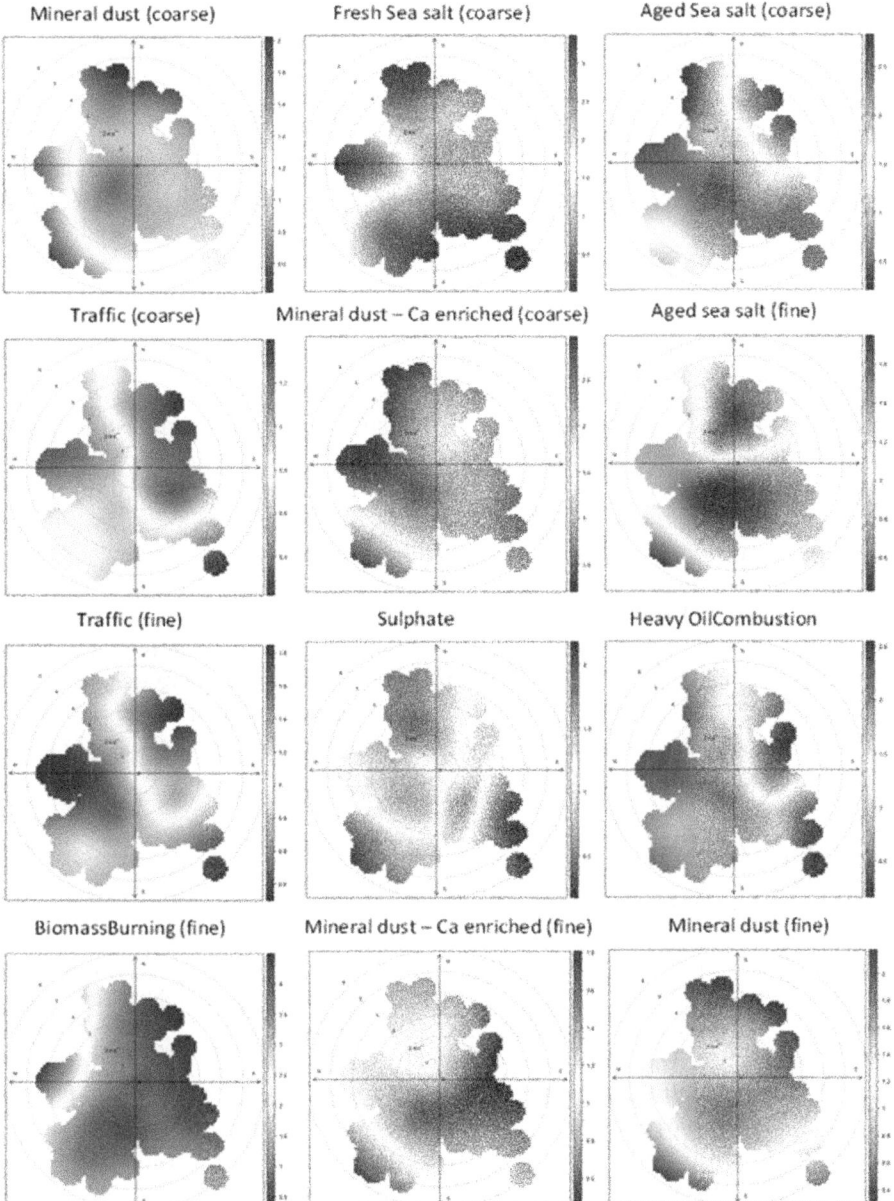

Figure 9   Polar plots of hourly source contributions (a.u.) in Athens (ATH-SUB).

in both fractions, fresh sea salt in the coarse fraction, secondary sulfate, heavy oil combustion and biomass burning in the fine fraction.

The traffic source is mainly non-exhaust source (NEX). The identification is confirmed by its time trend with peaks during traffic rush hours and by the source polar plot (Figures 9 and 10), which shows a stronger contribution from the eastern sector where a number of high speed and dense traffic ring road highways are located (Figure 11). The biomass burning (BB) source is characterized by a periodic pattern with peaks starting in the evening and lasting several hours (Figure 10). The polar plot shows an influence from the densely populated western suburbs and city center (Figure 11).

The secondary sulfate source does not exhibit a clear diurnal variation, indicating a regional rather than a local source, as confirmed by the polar plot (Figure 11). The smooth hourly time pattern is typical of secondary aerosols.

The heavy oil combustion (HOC) factor polar plot is similar to aged sea salt, pointing again towards aerosol transported from the sea (Figure 11).

3.4.3.3 Suggested Measures to Improve Air Quality

Measures of traffic management and emission reductions from vehicles in Athens are already in place but must be intensified. These already included an age limit of 23 years, which has been implemented for all urban, semi-urban and long distance buses. In addition, the limit of 11 years was set as the higher permissible age for buses in public transport. Under the provisions of the same law, economic incentives were given to the owners for the replacement of vehicles with new or used vehicles of lower age. Of the 5000 semi-urban and long distance buses licensed in Greece, 1846 buses have been replaced since 2004, of which 1746 were new and 100 were used with an age lower than 5 years. Table 25 summarises the measures proposed by AIRUSE for Athens based on PM speciation and source apportionment obtained in 2013 at the ATH-SUB site.

3.4.4 Porto (POR-TR)

3.4.4.1 Source Apportionment

For POR-TR (January 2013–January 2014), the best solution was found pooling PM_{10} and $PM_{2.5}$ samples in a single input matrix for PMF, comprising 226 samples, 27 strong species (EC, OC, levoglucosan, Na, Mg, Al, Si, P, S, Cl, K, Ca, Ti, Cr, Mn, Fe, Ni, Cu, Zn, Br, NO_3^-, Li, As, Rb, Sb and Ba), 6 weak species (V, NH_4^+, Cd, Sn, La and Pb) and setting PM as total variable with 400% uncertainty.[40] The distribution of residuals, G-space plots, Fpeak values and Q values were explored for solutions with numbers of factors varying between 6 and 9. The most reliable solution was found with 8 factors/sources, adding an extra 7% uncertainty and a minimum (base run) Q robust value of 7040.4, found over 30 runs (seed number 30), which exceeded the theoretical Q value (5800) by 21%. Species concentrations were reconstructed within 79–104%.

The following constraint was introduced: Pulling the difference of source contributions between PM_{10} and $PM_{2.5}$ to zero, only for those days and sources where $PM_{2.5}$ contribution was higher than PM_{10} contribution in the base run solution. The % of dQ was set at 5% for each constraint and the converged results used totally 2.9% dQ.

Table 25 Measures and priority degree proposed by AIRUSE for Athens based on PM speciation and source apportionment obtained in 2013 at ATH-SUB.

Sector	Measure	Competence	Description	Priority
Road traffic 23% of PM$_{10}$: Exhaust 10% Non-exhaust 8% Nitrate 5%	Low emission zones (LEZ)	National and local	Expand and prioritise the recently (2012) implemented unlimited access for electric and hybrid low emission cars in the so called "green ring", where regular vehicles are entering depending on odd/even days with respect to their number plate last digit.	***
	Parking	Local	Creation of large parking lots at main transport interfaces (train and metro stations) at the outskirts of the city (park and ride system) with incentives (low fares) in order to promote the combined use of car and public transport.	***
22% of PM$_{2.5}$: Exhaust 15% Non-exhaust 5% Nitrate 2%	Street cleaning	Local	• Tandem use of sweeping and, more importantly, water washing, especially during dry periods of the year and when African dust episodes are forecast. It is evident in Athens that non-exhaust traffic emissions lead to a major part of the coarse fraction of road dust that can be removed by street cleaning. • Street cleaning should be performed at the early morning, just before rush traffic hours and is specially indicated after African episodes.	***
	Promoting low-carbon and low-NO$_x$ vehicles and new technology vehicles	National and local	Implement further reductions in Road Tax and Import Tax for low emission vehicles (for NO$_2$ and PM). Incentives to withdraw aged private vehicles and replacement with modern (E5/E6) vehicles. Installation of particle filters on heavy duty commercial vehicles.	***
	Expand public transport network	Local	Continuous expansion of Metro lines (currently only 3) and improvement of Public Bus Network for a resource, environmentally friendly and faster public transportation (metro, train, and tram).	***
	Cycling paths and pedestrian lanes	Local	Facilitating alternative mobility by the construction of new cycling paths and the increase of the city surface reserved for pedestrians. Develop cycle parking and changing facilities; develop cycle sharing programmes; improved connectivity to public transport.	***

Table 25 Continued

Sector	Measure	Competence	Description	Priority
	Lowering number of cars in the urban areas	Local	Implementation of measures that attribute a direct fee to a car entering the city, such as road and congestion charges. Restrict vehicle circulation to neighbourhoods in specific areas of the city.	***
	Reducing road transportation for goods	National and local	Creating a terminal outside the Athens Metropolitan Area serviced by rail line to the Pireaus harbour while currently trucks travel for 50 km within the central axis of the Athens Metropolitan Area.	***
	Renewal of car/taxi/motorcycle fleet.	National	Subsidies and for increasing the share of hybrid, natural gas and LPG taxis. Promoting the replacement of 2-stroke motorcycles by 4-strokes.	***
	Reduced fares of public transport	National and local	Reduced fares for public transport during intensive Sahara dust intrusions or forecasted intense pollution episodes.	**
	Improving public fleet	Local	Increase the share of natural gas buses (currently at 35%). Enforce the measure of withdraw of old technology urban and regional buses.	**
	Vehicle and road maintenance	National and local	• Increase the frequency of inspection programmes to public vehicles to ensure that in-use engines continue to have functional controls and proper maintenance. • Maintaining roads in good repair to reduce the contribution of PM from road surface wear.	**
	Combat the illegal trade of adulterated fuel	National and local	Incidents of adulterated fuel circulation and use are still common in the Athens Metropolitan area and at national scale. Continuous controls are needed to eliminate this phenomenon.	**
	Industrial facilities	Local	Impose high standards for fuels and increase inspections to facilities. To update industrial activities inventory and to carry out a proper quantification.	**
	Stricter legislation for harbour	National and local	Docking at the Piraeus harbour is only permitted to vessels with engines operating with low sulfur content. These rules need to be enforced and monitored.	**
	Awareness	National Local	Integration of mobility awareness into school curricula.	*** ***

Source	Measure	Scale	Priority	Description
Shipping/heavy oil combustion 4% of PM_{10} 7% of $PM_{2.5}$	Stricter legislation for industrial heavy fuel oil users			Monitor with inspection checks the fuel efficiency of burners, boilers and power generators of small and medium scale industries operating machinery using heavy fuel oil.
Precursors of secondary particles **Sulfate:** 20% of PM_{10} 33% of $PM_{2.5}$ **Nitrate (non-traffic):** 10% of PM_{10} 4% of $PM_{2.5}$	Large Industry using fossil fuel for power generation or other industries	Local	**	• The introduction of natural gas in the national energy system is one of the largest investments ever carried out in Greece and it constitutes a major priority of the national energy policy. An important part of the infrastructure, mainly the high pressure transmission system and the medium pressure network, which is necessary for the transport of natural gas to the main regions of consumption, has been completed. • Expansion projects of Greek natural gas system are under way in order to link more cities and industries to the system (*e.g.* Aliveri, Megalopolis, *etc.*). Moreover, in the areas connected to the natural gas network, natural gas stations for feeding CNG vehicles have been created. • The high levels observed across Athens and in the whole of the country may be partly due to long range transport of secondary pollutants or gaseous precursors from outside Greece. • This is an area where policy makers must intensify efforts for resolving problems of transboundary pollution in Europe.
	Reduce NH_3 emissions	Local	***	• Application of "guidelines" for regional containment of ammonia resulting from agricultural, livestock and waste practices.
	Biomass combustion use	Local	***	• Communication campaign through the media and dissemination of "best practices". Central theme: the improper use of firewood in small household systems involves a number of negative effects on the environment and health. The practical suggestions relate to the choice of the type of plant and wood, the need for proper installation and maintenance and monitoring of the adequacy of combustion.
	Biomass combustion regulations	National and local	***	• Energy and emission classification for biomass burning appliances. Rules for new systems installation and for proper maintenance. Mandatory certification of residential combustion

Table 25 Continued

Sector	Measure	Competence	Description	Priority
			equipment. Census of domestic heating systems using wood for heating. Regulation for the use of equipment characterised by high emissions. Regulatory program to ban or restrict some wood burning devices in new homes, create financial incentives for old stove replacement, obligation to substitute old stoves upon re-sale of a house with energy efficient and emission controlled devices.	
Biomass burning 7% of PM_{10} 10% of $PM_{2.5}$	Reduction of low efficiency wood burning for residential heating	Local	• High price of diesel for residential heating during the economic crisis lead to the use of wood in a large scale in the densely populated areas of Athens leading to high pollution episodes during stagnation periods in winter. • Information material and training to discourage citizens from this inefficient form of energy are needed. • Introduction of natural gas and renewable energy sources. • Improvement of the thermal behavior of residential buildings. • Promotion of energy efficiency appliances and heating equipment. • News bulletins advising for reduction in wood burning during forecasted atmospheric stagnation periods.	
	Open biomass burning	Local	• Regulatory for open burning activities with particular reference to agriculture, forestry and construction. In general, it is forbidden but the regional government allows them with some restrictions and conditions.	***
Urban works/mineral 10% of PM_{10} 4% of $PM_{2.5}$	Storage and handling of dusty materials	Local	• Cover stock piles of bulk material during storage. • Covering the cargo beds of haul trucks with suitable covers to minimise wind-blown dust emissions. • Apply green curtains in areas with high fugitive dust emissions.	***
Inter-sectorial	**Environmental education and awareness raising**	Local	Communication campaigns through the media and dissemination of "best practices" should be favored in order to sensitise the population on the opportunity of the previous measures.	***

The 8 sources identified in Porto were: biomass burning (BB, Table 23), secondary nitrate (SNI, Table 7), heavy oil and secondary sulfate (HOS, Table 9), mineral (MIN, Table 8), sea salt (SEA, Table 12), industrial (IND, Table 10), vehicle non-exhaust (NEX, Table 6), and vehicle exhaust (VEX, Table 6).

BB comprises levoglucosan as well as OC (29% of OC explained variation), EC (11%) and K (17% of explained variation) (Tables 13–17 and 23). A high OC/EC ratio (2.9) is also characteristic of biomass burning emissions.[35] On average, the contribution of BB explains 17% of the $PM_{2.5}$ mass throughout the sampling campaign (Table 18), and the impact is especially high in the winter months owing to the generalized use of wood for residential heating. It should be noted that results from a survey questionnaire carried out in the early fall of 2010 to assess residential wood combustion (RWC) practices in the 18 districts of mainland Portugal revealed that emissions of $PM_{2.5}$ from RWC in the country represented 30% of the primary $PM_{2.5}$ emissions reported in official inventories. In the Porto district, which, in addition to the municipality with the same name, encompasses another 17 municipalities, it was estimated that about 250 kton of wood are burnt annually by householders for heating purposes, which contributes to the emission of 1.3 kton year^{-1} of $PM_{2.5}$.[55] The contribution of BB to PM is also higher in September. Several wildfires were registered in the Porto district in this particularly hot and dry month. According to the Institute for Nature Conservation and Forests,[56] this district recorded the highest number of occurrences and one of the largest burnt areas.

SNI factor explained about 16% of NH_4^+ and most of the variance of NO_3^- (58%) (Table 7). NO_x emissions from road traffic and industrial plants represent the primary source, although in winter, biomass burning could also contribute to the detection of NH_4^+ and NO_3^- in the particulate phase.[57] On an annual basis, the SNI contributions to PM_{10} and $PM_{2.5}$ were, on average, 3.2 and 1.4 µg m^{-3}, respectively (ratio $PM_{2.5}/PM_{10}$ of 0.44, Table 18), including a proportion of semi volatile organic aerosols, which easily condense on the high surface area of ammonium nitrate particles (Table 18). Although in winter ammonium nitrate formation is favored by low temperature and higher humidity and in summer it is volatilized more quickly owing to higher temperature, a marked seasonality is not observed. This is likely related to the fact that in summertime wildfires represent an additional source of nitrate precursors.

The HOS factor was traced by V and Ni (Tables 9 and 13–17), but also had a high contribution from SO_4^{2-} and NH_4^+. When the sampling site, in particular, and the city center, in general, is under the influence of NW winds, one of the major sources of these constituents could be the refinery, which began operating in 1970. Emissions from ships in the harbor, located at a short distance from the refinery, may represent another source. Source contributions for this factor were generally higher in summer than in winter owing to the higher photochemical activity during warmer months. Annual average contributions to PM_{10} and $PM_{2.5}$ were 3.5 (10%) and 3.3

(13%) µg m^{-3}, respectively. These constituents are concentrated in fine particles (ratio $PM_{2.5}/PM_{10}$ of 0.97) (Table 18).

The MIN factor was identified by the typical crustal species, such as Al, Si, Ca, Li, Ti, Rb, Ce, La (all with explained variation above 50%), Cr, Mn, Fe, Ba and K (above 20%; Tables 8 and 13–17). The MIN source represents a mixture of diverse contributions, mainly from fugitive and diffuse emissions (*e.g.* soil resuspension and constructions works). Saharan dust intrusions rarely reach the Porto area and, when this occurs, the input of desert particles is rather low. The overall composition of MIN reveals a significant enrichment in Ca, Fe and K when compared to the average crust composition,[41,42] indicating an anthropogenic component. Ca-rich particles may originate from concrete or limestone. Additionally K, beside from biomass burning, may result, together with Fe, from particle traffic emissions,[58] as it may be deposited on the roads and resuspended afterwards. The overall MIN contribution was 6.3 µg m^{-3} (18%) to PM_{10} and 3.8 µg m^{-3} (15%) to $PM_{2.5}$ levels (Table 18).

The SEA factor, traced by Na^+, Cl^- and Mg^{2+} (Tables 12 and 13–17), is associated with sea spray aerosol. It contributes, on average, 5.5 (16%) and 1.1 (4%) µg m^{-3}, respectively, to PM_{10} and $PM_{2.5}$. The $PM_{2.5}/PM_{10}$ ratio (0.21) reveals the dominance of these sea spray components in coarse particles (Table 18). Sea salt particles are more abundant in winter, when Atlantic winds pick up (Figure 14).

The IND factor, traced by Zn (70% of explained variation), also encompasses substantial percentages of Cd, Pb and Mn (Tables 10 and 13–17). Mean contributions during the AIRUSE campaign to PM_{10} and $PM_{2.5}$ concentrations were, for both size fractions, 1.2–1.3 µg m^{-3}, corresponding to 4% and 5% respectively (Table 18). No discernible seasonality is observed.

The NEX factor is traced by the aforementioned brake wear metals/metalloids (Cu, Ba, Cr, Fe, Sn and Sb), which individually explained 45–63% of the variation (Tables 6 and 13–17). NEX emissions accounted for 2.9 µg m^{-3} (8%) of PM_{10} and only 1.3 µg m^{-3} (5%) of $PM_{2.5}$ levels (Table 18). A $PM_{2.5}/PM_{10}$ ratio of 0.45 denotes the predominance of brake and tire wear components in coarse particles. Differently to the MIN source, NEX contributions were rather constant over the year with slightly higher values in winter.

VEX comprises organic particles (38% of OC explained variation) from tailpipes, as well as EC (60%) (Tables 5 and 13–17). The ratio OC/EC of 0.64 reveals a predominance of primary organic aerosol[59] from fuel incomplete combustion in vehicle engines. The presence of K (24% of variation), S (13%) and Br (24%) is also noted. Potassium is found in all unleaded fuels.[60] It is also used as an antifreeze inhibitor and as an additive in some oil types. Sulfur is a naturally occurring component of crude oil and is found in both gasoline and diesel. However, recent pollution reduction strategies have forced the reduction of the sulfur content of fuels to near-to-zero levels. Sulfur is also used in engine oil anti-wear additives.[61] Before leaded fuels were phased out, bromine was used to prepare 1,2-di-bro-moethane, which was used as an anti-knock agent. However, this use has declined as lead has gradually been removed from fuel. Bromine compounds are now being tested in batteries for electric cars, designed to produce zero emissions. Average source contributions from vehicle exhaust were around 8.0 $\mu g\ m^{-3}$ for both size fractions, corresponding to 23% of PM_{10} and 32% of $PM_{2.5}$ levels. The average $PM_{2.5}/PM_{10}$ ratio of 0.99 reveals an overwhelming contribution of exhaust particulate constituents to the lowest size fraction, which is related to the very fine mode of motor exhaust particles. Furthermore, besides representing the most significant source of particulate matter levels, the contribution of VEX emissions did not show significant variations between seasons, emphasizing a constancy in traffic patterns, which overlaps the seasonal weather and atmospheric dynamics.

The total contribution from traffic can be estimated as the sum $(VEX) + (NEX) + (0.6 \times SNI)$, which results in 12.8 and 10.3 $\mu g\ m^{-3}$ in PM_{10} (37%) and $PM_{2.5}$ (41%), respectively (Table 19). This indicates a clear preponderance of urban traffic emissions in Porto, at least in the vicinity of the sampling site. Oliveira *et al.*[62] estimated that direct vehicle emissions and road dust resuspension contributed with 44–66% to the fine aerosol and with 12 to 55% to the coarse particle mass at two contrasting sites in the center of the city of Porto (roadside and urban background), showing typically highest loads at roadside. The assumption that 60% of SNI originates from road traffic is based on NO_x emission estimates from the inventory of The North Regional Coordination and Development Commission for different sectors, sources, and activities.[63]

Finally, average source contributions to PM_{10} and $PM_{2.5}$ during the days in which the PM_{10} concentration is above 50 $\mu g\ m^{-3}$ are reported in Table 20. Traffic (as sum of VEX, NEX and 60% of secondary nitrate), biomass burning (as sum of BB and 16% of secondary nitrate) and mineral dust are the sources which give the most relevant contributions.

3.4.4.2 Validation with Hourly Elemental Data

Using hourly data from the Streaker-PIXE, six and five factors were identified in the fine and in the coarse fraction, respectively: traffic, aged sea salt, mineral dust and industry in both fractions; secondary sulfate (including a contribution from heavy oil) and biomass burning in the fine fraction; fresh sea salt in the coarse fraction.

The traffic source is mainly traced by Fe, Cu and, to a lesser extent Mn, in both the fine and coarse fractions; nevertheless, the coarse fraction profile of this source is also heavily characterized by Cr, Ni, Rb, Zr and Ba. The interpretation of these profiles as a traffic source is also reinforced by the daily time trends, showing peaks in both the fractions during the traffic rush hours (Figure 10); further, the source polar plots suggest a local emission for this source, that is near to the sampling site, which is actually characterized as a traffic site (Figure 11). The traffic source identified by the use of the

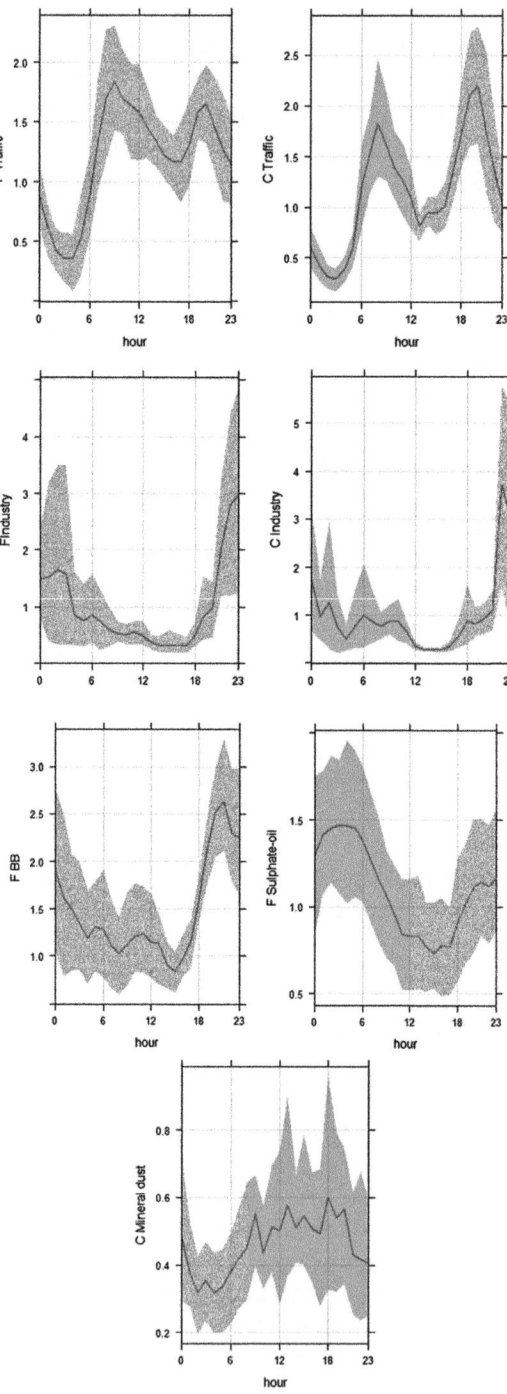

Figure 10   Daily cycle of the PMF sources (a.u.), in the fine (F) and coarse (C) fractions in Porto (POR-TR).

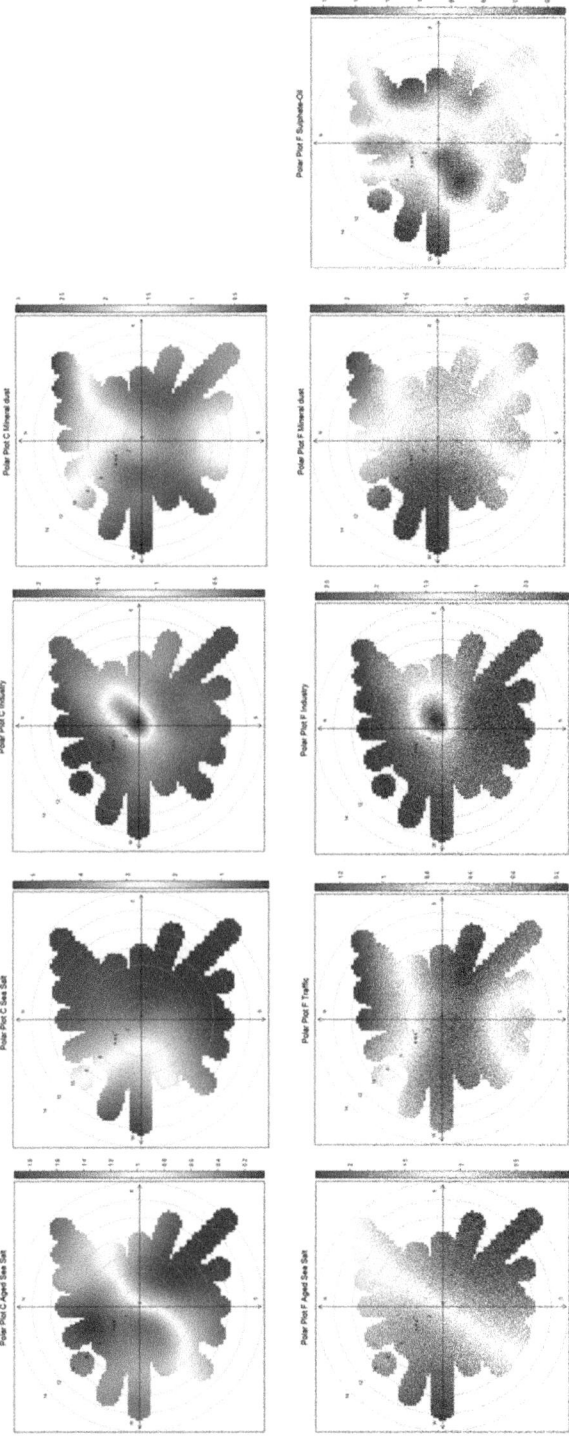

Figure 11 Polar plots of hourly source contributions (a.u.) in Porto (POR-TR), for the fine (F) and coarse (C) fractions.

Streaker corresponds to the vehicle non-exhaust source (NEX) resolved on the daily data; the difference between the fine and coarse source profiles shows that the road dust that is re-suspended by traffic and the contribution from direct brake and tire wear are higher in the coarse fraction.

As for the daily samples, the PMF on the Streaker hourly data identified, in both fractions, the contribution of the industry (IND) emissions: Zn and, to a lesser extent, Pb mainly characterize this source. The time trend for this source shows sharp peaks always occurring around 11 pm, probably linked to a specific phase of an industrial process (Figure 10). The polar plots for this source point to a location of the emission of this source close to the sampling site. Within the city and nearby there are several industries, most of them of small scale and not properly inventoried, whose welding, metalworking, foundry and metallurgical activities may contribute to the emission of non-ferrous metals.

The hourly time trend of BB (Figure 10) supports the interpretation of this factor as domestic heating: a sharp increase of the contribution of this source in the late afternoon, when most of the fireplaces are lit, and a slow decrease after some hours, as fires extinguish during the night. The polar plot shows a local origin for this source, with a prevailing influence from the south-western sector, where most of the Porto city is located. In addition, winds from this direction transport the plumes from the biggest suburban dormitory of the metropolitan area (Vila Nova de Gaia). This municipality, with more than 300 000 inhabitants, spreads over an area of about 168 km,^{2} encompassing many parishes with deeply rooted rural habits.

As for the daily samples, the PMF on the Streaker hourly data identified, in the fine fraction, a heavy oil and secondary sulfate (HOS), characterized by S and, to a lesser extent, by Ni and V. This factor has a daily trend with maxima during the night and minima during the day, mainly driven by the meteorological conditions, indicating a regional rather than a local source; this is also confirmed by the polar plot (Figure 11).

These characteristics (time trend and polar plot) are typical of aerosol with secondary origin. The polar plot does not show any clear prevailing emission direction; nevertheless during the Streaker sampling weeks, somehow relevant sources appear to be located NE and SW with respect to the sampling site. The northeast direction appears to also be linked to the industrial source (see above); in the SW direction there is the Douro River, with some ship traffic and anchorages on its mouth: however, the cargo vessel dock is located at the international harbor of Leixões, NW from the sampling site. Another major source to the northwest is the refinery, a crude oil industrial processing plant. The observed direction, and the daily time trend, could be the result of the recirculation of air masses over the coast, owing to the diurnal variation of the land/sea breezes combined with a channeling effect by the river.

The MIN source does not show any clear daily/time trend; nevertheless, the slight tendency to higher contributions during the day rather during the night (Figure 10) points to the possible emission of dust also during anthropogenic activities such as construction, demolition works and releases from buildings and other surfaces through weathering and other erosive processes. The polar plots show a minimum contribution for this source from the west, consistent with the geography of Porto, as in that direction the city faces the Atlantic Ocean.

In the Streaker hourly data, the PMF identified two sea salt sources: the aged sea salt (SEA) is present in both the fine and coarse fractions, whereas the fresh sea salt (FSS) is identified only in the coarse fraction. The polar plot for the FSS shows high contributions in correspondence with air masses coming straight from the Atlantic Ocean (west), while the SEA source appear linked to air masses originating from the ocean but transiting over the north-western and northern sectors, that is reaching POR-TR after passing over the land, sometimes over long distances (*e.g.* from north Portugal).

3.4.4.3 Suggested Measures to Improve Air Quality

Taking into account that the threshold limits of both $PM_{2.5}$ and PM_{10} are frequently exceeded in Porto, mitigation measures should be adopted. In Portugal, air quality management and monitoring is a responsibility of the Coordination and Regional Development Commissions (*Comissões de Coordenação e Desenvolvimento Regional*—CCDR) that have to take any necessary measures to ensure the respect of air quality legal limits in the national territory. Although a decreasing trend has been registered, in the last decade, $PM_{2.5}$ and PM_{10} have been exceeded in some sites of the metropolitan area. In order to set up mitigation measures, the Northern Region Commission (CCDR-N), together with public and private entities with responsibilities in the area, approved and published through Joint Legal Order no. 20762/2009, of 16 September, an Air Quality Improvement Program. The mitigation measures set out in this plan aimed at different sectors, such as traffic, industrial, domestic, civil works, agriculture and forestry and environmental awareness. Among the proposed policies and measures, the introduction of particulate filters in captive fleets, particularly in heavy duty vehicles with a strong share in total traffic, deserved particular attention. The adoption of the National Action Plan for Energy Efficiency (RCM no. 80/2008 of 20 May), principally in what respects residential combustion, was also targeted. However, CCDR-N has recently recognised that, owing to financial constraints, some of these plans have failed.

According to the AIRUSE results, measures for $PM_{2.5}/PM_{10}$ should especially focus on the traffic and biomass combustion sectors because they represent the major emission sources. However, the industrial sector should not be forgotten. These measures fall into two types: (a) municipal, achieved largely by local government, and (b) supra-municipal, often involving the central administration and covering several municipalities. Table 26 summarises some of the measures that can be adopted to reduce air pollution levels. These measures were prioritised from three stars (highest priority) to one star (lowest priority), depending on the source strength and on type of measure (municipal or supra-municipal).

Table 26 Measures and priority degree suggested by AIRUSE based on PM speciation and source apportionment obtained in 2013 at POR-TR.

Sector	Measure	Competence	Description	Priority
Road traffic	Improving private fleet	National and local	Support initiates of urban electric mobility; increase the share of electric vehicles, reduce the diesel share.	***
PM_{10}: 38% Exhaust 23%		Local	Introduction of high environmental standards when purchasing municipal vehicles.	***
Non-exhaust 9% Nitrate 6%		National	Subsidies and bonuses that either support the retrofitting of a vehicle, the scrapping of it or simply the transition to a new form of mobility; voluntary accelerated vehicle retirement programmes (also referred to as scrap, clunker, or old vehicle buy-back programmes); create a grant programme to co-finance the extra cost of replacing in-use diesel equipment and engines by retrofitting with certified technology or by purchasing new cleaner diesel engines or engines powered by alternative fuels or electricity.	***
$PM_{2.5}$: 39% Exhaust 31% Non-exhaust 5% Nitrate 3%	Lowering number of cars in the urban areas	Local	Implementation of measures that attribute a direct price to a car entering the city, such as road and congestion charges.	***
			Restrict vehicle circulation to neighbourhoods in specific areas of the city.	***
	Low emission zones (LEZ)	Local	Compulsory establishment of LEZ in all municipalities with >15 000 inhabitants.	***
	Improving public transportation	National and local	Continuous update and improvement for an economic, environmentally friendly and faster public transportation (train and tram). Reduction of congestion delays by using bus lanes.	***
		National and local	Technical improvements of the bus fleet (diesel filter and LPG engines), renewal of the municipal solid waste management fleet; improving resource efficiency by reducing fuel consumption and adopting fuel efficient technologies; achieving cleaner transport through cleaner diesel vehicles and cleaner forms of commuting.	***
		Local	Increase the number of routes and frequency of service in order to reduce waiting and exchange times. Expand the motorised transport.	

	Local	Make public transport more appealing (reduced crowding, better seats, cleaner vehicles, on-board internet access and other amenities). More efficient user information through signs, websites and mobile telephone applications that provide information on transit routes, fares and real-time vehicle arrival predictions.	*
	Local	Developing more integrated and seamless connections between bus/metro/train and parking.	***
	Local	Improved stops and stations, including shelter (enclosed waiting areas with heating in winter and cooling in summer), seating, washrooms, refreshments, Internet services, and other convenience features. This is a particularly important type of improvement in stations or stops where passenger traffic volumes often exceed the site design capacity.	***
	Local	Regular inspection programmes for public vehicles for proper maintenance.	***
	Local	Services targeting particular travel needs, such as express commuter buses, special event services and various types of shuttle services.	**
Random tests on vehicle emissions	Local	Random tests on vehicles (mainly diesel trucks) to ensure that smoke emissions are within acceptable limits.	**
Improve parking management	Local	Improved parking management with a focus on peak-hour pricing; restricting parking in the city centre with application of differential fees. High fares for parking in the city center for non residents and cheap fares for eco-cars.	***
Car-sharing	Local	Reduce single-occupancy car journeys by promoting car sharing with a proper connection to public transport services and cycling.	**
Cycling paths and pedestrian lanes	Local	Facilitating alternative mobility by the construction of new cycling paths and the increase of the city surface destined to pedestrian. Develop bike parking and changing facilities; develop bike sharing programmes; improved connectivity to public transport.	***
New park & ride	Local	New strategic parking lots at main transport interfaces (train, bus and tram stations) at the outskirts of the city in order to promote the combined use of car and public transport.	

Table 26 Continued

Sector	Measure	Competence	Description	Priority
Biomass burning PM$_{10}$: 12% PM$_{2.5}$: 18%	Awareness	National	Integration of mobility awareness into school curricula.	**
	Smoke from wood-burning fireplaces and heaters	National and local	• Mandatory certification of biofuel and residential combustion equipment. • Educational programmes (what to burn, how to burn, choosing the right appliance). • Regulatory programme to ban or restrict some wood burning devices in new homes. • Allow installation of only certified wood burning appliances. • Replace non-certified units upon property sale (removal of old stoves upon re-sale of a home). • Weatherization incentives. • Financial incentives for old wood stove and fireplace change outs. • Consider the possibility of enforceable emission reduction programme (mandatory opacity standard). • Establish a public awareness programme. • Set a voluntary curtailment during periods with predicted high PM levels (or update to mandatory).	***
		National and local	• Prohibit burning of materials not intended for use in wood-burning appliances. • Regulate moisture and ash content of wood.	***
	Biomass burning use	Local	Communication campaign through the media (web, radio, TV, newspapers) and dissemination of "best practices". Central theme: the improper use of firewood in small household systems involves a number of negative effects on the environment and health (including indoor air pollution) that, otherwise, may be at least partially contained by the correct use of biomass leading to a reduction in fuel consumption, lower emissions and increased safety in the home. The practical suggestions relate to the choice of the type of plant and wood, the need for proper installation and maintenance and monitoring of the adequacy of combustion. Educational programmes (what to burn, how to burn, choosing the right appliance).	***

Source	Scale	Measures	
Open biomass burning	Local	• Because of the generalised practice of agricultural biomass burning, regulations for open burning activities with particular reference to agriculture and forestry are necessary.	***
		• Development of a system for collecting firewood from to feed small controlled biomass heating plants. Create specific solid waste management programmes for garden and agriculture residues.	
		• More strict regulatory programmes and effective inspection applied to open biomass burning; mandatory curtailment.	
Wild fires	National and local	– Requirements for a burn authorization system.	
		– Requirements for smoke management plans by prescribed burners.	
		– Requirements for burn, no burn, and marginal burn days.	
		• Conception of a national plan of prevention and defence of forests against fires.	**
		• Incentives that promote effective cleaning of forests.	
Heavy oil + secondary PM_{10}: 10% $PM_{2.5}$: 13% Reduction of secondary sulfates precursors (SO_x and NH_3)	Local	• Promote the reconversion of oil burning power plants to gas power plants.	**
		• Enhanced surveillance of stationary sources.	
		• Survey of emission reduction systems in the industries of the Northern Region.	
		• Publication of a legal diploma with new emission limit values (ELV) for stationary sources.	
		• Permission to dock in Leixões harbour only given to vessel with engines operating with low sulfur content; best available technologies applied to marine ship flue gas must be required; use of plug-in shore power; alternative at-dock technologies (like channeling exhaust through barge-mounted control devices).	
Reduce NH_3 emis.	Local	Application of "guidelines" for regional containment of ammonia resulting from agricultural, zootechnics and waste practices.	***
NO_x reduction	Local	Introduction of centralised systems in buildings with more than 5 units. Favour the extensive use of valves for heat metering in residential housing.	***'

Table 26 Continued

Sector	Measure	Competence	Description	Priority
Local dust/urban works PM_{10}: 18% $PM_{2.5}$: 16%	Street flushing and washing	Local	Increase the frequency and/or the number of streets and sidewalks subjected to washing.	***
	Water spraying	Local	Apply water during construction operations (earthmoving, demolition, grading).	***
	Storage and handling of dusty materials	Local	Cover bulk material during storage; covering the cargo beds of haul trucks with a suitable closure to minimise wind-blown dust emissions and spillage; green curtains in areas with high diffusive dust emissions; reduction of vehicle speed.	***
	Unpaved areas	Local	Apply water, nano-polymers, gravel, or pave unpaved parking lots or roads; set additional control requirements for unpaved roads (*e.g.* vegetating, adopting speed limits).	**
Other sources	Commercial charbroiling	Local	Set requirements for commercial charbroiling operations (*e.g.*, emission control device).	
Inter-sectorial	Environmental education and awareness raising	Local	Communication campaigns through the media and dissemination of "best practices" should be favored in order to sensitise population on the opportunity of the previous measures. Integration of mobility awareness into school curricula.	***

3.4.5 Milan (MLN-UB)

3.4.5.1 Source Apportionment

For the city of Milan (January 2013–January 2014), the PMF has been applied to all 2013 days, pooling PM_{10} and $PM_{2.5}$ samples in a single input matrix for PMF; however, a further analysis only using the AIRUSE sampling period has been also performed. For Milan (MLN-UB) the best solution was found with 22 strong species (EC, OC, Levoglucosan, Al, Si, Cl, K, Ca, Ti, Cr, Mn, Fe, Ni, Cu, Zn, Br, Rb, Pb, Na^+, NH_4^+, SO_4^{2-}, NO_3^-), 2 weak species (V, Mg^{2+}) and setting PM as total variable with 400% uncertainty. The AIRUSE dataset comprised 253 samples *versus* 669 samples for the entire period. The distribution of residuals, G-space plots, Fpeak values and Q values were explored for solutions with number of factors varying between 6 and 10. The most reliable solution was found with 7 and 8 factors/sources respectively; the main difference between the selected solutions is related to the split of a source into two factors, owing to a greater number of samples in the second data set, which implies greater variability. For this reason, we discuss below the results of the application to the entire dataset. The best solution with 8 factors has a minimum (base run) Q robust value of 5589, found with 40 runs (seed number 25), which differs from the theoretical Q value (8856) by 37%. With the exception of Mg^{2+} (50%), and Na^+ (49%), species concentrations were reconstructed within 75–112%.

The 8 sources identified in Milan were: vehicle exhaust (VEX, Table 5) and non-exhaust (NEX, Table 6), mineral dust (MIN, Table 8), industrial (IND, Table 10), aged sea salt (SEA, Table 12), biomass burning (BB, Table 23), secondary nitrate (SNI, Table 7) and heavy oil combustion and secondary sulfate (HOS, Table 9), in agreement with previous studies in the Po Valley. To this goal, several constraints were added to the base run solution:

1. Setting to zero the presence of levoglucosan in two factors: vehicle non-exhaust and aged sea salt.
2. Pulling down maximally the presence of Na^+ in the SNI factor. The % dQ was set at 0.5%.
3. Pulling the value of 0.154 for the Na^+ of the sea salt profile pulling the ratio Cl/Na^+ to the literature value of 1.8. The % dQ was set at 5%.

The converged results used totally 3.9% dQ.

In general, considering the nature of the sampling site (urban background), the factors VEX and NEX are not well separated. In fact, crustal compounds are in the second factor but the main characterizing species of traffic are in each factor. Therefore, these include, as a sum of VEX and NEX, organic particles (8% of OC explained variation) from vehicles, EC (33%) as well as elements from mechanical wear (Cr 39%, Mn 33%, Fe 56%, Cu 16%, Zn 68% and Br 23%) and from direct resuspension (Al 3%, Si 25%, Ca 8% and Ti 19%) (Tables 5, 6 and 13–17). The ratio OC/EC of the sum of VEX and NEX is 0.9: considering the relationship between measured OC and EC (3.6), this value reveals a significant fraction of secondary organic aerosols from oxidation of primary volatile organic compounds (VOCs) from vehicles and other sources. Another secondary product from VEX is ammonium nitrate, which is easily formed in an environment with the meteorological conditions typical of the Po Valley; it was found in a separated factor. The total contributions from VEX and NEX were 6.3 $\mu g\ m^{-3}$ (16%) to PM_{10} and 4.4 $\mu g\ m^{-3}$ (17%) to $PM_{2.5}$ levels (Table 18). Source contributions of VEX and NEX are season-dependent with maxima in winter owing to its meteorological stability and to a decrease of vehicle traffic in the summer. The SNI factor explained about 63% of NH_4^+ and 78% of NO_3^- (Tables 13–17), as an oxidation product of local gaseous NO_x emissions (road traffic, biomass burning, domestic heating and industrial plants). From the Regional Emission Inventory, 67% of NO_x emissions is from road transport.[64] As expected, in winter ammonium nitrate formation is favored by low temperature and higher humidity while in summer it becomes volatilized more quickly owing to higher temperatures. On an annual basis, average contribution was 10.0 $\mu g\ m^{-3}$ in PM_{10} (26%) and 8.5 $\mu g\ m^{-3}$ in $PM_{2.5}$ (28%), including a proportion of semi volatile organic aerosols, which easily condense on the high surface area of ammonium nitrate particles (Table 18).

The total contribution from road traffic can therefore be estimated as the sum of VEX + NEX + (0.7×SNI), which results in 13 $\mu g\ m^{-3}$ in PM_{10} (34%), and 10 $\mu g\ m^{-3}$ in $PM_{2.5}$ (34%, Table 19). This indicates a tendency of a decrease in road traffic emissions in Milan, when compared to the 2002–2003 period,[65] when a mean contribution of 40% to $PM_{2.5}$ was estimated, and compared to the 2003–2007 period,[65] when a mean contribution to $PM_{2.5}$ of 63% was estimated. Moreover, this decrease is also in agreement with previous studies.[66]

The BB factor comprises organic particles (30% of OC and 75% of levoglucosan), EC (23%), the main percentage of K (64%) as well as other typical markers (Rb 53%, Cl 29%) (Tables 13–17 and 23). In this factor, the ratio OC/EC is 4.6; considering together the main combustion sources identified (traffic and biomass burning), this ratio becomes 2.5. Following the aforementioned statement done for VEX and NEX, and considering the ratio of measured OC and EC (3.6 as mentioned above), the new 2.5 ratio confirms the presence of secondary organic aerosols. The average source contribution from BB (Table 18) is 7.8 $\mu g\ m^{-3}$ (20%), and 5.3 $\mu g\ m^{-3}$ (18%), in PM_{10} and $PM_{2.5}$ respectively, with high season dependence, as expected for this source, with maxima in winter owing to domestic heating. Another secondary product of biomass burning emissions is ammonium nitrate, found in the SNI factor. Therefore, as in the foregoing, considering the Regional Emission Inventory, the total contribution from BB can be estimated as the sum BB + (0.13×SNI), which results in 9.1 $\mu g\ m^{-3}$ in PM_{10} (23%), and 6.4 $\mu g\ m^{-3}$ in $PM_{2.5}$ (21%, Table 19).

The MIN factor was identified by the typical crustal species, such as Al, Si, Ca, K, Fe, Ti, and Rb, all with explained variation above 45%, with the exception of K (22%), owing to its peculiarity as a BB marker (Tables 8 and

13–17), and it is interpreted as soil resuspension dust. On an annual basis, the average contribution was 2.6 µg m^{-3} in PM$_{10}$ (7%) and 1.5 µg m^{-3} in PM$_{2.5}$ (5%) (Table 18). As expected, the source contribution trend shows a seasonal dependence with maxima during the dry period (Figure 14).

The industrial factor (IND) explains 41, 40, 39, 34, 18, 16, 15 and 14% of the variance of Ni, Ca, Cr, Cu, Mn, Cl, OC and Fe, respectively (Tables 10 and 13–17). This source is representative of a mixture of different anthropogenic activities in the area of Milan, such as metallurgy or construction works. On the annual basis, the average contribution to PM$_{10}$ was 3.7 µg m^{-3} (9%), and to PM$_{2.5}$ was 1.4 µg m^{-3} (5%) (Table 18). IND contributions were rather constant across the year.

The HOS factor depends on the formation of secondary sulfate in the atmosphere from the photochemical oxidation of locally emitted gaseous sulfur oxides and from long range transport, and it is traced by SO$_4^{2-}$, NH$_4^+$ and OC (Tables 9 and 13–17). Condensation of semi-volatile organic compounds is suggested by the high content of OC (17%). In addition, this factor comprises the minor source related to residual oil combustion in industrial processes traced by the same previous components and with V (58% of explained variation) and Ni (51%). Source contributions were generally higher in summer than in winter owing to the higher photochemical activity during warmer months. The annual average contribution in Milan was 5.5 µg m^{-3} (14%) and 5.6 µg m^{-3} (19%), respectively, in PM$_{10}$ and PM$_{2.5}$ (Table 18).

The SEA, traced by Na$^+$ and Cl, is interpreted as sea salt particles including aged sodium nitrate and contributed on average 1.0 µg m^{-3} (3%) and 0.4 µg m^{-3} (1%) to PM$_{10}$ and PM$_{2.5}$ samples, respectively, substantially owing to long range transport events (Table 18).

Finally, average source contributions to PM$_{10}$ and PM$_{2.5}$ during the days in which the PM$_{10}$ concentration is above 50 µg m^{-3} are reported in Table 20. Traffic (40 and 39% of PM$_{10}$ and PM$_{2.5}$, respectively, as sum of VEX, NEX and 67% of secondary nitrates) and biomass burning (35 and 26% of PM$_{10}$ and PM$_{2.5}$, as sum of BB and 13% of secondary nitrates) are the sources that give the most relevant contribution.

In Figure 12, eight plots are shown for daily source contributions (owing to the lack of hourly data in Milan) in PM$_{2.5}$. In spite of the anisotropy of wind rising, the VEX and NEX plots show a homogeneous distribution, with higher concentrations when lowest wind speed occur, suggesting nearby sources with higher impact under stagnant atmospheric conditions. The BB plot shows again a homogeneous distribution as well as the SNI plot, with calm winds. This is a typical situation in a location such as the Po Valley. The HOS plot is less homogeneous indicating a contribution owing to the air mass transport. As aforementioned, the SEA contribution is essentially owing to the long range transport events, mainly from west sectors, in agreement with the intrusion of air from the Ligurian Sea. The IND plot reveals both a local contribution and a transport from the north-west sectors, where more industrialized areas are located. Considering the sampling site location in Milan, when it is downwind of the city, IND and MIN reveals

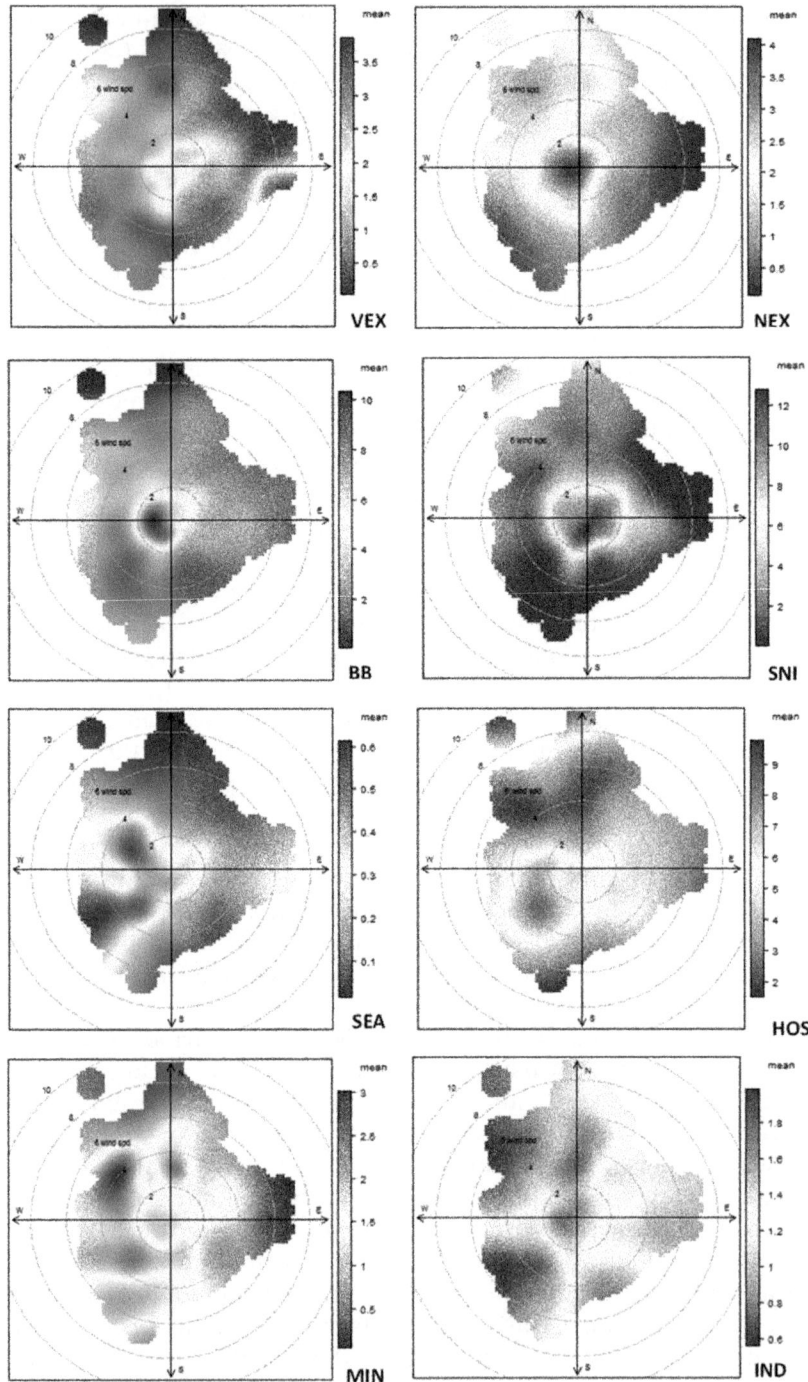

Figure 12 Polar plots of daily source contributions in Milan (MLN-UB).

major contributes from western sectors. The highest concentrations of HOS, IND, and SNI, occurring with the high wind speed from the north-west, have to be considered to be episodic events. Likewise for the easterly episode of the VEX factor.

3.4.5.2 Suggested Measures to Improve Air Quality

According to the national legislation, the regional agencies for the environmental protection (ARPA in each regions) are responsible for the Air Quality Monitoring; the regional/local measures to ensure the respect of the air quality legal limits are the responsibility of the Regional Administration, with the technical support of their ARPA, and they have to take any necessary measures to ensure the compliance with the air quality legal limits in the national territory.

The specific combination of the orographic and meteorological features, with intensive urban/industrial and agricultural emissions of the Po Valley contributes to the frequent occurrence of exceedances of PM_{10} and $PM_{2.5}$ limit values and targets. For this reason, mitigation measures should be adopted.

In the urban background site of Milan (MLN-UB), annual mean levels of PM_{10} and $PM_{2.5}$ during 2013 were 38.1 $\mu g\ m^{-3}$ and 30.5 $\mu g\ m^{-3}$, respectively. At this station, the annual PM_{10} limit value (2008/50/EC) was met, but the daily PM_{10} exceedances (80 for MLN-UB) and the target value for $PM_{2.5}$ (25 $\mu g\ m^{-3}$ at 01/01/2015) exceed the legislated limits. Furthermore, the annual limit value for NO_2 at the five traffic sites in Milan was not met, even if the trend of the last three years is for a slight decrease. The study year, 2013, was characterized by meteorology that disfavored dispersion and cleansing of the atmospheric pollution, as typically happens in the Po Valley.

According to our AIRUSE results, and to the emission inventories, road traffic is clearly the main source of PM_{10} and $PM_{2.5}$, followed by biomass burning and secondary particles, which have an important regional contribution. It is therefore evident that future efforts from air quality plans for the city of Milan must focus on the road traffic sector, and on biomass burning. The high penetration of natural gas in the domestic and residential heating of Milan accounts for a relatively low contribution of this source to PM levels. To reduce the contribution of the secondary particles, attention must focus on primary sources that are mainly in agriculture sectors, road traffic, biomass burning, and industrial sectors, already under effective control of their PM primary emissions. For several years, as the problem was recognised, regional AQ plans were developed, and the latter are the PRIA 2013–2020 (regional measures plan for the Air Quality). The application of those measures has led to a gradual improvement in the AQ but further efforts are still being made.

AIRUSE has therefore selected two main categories of mitigation measures (Table 27): priority (for road traffic, biomass burning and precursors of secondary particles) and ancillary (for other local sources). Moreover AIRUSE encourages horizontal (or inter-sectorial) measures which will have a medium-long-term beneficial effect on air quality.

Table 27 Measures and priority degree proposed by AIRUSE for Milan based on PM speciation and source apportionment obtained in 2013 at MLN-UB.

Sector	Measure	Competence	Description	Priority
Road traffic	Low emission zones (LEZ)	National and local	Adopting a national labelling scheme to enforce local emission-related traffic restrictions. Already applied in 220 EU cities. In Berlin a 58% reduction of exhaust particle emissions was achieved.	***
34% of PM$_{10}$: Exhaust 7% Non-exhaust 9% Nitrate 18%	Restrictions to entering the city centre	Local	Access to urban centre limited for vehicles in municipalities >40 000 inhabitants.	***
	Lowering number of cars in the urban areas	Local	Implementation of measures that attribute a direct price to a car entering the city, such as road and congestion charges Restrict vehicle circulation to neighbourhoods in specific areas of the city	***
34% of PM$_{2.5}$: Exhaust 6% Non-exhaust 9% Nitrate 19%	Transport (private and public)	Local	Facilitate the replacement of vehicles <Euro 3 diesel with Euro 6 adopting restrictions on their movement. Prohibition on movement for <Euro 2 diesel since 2015. Promotion of the LPG and hybrid engine vehicles.	***
	Support for cyclists	Local	Increase the city surface reserved for cyclists. Improvement in traffic planning for a shift in modal split from motor traffic towards public transport, cycling and pedestrian traffic.	***
	Improving public transportation	Local	Continuous update and improvement for an economic, environmentally friendly and faster public transportation (metro, train, and tram).	***
	Cycling paths and pedestrian lanes	Local	Facilitating alternative mobility by the construction of new cycling paths and the increase of the city surface destined to pedestrians. Develop bike parking and changing facilities; develop bike sharing programmes; improved connectivity to public transport.	***
	Promoting low-carbon and low-NO$_x$ vehicles and diesel retrofit	National and local	Incentive for the installation of particle filters in heavy duty commercial vehicles. Development of electric vehicles. Infrastructure.	***
	Reducing road transportation for goods	National and local	Creating a train–harbor interface for transport of goods, which is currently carried out by trucks. Defining of a more sustainable logistic.	***

Car-sharing and motorcycle-sharing	Local	Reduce single-occupancy car journeys by promoting car sharing and suitable motorcycle sharing, linking car sharing to public transport services and cycling.	***
Mobility card	Local	Incentives to alternative mobility with a "Mobility Card" prepaid for the public transport system, through the destruction of an old and polluting vehicle.	**
Promoting innovative fuels	Local	Development of studies on innovative fuels, such as dual-fuel systems.	**
Vehicle and road maintenance	National and local	Control of the inspection programmes to public vehicles to ensure that in-use engines continue to have functional controls and proper maintenance.	**
		Maintaining roads in good repair to reduce the contribution of PM from road surface wear.	***
Biomass burning 24% of PM$_{10}$ 21% of PM$_{2.5}$			
Awareness	National	Integration of mobility awareness into school curricula.	***
Biomass combustion use	Local	Communication campaign through the media and dissemination of "best practices". Central theme: the improper use of firewood in small household systems involves a number of negative effects on the environment and health. The practical suggestions relate to the choice of the type of combustion system and wood, the need for proper installation and maintenance and monitoring of the adequacy of combustion.	***
Biomass combustion regulations	Local	Rules for new systems installation and for proper maintenance. Census of domestic installations using wood for heating (including existing ones) in order to contain emissions. In 2015, following the introduction of the classes of emission, regulate the use of equipment characterised by high emissions.	***
Biomass combustion regulations	National and local	Energy and emission classification for biomass burning appliances.	***
Open biomass burning	Local	Regulations for open burning activities with particular reference to agriculture, forestry and construction. In general, it is forbidden but the regional government allows them with some restrictions and conditions.	***
Biomass combustion use	Local	Prohibition to use biomass burning in domestic heating for municipalities <300 m asl and with domestic plant efficiency less than 63% (by 2018 extension to the entire region).	***

Table 27 Continued

Sector	Measure	Competence	Description	Priority
	Biomass combustion regulations	National and local	Updating of the regulations regarding the deployment of biomass burning plants for energy production. On the land, the use of biomass for energy production must be steered towards producing heat, favouring the use of heat in the district heating network.	***
Other sources				
Precursors of secondary particles Secondary particles: 19% of PM_{10} 25% of $PM_{2.5}$	Reduce NH_3 emissions	Local	Application of the "guidelines" for regional containment of ammonia resulting from agricultural, livestock and waste practices.	***
Industrial 9% of PM_{10} 5% of $PM_{2.5}$	Industrial facilities	Local	Impose high standards for fuels and BATs (channeled and fugitive emissions), and increase inspections to facilities.	***
	Urban waste incineration	National and local	Favour the extensive application of incinerator for domestic and industrial district heating. Introduction of target values and use of BATs in authorizations.	**
Urban works/mineral 7% of PM_{10} 5% of $PM_{2.5}$	Water spraying	Local	Apply water during construction operations (earthmoving, demolition, grading), within a radius of 500 m and at least twice a day. Clean vehicle wheels leaving the site.	**
	Storage and handling of dusty materials	Local	Cover bulk material during storage; covering the cargo beds of haul trucks with a tarp or other suitable closure to minimise wind-blown dust emissions and spillage; green curtains in areas with high diffusive dust emissions; reduction of vehicle speed and eco-driving test for workers.	**
Inter-sectorial	Environmental education and awareness raising	Local	Communication campaigns through the media and dissemination of "best practices" should be promoted in order to sensitise population on the opportunity of the previous measures.	

4 Comparison Among Case Study Cities

The total impact of traffic as well as the proportion between different sub-sources may change significantly from one site to another, justifying different mitigation measures and strategies. For the NEX source rather dissimilar chemical profiles were found comparing different cities. Although the enrichment in Fe is common to all the cities, the main component of NEX can be either Ca (in BCN-UB), EC (in POR-TR and MLN-UB), OC (in FI-UB), or S (in Athens-SUB). These differences can be owing to:

- The proximity to the source: at the traffic site (POR-TR) the NEX source is dominated by brake wear (EC).
- The climatic conditions: Ca is higher in drier regions (BCN-UB and ATH-SUB) owing to the enhanced resuspension, when compared to POR-TR and MLN-UB.
- Type of materials used for brakes and road pavement (the higher OC in Florence might be owing to higher road wear compared to other cities).
- The "rotational ambiguity" of receptor modeling, which may produce false bias.

Other important differences are the absence of OC in POR-TR (again likely owing to the dominance of brake wear particles) and in ATH-SUB, and the high abundance of NO_3^- in POR-TR and NH_4^+ in Athens. Note that some elements are absent in some cities since they were not used for the source apportionment study.

The VEX source shows at all cities an almost total carbonaceous composition (Table 5), with the sum of OC + EC approximately 90–98% of the mass. The ratio OC/EC varies widely among different cities: the lowest at POR-TR owing to its proximity to the source and the consequent lower proportion of secondary OC. The value varies within 1.8–3.7 at the UB sites, probably linked to the distance from main roads (BCN-UB, MLN-UB and FI-UB). A much higher value (16.4) is observed in ATH-SUB owing to the reduced share of diesel vehicles in the local fleet.

Based on the regional NO_x emission inventory of each AIRUSE city, a locally based proportion of the secondary nitrate contribution was apportioned to road traffic (within 30–80%). The composition of this factor (Table 7) is very similar among the different cities, NO_3^- being the main component (28–50%). Nitrate is usually neutralized by ammonium (5–16%) although in the case of ATH-SUB by Na (11%). EC is also present in significant concentrations (>1%) although not in the case of POR-TR where other primary elements can be observed in this profile (Si, Al, Ca, Fe and K).

The total contribution from road traffic emissions to PM_{10} varies significantly in absolute terms (4.8–12.7 µg m^{-3}) with the maximum found at the POR-TR traffic station; nevertheless, the contribution to PM_{10} is quite similar among the Mediterranean countries (23–38%). Similarly. in $PM_{2.5}$ absolute annual contributions vary within the range 2.5–10.3 µg m^{-3} but the percentage is quite constant within 22–39% (Table 9). These results show unequivocally that road traffic is the main source of PM_{10} (at all sites) and $PM_{2.5}$ (first source in MLN-UB, FI-UB, and POR-TR) except in ATH-SUB and BCN-UB where it is the second most important after secondary sulfate (which however it is not identified with a specific source).

During exceedance days (or high pollution days), the total contribution from traffic increases in PM_{10} and $PM_{2.5}$, respectively, from 33 and 28% to 45 and 42% in BCN-UB, from 31 and 31% to 43 and 32% in FI-UB, from 34 to 39% in MLN-UB, and decreases from 23 and 22% to 9 and 11% in ATH-SUB and from 38 and 39% to 36 and 35% in POR-TR.

Among different traffic sources, VEX is still in general the highest contributor to PM_{10} (Table 9). The only exception is in MLN-UB for $PM_{2.5}$ where the highest contribution is from traffic-related SNI. The second most important traffic-source is NEX, whose importance has been increasing during the last decade owing to the lack of mitigation measures. However the difference (in PM_{10}) between VEX, NEX and SNI contributions is not significant, which can be generally translated into an equal contribution from the three sub-sources at the UB and SUB sites. At the TR site (POR-TR) the VEX contribution is significantly higher than the other two sub-sources (by a factor >2).

In $PM_{2.5}$, the share between different sub-sources varies considerably depending on the site (Table 9). At POR-TR, BCN-UB, FI-UB and ATH-SUB the emissions from VEX have the highest contribution, while in MLN-UB SNI is the major source owing to the regional stagnation in the Po Valley that favors nitrate formation. The NEX contribution is significantly lower in $PM_{2.5}$ owing to the coarser size distribution; only in MLN-UB the VEX and NEX contributions in $PM_{2.5}$ are similar.

The daily variation of source contribution at all sites can be seen in Figures 13 and 14.

Fresh sea salt was identified at POR-TR, FI-UB and ATH-SUB, while aged sea salt was found at MLN-UB, FI-UB and BC-UB. As shown in Table 12, the concentration of nitrate (µg/µg) is considerably higher in the aged profiles, by one order of magnitude when compared to fresh profile. In FI-UB, where both fresh and aged factors were found, nitrate is absent in the fresh sea salt and Cl is absent in the aged factor, which shows also particularly high level of OC indicating the mixing of the aged sea salt with anthropogenic plumes. In all factors the absence of ammonium indicates clearly the neutralization of nitrate by sodium.

Mean contributions of sea salt (Table 19) in PM_{10} were revealed to be generally higher (by 25%) than what was found with the chemical speciation (2.5 *vs.* 1.5 µg m^{-3} in BCN-UB, 5.5 *vs.* 4.4 µg m^{-3} in POR-TR, 1.8 *vs.* 0.6 µg m^{-3} in FI-UB, 1.0 *vs.* 0.7 µg m^{-3} in MLN-UB, and 1.0 *vs.* 1.6 µg m^{-3} in ATH-SUB) owing to the involvement of coarse nitrate and water. In $PM_{2.5}$ the PMF contribution matches with the chemical data (slope $= 1$, $r^2 = 0.95$). Consequently, the comparison of PMF contributions reflects the same conclusions drawn in the PM speciation section with lower levels of sea salt at

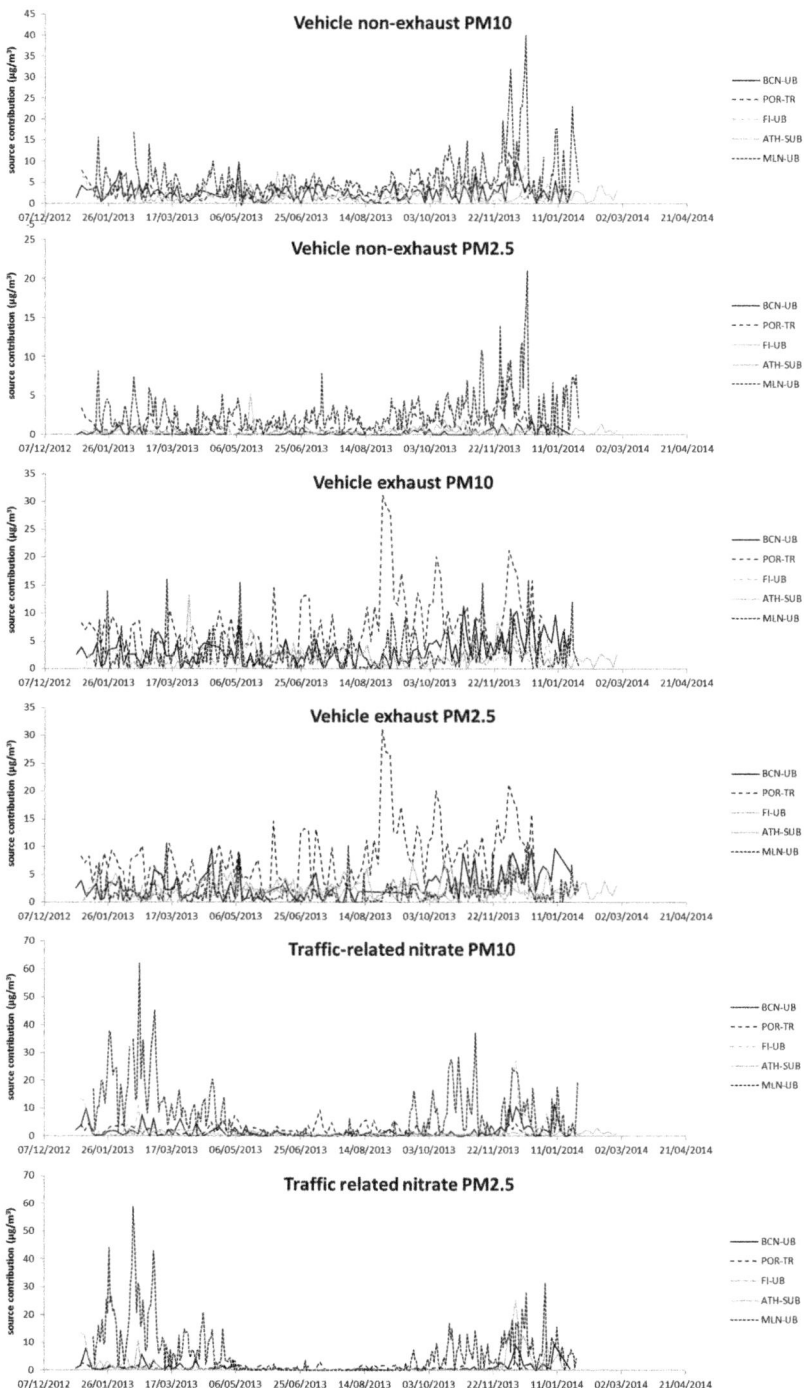

Figure 13 Daily contributions for traffic sources at the five AIRUSE cities.

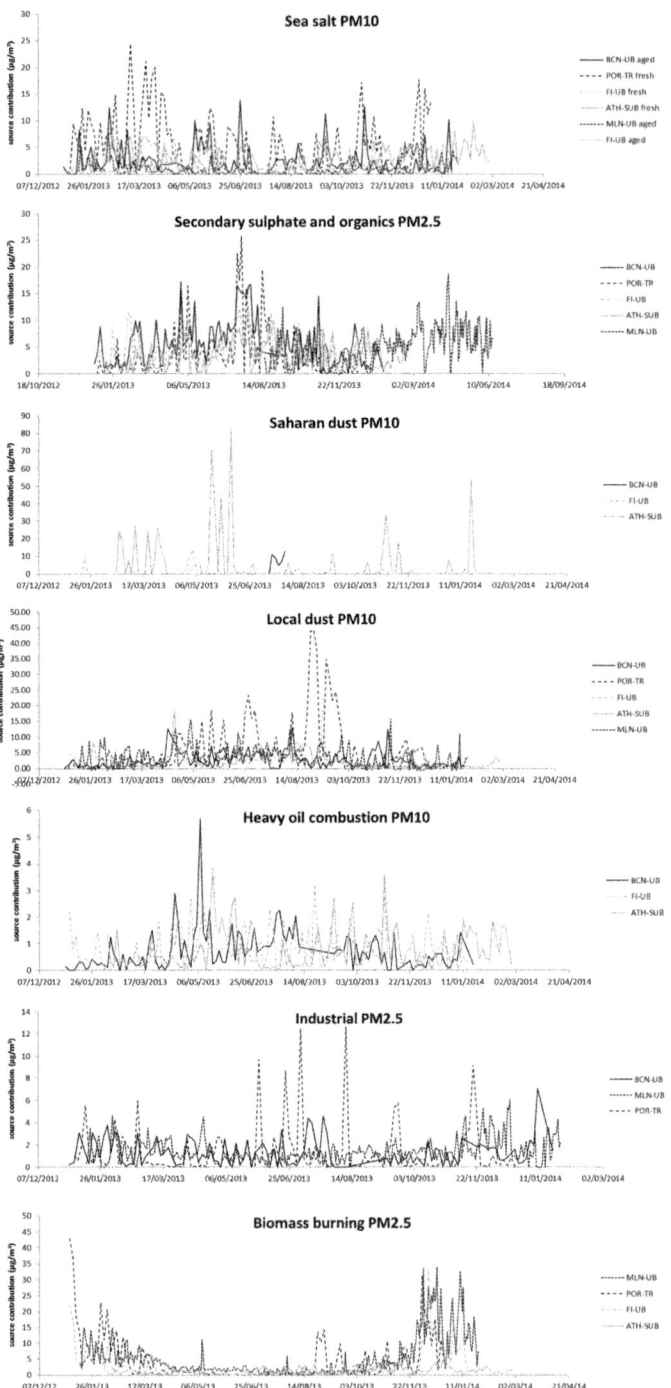

Figure 14 Daily contributions for non-traffic sources at the five AIRUSE cities.

the inland Italian cities (FI and MLN) and higher at the Mediterranean coastal sites, with the highest contribution observed at the Atlantic site (POR). Daily source contributions can be seen in Figure 14.

During exceedance days (or high pollution days), the sea salt contribution in PM_{10}–$PM_{2.5}$ generally decrease from 6–3% to 2–1% in BCN-UB, from 16–5% to 3–0% in POR-TR, from 4–0% to 0–1% in FI-UB, or does not change in MLN-UB and ATH-SUB.

The factor SSO was found at all cities with a rather constant composition, revealing a mainly regional origin of these aerosols across the Mediterranean (Table 9). At two sites (POR-TR and MLN-UB), this factor was combined with the heavy oil combustion, thus adding NO_3^-, V and Ni, which in the other cities (BCN-UB, FI-UB and ATH-SUB) appears as an independent source.

SSO aerosols are at all sites in the fine mode but the variability of contributions across the Mediterranean does not show the same pattern observed for the concentrations of sulfate. The contributions were progressively increasing from POR-TR (3.4 µg m^{-3} in $PM_{2.5}$), ATH-SUB (3.8 µg m^{-3}), FI-UB (4.1 µg m^{-3}) to MLN-UB and BCN-UB (5.6 µg m^{-3}) (Table 18). The reason for this higher contribution might be related to the inclusion of some local sources, as suggested by the presence of EC primary combustion particles only at these two sites. The seasonal trend is clear with maxima in the warmer months, probably owing to the enhanced photochemical activity; this is less clear in MLN-UB where high concentrations in fall and winter may be owing to aqueous phase formation in the fogs that occur frequently at this time of year (Figure 14).

POR-TR registered the highest daily peaks, when the site is under the influence of NW winds, probably owing to the emissions from the Porto Refinery, a plant carrying out crude oil industrial processing for the petrochemical industry.

During exceedance days (or high pollution days), the SSO contribution in PM_{10}–$PM_{2.5}$ always decreases from 26–37% to 19–22% in BCN-UB, from 10–13% to 5–2% in POR-TR, from 21–30% to 6–6% in FI-UB, from 14–19% to 9–11% in MLN-UB, and from 20–33% to 2–5% in ATH-SUB.

The annual mean Saharan dust contribution during AIRUSE sampling days was estimated as 0.3 µg m^{-3} (1%) in BCN-UB, 0.7 µg m^{-3} (4%) in FI-UB, and 3.0 µg m^{-3} (14%) in ATH-SUB (Table 18). This large difference is owing to the Southern location of Athens, and the severity of some Saharan dust episodes in the eastern part of the basin. Saharan dust inputs in the western side of the Mediterranean are considerably higher between May and October, and in March, when compared to the rest of the year. On the contrary, such inputs are clearly higher between November and May in the eastern part of the Mediterranean.[67] An intermediate outcome is observed for central locations in the Mediterranean, where only slightly higher summer contributions are detected.[1] For $PM_{2.5}$ the SAH contribution was estimated only at FI-UB (0.2 µg m^{-3}) and ATH-SUB (0.7 µg m^{-3}), with the $PM_{2.5}$/PM_{10} ratio equal to 0.2 in both cases.

Concerning PM exceedances, the relative burden of Saharan dust increases during exceedance days only in ATH-SUB. In ATH-SUB, Saharan dust is on average the main source of PM_{10} (52%) and $PM_{2.5}$ (45%) when the daily limit value of 50 μg m^{-3} is exceeded. No contributions from SAH were found in FI-UB and BCN-UB during days with PM_{10} above 50 μg m^{-3} and 40 μg m^{-3}, respectively.

Besides the long-range transported mineral dust from the Sahara, a significant part of mineral dust was found to be locally emitted in all cities (local dust, LDU). The ratio SAH/LDU is usually very low: 0.0 in POR-TR and MLN-UB, 0.1 in BCN-UB and 0.3 in FI-UB, indicating that at these sites Saharan dust is 0–23% of measured mineral dust, with the rest emitted by human activities or of local origin. Only in ATH-SUB the SAH contribution was higher (ratio SAH/LDU = 1.4) than the local dust owing both to the geographical position of Athens and the suburban location of the measurement site (*i.e.* lower anthropogenic contribution than at UB sites).

The chemical profile of LDU is the one shown in Table 8 (the MIN profile is used). Although the major components are similar at all sites (Si, Al, Ca, Fe, OC and K), some differences can be observed. The highest ratio Ca/Al is found in FI-UB (6.3) owing to the perfect PMF separation from SAH contribution (ratio Ca/Al = 0.4, similarly to earth crust). The other cities show Ca/Al ratios in the mineral dust factors varying from 0.5 (POR-TR) to 2.2 (BCN-UB), which is influenced by the SAH contribution as well as the local geology and share of Ca-rich local emissions. The ratio Si/Al varies within 2.0 and 3.8. The ratio OC/Ca is generally above the stoichiometric ratio in calcite (0.3) revealing additional sources of organic carbon such as biogenic OC and/or road dust, mostly in MLN-UB where the complex atmospheric dynamic impedes the separation of sources.

LDU contributions to PM_{10} range within 7–12% at SUB and UB sites (2.1–2.6 μg m^{-3}) and increase to 6.4 μg m^{-3} (18%) at the TR site, revealing a contribution from road dust resuspension. In $PM_{2.5}$ the SUB-UB range was 0.3–1.5 μg m^{-3} (2–5%) and 3.9 μg m^{-3} (15%) at the TR site. The relative contribution (%) does not increase during exceedances days, with the exception of POR-TR, where it rises to 28% and 20% for PM_{10} and $PM_{2.5}$ respectively.

The daily variation of LDU contributions shows generally higher values from spring to autumn for all cities (Figure 14). Above the background contributions, sporadic peaks are also found, mostly at POR-TR, probably related to road dust emissions, not completely included in the NEX factor.

As already mentioned, the contribution of heavy oil combustion was separated only at the cities nearest to the Mediterranean shipping routes (BCN, FI and ATH) in spite of the fact that V and Ni concentrations in POR-TR were twice as high as in FI-UB. In POR-TR and MLN-UB, the HOC source is mixed with the SSO in the combined source HOS. HOC particles are commonly composed by EC, OC and S, explaining the high variance of V and Ni. The difference between AIRUSE sites concerns only specific trace elements, such as Zn, Sn, Ba (in BCN-UB), Ba and Se (in FI-UB), Sr and Sb (in ATH-SUB), although only a small amount of variance of these elements in explained by HOC (Table 11).

The annual contribution of HOC is practically the same in BCN-UB, FI-UB and ATH-SUB (0.9 µg m^{-3} in PM$_{10}$ and 0.8 µg m^{-3} in PM$_{2.5}$). The contribution in POR-TR and MLN-UB could not be separated from HOS.

The daily variation of HOC contributions, as estimated by PMF, is shown in Figure 14. In FI-UB and BCN-UB, higher contributions are observed in summer owing to the higher air circulation, which favors their transport and distribution across the regional area, while in ATH-SUB no clear seasonal trend is observed for HOC, probably because of the variety of the sources that may contribute to this factor (residential heating and shipping). During exceedance days (or high pollution days), the HOC contribution in PM$_{10}$–PM$_{2.5}$ does not change significantly.

The impact of industrial emissions was identified only in only three cities: BCN-UB, POR-TR and MLN-UB. Florence and Athens are in fact the less industrialized cities among the AIRUSE consortium. In the three industrialized cities, OC and Fe are commonly present as main components (Table 10). Beside OC and Fe, in BCN-UB and POR-TR the presence of Zn, Pb, S, Cu, Cd, Sb and Mn indicate high temperature metal processing, pointing at the smelters located SW of BCN-UB and east of POR-TR. In MLN-UB the industrial source presents a different chemical profile with NO$_3$, EC, Ca, Cl as main components (together with OC and Fe) and a high variance of Cr, Ni, Cu and Mn, suggesting a more mixed origin including metallurgy and construction activities.

The impact of industrial emissions upon PM$_{10}$ and PM$_{2.5}$ is similarly low at MLN, POR and BCN, ranging within 1.2–1.7 µg m^{-3} as annual means and with a PM$_{2.5}$/PM$_{10}$ ratio close to 1. No typical seasonal trend is observable at any site (Figure 14). The contribution was rather constant throughout the year (around 1 µg m^{-3} daily) in MLN-UB and BCN-UB while elevated peaks (up to 13 µg m^{-3} as a daily mean) are registered in PR-TR mostly in the warmer months.

During exceedance days (or high pollution days), the IND contribution in PM$_{10}$–PM$_{2.5}$ slightly increases from 11–12% to 17–19% in BCN-UB, and decreases from 4–5% to 2–1% in POR-TR, and from 9–5% to 4–3% in MLN-UB.

The PMF identified a biomass burning source only in four of the five cities owing to the low levels of levoglucosan in BCN-UB (22 ng m^{-3} as annual mean). In all other cities, levoglucosan is the main tracer in the chemical profile of the biomass burning source identified by PMF (except in ATH-SUB, where levoglucosan was not used as input specie for PMF owing to the high S/N ratio). Levoglucosan represents 4–8% of PM mass (Table 22) emitted by biomass burning; OC and EC are the major components in the BB profile (12–65% and 4–14%, respectively). In spite of these quite large ranges (which can be owing to the rotational ambiguity of

PMF), the OC/EC ratio can be used as a more robust diagnostic of BB composition. The OC/EC ratio in BB aerosols varies from 2.6 (ATH-SUB), 2.9 (POR-TR), 4.6 (MLN-UB) to 6.1 (FI-UB) which may be explained by a higher proportion of secondary organic aerosols in MLN-UB and FI-UB or by different wood types and combustion appliances. In addition, K (probably the soluble fraction) traces BB aerosols, representing 2–4% of the mass. Other components can be observed, although more sporadically, such as Cl, S, Zn, Pb, NH_4^+ and NO_3^- (Tables 13–17 and 23). BB contributions reproduce quite well the gradients found for levoglucosan among the AIRUSE cities. Although levoglucosan has been detected in some samples from BCN, BB could not be assigned as a significant contributor to PM. In the other cities, an annual mean of 1.2–1.4 µg m^{-3} (7–11%) is estimated in ATH-SUB, 2.9–3.0 µg m^{-3} (15–21%) in FI-UB, 4.2–4.4 µg m^{-3} (12–17%) in POR-TR, up to 7.8–5.3 µg m^{-3} (20–18%) in MLN-UB. Therefore, this reveals quite a contrasting impact of BB emissions across the Mediterranean depending on the type of fuel and combustion device used in each region for residential heating. Differently from other cities, Barcelona is well supplied with natural gas for residential heating; Florence is also well supplied with natural gas but the neighborhoods on the hill are often provided with chimneys. Even in Milan the use of natural gas for heating is very extensive, however, also owing to the current economic crisis, many citizens are equipped with small pellet stoves. In ATH the BB source is also associated with tracers of waste combustion, such as As, Cd, Sb and Pb, with explained variance ranging between 12 and 72% as citizens of Athens have turned to alternative heating fuels, such as wood owing to the economic crisis and the increased prices of diesel oil, which was the most common means of residential heating in Greece. In many cases, treated wood or even combustible wastes are now used as fuel.

As previously shown for the traffic source, another factor identified by PMF was secondary nitrate (SNI). Although in urban environments nitrate mainly arises from NO_x from traffic, a substantial fraction can also derive from biomass burning emissions. Therefore, for each city, the corresponding share of NO_x owing to biomass burning can also be applied to SNI. Based on this approach, percentages of 16 and 13 were adopted in POR-TR and MLN-UB, respectively, to account for SNI from biomass burning. In FI-UB, on the basis of the emission inventory, about 10% of NO_x emissions is owing to domestic heating, with only 2% attributable to stoves and chimneys (http://servizi2.regione.toscana.it/aria/); it is however suspected that these data underestimate the contribution of domestic heating BB to NO_x. Thus, the total contribution from BB in FI-UB can be appraised as the sum $BB + (0.02 \times SNI)$, which results in 3.1 and 2.9 µg m^{-3} in PM$_{10}$ (16%) and PM$_{2.5}$ (21%), respectively. In POR-TR, the total contribution from BB was estimated to be 4.7 µg m^{-3} (13% of PM$_{10}$) and 4.7 µg m^{-3} (18% of PM$_{2.5}$). In MLN-UB, the total contribution from BB represented 9.1 µg m^{-3} in PM$_{10}$ (24%) and 6.4 µg m^{-3} in PM$_{2.5}$ (21%).

The impact of BB emissions is especially high in the winter months (Figure 14), owing to the generalized use of wood for residential heating.[55] The contribution of BB to PM in POR-TR was also higher in September. Several wildfires were registered in the Porto district in this particularly hot and dry month. In MLN-UB the specific stagnant and reduced boundary layer depths induced by the typical meteorology of the Po Valley also enhance BB contributions during winter months.

The percentage contribution from biomass burning to PM_{10} and $PM_{2.5}$ generally increases on exceedance days in POR-TR, MLN-UB and FI-UB. The percentages increase in fact from 24–21% to 35–26% in MLN-UB, from 13–18% to 25–36% (POR-TR), from 16–21% to 30–32% (FI-UB). Conversely, in ATH-SUB, during exceedances days the contribution from BB is substantially reduced (from 7–11% to 1–2%).

Recent research attention has been focused on the importance of other sources/process contributing to the non-fossil OC in urban ambient air $PM_{2.5}$, namely food cooking (COA) and enhanced biogenic secondary aerosols. These two sources/processes can be separated through the application of an online Aerosol Mass Spectrometer (AMS) and off-line radiocarbon analysis on PM filters, which could not be performed within AIRUSE.

In UK cities, Allan *et al.* (2010) identified important contributions of COA (19–30%) to primary organic aerosols. In London, Yin *et al.* (2015) found cooking contributions up to 4% of $PM_{2.5}$. In Zurich, Canonaco *et al.*[68] found a mean contribution of 7.5% of non-refractory (NR) PM1, with a clear peak at noon. In Paris, contributions from cooking were found to reach up to 35% of NR-PM1 during meal hours only in the core of the city. In Barcelona, cooking was estimated to be responsible of 5% of NR-PM1 in March 2009. In Milan, AMS was applied offline on PM_{10} and $PM_{2.5}$ filters during 2013, revealing contributions of around 3% $PM_{2.5}$ (A. Prevot, Personal communication). In Athens, COA was not found as a separate source but mixed with traffic emission in the HOA-2 factor, contributing 17% of OA.[68]

Urban anthropogenic emissions can also lead to an enhancement in secondary organic aerosol (SOA) formation from naturally emitted precursors. Even determining a fraction of modern VOC that is actually emitted through anthropogenic activities does not capture the whole human influence on the organic aerosol budget, as it ignores any possible enhancement, through anthropogenically emitted compounds, of SOA formation from true biogenic precursors. In fact, the high concentrations of specific anthropogenic pollutants (*e.g.* NO_x, SO_2 and O_3) may enhance the formation of secondary organic aerosols from biogenic precursors (BSOA), giving rise to eBSOA.[3] Thus, in this late case, although OC has a biogenic origin from volatile organic compounds (BVOCs), the formation of eBSOA is anthropogenically driven. The interaction between the systems BVOCs-O_3-NO_x-SOA is very complex.[69] Thus, several studies have reported higher yields of eBSOA formation from isoprene in the presence of relatively high concentrations of NO_x and lower SOA formation for very high NO_x levels.[70] In any case, one of the major BSOA formation pathways, especially in high NO_x and O_3 regions, seems to be related with the nocturnal oxidation of VOCs by the nitrate radical (NO_3, a product of the $NO_2 + O_3$ interaction) reaction.[3]

5 Conclusions

Table 18 shows the annual average source contributions in simplified pie charts, where similar sources are combined in fewer categories (*e.g.* traffic). Road traffic (as the sum of vehicle exhaust, vehicle non-exhaust and traffic-related secondary nitrate) is unequivocally the most important source of PM_{10} (at all sites) and $PM_{2.5}$ (at MLN-UB, FI-UB, and POR-TR) while for $PM_{2.5}$ at ATH-SUB and BCN-UB it is the second most important after secondary sulfate and organics (which, however, does not identify one specific source and likely receives significant transboundary contribution). The total annual mean contribution from road traffic to PM_{10} is commonly high (23–37%) at all AIRUSE monitoring sites varying, in absolute terms, from 4.8 µg m^{-3} (ATH-SUB) to 13.0 µg m^{-3} (MLN-UB). Similarly, in $PM_{2.5}$, traffic emissions increase concentrations by 22–40% (2.5–10.3 µg m^{-3} as an annual mean, Table 18).

The second most important "source" of PM_{10} (20–26%) is secondary sulfate and organics at BCN-UB, FI-UB and ATH-SUB, while this only represents 14% of PM_{10} in MLN-UB and 10% of PM_{10} in POR-TR (Table 18). The relative importance of secondary sulfate and organics is higher in $PM_{2.5}$ (19–37% at SUB and UB sites and 13% in POR-TR). The contributions (in $PM_{2.5}$) progressively increase from POR-TR (3.4 µg m^{-3}), ATH-SUB (3.8 µg m^{-3}), FI-UB (4.1 µg m^{-3}) to MLN-UB and BCN-UB (5.6 µg m^{-3}).

Another important source of PM_{10} is biomass burning (13% in POR-TR, 16% in FI-UB, and 24% in MLN-UB), although it is only 7% in ATH-SUB and negligible in BCN-UB. In $PM_{2.5}$, BB is the second most important source in MLN-UB (21%) and in POR-TR (18%), the third in FI-UB (21%) and ATH-SUB (11%), but again negligible (<2%) in BCN-UB. This large discrepancy among cities is mostly owing to the degree of penetration of wood (and its derivatives) as fuel for residential heating. In Barcelona, natural gas is very well supplied across the city and used as fuel in 96% of homes, while in other cities, PM levels increase on an annual basis by 1–6 µg m^{-3} owing to this source.

Other significant anthropogenic sources are:

- Local dust, 7–12% of PM_{10} at SUB and UB sites and 18% at the TR site, revealing a contribution from road dust resuspension. In $PM_{2.5}$ percentages decrease to 2–7% at SUB-UB sites and 15% at the TR site.
- Industries, mainly metallurgy contributing 4–11% of PM_{10} (5–12% in $PM_{2.5}$), but only at BCN-UB, POR-TR and MLN-UB. No clear impact of industrial emissions was found in FI-UB and ATH-SUB.

- Remaining secondary nitrate, emitted from multiple sources, such as industries, shipping and power generation, and contributing 2–10% of PM_{10} and 1–6% of $PM_{2.5}$.
- Natural contributions consist of sea salt (16% of PM_{10} in POR-TR but only 2–7% in the other cities) and Saharan dust (14% in ATH-SUB) but less than 4% in the other cities.
- Other sources of non-fossil OC, such as food cooking and enhanced biogenic secondary aerosols could not be separated owing to the lack of AMS techniques.

During high pollution days, road traffic is the largest source of PM_{10} and $PM_{2.5}$ at all sites (UB and TR): 35–45% to PM_{10} and 32–42% to $PM_{2.5}$, except at ATH-SUB (9 and 11% respectively) owing to the suburban location of this monitoring site (more distant from urban emissions). At ATH-SUB the highest contribution is from Saharan dust (52 and 45%, respectively). Biomass burning is the second most important source during high pollution episodes at FI-UB, POR-TR and MLN-UB (25–30% of PM_{10} and 26–36% of $PM_{2.5}$). During those days, there are also quite important industrial emissions in BCN-UB (17–19%) and local dust in POR-TR (28–20%).

References

1. J. Pey, X. Querol, A. Alastuey, F. Forastiere and M. Stafoggia, *Atmos. Chem. Phys.*, 2013, **13**, 1395.
2. A. M. M. Manders, M. Schaap, X. Querol, M. F. M. A. Albert, J. Vercauteren, T. A. J. Kuhlbusch and R. Hoogerbrugge, *Atmos. Environ.*, 2010, **44**(20), 2434.
3. C. R. Hoyle, M. Boy, N. M. Donahue, J. L. Fry, M. Glasius, A. Guenther, A. G. Hallar, K. Huff Hartz, M. D. Petters, T. Petäjä, T. Rosenoern and A. P. Sullivan, *Atmos. Chem. Phys.*, 2011, **11**, 321.
4. A. Hodzic, S. Madronich, B. Bohn, S. Massie, L. Menut and C. Wiedinmyer, *Atmos. Chem. Phys.*, 2007, **7**, 4043.
5. X. Querol, A. Alastuey, C. R. Ruiz, B. Artinano, H. C. Hansson, R. M. Harrison, E. Buringh, (. . .) and J. Schneider, *Atmos. Environ.*, 2004, **38**, 6547.
6. J. Kukkonen, M. Pohjola, R. S. Sokhi, L. Luhana, N. Kitwiroon, L. Fragkou, M. Rantamaki, (. . .) and S. Finardi, *Atmos. Environ.*, 2005, **39**(15), 2759.
7. M. Lianou, M.-C. Chalbot, I. G. Kavouras, A. Kotronarou, A. Karakatsani, A. Analytis, K. Katsouyanni, (. . .) and G. Hoek, *Environ. Sci. Pollut. Res.*, 2011, **18**, 1202.
8. EEA, Air quality in Europe — 2013 report. EEA Report No 9/2013.
9. B. Brunekreef and B. Forsberg, *Eur. Respir. J.*, 2005, **26**, 309.
10. M. Eeftens, R. Beelen, K. De Hoogh, T. Bellander, G. Cesaroni, M. Cirach, C. Declercq, (. . .) and G. Hoek, *Environ. Sci. Technol.*, 2012, **46**(20), 11195.

11. S. Medina, A. Plasencia, F. Ballester, H. G. Mucke and J. Schwartz, *J. Epidemiol. Commun. Health*, 2004, **58**(10), 831.
12. X. Meng, C. Wang, D. Cao, C.-M. Wong and H. Kan, *Atmos. Environ.*, 2013, **77**, 149.
13. I. Romieu, N. Gouveia, L. A. Cifuentes, A. P. de Leon, W. Junger, J. Vera, V. Strappa, V. Hurtado-Díaz, Miranda-Soberanis, L. Rojas-Bracho, L. Carbajal-Arroyo and G. Tzintzun-Cervantes, *Res. Rep. – Health Eff. Inst.*, 2012, **171**, 5.
14. A. D'Alessandro, F. Lucarelli, P. A. Mandò, G. Marcazzan, S. Nava, P. Prati, G. Valli, R. Vecchi and A. Zucchiatti, *J. Aerosol Sci.*, 2003, **34**, 243.
15. F. Amato, A. Alastuey, A. Karanasiou, F. Lucarelli, S. Nava, G. Calzolai, M. Severi, S. Becagli, V. L. Gianelle, C. Colombi, C. Alves, D. Custódio, T. Nunes, M. Cerqueira, C. Pio, K. Eleftheriadis, E. Diapouli, C. Reche, M. C. Minguillón, M. Manousakas, T. Maggos, S. Vratolis, R. M. Harrison and X. Querol, *Atmos. Chem. Phys. Discuss.*, 2016, **15**, 23989–24039.
16. F. Lucarelli, G. Calzolai, M. Chiari, M. Giannoni, D. Mochi, S. Nava and L. Carraresi, *Nucl. Instrum. Methods Phys. Res.*, 2014, **B 318**, 55.
17. X. Querol, A. Alastuey, S. Rodríguez, F. Plana, E. Mantilla and C. R. Ruiz, *Atmos. Environ.*, 2001, **35**, 845–858.
18. X. Querol, J. Pey, M. Pandolfi, A. Alastuey, M. Cusack, T. Moreno, M. Viana, N. Mihalopoulos, G. Kallos and S. Kleanthous, *Atmos. Environ.*, 2009, **43**, 4266.
19. C. A. Pio, M. M. Ramos and A. C. Duarte, *Atmos. Environ.*, 1998, **32**, 1979.
20. C. A. Pio, M. Cerqueira, R. M. Harrison, T. Nunes, F. Mirante, C. Alves, C. Oliveira, A. Sanchez de la Campa, B. Artiñano and M. Matos, *Atmos. Environ.*, 2011, **45**, 6121.
21. C. A. Pio, L. M. Castro, and M. O. Ramos, *Proceedings of the Sixth European Symposium: Physico-Chemical Behaviour of Atmospheric Pollutants*, ed. G. Angeletti and G. Restelli, Report EUR 15609/2 EN, European Commission, 1994, p. 706.
22. G. D. Thurston and J. D. Spengler, *Atmos. Environ. (1967)*, 1985, **19**(1), 9.
23. R. C. Henry and G. M. Hidy, *Atmos. Environ. (1967)*, 1979, **13**(11), 1581.
24. P. Paatero and U. Tapper, *Environmetrics*, 1994, **5**, 111.
25. A. V. Polissar, P. K. Hopke, P. Paatero, W. C. Malm and J. F. Sisler, *J. Geophys. Res.: Atmos.*, 1998, **103**(D15), 19045.
26. F. Amato, M. Pandolfi, A. Escrig, X. Querol, A. Alastuey, J. Pey, N. Perez and P. K. Hopke, *Atmos. Environ.*, 2009, **43**, 2770.
27. P. Paatero and P. K. Hopke, *Anal. Chim. Acta*, 2003, **490**(25), 277.
28. SEC, 2011. 208 final. Secretary-General of the European Commission. Commission Staff Working Paper establishing guidelines for demonstration and subtraction of exceedances attributable to natural sources under the Directive 2008/50/EC on ambient air quality and cleaner air for Europe. Brussels, 15.02.2011.

29. D. C. Carslaw, The openair manual – open-source tools for analysing air pollution data, Manual for version 0.7-0, King's College, London, 2012.

30. D. C. Carslaw and K. Ropkins, *Environ. Modell. Softw.*, 2012, **27–28**, 52.

31. X. Querol, A. Alastuey, M. Pandolfi, C. Reche, N. Pérez, M. C. Minguillón, T. Moreno, M. Viana, M. Escudero, A. Orio, M. Pallarés and F. Reina, *Sci. Total Environ.*, 2014, **490**(15), 957.

32. WHO, *Health Effects of Black Carbon*, 2012, ISBN: 978 92 890 0265 3. See at: http://www.euro.who.int/__data/assets/pdf_file/0004/162535/e96541. pdf.

33. H. Puxbaum, A. Caseiro, A. Sanchez-Ochoa, A. Kasper-Giebl, M. Claeys, A. Gelencser, M. Legrand, (. . .) and C. A. Pio, *J. Geophys. Res.: Atmos.*, 2007, **112**(23), D23S05.

34. P. M. Fine, G. R. Cass and B. R. T. Simoneit, *Environ. Sci. Technol.*, 2001, **35**(13), 2665.

35. C. Gonçalves, C. Alves, M. Evtyugina, F. Mirante, C. Pio, A. Caseiro, C. Schmidl, H. Bauer and F. Carvalho, *Atmos. Environ.*, 2010, **44**, 4474.

36. S. Nava, F. Lucarelli, F. Amato, S. Becagli, G. Calzolai, M. Chiari, M. Giannoni, R. Traversi and R. Udisti, *Sci. Total Environ.*, 2015, **511**, 11.

37. B. R. T. Simoneit, J. J. Schauer, C. G. Nolte, D. R. Oros, V. O. Elias, M. P. Fraser, W. F. Rogge and G. R. Cass, *Atmos. Environ.*, 1999, **33**(2), 173.

38. U. S. EPA Positive Matrix Factorization (PMF) 5.0 Fundamentals and User Guide. EPA/600/R-14/108 April 2014 www.epa.gov.

39. P. Paatero and P. K. Hopke, *Chemometrics*, 2008, **23**(2), 91.

40. X. Querol, A. Alastuey, M. Viana, T. Moreno, C. Reche, M. C. Minguillón, A. Ripoll, (. . .) and R. Fernández Patier, *Atmos. Chem. Phys.*, 2013, **13**(13), 6185.

41. B. Mason, *Principles of Geochemistry*, Wiley, New York, 3rd edn, 1966.

42. K. A. Rahn, *Atmos. Environ.*, 1976, **10**, 597.

43. F. Amato, M. Viana, A. Richard, M. Furger, A. S. H. Prévôt, S. Nava, F. Lucarelli, N. Bukowiecki, A. Alastuey, C. Reche, T. Moreno, M. Pandolfi, J. Pey and X. Querol, *Atmos. Chem. Phys.*, 2011, **11**, 2917.

44. M. C. Minguillón, A. Schembari, M. Triguero-Mas, A. de Nazelle, P. Dadvand, F. Figueras, J. A. Salvado, J. O. Grimalt, M. Nieuwenhuijsen and X. Querol, *Atmos. Environ.*, 2012, **59**, 426.

45. J. C. Chow, J. G. Watson, D. H. Lowenthal and K. L. Magliano, *J. Air Waste Manage. Assoc.*, 2005, **55**, 1158.

46. A. L. Corrigan, L. M. Russell, S. Takahama, M. Aijala, M. Ehn, H. Junninen, J. Rinne, T. Petaja, M. Kulmala, A. L. Vogel, T. Hoffmann, C. J. Ebben, F. M. Geiger, P. Chhabr, J. H. Seinfeld, D. R. Worsnop, W. Song, J. Auld and J. Williams, *Atmos. Chem. Phys.*, 2013, **13**, 12233.

47. M. Jöller, T. Brunner and I. Obernberger, *Fuel Process. Technol.*, 2007, **88**, 1136–1147.
48. V. Bernardoni, R. Vecchi, G. Valli, A. Piazzalunga and P. Fermo, *Sci. Total Environ.*, 2011, **409**, 4788.
49. J. H. Seinfeld, *Atmospheric Chemistry and Physics of Air Pollution*, Wiley, New York, 1986.
50. H. B. Singh, *Composition, Chemistry and Climate of the Atmosphere*, Van Nostrand Reinhold, New York, 1995.
51. H. Zhuang, C. K. Chan, M. Fang and A. S. Wexler, *Atmos. Environ.*, 1999, **33**(26), 4223.
52. K. Eleftheriadis, D. Balis, I. Ziomas, I. ColBeck and N. Manalis, *Atmos. Environ.*, 1998, **32**(12), 2183–2191.
53. A. Karanasiou, P. Siskos and K. Eleftheriadis, *Atmos. Environ.*, 2009, **43**(21), 3385.
54. A. Saffari, N. Daher, C. Samara, D. Voutsa, A. Kouras, E. Manoli *et al.*, *Environ. Sci. Technol.*, 2013, **47**(23), 13313.
55. C. Gonçalves, C. Alves and C. Pio, *Atmos. Environ.*, 2012, **50**, 297–306.
56. ICNF (2014) Relatório provisório de incêndios Florestais – 2014. Instituto de Conservação da Natureza e das Florestas. Departamento de Gestão de Áreas Classificadas, Públicas e de Proteção Florestal.
57. A. I. Calvo, V. Martins, T. Nunes, M. Duarte, R. Hillamo, K. Teinilä, V. Pont, A. Castro, R. Fraile, L. Tarelho and C. Alves, *Atmos. Environ.*, 2015, **116**, 72.
58. S. Lawrence, R. Sokhi, K. Ravindra, H. Mao, H. D. Prain and I. D. Bull, *Atmos. Environ.*, 2013, 77, 548.
59. C. A. Alves, J. Gomes, T. Nunes, M. Duarte, A. Calvo, D. Custódio, C. Pio, A. Karanasiou and X. Querol, *Atmos. Res.*, 2015, **153**, 134.
60. M. T. Spencer, L. G. Shields, D. A. Sodeman, S. M. Toner and K. A. Prather, *Atmos. Environ.*, 2006, **40**, 5224.
61. J. Fitch, Copper and Your Diesel Engine Oils. Practicing Oil Analysis Magazine. September 2004.
62. C. Oliveira, C. A. Pio, A. Caseiro, P. Santos, T. Nunes, H. Mao, L. Luahana and R. Sokhi, *Atmos. Environ.*, 2010, **44**, 3147–3158.
63. DAO-UA/CCDR-N (2009) Melhoria do actual inventário de emissões de poluentes atmosféricos da região norte. Relatório Final. Departamento de Ambiente e Ordenamento – Universidade de Aveiro; Comissão de Coordenação e de Desenvolvimento Regional do Norte.
64. ARPA Lombardia, 2010, INEMAR, http://www.ambiente.regione. lombardia.it/inemar/webdata/main.seam.
65. G. Lonati, C. Colombi and S. Cernuschi, PM2.5 source apportionment in Milan by UNMIX receptor model. 2012. Proceeding on EAC2012.
66. G. Pirovano, C. Colombi, A. Balzarini, G. M. Riva, V. Gianelle and G. Lonati, *Atmos. Environ.*, 2015, **106**, 56–70.
67. B. R. Larsen, Overview of the Lombardy region (I) Source Apportionment Study, *Geophys. Res. Abstr.*, 2009, **11**, EGU2009-11091.

68. F. Canonaco, M. Crippa, J. G. Slowik, U. Baltensperger and A. S. H. Prévôt, *Atmos. Meas. Tech.*, 2013, **6**(12), 3649.
69. E. Kostenidou, K. Florou, C. Kaltsonoudis, M. Tsiflikiotou, S. Vratolis, K. Eleftheriadis and S. N. Pandis, *Atmos. Chem. Phys. Discuss.*, 2015, **15**, 3455.
70. S. N. Pandis, S. E. Paulson, J. H. Seinfeld and R. C. Flagan, *Atmos. Environ., Part A*, 1991, **25**, 997.

PM$_{10}$ Source Apportionment in Five North Western European Cities— Outcome of the Joaquin Project

DENNIS MOOIBROEK, JEROEN STAELENS,* REBECCA CORDELL, PAVLOS PANTELIADIS, TIPHAINE DELAUNAY, ERNIE WEIJERS, JORDY VERCAUTEREN, RONALD HOOGERBRUGGE, MARIEKE DIJKEMA, PAUL S. MONKS AND EDWARD ROEKENS

ABSTRACT

The aim of this study was to identify and quantify sources contributing to particulate matter (PM$_{10}$) at four urban background sites and an industrial site in North West Europe using a harmonized approach for aerosol sampling, laboratory analyses and statistical data processing. Filter samples collected every 6th day from April 2013 to May 2014 were analysed for metals, monosaccharide anhydrides, elemental and organic carbon, water-soluble ions and oxidative potential. The receptor-oriented model EPA-PMF 5.0.14 was used to carry out a source apportionment using the pooled data of all sites. A solution with 13 factor profiles was found which could be aggregated into eight groups: secondary aerosol; furnace slacks, road wear and construction; sea spray; mineral dust; biomass burning; industrial activities; traffic emissions and brake wear; and residual oil combustion. The largest part of PM$_{10}$ (40–48%) was explained by nitrate-rich and sulphate-rich secondary aerosol, followed by (aged) sea spray (11–21%). Clear traffic and biomass burning profiles were also found. Conditional probability function plots were used to indicate the likely directions of the sources,

*Corresponding author.

Issues in Environmental Science and Technology No. 42
Airborne Particulate Matter: Sources, Atmospheric Processes and Health
Edited by R.E. Hester, R.M. Harrison and X. Querol
© The Royal Society of Chemistry 2016
Published by the Royal Society of Chemistry, www.rsc.org

while air mass back-trajectories were analysed using the HYSPLIT model. A better understanding of the composition and sources of particulate matter can facilitate the development of health-relevant air quality policies.

1   Introduction

Northwest Europe is still considered as a hot spot for air pollution with high ambient concentrations of, amongst others, particulate matter (PM) and nitrogen oxides (NO$_x$).[1] Particulate matter is a heterogeneous mixture of components resulting from multiple natural and anthropogenic sources, including sea salt, naturally suspended dust, pollen, volcanic ash, combustion processes, industrial activities, vehicle tyre and break wear, and road surface wear. Epidemiological studies attribute the most important health impacts of air pollution to PM,[2] although currently it is still unclear which specific properties (such as size and chemical composition) or sources of particles are most relevant to health effects.[3] Ambient PM concentrations vary substantially between and within regions, as indicated for example by routine air quality monitoring networks.[1] In urban areas, in addition to background PM concentrations often imported, traffic-related emissions and domestic heating can significantly contribute to ambient PM levels.[1]

Current monitoring efforts generally focus on the mass concentration of PM$_{10}$ and PM$_{2.5}$ in line with current air quality legislation,[4] but these data generally do not allow the assessment of differing sources. To facilitate the development of health-relevant air quality policies a better understanding of the sources and composition of PM is required. Information about the pollution sources is also required to identify if exceedances are owing to either natural sources (including road salting and sanding) or anthropogenic sources.[4] Using source apportionment additional information can be derived about pollution sources and the amount they contribute to pollution levels.[5] To accomplish this, three main approaches exists: emission inventories, source-oriented models and receptor-oriented models. In this study we have used the receptor-oriented model US-EPA-PMF (Positive Matrix Factorization) to identify and quantify the most relevant sources in NW Europe. The PMF model is based on uncertainty-weighted factor analysis and relies on pollutant measurements. This approach has the advantage of scaling each data point individually using an uncertainty matrix, so that data with a higher precision have a larger influence on the solution.[6]

Information on the composition of PM in NW Europe has been reported before, for example for the north of Belgium,[7] the Netherlands[8] and the United Kingdom,[9] but comparison of the findings between sites or regions is hampered by differences in study periods, analytical methods and modelling and reporting methods. Therefore, the aim of this study was to quantitatively identify sources contributing to PM$_{10}$ at five sites in NW Europe using a harmonized approach for aerosol sampling, laboratory analyses and statistical data processing. PM_{10} was sampled at four urban background sites and one industrial site and the chemical composition and oxidative potential of PM_{10} were measured. The data of the five sites were pooled to carry out a source apportionment using PMF.

2 Site Description and Chemical Characterisation of PM_{10}

2.1 Sites

Aerosol samples were collected at five sites in NW Europe: Amsterdam (site code: AD), Wijk aan Zee (WZ) (The Netherlands), Antwerp (AP; Belgium), Leicester (LE; United Kingdom) and Lille (LL; France). Table 1 summarizes the characteristics of the sites; detailed site descriptions are given elsewhere.[10] Site WZ is an industrial monitoring site at about 30 km from Amsterdam. The four other sites are considered to be urban background sites for PM_{10} monitoring, although there is considerable influence of local traffic. Traffic data for AD,[11] LE[12] (2013) and AP[13] (February and October 2013) show that traffic intensity was highest near AP, which was also closest to a main road. Air quality monitoring at AD, AP and LE indicated a clear traffic-related diurnal variation in black carbon and ultrafine particles at these sites.[10]

Table 1 Location and characteristics of the five PM_{10} sampling sites (traffic intensity data for 2013).

City	Code	Latitude	Longitude	Description
Amsterdam	AD	52°21′35″ N	4°51′59″ E	Near Vondelpark, 64 m south from a main road (Overtoom, 15 000 vehicles day^{-1}).
Antwerp	AP	51°12′35″ N	4°25′55″ E	In Borgerhout, 30 m north from of a main road (Plantin en Moretuslei, 29 500 vehicles day^{-1}).
Leicester	LE	52°37′12″ N	1°07′38″ W	At the main campus of the University of Leicester, 140 m northeast from a main road (Welford Road, 22 500 vehicles day^{-1}).
Lille	LL	50°37′41″ N	3°05′25″ E	At a school campus in Lille Fives, 35 m north from a local road (rue du Vieux Moulin, no traffic data available).
Wijk aan Zee	WZ	52°49′40″ N	4°60′23″ E	On a parking lot, 70 m north from a local road (Burgemeester Rothestraat, no traffic data available).

Wind speed and direction were recorded at meteorological monitoring stations located at a distance of 3–7 km from the aerosol sampling sites (AD: Schiphol airport, AD: Antwerp Luchtbal, LE: Groby Road, LL: Sequedin, WZ: IJmuiden), as detailed elsewhere.[10] As the wind was not measured at the receptor sites, the available data can be considered to be representative for regional wind conditions but do not account for the potential influence of *e.g.* high buildings on wind conditions in urban environments.

2.2 PM_{10} Sampling and Gravimetric Analysis

Sampling was carried out for 14 months (426 days) from 1 April 2013 to 31 May 2014, except for LL where the measurements started 2 months later (5 June 2013 to 31 May 2014). The samples were collected daily (24 hour exposure) onto 47 mm quartz filters (Pall Tissuquartz™ filters, 2500 QAT-UP) using a sequential sampler (Derenda PNS16 at AD and WZ and Leckel SEQ47/50 at AP, LE and LL) with a PM_{10} inlet running at 2.3 m^3 h^{-1} for 24 h per filter. Flows were checked every 14 days when changing the filter compartments. Filters were weighed before and after sampling in order to determine total PM_{10} mass according to the (stricter) standard for $PM_{2.5}$.[14] Before and after sampling, filters were conditioned at 20 ± 1 °C and $50 \pm 5\%$ relative humidity for 48 h, weighed, left for a further 24 h and re-weighed. After gravimetric analysis, filters were stored at -18 °C. Ambient nitrogen oxide concentrations were measured continuously at the five sites by chemiluminescence NO-NO_2-NO_x analysers (API 200A at AD and WZ and Thermo 42i at AP, LE and LL).

2.3 Chemical Analysis

One in six filters was used for chemical analysis. Six punches of 1×1 cm^2 (puncher manufactured by Sunset Laboratory Inc., USA) were taken per filter and transported in cooling boxes to the different laboratories. Three punches were analysed for metals, monosaccharide anhydrides (MAs) and elemental and organic carbon (EC and OC), respectively. The other three punches were used to determine water-soluble ions and the remaining filter part was analysed for oxidative potential (OP). Each type of analysis was carried out by a single laboratory for all the samples. After microwave-assisted acid digestion (HNO_3/H_2O_2), calcium (Ca), iron (Fe), potassium (K) and zinc (Zn) were quantified by inductively coupled plasma optical emission spectroscopy (ICP-OES) and aluminium (Al), titanium (Ti), vanadium (V), chromium (Cr), manganese (Mn), nickel (Ni), copper (Cu), arsenic (As), molybdenum (Mo), cadmium (Cd), antimony (Sb), barium (Ba) and lead (Pb) by inductively coupled plasma mass spectrometry (ICP-MS). The HNO_3/H_2O_2 digestion leads to incomplete recovery of Al and Ti because HF is needed to dissolve aluminium silicates, but was chosen because it results in a lower limit of detection (LOD) for most elements. The MAs levoglucosan, mannosan and galactosan were quantified using a validated gas chromatography mass spectrometry (GC-MS) method described in detail previously.[15] The EC/OC content was analysed according to CEN/TR 16243[16] using a laboratory organic/elemental carbon aerosol analyser (Sunset Laboratory Inc., USA) and the NIOSH protocol, which was considered to be most suitable for the traffic-influenced PM_{10} samples in this study. The water-soluble ions were determined according to CEN/TR 16269.[17] After ultrasonic water extraction, potassium (K^+), calcium (Ca^{2+}), magnesium (Mg^{2+}) and sodium (Na^+) were analysed by ICP-OES and ammonium (NH_4^+), chloride (Cl^-), nitrate (NO_3^-) and sulfate (SO_4^{2-}) by ion chromatography with conductivity detection (IC-CD). The method used for the detection of the oxidant (radical) generation capacity of PM was electron paramagnetic resonance (EPR) spectroscopy using the spin trap 5,5-dimethyl-1-pyrroline-*N*-oxide (DMPO). The OP was determined directly on the filter material without prior PM extraction.[18]

In addition to the filters collected every 6th day from April 2013 to May 2014, the analyses were also carried out for 6–9 filters taken every 2nd day for three sites (AD: $n = 6$ in April 2013; AP: $n = 8$ in September 2013; LE: $n = 9$ in March 2014) and for 17 filters per site taken on days with regionally enhanced PM_{10} levels (10 April and 25 September 2013; 30 January, 5–15 March, 30 March to 3 April, and 24 April 2014). Over the study period 68 field blank filters, which were kept in the sampling devices for 14 days without exposure to sampled air, were also analysed as part of the quality control.

3   Source Apportionment Using Positive Matrix Factorization

3.1   Data Preparation and Uncertainty Matrix

For the five sites, in total 434 valid filters were available for the source apportionment analysis ($n = 94$ for AD, 95 for AP, 89 for LE, 72 for LL and 84 for WZ). In addition, the analysis included 31 filters that were collected during short sampling campaigns using a mobile trailer (Leckel SEQ47/50) at a second urban background site in Amsterdam (6 km from AD, May 2013, $n = 8$), Antwerp (1 km from AP, Oct 2013, $n = 12$) and Leicester (1 km from LE, April 2014, $n = 11$).[19] These filters were analysed as described above. In the present study we focus on the five main sites (Table 1) because at the three temporary sites only a few samples were collected with a PM_{10} composition that was generally similar to those of the main site nearby (AD, AP, LE).

Ambient concentrations of the measured variables per filter were calculated based on the exposed surface area of the filter, which was considered to be a constant value per sampling device, and the sampled air volume measured per sampling device and day. No correction was carried out for the concentrations of blank filters. The amount of missing chemical data was generally negligible, except for Al, Ca, Ti and Zn. Owing to sample contamination in the laboratory, results of these elements were lacking for 171 filters (39%). The missing data were estimated by multiple imputation,[20] a technique frequently applied in environmental research and biostatistics.[21,22]

Some variables have less additional informative value in the PMF analysis and are therefore downscaled or excluded. In this study the following classification was used: (1) the PM_{10} mass from the weighed filters was included as default mass variable and hence automatically downscaled; (2) Al, Ca, Ti and Zn were downscaled owing to the large number of missing values; (3) total K was excluded because the concentrations were often lower than for the water-soluble fraction K^+, likely owing to the low levels of ambient K and K^+ compared to the LOD of the analytical methods; and (4) OP, NO and NO_2 were downscaled because in the present study they were not considered as driving variables, however the distribution of their concentrations in the profiles is of interest, *e.g.* to test the validity of a traffic profile. This resulted in a data set of 465 samples with 33 variables, including PM_{10} mass, 25 of which were categorized as strong.

The coefficient of variation (CV) and LOD of each variable were estimated based on the relative standard deviation of repeated analyses of blanks and exposed filters. The CVs and LODs were used to calculate the uncertainties (unc) for a given concentration level (conc) with the following equation:

$$unc = \sqrt{(0.5 \cdot LOD)^2 + (CV \cdot conc)^2} \tag{1}$$

Chemical results below LODs were not replaced by a fixed value. However, uncertainties for values below the LOD were set equal to the LOD. For imputed values the uncertainties were increased by a factor of four. This ensured that imputed values and values below the LOD had less weight in obtaining the final solution.

3.2 Positive Matrix Factorization

The source apportionment was carried out with US-EPA-PMF 5.0.14, a programme that uses ME-2 as the underlying engine to solve the PMF problem.[23,24] The calculations were performed in robust mode, in which outliers have less influence on the obtained solution. The data of all sites were pooled prior to the PMF analysis. This approach increases the size of the data set and hence decreases the significance of random errors and errors owing to rotational ambiguity.[25] However, this approach assumes that there is little variance between the composition of the source profiles between the sites. In other words, the main variability is assumed to be found in the contribution of each factor at the different sites.

When applying the PMF technique, the number of factors has to be selected. An indicator for selecting this number is the ratio between the (robustly) calculated and expected values of the sum-of-squares object function (Q) that is minimized in the PMF. Runs were carried out with 2 to 20 factors to evaluate the robust to expected Q value (Q ratio) and to examine the composition of the obtained factors. A solution with 13 factor profiles was selected because this yielded the most physically interpretable results. The Q ratio for this solution was approximately 6, which is larger than the expected ratio of 1. Part of this increase can be explained by the decision to pool the data from all receptor sites, as multiple-site data will have a higher Q ratio owing to the variation in source composition between the sites. Brown et al.[26] reported Q ratios between 2 to 5.5 for single-site data, which is fairly similar to the Q ratio in the present study. Furthermore, even while performing calculations in robust mode, it might be difficult to assess if the calculated Q value is as expected or too large. In this case, it might be helpful to examine the scaled residuals, *i.e.* the residuals divided by their error estimates. In a well-fit model these scaled residuals should be symmetrically distributed and lie in the -3 to $+3$ range. The majority of all scaled residuals were generally within this range, except for K^+, Ba, Fe, Mo, Ni, Sb and the MAs, for which the distribution of the scaled residuals was skewed to the right. This may be owing to high concentrations at one or two sites, probably by extending the data set of the predefined filter samples taken every 6th day with the samples on days with enhanced PM_{10} values. To examine the effect of the uncertainties an indicative analysis was performed, with 13 factors and an 'added modelling uncertainty' of 15%. This decreased the Q values and made the Q ratio approach unity, but only had a small influence on the composition of the resulting factors. It was therefore decided to evaluate the 13 factor solution without added modelling uncertainty.

Even though a global minimum of the Q function is found by the least squares fitting process, there may not be a unique solution because of rotational ambiguity. The addition of constraints such as non-negativity can reduce the rotational freedom in the model, but non-negativity alone does not generally produce a unique solution.[27] The PMF algorithm provides a 'peaking parameter' called FPEAK allowing the user some control over the rotations. If a value of zero is used for this parameter, the algorithm will produce a more central solution. The use of non-zero values allows sharper peaks to be obtained, which are to be expected in source profiles, and will limit the rotational freedom.[6,28] Several runs with different positive and negative FPEAK values were performed to get more insight into the rotational freedom of the solution. Judging from the change in Q values, there was not much improvement over the 'central' solution (FPEAK $= 0$), which consequently has been used in the rest of this study.

The degree of spatial uniformity of the calculated source contributions between the receptor sites was determined using the coefficient of divergence (COD).[22,29] If the source contributions are similar at all sites, the COD approaches zero. In case the source contributions are very different, *e.g.* highly impacted by local sources, the COD approaches unity.

Conditional probability function (CPF) plots were used to indicate directions in which sources are likely to be located.[22,30] The CPF was calculated using the upper 25th percentile of the source contributions from the PMF analysis and the wind direction values measured on meteorological stations near the sites. To match the hourly wind data, each daily contribution was assigned to each hour of a given day.[31] Back-trajectories of air masses during specific periods were analysed using the Hybrid Single-Particle Lagrangian

Integrated Trajectory (HYSPLIT) model.[32] By performing a backwards trajectory the model follows a single air parcel backwards in time, providing information about the origin of a local or regional source. For the interpretation of the CPF plots, it should be noted that the wind data used are representative for area sources but might be less representative for local sources in urban areas (*e.g.* traffic) because of the potential influence of high buildings on wind speed and direction. For trajectory analyses, however, regional meteorology is most relevant.

The European air quality directive[4] has set two limit values for PM_{10}, one of which is that the daily mean PM_{10} value may not exceed 50 μg m^{-3} more than 35 days in a year. The contribution matrix calculated by PMF contains factor contributions for each sample day included in the analysis. We have selected the days with known PM_{10} composition on which the PM_{10} concentration was above 50 μg m^{-3} to determine the contribution of the calculated sources on these exceedance days.

The PMF model version used contains three methods for estimating the uncertainty in a PMF analysis owing to random errors and rotational ambiguity: classical bootstrap (BS), displacement of factor elements (DISP), and bootstrap enhanced by displacement of factor elements (BS-DISP). The three methods are considered complementary; depending on the situation one method might provide better results than the other methods.[25] Therefore, the uncertainty of the 13-factor solution was analysed according to the three methods. In our evaluation NO_3^-, Cl^-, SO_4^{2-}, Na^+, NH_4^+, Ca^{2+}, Cr, Cu, Fe, Mn, Pb, V, levoglucosan and EC were included as 'key species' for the BS-DISP method. A general problem in comparing uncertainty analysis results is that species concentrations differ in magnitude. In our study this is for example evident by including the OP of PM_{10}, with much higher values compared to other species. To compare the results among species, the so-called interval ratios of the BS, DISP and BS-DISP methods were used.[25,26]

4 Results and Discussion

4.1 *PM₁₀ Mass Concentrations*

During the common sampling period at the five sites (1 June 2013 to 31 May 2014), the mean PM_{10} concentration was highest at the site in Wijk aan Zee (WZ; annual mean of 25.0 μg m^{-3}) and Antwerp (AP; 24.5 μg m^{-3}), intermediate in Lille (LL; 22.4 μg m^{-3}) and Amsterdam (AD; 20.3 μg m^{-3}) and lowest in Leicester (LE; 16.0 μg m^{-3}). Data coverage during this period was 90–97% depending on the site. The number of exceedances of the EU day limit value[4] for PM_{10} (50 μg m^{-3}) was highest at AP (20 days year^{-1}) and WZ (16), moderate at LL (12) and lowest at AD (8) and LL (6). Exceedances of the day limit value mainly occurred in March and April. Sampling started 2 months earlier at all sites other than LL, and the mean PM_{10} values for this 14 month period (Figure 1, left boxplot for each site) differed by less than 2% from the annual means.

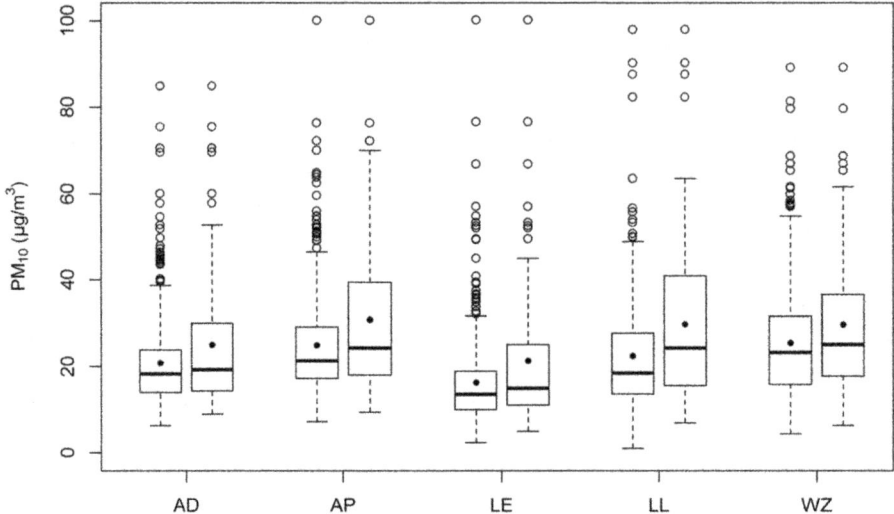

Figure 1 Boxplots of daily PM_{10} ($\mu g\ m^{-3}$) at the five sites (see Table 1) according to all valid filters (left boxplot per site) and according to the filters used for source apportionment (right boxplot per site) (dot: mean value; line in box: median value; box: 25th and 75th percentiles; whiskers: 1.5 times the interquartile range from the box).

In this study a subset of daily filters was selected for chemical analyses and source apportionment, so it is relevant to evaluate the representativeness of the selected days. The data set mainly consisted of filters collected every 6th day; the mean PM_{10} concentration for these data differed less than 2% from the 12/14 month mean per site. By expanding the data set with 6–9 filters per site taken every 2nd day and with 17 filters per site for days with regionally enhanced PM_{10} levels, the PM_{10} distribution per site was shifted upwards (Figure 1, right boxplot per site). Figure 2 presents the time variation of PM_{10} per site for the source apportionment data used in this study. The relative similarity across the five sites over time indicates a regional pattern of PM_{10}. A detailed description of all the measurement results, including time series and site comparisons per individual PM_{10} composition variable, is reported elsewhere.[19]

4.2 Identification and Temporal Variation of the Calculated Factors

Figure 3 shows the normalized factor profiles, indicating the fraction of all the measured variables accounted for by each factor. The chemical fingerprint of factor 1 consists primarily of markers associated with nitrate-rich secondary inorganic aerosols (NO_3^- and NH_4^+). Based on the calculated contributions of this profile we observed a seasonal pattern (results not shown) consisting of higher concentrations during the autumn/winter

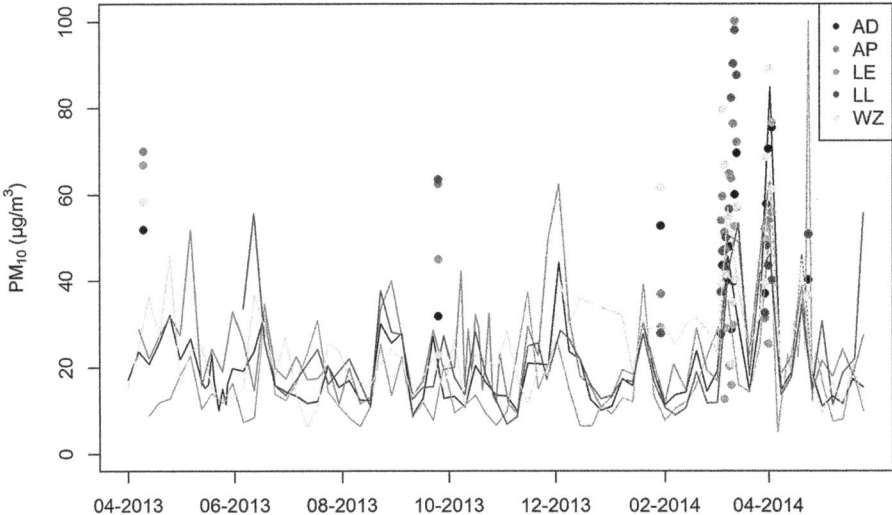

Figure 2 Time variation of daily PM$_{10}$ (μg m^{-3}) at the five sites according to the filters used for source apportionment. Full lines present data of the 1 in 6th day and 1 in 2nd day filters, dots show the additionally analysed filters for days with regionally enhanced PM$_{10}$ ($n = 17$ per site).

period compared to the summer/spring period. It is well known that temperature greatly affects the equilibrium of ammonium nitrate (NH$_4$NO$_3$), ammonia (NH$_3$) and nitric acid. Low temperature during the autumn/winter period favours the formation of the particulate form.[22] Nitrate is formed by the oxidation of NO$_2$ emitted by combustion processes, such as vehicle engines. This process is slow, which explains the small contribution of NO$_2$ and other combustion indicators associated with this factor. Nevertheless, a limited contribution of OC is found in this factor. This observation is however not unrealistic because part of the OC is also thought to be semi-volatile.[33] Similar results have been reported for the Netherlands.[34,35] We categorized factor 1 as 'nitrate-rich secondary aerosol (SA)'. Factor 2 mainly contains secondary sulfate, as indicated by the presence of SO$_4^{2-}$ and NH$_4^+$, and a limited contribution of OC. Secondary sulfate is formed from atmospheric oxidation of SO$_2$ and is associated with long-range transport. It exhibits a seasonal variation with higher contributions during the summer period.[22] Factor 2 is categorized as 'sulfate-rich SA'.

Factor 3 contains the majority of the Ca contributions. In general, Ca is commonly associated with mineral dust, building activities and fertilizers. However, Ca has also been related to vehicular emissions and iron and steel plants.[36–38] Iron and steel plants commonly produce furnace slacks, *i.e.* the glass-like by-product left over after a desired metal has been separated (*i.e.*, smelted) from its raw ore. It consists mainly of Ca, silicon (Si), Mg, and Al oxides. Furnace slacks are commonly used in asphalt which is a composite material used to surface roads, parking lots and airports. Additionally,

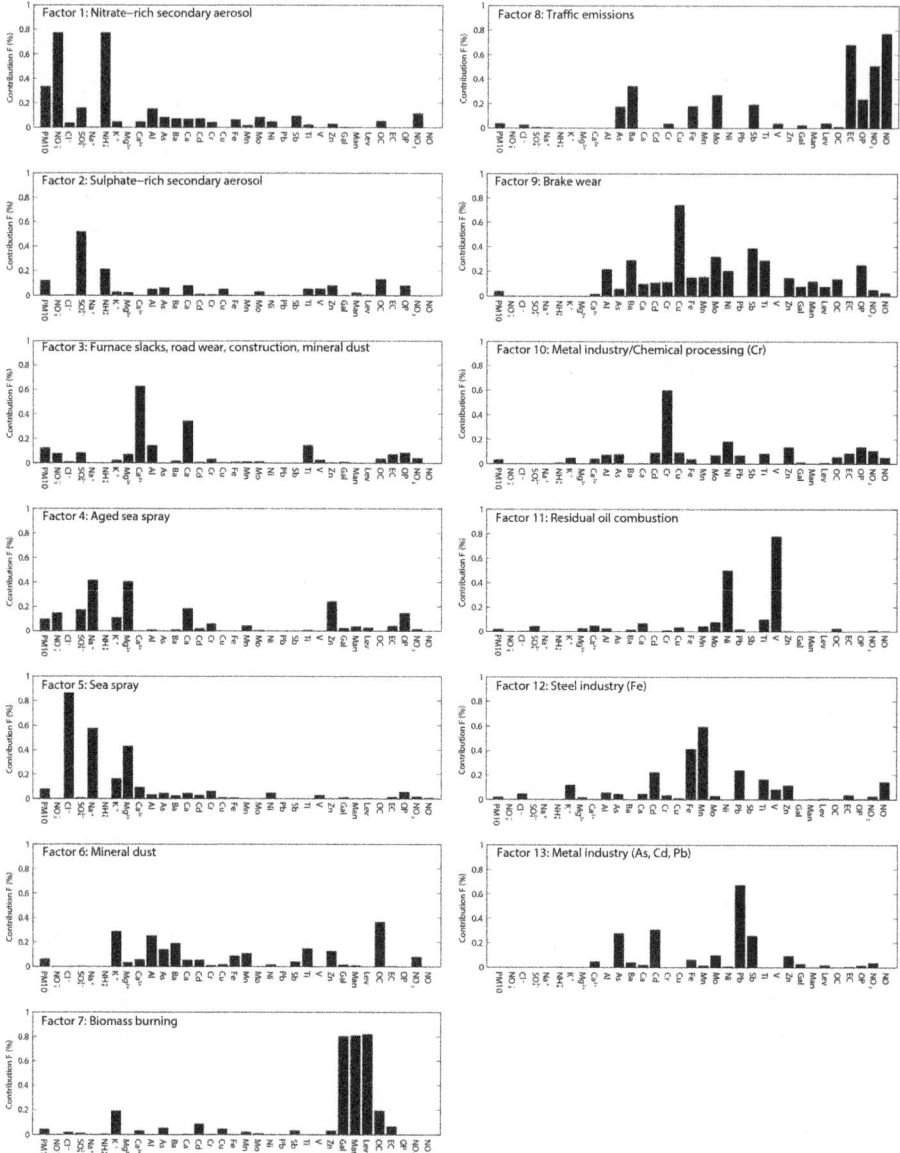

Figure 3 Normalized US-EPA-PMF source profiles based on the chemical composition, the oxidative potential (OP) of PM_{10} and the ambient NO and NO_2 concentration (average profiles for the five sites).

building/concrete debris is also used as the mineral aggregate in asphalt. In another source apportionment study in Amsterdam[39] a crustal component based on a high loading of Ca was found. Despite the large contribution of Ca in that study[39] no other typical crustal elements, such as Al and Ti, were detected, and the profile was not associated with any wind direction, suggesting it was not emitted by a single major point source in or near Amsterdam. Valius[39] associated this profile to a distinctive composition profile of the resuspended road dust near the measurement site. The different composition profile was thought to be a consequence of the widespread use of road structures made from asphalt/concrete in Amsterdam. In the present study, contributions of factor 3 are found at all sites, with WZ and AP showing the highest contributions. The high contributions found at WZ might be primarily associated with the storage/transhipment of furnace slacks from the nearby steel melting furnace. However, this does not explain the high contributions at AP. Compared to the other sites, AP is most closely located to a main road with high traffic intensity, so part of the contributions may be caused by the wear and tear on the asphalt/concrete roads owing to traffic. However, factor 3 does not show a seasonal variation associated with traffic emissions so not all contributions can be related to road surface wear. It is plausible that part of the contribution of this factor is associated with mineral dust and construction/building. Factor 3 is categorized as 'furnace slacks, road wear, construction, mineral dust'.

Factor 4 is characterized by high amounts of Na^+, Mg^{2+} and some SO_4^{2-}. Both Na^+ and Mg^{2+} are associated with sea spray. The factor does not contain Cl^-, which can be explained by chloride depletion, commonly seen in aged sea spray.[40,41] This depletion is directly correlated with the retention time. During this time chloride can be removed from sea salt particles in the air when it reacts to HNO_3 as well as H_2SO_4. In both cases Cl^- becomes associated with the gaseous HCl. The amount of (secondary) NO_3^- and SO_4^{2-} associated with this profile suggest a longer retention time than fresh sea spray. Factor 4 is categorized as 'aged sea spray'. In factor 5 the majority of Na^+, Mg^{2+} and Cl^- concentrations were found, which are associated with sea salt particles. Traces of K^+ and Ca^{2+} are also associated with these particles. The lack of NO_3^- and SO_4^{2-} in this factor suggests a short retention time in the air. This is supported by the high amount of Cl^-, showing limited signs of Cl^- depletion. The Cl^-/Na^+ ratio in this profile is 1.8, corresponding with that found in an earlier study.[22]

Factor 6 is identified by the presence of Al, Ti, Fe and Ca, species that are all associated with mineral dust. This factor also contains a significant portion of OC, which is an indication of mixing of dust and organic matter contributed by biogenic sources (*e.g.* plant detritus or other plant fragments) during resuspension.[41-43] The profile also contains a high amount of K, which can be attributed to sources like mineral dust, combustion and industrial activities. For WZ and AD high source contributions were found at Easter in 2014 (20 April). Factor 6 is categorized as 'mineral dust'. Factor 7 contains almost all available concentrations for the MAs, which are considered to be biomarkers for biomass or wood burning. A small amount of K^+ is also found in this factor and is commonly associated with biomass burning.[44] The factor shows some OC, but the attributed concentration is smaller than the sum of the apportioned MA concentrations. Since the MAs are part of the OC fraction, this indicates that not enough OC is apportioned to this profile. The highest amounts of biomass burning are found at LL, initially showing a seasonal pattern with high contributions during wintertime. Both AD and WZ show distinct peaks on 30 January and 20 April 2014. The latter is associated with bonfires during Easter commonly seen in the east of the Netherlands as well as Germany. As mentioned, factor 6 (mineral dust) contains high amounts of OC and K^+ and also has a high peak during Easter. Therefore, part of the OC and K^+ in factor 6 is likely related to biomass burning.

Factor 8 contains high amounts of EC, NO_2 and NO, which are commonly associated with primary traffic emissions. In this factor, these species represent the tail-pipe emissions. To a lesser extent, this profile also contains Ba, Mo and Sb. Barium emissions are linked to tyre wear[45] whereas Mo is used as an additive to lubricants and is also released by the combustion of fossil fuels.[46] The highest contribution of this profile is found at AP, which is a traffic-exposed background site. A clear weekend/weekday variation for this profile is found as well as increased contributions during autumn and winter. Factor 8 is categorized as 'traffic emissions'.

Factor 9 is associated with a range of metals (Fe, Cu, Ba, Mn, Mo and Sb) and some OC. The presence of high amounts of Cu along with these other metals suggests this factor is mostly associated with brake wear emissions.[22,47] As this factor is coupled with Al and some Ca and Ti, this would suggest resuspension of traffic-related crustal material. Similar to the traffic emissions a clear weekend/weekday variation for this profile is found at AP as well as increased contributions during autumn and winter. Factor 9 is categorized as 'brake wear'. High contributions of Cr in factor 10 are found at AD, AP and WZ, in contrast to the low contributions at LE and LL. Chromium is associated with several sources, such as traffic and industrial activities. The lack of other markers for traffic or industry hampers the identification of factor 10, which is dubbed as 'metal industry/chemical processing (Cr)'.

In factor 11 significant fractions of Ni and V are found, which are known tracers for the combustion of crude oil.[48] Typical oil combustion sources are shipping,[49] municipal district heating power plants and industrial power plants using heavy oil.[22] While the profile does contain the majority of V, it contains only about 30% of the Ni concentration, suggesting there is at least one other Ni source. Factor 11 is called 'residual oil combustion'. Factor 12 is characterized by high amounts of Fe, Mn and Cd, species commonly associated with the steel industry. The contributions are high at WZ, a site impacted by the presence of a steel industry nearby. We dubbed this profile as 'steel industry (Fe)'. Factor 13 contains high contributions of As, Cd and Pb, which are commonly associated with the metal industry, but are also related to other sources. An example is the bioaccumulation of metals in plants, which are released in the atmosphere by biomass burning. Factor 13 is categorized as 'metal industry (As, Cd, Pb)'. In this analysis no PM_{10} mass was apportioned to the profile metal industry (As, Cd, Pb), despite the distinct chemical fingerprint. This occasionally happens when there is some

Table 2 Linear regressions and coefficient of determination (R^2) between the measured and predicted (apportioned) PM_{10} mass per site.

Site	Linear regression equation	R^2
AD	Predicted mass $= 0.37 + 0.98 \cdot$ total mass	0.98
AP	Predicted mass $= 0.71 + 0.98 \cdot$ total mass	0.98
LE	Predicted mass $= 2.94 + 0.84 \cdot$ total mass	0.81
LL	Predicted mass $= 1.70 + 0.94 \cdot$ total mass	0.94
WZ	Predicted mass $= 0.29 + 0.99 \cdot$ total mass	0.98
Full data set	Predicted mass $= 2.01 + 0.92 \cdot$ total mass	0.91

rotational ambiguity in the solution as well as very low PM_{10} concentrations with high uncertainties associated with this source.

Table 2 shows the relationships between the measured and apportioned PM_{10} mass per site and for the full data set. The lowest coefficient of determination was found for LE $(R^2 = 0.81)$, suggesting that the current suite of profiles and corresponding contributions are not an optimum match for the actual situation at LE.

4.3 Spatial Variation of the Source Profiles

Figure 4 shows the relative source profile contributions per sampling site as a percentage of the total PM_{10} mass concentration. The 13 profiles in Figure 4 have been aggregated to eight groups plus an unexplained fraction. The unexplained fraction is defined as the difference between the sum of the mean apportioned and measured PM_{10} mass concentration at each site. Table 3 compares the individual source profiles between the sites, expressed in mass concentration units $(\mu g\ m^{-3})$. The largest part of PM_{10} (40–48%) was explained by SA, which was mainly owing to the nitrate-rich SA factor. The two SA profiles were estimated to contribute similar mass concentrations at the four sites on the European continent $(12–13\ \mu g\ m^{-3})$, but had clearly lower values at the UK site LE $(8.4\ \mu g\ m^{-3})$. Nitrate-rich SA accounted for 27% (LE) to 37% (LL) of the PM_{10}, while sulfate-rich SA contributed 9–13%. The high contribution of nitrate-rich SA, with NH_4NO_3 as main compound, indicates that decreasing the emissions of its precursor gases NO_x and NH_3 can meaningfully reduce the PM concentrations. For example, a model study found that a reduction of NH_3 emissions by 50% could lead to a 24% reduction of the $PM_{2.5}$ concentrations in NW Europe, mainly driven by reduced formation of NH_4NO_3.[50]

The second-most important source profiles were related to sea spray (11–21% of PM_{10}). The fresh and aged sea spray profiles contributed most PM_{10} in absolute term at WZ $(6.3\ \mu g\ m^{-3}$ or 21%), which is the site closest to the North Sea. The absolute contribution was lower and similar at the other four sites $(3–4\ \mu g\ m^{-3})$, even though at AD and LE sea spray was relatively more important (17–20%) than at AP and LL (11–13%) because of the lower total PM_{10} levels at AD and LE. The source profile furnace slacks, road wear and construction was on average the third most important, but there was

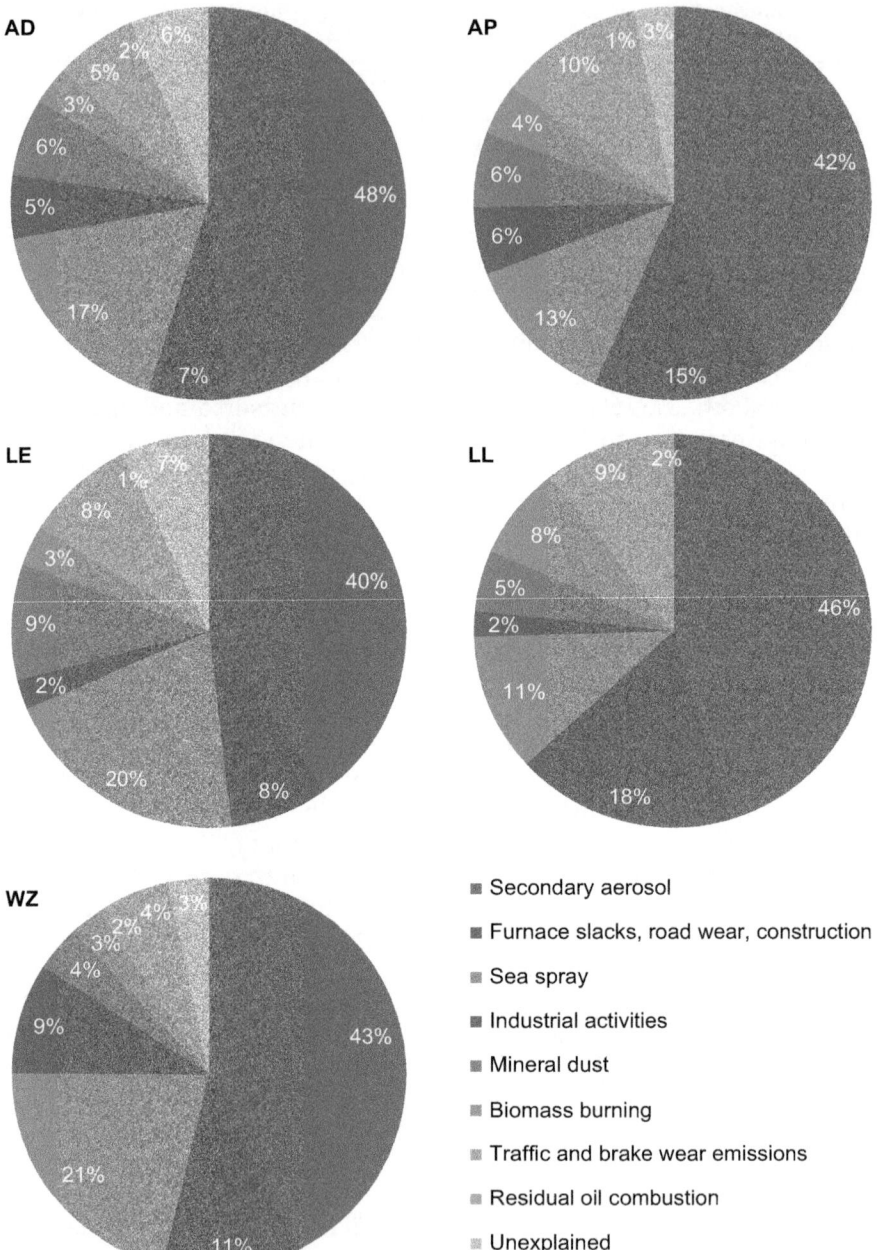

Figure 4   Source contributions (% of the apportioned PM$_{10}$ concentration per site) per site.

more variation between the receptor sites both in the absolute as well at the relative contributions. The estimated mass contribution of furnace slacks, road wear and construction was lower at AD and LE (1.7 μg m^{-3} or 7–8%)

Table 3 Contribution of source profiles ($\mu g \ m^{-3}$) to the measured PM_{10} per site.

Source profile	Contribution to PM_{10} ($\mu g \ m^{-3}$)				
	AD	AP	LE	LL	WZ
Secondary aerosol	12.1	12.9	8.4	13.6	12.6
Nitrate-rich secondary aerosol	9.1	9.9	5.8	10.9	8.8
Sulfate-rich secondary aerosol	3.0	3.0	2.6	2.7	3.8
Furnace slacks, road wear, construction	1.7	4.5	1.7	5.3	3.4
Sea spray	4.3	3.9	4.3	3.4	6.3
Aged sea spray	2.5	2.5	2.4	1.9	2.8
Sea spray	1.8	1.4	2.0	1.5	3.4
Mineral dust	1.6	1.9	2.0	1.5	1.1
Biomass burning	0.8	1.3	0.7	2.3	0.8
Industrial activities	1.3	1.7	0.5	0.6	2.7
Metal industry/chemical processing (Cr)	1.1	1.3	0.4	0.4	1.1
Steel industry (Fe)	0.2	0.4	0.2	0.2	1.6
Metal industry (As, Cd, Pb)	0.0	0.0	0.0	0.0	0.0
Traffic emissions and brake wear	1.2	3.2	1.7	2.7	0.7
Traffic emissions	0.5	1.6	0.9	1.2	0.5
Brake wear	0.7	1.5	0.8	1.5	0.2
Residual oil combustion	0.5	0.4	0.2	0.5	1.1
Unexplained	1.6	1.0	1.5	0.0	1.0
Total measured PM_{10}	25.0	30.8	21.1	29.8	29.6

than at WZ (3.4 $\mu g \ m^{-3}$ or 11%) and highest at AP and LL (4.5–5.3 $\mu g \ m^{-3}$ or 15–18%).

The three source profiles for industrial activities were most important at AD and AP, and in particular at WZ. Metal industry/chemical processing (Cr) was estimated to contribute similarly at these three sites (1 $\mu g \ m^{-3}$ or 4%), with lower levels at LE and LL, suggesting that this source had a minor impact on PM_{10} concentrations at LE and LL. The steel industry (Fe) profile was only important at WZ (1.6 $\mu g \ m^{-3}$ or 5%). As mentioned earlier, the PMF analysis did not apportion any PM_{10} mass to the metal industry (As, Cd, Pb) profile. However, the PMF output also provides a scaled contribution matrix, which revealed that the largest scaled contribution for the metal industry (As, Cd, Pb) profile was at the receptor site in Antwerp. This indicates the physical relevance of this profile, because upwind of AP (255–225 °N) an industrial site is located that is known to emit As, Cd and Pb.

The biomass burning profile was higher at LL (2.3 $\mu g \ m^{-3}$) and AP (1.3 $\mu g \ m^{-3}$) than at the other sites (0.7–0.8 $\mu g \ m^{-3}$). The mineral dust profile contributed on average 1.1 (WZ) to 2.0 $\mu g \ m^{-3}$ (LE), in agreement with the value of 2 $\mu g \ m^{-3}$ reported for the regional mineral contribution to total PM_{10} in most of Northern and Central Europe.[51]

The contribution by traffic-related emissions was split up over different source profiles. The PMF analysis resulted in two distinct traffic-related profiles: primary traffic emissions (exhaust) and brake wear, which were highest at LL (2.7 $\mu g \ m^{-3}$) and AP (3.2 $\mu g \ m^{-3}$). For AP this seems logical since the sampling site has the closest distance to a main road with high traffic intensity. The high contribution of brake wear at LL is somewhat conspicuous as there are no traffic lights near the site. However, a shunting site for trains is located near the site. Shunting in general requires some stop and go actions from trains, so part of the brake wear at LL might be associated with this activity.

In addition to traffic emissions and brake wear, other profiles in this study also contribute to traffic-related PM_{10}, *e.g.* the secondary nitrate-rich SA profile, the furnace slacks, road wear and construction profile and the mineral dust profile. Hence, the sum of the traffic emissions and brake wear profiles in the current study underestimates the actual contribution of road traffic to ambient PM_{10}. For example, for five cities in Flanders the annual PM_{10} concentration in 2013–2014 was on average 7 µg m^{-3} (or 38%) higher in busy street canyons than at a background location near the city.[52] This local contribution was mostly owing to two approximately equally important traffic-related contributions: the contribution of EC and organic matter in exhaust gases on the one hand and the contribution of mineral dust by soil dust resuspension and the wear of vehicles (tires, brakes, bodywork) and the road surface on the other hand.[52]

The spatial variation between the sites was also evaluated using the COD values per source profile (Figure 5). We found that the steel industry (Fe) profile had the largest COD, indicating very different source contributions between the sites. The highest contributions for the steel industry profile

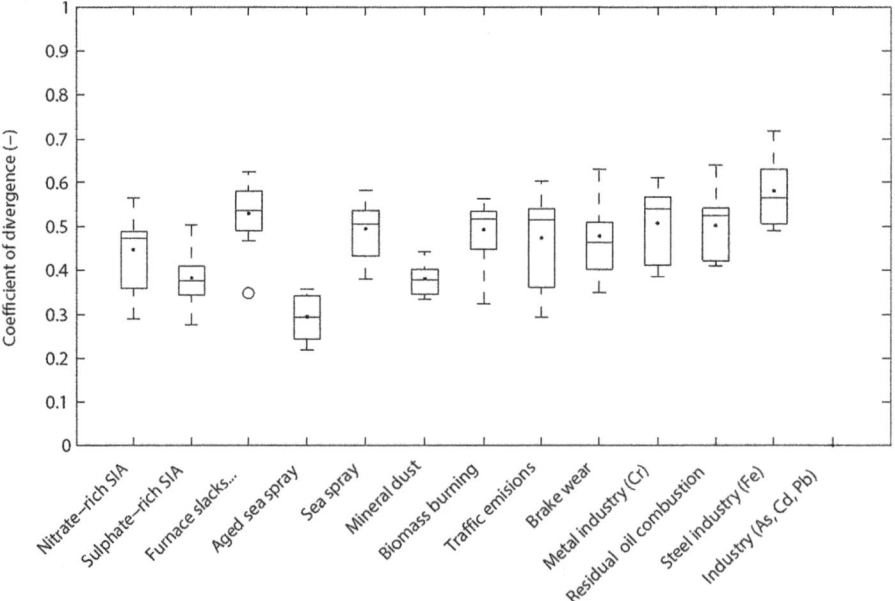

Figure 5   Coefficient of divergence between the five sampling sites per PM_{10} source profile (for explanation of boxplots see Figure 1).

were found at WZ, which is directly impacted by steel industry located south of the receptor site. The factor with the lowest spatial variability was aged sea spray, closely followed by sulfate-rich SA. The small spatial variation reflects the importance of long-range transport into and across the monitored area.[22] Smaller CODs for sulfate-rich SA compared to the nitrate-rich SA have been reported.[53] In the present study the nitrate-rich SA contribution at LE is considerably smaller than at the other sites, indicating a difference in precursor concentrations.

The largest CODs for the traffic emissions profile were found between WZ and the four other sites. This indicates that the contribution from this source is primarily local in the urban environment and the spatial variability is very similar across urban sites in NW Europe. Interestingly, the brake wear source showed a similar behaviour. However, the spatial variability for brake wear between the sites other than WZ was larger compared to traffic emissions. The reason may be that AP, LE and LL are impacted from nearby traffic junctions, leading to a higher contribution of brake-related emissions. Aged sea spray showed the smallest COD values, indicating that this can be considered as an area source. The fresh sea spray profile had larger CODs; the largest contribution of this profile was found at locations relatively close to the sea. The CODs for mineral dust were lower than for sea spray, which is surprising because there are many local and global sources related to mineral dust compared to the single source for sea salt.

The highest COD values for the factors furnace slacks, road wear, construction and mineral dust were found between AD, LE and WZ. This suggests the contribution at these sites is a combination of several local sources with a location-specific source contribution ratio (*e.g.* WZ: primarily furnace slacks; LE: primarily building/construction). The importance of local source contributions was also observed for black carbon and ultrafine particles measured at the same sites.[10] The low COD values for the factors furnace slacks, road wear, construction and mineral dust between AP and LL suggests that the associated sources in these two areas may still be local but similar in composition.

4.4 Wind-directional and Trajectory Analysis of the Source Profiles

Figure 6 shows CPF plots for the contributions of some example source profiles and sites. In these plots the largest peaks points towards the wind sector responsible for the highest contribution at a site. A high contribution of nitrate-rich SA was associated with wind coming from the northeast-east at all the sites, as illustrated for AP (Figure 6a). The contributions of sulfate-rich SA showed a similar pattern as for the nitrate-rich profile (not shown). In general, eastern winds bring in more air pollution compared to western winds. The contributions of sources from Eastern Europe are accumulated leading to higher PM concentrations compared to the relatively cleaner air from the west travelling primarily over the ocean.

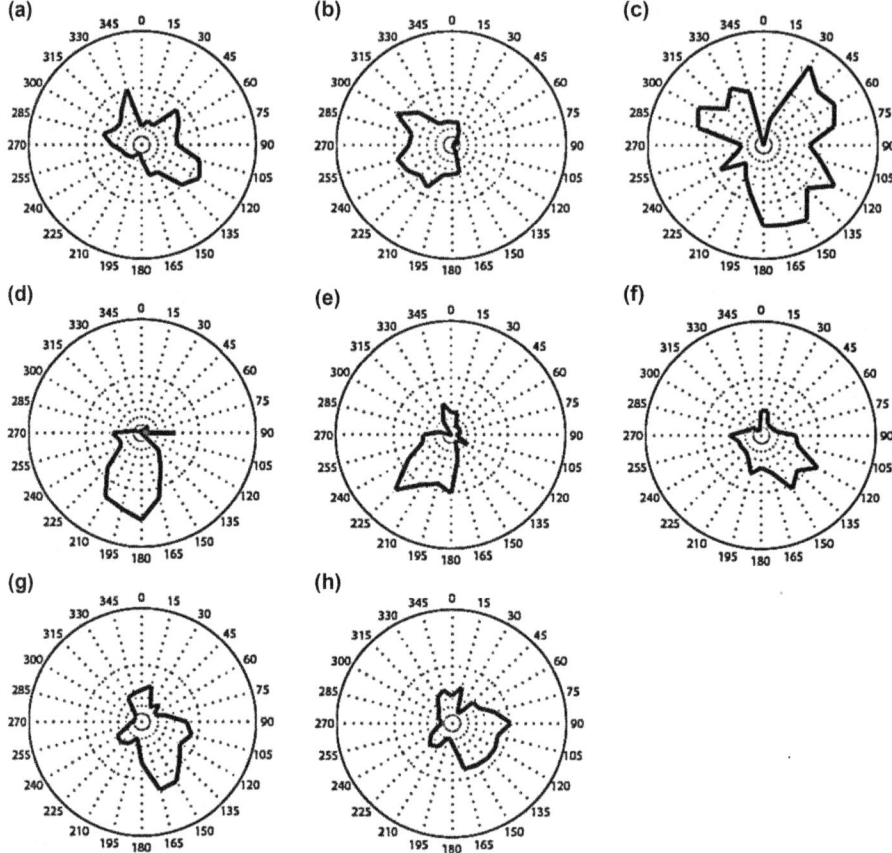

Figure 6   Examples of conditional probability function plots of the contributions of the PM_{10} source profile (a) nitrate-rich secondary aerosol at AP, (b) sea spray at AD, (c) biomass burning at LL, (d) steel industry at WZ, (e) residual oil combustion at WZ, (f) mineral dust at LE, (g) traffic emissions at AP and (h) brake wear at AP.

For the sea spray and aged sea spray profiles, the four sites on the European mainland had the highest contributions from the west, corresponding with the location of the North Sea. This is shown for sea spray at AD (Figure 6b). The highest contributions for (aged) sea spray were found at WZ (Table 3), which is the site closest to the North Sea. For the UK site LE, the highest contribution of sea salt came from the southwest, while the aged sea spray profile seemed to behave like an area source with equal contributions from every direction (not shown).

Figure 7 shows the CPF plots of the Cr profile at AD, AP, LE and LL in combination with a selection of the largest registered sources of Cr in NW Europe (data from 2013). Note that a scaling is used in the CPF plots, so that the size of the peaks in the plots does not correspond with the actual concentrations from a particular wind sector. As mentioned, the highest

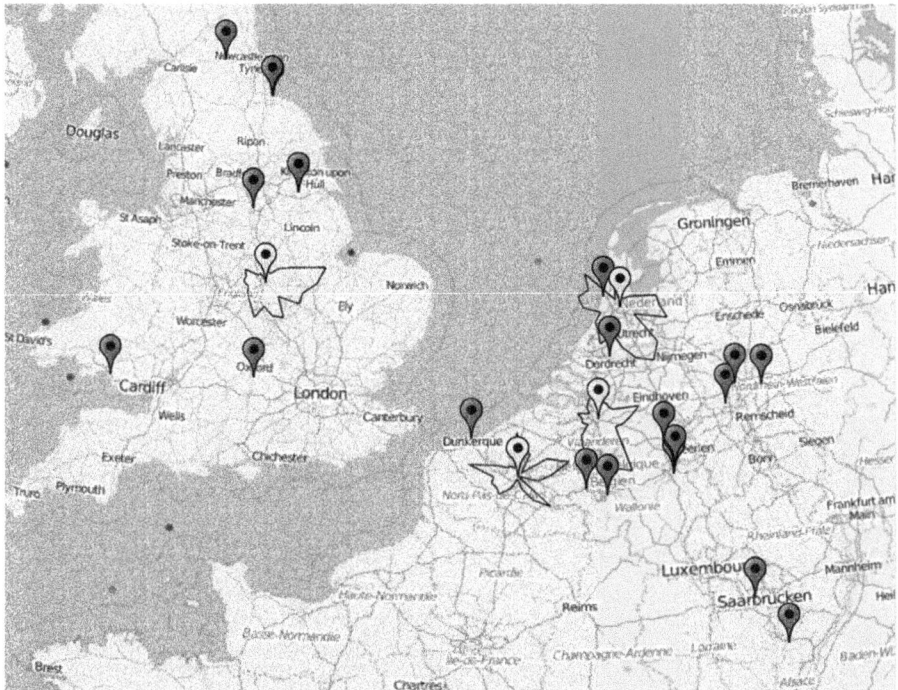

Figure 7 Conditional probability function plots of the contributions of the metal industry/chemical processing (Cr) profile at four sites, indicated by the white symbols (AD, AP, LE and LL). Grey symbols indicate some of the largest registered Cr sources according to the European Pollutant Release and Transfer Register (E-PRTR, http://prtr.ec.europa.eu, data of 2013).

concentrations of the metal industry/chemical processing (Cr) profile were found at AD, AP and WZ (Table 3). The largest contribution of the metal industry/chemical processing (Cr) profile was found on 3 March 2014 at AD. The back-trajectories for this day indicate that the air picks up several emissions from Cr sources located south of AD (Figure 8). The impact of Cr sources at LE seems to be coming primarily from the European mainland (Figure 7).

For biomass burning, the highest concentrations at AD and WZ during the monitoring period were found from the east (not shown). This is related to the Easter bonfires on 20 April 2014 in large parts in the east of Netherlands and Germany, as indicated by the HYSPLIT backward trajectories shown for AD (Figure 9). The biomass burning profile had the highest concentrations at LL (Table 3) and showed contributions from several wind directions (Figure 6c). This indicates that multiple sources associated with biomass burning are affecting the receptor site LL.

The steel industry (Fe) profile was a local source associated with WZ only (Table 3). The highest contributions for steel industry (Fe) were found from the south of WZ (Figure 6d), coinciding with the location of a major steel

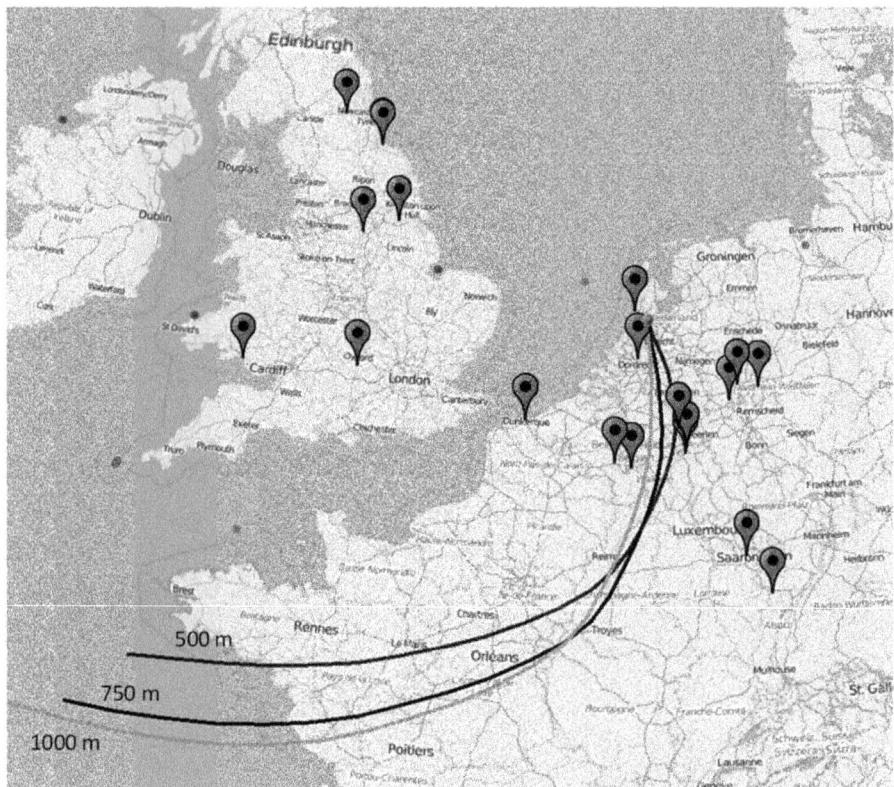

Figure 8   HYSPLIT model analysis with backward trajectories ending at site AD on 3 March 2014, 23:00 UTC. Grey symbols: see Figure 7. The three lines have different starting heights above ground level (500, 750 and 1000 m).

industry in the IJmond industrial area. For the source profile furnace slacks, road wear, construction and mineral dust, the major contribution at WZ was also coming from the south. This finding corresponds well with the assumption that the concentrations in this profile were partly contributed by furnace slacks, which are a by-product of steel manufacturing. The plots for the residual oil combustion profile at AD, AP and WZ point towards shipping emission sources, *e.g.* for WZ there was a strong influence of the port of IJmuiden SW of the site (Figure 6e).

The mineral dust profile showed the highest contributions from the east for the sites on the European mainland and from the S-SE for LE (Figure 6f). For the traffic emissions profile, the majority of the concentrations were contributed by local sources. Although the wind data used in this study are not necessarily representative for the local situation at the aerosol sampling sites, the traffic emission contributions at *e.g.* AP point towards the location of a traffic junction near the site (Figure 6g). The same holds true for the brake wear profile points at AP (Figure 6h).

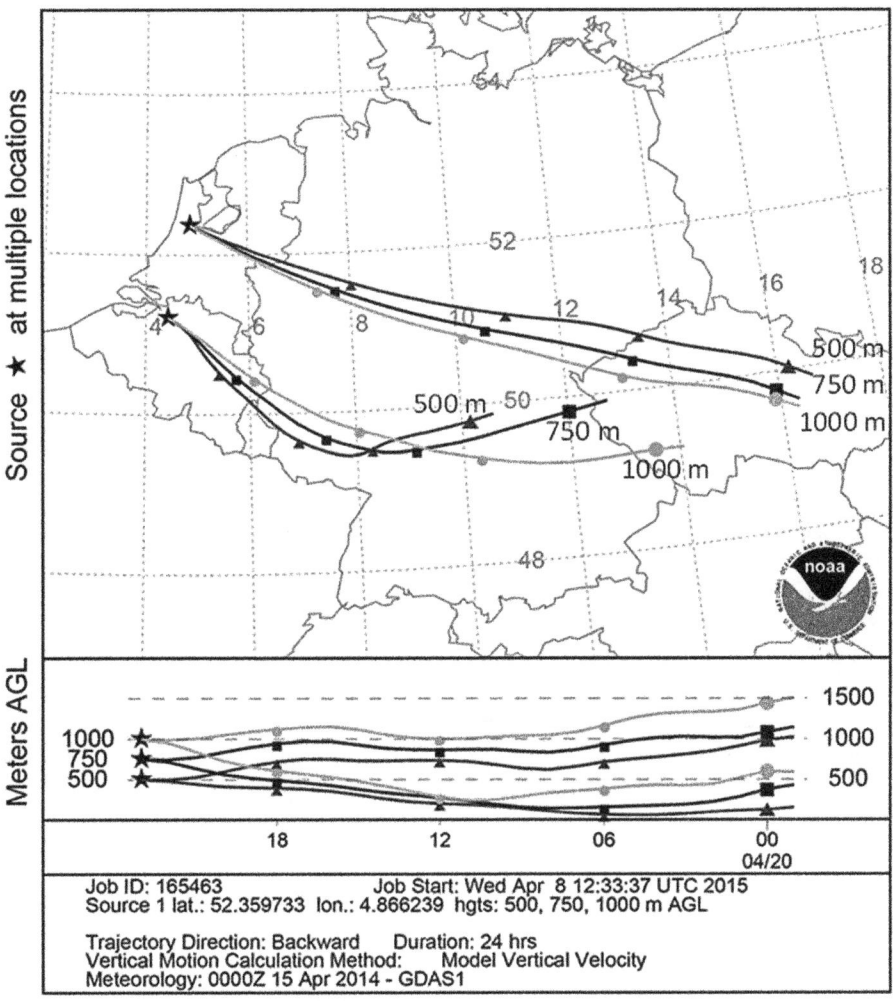

Figure 9 HYSPLIT model analysis with backward trajectories ending at sites AP and AD on 20 April 2014, 23:00 UTC. Lines: see Figure 8.

4.5 Source Profiles on Days Exceeding the Daily Limit Value

Using the calculated PMF source profiles, the dominant source contributions were evaluated for days with known PM$_{10}$ composition on which the concentrations exceeded 50 µg m^{-3} ($n = 8$ for AD, 18 for AP, 7 for LE, 10 for LL and 11 for WZ). The calculated source with the highest contribution was found to be nitrate-rich SA at all sites, as illustrated for AP and WZ (Figure 10). The contribution of this source during exceedance days was on

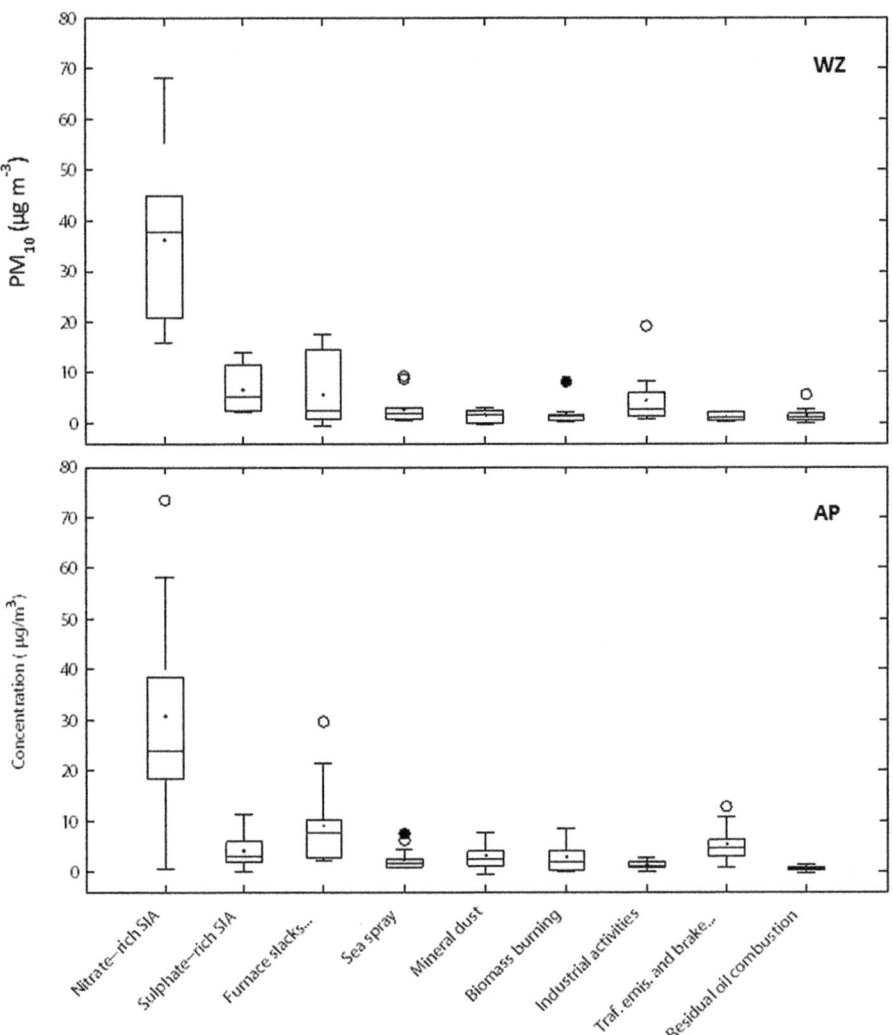

Figure 10 Apportioned PM$_{10}$ concentration per aggregated source profile on days with analysed PM$_{10}$ composition on which PM$_{10}$ exceeded 50 µg m^{-3} at sites WZ and AP.

average 35 ± 18 µg m^{-3} for all sites. The highest mean contribution during exceedance days was found at AD (43 ± 13 µg m^{-3}) and the lowest at LE (32 ± 12 µg m^{-3}). This is in line with other studies in *e.g.* Flanders[52] and the Netherlands.[34] The majority of exceedance days occurred during spring. This is likely related to increased emissions of NH$_3$ when manure is spread on agricultural lands, and the subsequent increase in the formation of NH$_4$NO$_3$. There were also exceedance days with relatively low contributions of nitrate-rich SA, in the winter period when NH$_3$ emissions are lower.[50] This can probably be attributed to meteorological conditions that are unfavourable to air dispersion. In winter, a lower inversion layer is generally observed, resulting in a lower mixing height and higher PM concentrations.

Sulfate-rich SA remained fairly constant during exceedance days with a mean contribution of 6 ± 4 µg m^{-3} for all sites. This suggests that SO$_2$ is not driving increased PM levels, in contrast to NH$_3$.[50] For AP and LL the second most important source during exceedances days was the furnace slacks, road wear and construction source, with contributions of 9 ± 7 µg m^{-3} (AP) and 12 ± 14 µg m^{-3} (LL). The contribution to PM$_{10}$ of the combined sea salt profiles was fairly constant on all sites with an average of 3 ± 2 µg m^{-3}. The contribution of other sources, such as mineral dust, biomass burning, industrial activities, traffic and residual oil combustion, was small and fairly constant. For some exceedance days, however, for example at WZ, local sources had a meaningful impact on PM$_{10}$.

4.6 Estimated Uncertainty of the PMF Analysis

The bootstrap (BS) method showed that some high elemental concentrations (*e.g.* Fe and Cr at WZ) critically influenced the bootstrap results. If the peak concentrations for Fe are not present in the bootstrap data set, the steel industry factor is not found and cannot be matched to the base run factor (74% BS mapping). The same phenomenon is observed for Cr (79% mapping). All other factors have approximately 95–100% mapping from the base runs. The displacement of factor elements method (DISP) indicated that there was no significant rotational ambiguity and that the solution was sufficiently robust to be used. The decrease in Q was small (d$Q_{DISP} < 0.1\%$), indicating that a true global minimum of Q was likely found in the base run.[24] Furthermore, no factor swaps were observed in the DISP method, indicating the solution has no or few data errors so that the solution can be considered to be well defined.[25] For the bootstrap enhanced by displacement of factor elements (BS-DISP) method, 71% of the runs were accepted, indicating some uncertainty in the chosen solution. However, the number of swaps can be considered as not fatal to the analysis when positive BS and DISP results as well as clear interpretable factors are found.[26]

The interval ratios of the BS, DISP and BS-DISP methods[25,26] are reported in detail elsewhere.[19] For the key species of both the nitrate-rich and sulfate-rich SA profiles, the interval ratios for the three uncertainty methods were low, indicating a very stable apportionment of NO$_3^-$, NH$_4^+$ and SO$_4^{2-}$. Much higher interval ratios and a higher variation between the ratios of the three methods indicated instability of OC in the sulfate-rich SA. For the profile furnace slacks, road wear, construction and mineral dust, Ca and Ca^{2+} seemed the main key species. Other key species in this factor such as Al and Ti had higher interval ratios, probably owing to the large amount of missing data for these species. The key species for both fresh (Na$^+$, Mg^{2+} and Cl$^-$) and aged sea salt (Na$^+$, Mg^{2+}) showed low interval ratios for all three error estimation methods, indicating that the apportionment of this profile is very stable. For K$^+$ and OC in the mineral dust profile, the higher interval ratios for BS-DISP were high compared to the other methods, indicating that the apportionment of these species shows instability. The apportionment of the biomass burning profile, with low interval ratios for the key species levoglucosan, galactosan and mannosan, was very stable. For the distinguished traffic-related factors, the primary traffic emissions profile with key species EC, NO and NO_2, is the most stable solution. For the brake wear profile, Cu appeared to be the main key species, whereas other key species such as Ba, Mo and Sb had higher ratios. The difference between the interval ratios of the different methods indicates that the apportionment of the brake wear profile was not completely stable. This can be explained by the data pooling for sites with differences in the amount and composition of brake wear contributions, *e.g.* owing to differences in traffic intensities, road infrastructure (*e.g.* traffic lights) and the presence of other sources (*e.g.* train shunting facility). Some instability issues were also found for the industry factors (Cr, Fe, As and Cd) and the residual oil combustion factor (Ni, V). In this study the contributions from the industry factors as well as the residual oil combustion differ among the receptor sites and have different local impacts. We established earlier that steel industry has a major impact at WZ whereas higher contributions from residual oil combustion can be seen at sites impacted by shipping emissions.

5 Conclusions

The aim of this study was to identify and quantify the most relevant sources of PM_{10} in NW Europe using positive matrix factorization (PMF). Based on PM_{10} sampling at four urban background sites and one industrial site, an overview of global sources impacting the studied sites was established. A solution with 13 factor profiles was selected, which could be aggregated to eight groups: secondary aerosol; furnace slacks, road wear and construction; sea spray; mineral dust; biomass burning; industrial activities; traffic emissions and brake wear; and residual oil combustion. The largest part of PM_{10} (40–48%) was explained by nitrate-rich (27–37%) and sulfate-rich (9–13%) secondary aerosol. The second-most important source profiles were related to sea spray and aged sea spray (11–21%). These findings correspond with previous source apportionment studies in the Netherlands.[22,35] The source profile furnace slacks, road wear and construction was on average the third most important, but there was more variation between the receptor sites. In addition, clear traffic and biomass burning source profiles were found. Decreasing the emissions of precursor gases of secondary inorganic aerosol, and particularly NH_3 and NO_x, can contribute to decreasing the ambient PM_{10} concentrations in the study region.

The relationship between the measured and apportioned PM_{10} mass per site was good ($R^2 > 0.94$), except for the site in Leicester ($R^2 = 0.81$), suggesting that the current suite of profiles and corresponding contributions are not an optimum match for this site. Combining the data from sites still gives information about local sources, provided the contribution is strong enough (*e.g.* traffic emissions near the urban sites, the steel industry near the site in Wijk aan Zee). The results of the error estimations for the 13-factor solution show that the apportionment of some factors (secondary aerosol, sea salt, biomass burning and traffic emissions) is stable across the five receptor sites. Other profiles (*e.g.* industrial, brake wear) are subjected to some instability in the PMF analysis, indicating that either the error uncertainties are not accurate and/or the contributions of these factor profiles vary per site or over time. It is recommended to complement this pooled PMF analysis with site-specific source apportionment modelling.

Acknowledgements

This study was carried out in the framework of the Joint Air Quality Initiative (Joaquin), an EU cooperation project supported by the INTERREG IVB North West Europe programme (http://www.nweurope.eu). We thank all the persons who contributed to the PM₁₀ sampling and chemical analyses.

References

1. *Air quality in Europe – 2014 report*. EEA Report No 5/2014, European Environment Agency, Luxemburg, 2014.
2. *Air quality guidelines. Global update 2005. Particulate matter, ozone, nitrogen dioxide and sulfur dioxide*. World Health Organization, Regional Office for Europe, Copenhagen, 2006.
3. F. J. Kelly and J. C. Fussell, *Atmos. Environ.*, 2012, **60**, 504.
4. *Directive 2008/50/EC of the European Parliament and of the Council of 21 May 2008 on ambient air quality and cleaner air for Europe. Off. J. Eur. Union*, 11.6.2008, 1–44.
5. C. A. Belis, B. R. Larsen, F. Amato, I. El Haddad, O. Favez, R. M. Harrison, P. K. Hopke, S. Nava, P. Paatero, A. Prevot, U. Quass, R. Vecchi and M. Viana. *European guide on air pollution source apportionment with receptor models*, European Commission, Joint Research Centre, Institute for Environment and Sustainability, report EUR 26080 EN. Publications Office of the European Union, Luxembourg, 2014.
6. P. K. Hopke, *A Guide to Positive Matrix Factorization*, Available at: http://people.clarkson.edu/~hopkepk/PMF-Guidance.htm, 2003.
7. J. Vercauteren, C. Matheeussen, E. Wauters, E. Roekens, R. van Grieken, A. Krata, Y. Makarovska, W. Maenhaut, X. Chi and B. Geypens, *Atmos. Environ.*, 2011, **45**, 108.
8. E. P. Weijers, M. Schaap, L. Nguyen, J. Matthijsen, H. A. C. Denier van der Gon, H. M. ten Brink and R. Hoogerbrugge, *Atmos. Chem. Phys.*, 2011, **11**, 2281.
9. Air Quality Expert Group, *Fine Particulate Matter (PM₂.₅) in the United Kingdom*, Department for Environment, Food and Rural Affairs, London, UK, 2012. Available at https://www.gov.uk/government/uploads/system/

uploads/attachment_data/file/69635/pb13837-aqeg-fine-particle-matter-20121220.pdf.

10. *Monitoring of ultrafine particles and black carbon. Joint Air Quality Initiative (Joaquin), work package 1 action 1 and 3.* Flanders Environment Agency, Aalst, 2015. Available at http://joaquin.eu.

11. Nationaal Samenwerkingsprogramma Luchtkwaliteit (NSL), The Netherlands. Available at: https://www.nsl-monitoring.nl/viewer.

12. Department for Transport, London, UK. Available at http://www.dft.gov.uk/traffic-counts.

13. *Intra-urban variability of ultrafine particles in Antwerp (February and October 2013).* Flanders Environment Agency, Erembodegem, 2014. Available at http://www.vmm.be.

14. EN 14907. *Ambient air quality – Standard gravimetric measurement method for the determination of the $PM_{2.5}$ mass fraction of suspended particulate matter*. European Committee for Standardization, Brussels, 2005.

15. R. L. Cordell, I. R. White and P. S. Monks, *Anal. Bioanal. Chem.*, 2014, **406**, 5283.

16. CEN/TR 16243. *Ambient air - Guide for the measurement of elemental carbon (EC) and organic carbon (OC) deposited on filters.* European Committee for Standardization, Brussels, 2011.

17. CEN/TR 16269. *Ambient air - Guide for the measurement of anions and cations in $PM_{2.5}$.* European Committee for Standardization, Brussels, 2011.

18. A. Yang, A. Jedynska, B. Hellack, I. Kooter, G. Hoek, B. Brunekreef, T. A. J. Kuhlbusch, F. R. Cassee and N. A. H. Janssen, *Atmos. Environ.*, 2014, **83**, 35.

19. *Composition and source apportionment of PM_{10}. Joint Air Quality Initiative (Joaquin), work package 1 action 1 and 2.* Flanders Environment Agency, Aalst, 2015. Available at http://joaquin.eu.

20. D. B. Rubin, *Biometrika*, 1976, **63**, 581.

21. P. K. Hopke, *Chemom. Intell. Lab. Syst.*, 1991, **10**, 21.

22. D. Mooibroek, M. Schaap, E. P. Weijers and R. Hoogerbrugge, *Atmos. Environ.*, 2011, **45**, 4180.

23. G. Norris, R. Vedantham, K. Wade, S. Brown, J. Prouty, C. Foley, *EPA Positive Matrix Factorization (PMF) 3.0 Fundamentals and User Guide*, U.S. Environmental Protection Agency, Washington, DC, 2008, EPA 600/R-08/108.

24. G. Norris, R. Duvall, S. Brown, S. Bai, *EPA Positive Matrix Factorization (PMF) 5.0 Fundamentals and User Guide*, U.S. Environmental Protection Agency, Washington, DC, 2014, EPA/600/R-14/108.

25. P. Paatero, S. Eberly, S. G. Brown and G. A. Norris, *Atmos. Meas. Tech.*, 2014, **7**, 781.

26. S. G. Brown, S. Eberly, P. Paatero and G. A. Norris, *Sci. Total Environ.*, 2015, **518–519**, 626.

27. P. Paatero, P. K. Hopke, X. H. Song and Z. Ramadan, *Chemom. Intell. Lab. Syst.*, 2002, **60**, 253.

28. P. Paatero, *User's Guide for Positive Matrix Factorization Programs PMF2 and PMF3, Part 2: Reference, Tutorial*, 2000.
29. E. Kim, P. K. Hopke, J. P. Pinto and W. E. Wilson, *Environ. Sci. Technol.*, 2005, **39**, 4172.
30. E. Kim and P. K. Hopke, *Atmos. Environ.*, 2004, **38**, 4667.
31. E. Kim, P. P. Hopke and E. S. Edgerton, *J. Air Waste Manage. Assoc.*, 2003, **53**, 731.
32. R. R. Draxler and G. D. Hess, *Austr. Meteorol. Mag.*, 1998, **47**, 295.
33. N. M. Donahue, A. L. Robinson and S. N. Pandis, *Atmos. Environ.*, 2009, **43**, 94.
34. M. Schaap, E. P. Weijers, D. Mooibroek, L. Nguyen, *Composition and origin of particulate matter in the Netherlands*. Bilthoven, the Netherlands, PBL report 500099007, 2010.
35. D. Mooibroek, E. van der Swaluw, R. Hoogerbrugge. *A reanalysis of the BOP dataset: Source apportionment and mineral dust.* National Institute for Public Health and the Environment, Bilhtoven, RIVM Rapport 680356001, 2012.
36. S. H. Cadle, P. A. Mulawa, J. Ball, C. Donase, A. Weibel, J. C. Sagebiel, K. T. Knapp and R. Snow, *Environ. Sci. Technol.*, 1997, **31**, 3405.
37. D. S. Lee and J. M. Pacyna, *Atmos. Environ.*, 1999, **33**, 1687.
38. M. J. Kleeman, J. J. Schauer and G. R. Cass, *Environ. Sci. Technol.*, 2000, **34**, 1132.
39. M. Valius, *Characteristics and sources of fine particulate matter in urban air*, PhD thesis, University of Kuopio, Finland, 2005.
40. H. Beuck, U. Quass, O. Klemm and T. A. J. Kuhlbusch, *Atmos. Environ.*, 2011, **45**, 5813.
41. A. Waked, O. Favez, L. Y. Alleman, C. Piot, J.-E. Petit, T. Delaunay, E. Verlinden, B. Golly, J.-L. Besombes, J.-L. Jaffrezo and E. Leoz-Garziandia, *Atmos. Chem. Phys.*, 2014, **14**, 3325.
42. J. G. Watson and J. C. Chow, *Sci. Total Environ.*, 2001, **276**, 33.
43. N. J. Kuhn, *Earth Surf. Processes Landforms*, 2007, **32**, 794.
44. V. Bernardoni, R. Vecchi, G. Valli, A. Piazzalunga and P. Fermo, *Sci. Total Environ.*, 2011, **409**, 4788.
45. A. J. Fernández-Espinosa and M. Ternero-Rodríguez, *Anal. Bioanal. Chem.*, 2004, **379**, 684.
46. *Geochemical Atlas of Europe. Part 1: Background Information, Methodology and Maps*, ed. R. Salminen, M. J. Batista, M. Bidovec and A. Demetriades, ISBN 951-690-921-3, p. 525 (printed) & 951-690-913-2 (electronic version), 2005.
47. H. A. C. Denier van der Gon, J. H. J. Hulskotte, A. J. H. Visschedijk and M. Schaap, *Atmos. Environ.*, 2007, **41**, 8697.
48. Olmez, A. E. Sheffield, G. E. Gordon, J. E. Houck, L. C. Pritchett, J. A. Cooper, T. G. Dzubay and R. L. Bennett, *J. Air Waste Manage. Assoc.*, 1988, **38**, 1392.
49. M. Pandolfi, Y. Gonzalez-Castanedo, A. Alastuey, J. D. de la Rosa, E. Mantilla, A. S. de la Campa, X. Querol, J. Pey, F. Amato and T. Moreno, *Environ. Sci. Pollut. Res. Int.*, 2011, **18**, 260.

50. A. Backes, A. Aulinger, J. Bieser, V. Matthias and M. Quante, *Atmos. Environ.*, 2015, **126**, 153.
51. X. Querol, A. Alastuey, M. M. Viana, S. Rodriguez, B. Artiñano, P. Salvador, S. Garcia do Santos, R. Fernandez Patier, C. R. Ruiz, J. de la Rosa, A. Sanchez de la Campa, M. Menendez and J. I. Gil, *J. Aerosol Sci.*, 2004, **35**, 1151.
52. *Chemkar PM$_{10}$ – Second city campaign: Chemical characterisation of particulate matter in Mechelen, Leuven, Kortrijk, Hasselt en Aalst, 2013-2014* [in Dutch]. Flanders Environment Agency, Aalst, 2015. Available at http://www.vmm.be.
53. E. Kim and P. K. Hopke, *Atmos. Environ.*, 2008, **42**, 6047.

PM$_{2.5}$ Source Apportionment in China

MEI ZHENG,* CAIQING YAN AND XIAOYING LI

ABSTRACT

China has been facing a severe air pollution challenge in recent years. It is known that fine particulate matter is closely linked to haze. It is very important to have a good understanding of the formation mechanisms and sources of haze in China. This study provides long-term variation trends of meteorology and emissions during the past decades, reviews methodologies used in source apportionment of fine particulate matter based on published literature, and presents most recent source apportionment results from different cities in China, especially Beijing, the capital of China. Directions and key challenges in current source apportionment research are also discussed and suggestions are provided.

1 Introduction

With the rapid development of its economy and urbanization, China is now facing serious problems of air pollution. Recently, there has been growing public concern about air pollution, its sources and health impact in China after experiencing frequent, long-lasting, and wide area-covered hazy days associated with high PM$_{2.5}$ concentrations (fine particulate matter with aerodynamic diameter less than 2.5 μm). A number of studies have been conducted to investigate the formation mechanisms and sources of haze in China.[1-3]

---

*Corresponding author.

Issues in Environmental Science and Technology No. 42
Airborne Particulate Matter: Sources, Atmospheric Processes and Health
Edited by R.E. Hester, R.M. Harrison and X. Querol
© The Royal Society of Chemistry 2016
Published by the Royal Society of Chemistry, www.rsc.org

Owing to the significant reduction in visibility and increased health threat on hazy days, there is an urgent need to implement efficient and effective controls for air pollution, especially for $PM_{2.5}$ during haze episodes. Currently, source apportionment is the major tool used by the Chinese government to identify and quantify the major sources of these fine particles, and to provide the scientific basis for control measures. The primary goal of this study is to introduce current source apportionment research in China and its challenges.

2 Time Trends

2.1 Visibility Trends

Haze refers to a phenomenon with visibility lower than 10 km and relative humidity (RH) lower than 90%.[4] The major haze regions in China include the North China Plain, the Pearl River Delta region in southern China, and the Yangtze River Delta in eastern China, which are relatively developed regions with active anthropogenic emissions.[5,6]

As China added $PM_{2.5}$ as a new National Air Quality Standard in 2012 and started to monitor it in some cities in 2013, there is a very short record of $PM_{2.5}$ in this country. Therefore, the long-term trend in $PM_{2.5}$ mass concentration and haze days can only be indirectly inferred using long-term visibility/visual range data from limited *in situ* measurements[4,5,7] and satellite observation (*e.g.*, aerosol optical density or absorbing aerosol index).[6]

Long-term trend studies on visibility in major cities or regions have been conducted using meteorological data based on surface observations as well as data from the US National Climatic Data Center. Che *et al.*[8] analyzed the visibility trend in the past 25 years over China, and found there was a decreasing trend on average, especially in the eastern region. Chen *et al.*[4] emphasized that visibility exhibited a descending trend in both urban and rural sites in North China from 1960 to 2012, with more haze days in urban sites and fewer in rural sites.

Statistical analyses of long-term visual range and satellite observations suggest that there is an increasing trend in the outbreak of haze events in eastern and northeastern China.[4,5]

2.2 Emission Trends

Increased anthropogenic emissions with the development of industrialization and urbanization has been considered as one of the major factors that contribute to the frequent haze events in eastern China. With increasing energy consumption, electricity generation and vehicle fleet, emissions of various air pollutants have also increased.[9]

Recent emission trends of primary $PM_{2.5}$, and its gaseous precursors, such as sulfur dioxide (SO_2), nitrogen oxides (NO_x), and non-methane volatile organic compounds (NMVOC), in East Asia have been reported by Wang *et al.*[10] Their results suggested that during 2005–2010, the emissions of SO$_2$ and PM$_{2.5}$ decreased by about 15 and 12% in East Asia, respectively. However, NO$_x$ and NMVOC increased by 25 and 15%, respectively. The projections of future emissions indicate that gaseous precursor (*e.g.* NO$_x$, SO$_2$, NMVOC) emissions in East Asia would increase by about one-quarter by 2030 (compared to the level in 2010), with current regulations and implementation, while primary PM$_{2.5}$ emission is expected to decrease by 7%. However, with stringent new energy-saving policies, both emissions of PM$_{2.5}$ and its gaseous precursors will decrease.

2.3 Meteorology Trends

Besides emissions, meteorology plays an important role in haze formation. Unfavorable meteorological conditions (*e.g.*, high RH and weak surface winds) along with high emissions have been identified to induce heavy-haze pollution in China, such as in the Beijing–Tianjin–Hebei region over the past two decades.[11] For example, meteorological conditions such as the weak east-Asian winter monsoon, southerly wind anomalies in the lower troposphere, abnormally weak surface wind, and high RH were more responsible for the severe haze pollution events in January 2013 rather than the fast increase of emissions.[11] Urbanization is another factor that impacts wind speed.

The long-term record shows a decreasing trend of surface wind speed during recent decades over China.[12,13] Guo *et al.*[12] analyzed the long-term trend of near-surface wind speed in China with a dataset including 652 stations, and found that annual and seasonal mean wind speed weakened during the period from 1969 to 2005, especially in northern China, which is not favorable for dispersion of atmospheric pollutants.

3 Sources of PM$_{2.5}$ in China

3.1 Methods for Source Apportionment

Emission inventories, 3D air quality models and receptor models are three major ways for providing source information, of which the receptor model is the most commonly applied in China. A brief summary of published source apportionment studies in China (before 2012, especially receptor model-based studies) including ambient sampling, chemical analysis and receptor models[14] is presented here.

Sampling methods vary widely from study to study. In general, quartz and Teflon filters are the two most used filter materials, accounting for 73% of all filter types, which are used for organic and inorganic tracer analysis, respectively. The samplers with low volume (flow rate <30 L min^{-1}) are widely applied in these studies (especially a four-channel sampler at 16.7 L min^{-1}). Multiple channels can provide filters for different types of chemical analysis.

Medium-volume samplers are the second to be commonly used in China, accounting for 28% of all samplers.

Various analytical methods are used for analysis of $PM_{2.5}$ chemical components. For elements, X-ray fluorescence (XRF), inductively coupled plasma-atomic emission spectrometry (ICP-AES), and inductively coupled plasma-mass spectrometry (ICP-MS) are major tools, with ICP-MS more commonly used in recent years. Ion chromatography (IC) is the major method for analysis of ions. For organic carbon (OC) and elemental carbon (EC), thermal optical transmittance (TOT) and thermal optical reflectance (TOR) are equally important. Gas Chromatography-Mass Spectrometry (GC-MS) is the most commonly used method for organic speciation analysis.

Receptor models are primarily applied for quantifying sources for $PM_{2.5}$ in China, including Chemical Mass Balance (CMB), Positive Matrix Factorization (PMF), Principal Component Analysis (PCA), multivariate receptor (UNMIX) and Factor Analysis (FA). Therein, the latter four receptor models, which are based on multivariate statistical techniques, are defined as PPUF. Then all these receptor models can be further divided into two main types based on the source profile requirements. CMB needs source profiles while PPUF does not. Based on the tracers used in the receptor model, these methods are then classified into six categories including CMB-I (inorganic tracer), CMB-O (organic tracer), CMB-C (inorganic and organic tracers combined), PPUF-I, PPUF-O and PPUF-C. PPUF-I is mostly used in source apportionment studies in China. The details can be found in ref. 14.

3.2 Source Apportionment Results in China

3.2.1 Nationwide Source Apportionment Study.
Extensive studies have been conducted in recent years to investigate the sources and formation mechanisms of haze in China.[15-23]

Major sources of haze vary with seasons. For example, agricultural burning could be the major source of haze in summer,[16,17] while coal combustion is more responsible for haze during wintertime.[24,25] Severe haze pollution occurs frequently in winter, due to not only a substantial increase of coal combustion emissions for heating and cooking but also the frequent occurrence of fog episodes.[26] In the autumn harvest season, open biomass burning could be important as well. Based on joint observations in five cities (Shanghai, Hangzhou, Ningbo, Suzhou and Nanjing) of the Yangtze River Delta, Cheng et al.[27] showed that open biomass burning contributed to 37% of $PM_{2.5}$, 70% of OC and 61% of EC, and suggested that the impact of open biomass burning was regional, owing to the substantial inter-province transport of air pollutants. They also estimated that levels of $PM_{2.5}$ exposure could be reduced by 47% for the Yangtze River Delta if open biomass burning is forbidden and a significant health benefit is expected.

Since the new $PM_{2.5}$ national ambient air quality standard was introduced in China (GB 3095-2012), immediate actions followed. For example, environmental monitoring centers are required to routinely monitor $PM_{2.5}$

concentration in 74 cities since 2013. However, only 3 out of 74 cities could meet the standard in 2013.[28]

Nine cities, including Beijing, Tianjin, Shijiazhuang, Nanjing, Shanghai, Hangzhou, Ningbo, Guangzhou, and Shenzhen, have released their source apportionment results to the public through websites (see Figure 1 and Table 1). It is clear that there are some similarities in the source types identified in different cities. Up to eight types of emission sources have been identified and quantified in each city, including three commonly found sources, such as vehicular exhaust, road dust, and industry processes. Coal combustion has also been quantified in most of the cities except Ningbo and Nanjing, which might be partly owing to the different source apportionment methods and tracers used. Other sources, such as biomass burning, ship emission, marine aerosol, residential combustion, and agricultural emissions, have also been identified in some cities.

From Figure 1, contributions from vehicles, road dust, and industry sources to $PM_{2.5}$ mass concentrations in the above-mentioned nine cities were in the range of 20–41%, 10–30%, and 12–47%, respectively. Road dust in northern cities (14–30%) was slightly higher compared to that in the southern cities (10–12%). However, vehicular emission contributed more in the southern cities (especially in Shenzhen, up to 41%) than that in the northern cities (*i.e.*, Beijing, Tianjin, and Shijiazhuang, around 20%). Coal combustion accounted for 13–29% of $PM_{2.5}$ mass concentrations in cities where a coal combustion source has been identified. It should be noticed that annual average contribution from coal combustion in the northern cities (Beijing–Tianjin–Shijiazhuang, representative of Beijing–Tianjin–Hebei area) was significantly higher compared to that in the southern cities.

There were some differences in sample collection, chemical analysis, as well as methods for source apportionment of $PM_{2.5}$ in the above nine cities. For most cities, receptor models such as PMF and CMB were the key tools, as stated in Section 3.1. In some cities, a combination of receptor models with 3D air quality models was applied. These $PM_{2.5}$ source apportionment results for different cities were conducted by each local environmental protection agency with collaboration with different research groups, therefore, direct comparison of source apportionment results among cities should be made with caution.

3.2.2 Contributions from Vehicular Exhaust in Urban Areas. Fast increase of the number of on-road vehicles can be seen in China especially in mega cities. The number of on-road vehicles in China reached 154 million in 2014, of which 5.37 million vehicles are in Beijing.[39] The emissions of carbon monoxide (CO), hydrocarbon (HC), NO_x, and particulate matter (PM) from vehicles all increased by about 10-fold from 1980 to 2008.[39]

Vehicular exhaust is identified as one of the most important sources of air pollution in China, especially in Beijing.[40,41] However, the contributions of vehicular exhaust to $PM_{2.5}$ in Beijing estimated by different groups vary greatly and its quantification is still a great concern as Beijing implements

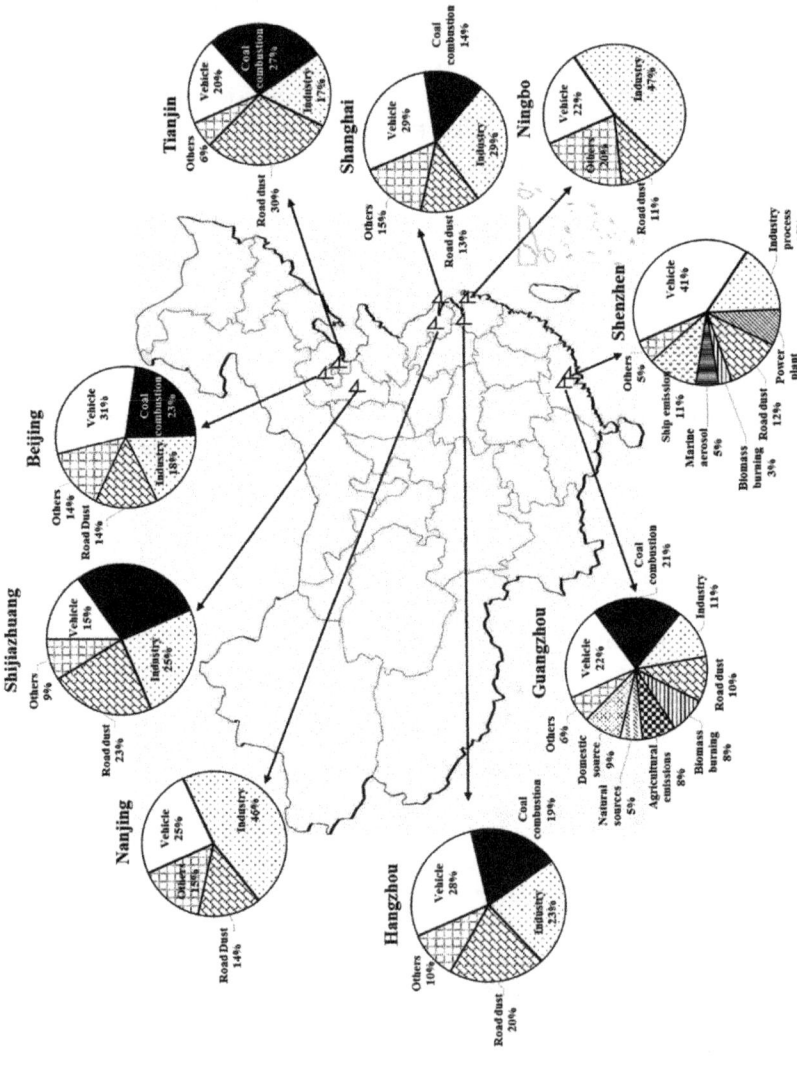

Figure 1   Source apportionment results of nine cities in China (data source is shown in Table 1).

Table 1 Source apportionment results of nine cities in China.

City	PM	Source type	Results	Methods	Data source	Source
Beijing	PM$_{2.5}$	Regional transport	28–36%	Not given	Beijing Municipal Environmental Protection Bureau	29
		Local emission	64–72%			
		Vehicle	31%			
		Coal combustion	22%			
		Industry	18%			
		Road dust	14%			
		Others	14%			
Tianjin	PM$_{2.5}$	Regional transport	22–34%	Air quality model, CMB, Emission inventory	Tianjin Environmental Protection Bureau	30
		Local emission	66–78%			
		Vehicle	20%			
		Coal combustion	27%			
		Industry	17%			
		Road dust	30%			
		Others	6%			
	PM$_{10}$	Regional transport	10–15%	Air quality model, CMB, Emission inventory	Tianjin Environmental Protection Bureau	
		Local emission	85–90%			
		Vehicle	14%			
		Coal combustion	23%			
		Industry	14%			
		Road dust	42%			
		Others	7%			
Shijiazhuang	PM$_{2.5}$	Regional transport	23–30%	Air quality model, CMB, Emission inventory	Shijiazhuang Environmental Protection Bureau	31
		Local emission	70–77%			
		Vehicle	15%			
		Coal combustion	29%			
		Industry	25%			
		Road dust	23%			
		Others	9%			
	PM$_{10}$	Regional transport	85–90%			
		Local emission	10–15%			
		Vehicle	13%			
		Coal combustion	25%			
		Industry	21%			
		Road dust	38%			
		Others	5%			

Table 1 Continued

City	PM	Source type	Results	Methods	Data source	Source
Nanjing	PM$_{2.5}$	Vehicle	25%	Air quality model	Nanjing Environmental Protection Bureau	32
		Industry	46%			
		Road dust	14%			
		Others	15%			
Ningbo	PM$_{2.5}$	Vehicle	22%	Not given	News from Ningbo Ecological Office	33
		Industry	47%			
		Road dust	11%			
		Others	20%			
	PM$_{2.5}$	Vehicle	15%	CMB		34
		Coal combustion	14%			
		Steel	4%			
		Road dust	20%			
		Cement of construction	1%			
		Sea salt	4%			
		Sulfate	17%			
		Nitrate	10%			
		SOC	9%			
		Others	6%			
	PM$_{10}$	Vehicle	12%			
		Coal combustion	16%			
		Steel	6%			
		Road dust	23%			
		Cement of construction	2%			
		Sea salt	5%			
		Sulfate	13%			
		Nitrate	9%			
		SOC	6%			
		Others	8%			
Hangzhou	PM$_{2.5}$	Regional transport	18–38%	Not given	Hangzhou Environmental Protection Bureau	35
		Local emission	62–82%			
		Vehicle	28%			
		Coal combustion	19%			
		Industry	23%			
		Road dust	20%			
		Others	10%			

City		Sources			Organization	Ref.
Shanghai	PM$_{2.5}$	Regional transport	16–36%	Not given	Shanghai Environmental Protection Bureau	36
		Local emission	64–84%			
		Mobile source[a]	29%			
		Coal combustion[b]	14%			
		Industry[c]	29%			
		Road dust[d]	13%			
		Others[e]	15%			
Guangzhou	PM$_{2.5}$	Vehicle	22%	Not given	Guangzhou Environmental Protection Bureau	37
		Coal combustion	21%			
		Industry	12%			
		Road dust	10%			
		Domestic source	9%			
		Biomass burning	8%			
		Agriculture	8%			
		Natural sources	5%			
		Others	6%			
Shenzhen	PM$_{2.5}$	Vehicle	41%	Not given	Human Settlements and Environment Commission of Shenzhen Municipality	38
		Industry process	15%			
		Power plant	8%			
		Road dust	12%			
		Biomass burning	3%			
		Marine	5%			
		Shipping	11%			
		Others	5%			

Note:
[a]Emissions from vehicles, ships, airplanes and off-road vehicles.
[b]Emissions from coal combustion based power plants, heating boilers and industrial furnaces.
[c]Emissions from other coal combustion based heating boilers, industrial furnaces and industry processes, as well as petroleum industry, chemical industry, equipment manufacture, surface spray, and printings.
[d]Emissions from bare surface, construction, and soils.
[e]Biomass burning, agriculture, cooking, residential emissions, as well as natural emission from sea-salts and plants.

control policy for mobile sources based on their plate numbers. Huang *et al.*[42] combined a few methods to estimate the contribution of traffic to $PM_{2.5}$ in Beijing in January 2013, and their results indicated that about 5.6% of $PM_{2.5}$ was from traffic during that period.

Tracers used for estimating vehicular exhaust vary widely among studies. In one study, the factor with high nitrate, EC, Cu, Zn, Cd, Pb, Mo, Sb, and Sn in PMF was identified as the source from traffic and waste incineration.[41] Some used certain organic species (*e.g.* hopanes) and EC as tracers for vehicular exhaust.[43] Many studies identified vehicular exhaust by Pb and Zn in PCA and UNMIX analyses. It is a complex issue as vehicular exhaust should include its primary as well as secondary contributions.[44] In addition, motor vehicle type and fuel type in China have a wide range, and knowledge of emission profiles is very limited. The above factors all add difficulty for accurate estimation of the vehicular source contribution. This area still needs further research.

3.2.3 Contributions of Secondary Components to Fine PM During Haze Episodes. When haze occurs, it is often found that mass concentrations of secondary compositions such as SNA (sum of sulfate, nitrate and ammonium) increase quickly, which are major contributors to severe haze, based on offline $PM_{2.5}$ analysis,[45] and online non-refractory PM_1 analysis by an Aerosol Chemical Speciation Monitor.[25] SNA arise mainly from secondary formation.

SNA formed predominantly through gas-phase oxidation of anthropogenic SO_2 and NO_x contributed up to 60% of the total PM_{10} mass observed during the summer haze periods in Beijing in 2006.[46] More recent studies of $PM_{2.5}$ also confirm the importance of secondary components in haze formation in urban Beijing and other haze-impacted regions in China.[1,42]

Secondary organic aerosol (SOA) produced through photochemical processes of anthropogenic volatile organic compounds (VOCs) shows increased importance during haze episodes as well. Huang *et al.*[42] reported that SOA could account for 16–30% of $PM_{2.5}$ mass and 40–70% of organic aerosol mass in urban sites of Beijing, Shanghai, Guangzhou and Xi'an in January 2013 when daily average $PM_{2.5}$ mass concentrations ranged from 69 to 345 μg m^{-3}.

Chemical components in $PM_{2.5}$ are directly linked to visibility reduction. Studies show that sulfate and organic matter can each contribute to about one third of total light extinction and they play a key role among all of the components that impair visibility.[47,48]

Certain meteorological conditions, such as higher RH, lower planetary boundary layer (PBL) height, and lower wind speed, could favor haze formation. The SOR (sulfur oxidation ratio) and NOR (nitrogen oxidation ratio) show an increasing trend with the increase of RH. Therefore, controlling emissions of SO_2 and NO_x is crucial, especially for days with higher RH.[46]

In addition, studies suggest that there is positive feedback between PBL height and aerosol loading.[49] The substantial high concentration of aerosol tends to depress the development of the PBL by decreasing solar radiation at the surface, while the lower PBL will in turn weaken the dispersion of pollutants, leading to a higher aerosol loading. This positive feedback loop (more aerosols/lower PBL height/more aerosols) may result in heavier haze at ground level and prevent vertical dispersion of particles (Zhang *et al.*, 2009).[50]

3.3 PM$_{2.5}$ Source Apportionment in Beijing, China

Beijing is experiencing one of the highest PM$_{2.5}$ concentrations in the world. It is surrounded by mountains in the west, north and northeast, and faces the Bohai Sea in the southeast. The center of Beijing (39°54'N, 116°23'E) is located in the northwest edge of the North China Plain, adjacent to Tianjin, and surrounded by Hebei Province in the south. Some cities in Hebei Province remain on the top list of the most polluted cities in China.

3.3.1 Contribution of Pollutants from South to Beijing. Time series of PM$_{2.5}$ concentration in Beijing show a typical cycle (from low to high concentration with a period of about 4–7 days) especially in the autumn and winter seasons. This was first reported by Jia *et al.*[51] and named as the "sawtooth cycle". The sawtooth cycle is usually controlled by the synoptic cycle, especially the passage of the cold front, and PM$_{2.5}$ concentration can increase with air back-trajectories rotating from the northwest to the west and south. One sawtooth cycle contains a smoothly increasing baseline, dominated by SNA, with daily cycles superimposed.

Another feature of PM$_{2.5}$ in Beijing, which is quite different from other cities in China, is that its concentration can increase from a very low level at the very beginning to several hundred micrograms per cubic meter.[1,2] This increase can be further divided into two types: one with a gradual increase or "sustained growth", and the other with a sharp increase within several hours or "explosive growth".[2]

An important concern for PM$_{2.5}$ in Beijing is to correctly identify contributions from local and regional sources. Regional transport of pollutants has sometimes been found to contribute greatly to PM$_{2.5}$,[52,53] dust and SO$_2$[54] in Beijing. Based on the estimate from the sawtooth, the regional contribution averages about 50% and can range from 10–20% during northwesterly flow to 70% during southerly flow in Beijing.[51] To effectively reduce PM$_{2.5}$ concentrations in urban areas, it is necessary and important to coordinate local and regional efforts.

The Environmental Protection Bureau of Beijing has released source apportionment result of PM$_{2.5}$ in 2014, together with the relative contribution from local emissions and regional transport (see Figure 2). The annual average PM$_{2.5}$ mass concentration in 2014 in Beijing was 86 μg m^{-3}. The results show that the local source contributed to the majority of PM$_{2.5}$ in Beijing (64–72%) while the regional transport was less than 40%. However, details of the methods for quantifying local emission and regional transport were not provided. Similar to Beijing, the Environmental Protection Bureau of Shanghai also released source apportionment result of $PM_{2.5}$ in 2014 (see Figure 2). The relative contributions from local emissions (less than 40%) and regional transport (64–84%) in Shanghai were quite similar to those in Beijing. Additionally, based on these results, it can be seen that higher coal combustion was found in Beijing (23%) compared to Shanghai (14%), but there was a greater industry contribution in Shanghai (29%) compared to that in Beijing (18%).

3.3.2   Source Apportionment Studies in Beijing.   Inorganic tracers were previously applied to apportion $PM_{2.5}$ sources in Beijing.[41,55–57] However, the inorganic tracer method cannot characterize some important sources that emit mostly organic matter, such as meat cooking and diesel/gasoline engine exhaust, when Pb is no longer added in gasoline. Particulate organic matter accounts for a large fraction of $PM_{2.5}$. Organic tracers have been applied in several fine particle source apportionment studies in ref. 43 and 58–69. Contributions of sources like coal combustion, diesel/gasoline emission, and other sources have been quantified using organic tracers. Additionally, radiocarbon analysis is powerful in distinguishing two types of carbonaceous sources, fossil fuel carbon and contemporary carbon. Sun *et al.*[24] used radiocarbon (^{14}C) analysis to determine the sources of carbonaceous aerosol in Beijing.

Lv *et al.*[61] reviewed over sixty $PM_{2.5}$ source apportionment related studies conducted in Beijing from 2000 to 2012, and reported that the annual average contribution from vehicle exhaust, dust, industry, biomass burning, coal combustion and secondary source in the past decade was about $7.5 \pm 3.9\%$, $14.6 \pm 5.9\%$, $9.1 \pm 7.0\%$, $9.8 \pm 2.9\%$, $15.4 \pm 7.7\%$,   and $29.1 \pm 11.0\%$, respectively. This study also compared source apportionment results (the annual average) before and after 2005. For example, the vehicle exhaust contribution increased from 7% before 2005 to 11% of $PM_{2.5}$ after 2005. Industry contribution increased from 7% (before 2005) to 16% (after 2005). However, the annual average contribution from coal combustion remained the same (15%). For the seasonal variation, among the major sources, secondary source and vehicle exhaust exhibited greater contributions in summer, while dust peaked in spring, biomass burning in autumn and winter, and coal combustion in winter.

Several year-long $PM_{2.5}$ source apportionment studies[62–65] conducted from 2012 to 2014 have been published recently. The reported contributions from vehicle, dust, industry, biomass burning, coal combustion and secondary source were 6–16%, 6–23%, ~6%, ~8.5%, 15–19% and 26–42%, respectively. According to Section 3.3.1, the Environmental Protection Bureau of Beijing released its official $PM_{2.5}$ source apportionment results for the year 2013. Amongst the local emission (64–72%), the annual average contribution of vehicle exhaust, coal combustion, industry, and dust to $PM_{2.5}$ was 31%, 22%, 18% and 14%, respectively. In addition, some studies investigated sources of $PM_{2.5}$ during the haze episodes. For example, Huang *et al.*,[42]

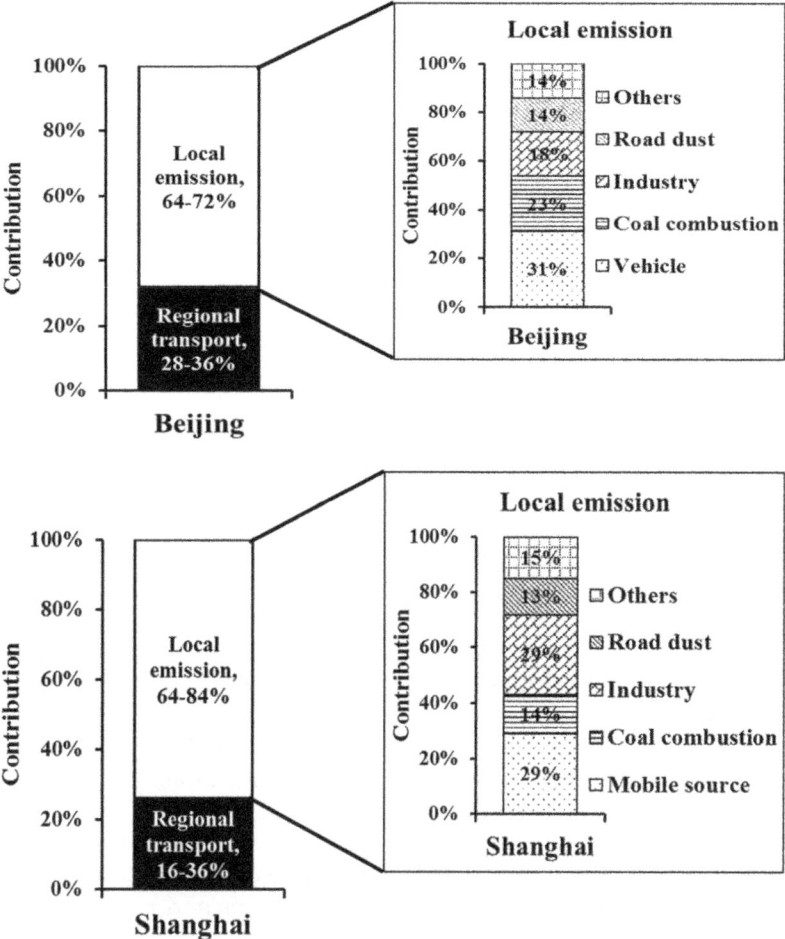

Figure 2 Local and regional contributions to PM$_{2.5}$ (A) in Beijing and (B) in Shanghai (data source: Beijing Municipal Environmental Protection Bureau, http://www.bjepb.gov.cn/bjepb/372794/index.html; Shanghai Municipal Environmental Protection Bureau, http://www.envir.gov.cn/law/bulletin/2014/2014bteng.pdf; note: Mobile sources in Shanghai include a variety of vehicles such as automobiles, aircraft, shipping and railways).

conducted source apportionment of PM$_{2.5}$ during the severe haze episode in January 2013, reporting a high contribution from coal combustion (26.1%) and secondary organic and inorganic formation (25.8 and 25.3%, respectively), while vehicular emission and biomass burning showed similar contributions to PM$_{2.5}$ (5.6%), but smaller than dust (10%). Sun *et al.* (2014) conducted source apportionment of organic aerosol (OA) and found that secondary source contributed to 55% of the total OA, followed by primary coal combustion source (19%), hydrocarbon-like OA (14%) and cooking OA (12%).

It should be noted that abundant coal is used in Beijing, a mega city located in North China, in winter for heating and cooking and researchers pointed out that coal burning is still a major source of fine PM,[41,55] and also of OC.[42,60] Although various measures have been taken to clean up the air, extensive residential coal burning still exists in rural and semi-rural areas in Beijing, especially in the south, and can contribute significantly to $PM_{2.5}$ in Beijing. Zhao *et al.*[66] estimated that 3×10^6 tons of coal were consumed and more than 3000 tons of PM were emitted in 2013 in these semi-rural areas of Beijing.

4   Future Research

4.1   Health Effects

A number of studies have reported that those living nearer to urban roads or highways often have higher risks for mortality, morbidity or other indicators for adverse health effects, such as heart rate variability.[67-69] Industrial sources were also highly related to mortality, especially for those living near to a steel plant.[70-72] However, research related to assessing the health effects of fine particles from different sources is still limited in China.

Based on a panel of 40 healthy university students, Wu *et al.*[73] examined the cardiopulmonary health effects of PM from different pollution sources in Beijing, China. Potential key sources responsible for cardiopulmonary health effects of PM were identified as SNA and dust/soil (inflammatory biomarkers), coal combustion and metallurgical emissions (blood pressure), and dust/soil and industry (pulmonary function).

Fine PM emitted from various sources has different physicochemical characteristics as well as potential toxicity and health impacts, which should be investigated in the future, especially some major and typical sources in China, such as coal combustion, vehicular exhaust and biomass burning. We suggest that more research is needed to study what sources and chemical components are most responsible for the adverse effects of fine PM in China.

4.2   Online Source Apportionment

Previous studies have shown that some haze episodes could come and go within only 24 hours. For example, in January 2013, $PM_{2.5}$ concentrations could increase from 35 to 500 µg m^{-3} within only 8 hours.[2] The low time resolution (usually 24 hours or more) and detection limit associated with regular offline filter measurement could not meet the need for understanding sources and processes of haze episodes occurring in such a short time period. High time resolution data of $PM_{2.5}$ and its major components (*e.g.*, sulfate, nitrate, ammonium, other water-soluble ions, OC, EC, and heavy metals) are essential to study rapid changes of fine PM sources. Developing novel instrumentation and online source apportionment is an urgent requirement for current source apportionment studies in China.

Online or semi-continuous instruments for different components in PM$_{2.5}$ have been developed. For example, for water-soluble ion measurement, online instruments include Particle-Into-Liquid Sampler (PILS), Gas and Aerosol Collector (GAC), Monitoring for AeRosols and Gases (MARGA), and ambient ion monitor (URG9000D). High time resolution EC and OC can be provided by semi-continuous Sunset OCEC field analyzer. Recently, new techniques for quasi online compound specific measurements of organic aerosol particles (*e.g.*, Aerosol Collection Module interfaced with Gas Chromatograph-Mass Spectrometer or Flame Ionization Detector system) have also been developed.[74] Quasi-online settings for elemental analysis, such as streaker and PIXE, and online XRF (*e.g.*, Xact) can provide online measurements of about 20 metals simultaneously (Dall'Osto *et al.*, 2013; Gao *et al.*, 2016; Park *et al.*, 2013; Prati *et al.*, 1998; Richard *et al.*, 2010).[75–79] When information of different chemical species in PM$_{2.5}$ is available, receptor models such as CMB and PMF can be used to quantify the source contribution of fine PM. However, publications on source apportionment combining multiple online instruments are still very limited.

Online aerosol mass spectrometry has been widely deployed and used for PM source apportionment studies. There are mainly two types of online aerosol mass spectrometry, one for bulk measurement and the other for single particle measurement. Instruments such as the aerosol mass spectrometer (AMS) and aerosol chemical speciation monitor (ACSM) are used for online aerosol bulk measurement. By coupling with receptor models such as PMF, AMS data can be used to obtain quantitative source contributions for organic aerosol. However, since AMS cannot measure refractory components of PM, many important tracers such as EC, crustal elements, and heavy metals are not available. Aerosol time-of-flight mass spectrometry (ATOFMS) and single particle aerosol mass spectrometry (SPAMS, which is manufactured in China and has been available since 2010) are two commercial single particle instruments that can measure the size distribution, chemical composition and mixing state of a single particle at the same time. Source apportionment by single particle aerosol mass spectrometry is done by first clustering all particles detected and then determining possible sources or chemical processes of these major clusters based on their spectra. The PMF-ATOFMS method has also been developed in recent years. Particles can be first classified by ART-2a or K-means method and then serve as the input into the PMF model. Another method is to take raw data (signal intensity) from single particle aerosol mass spectrometry directly as inputs into the PMF model. With the single particle mass spectrometry technique, source categories of fine PM, such as biomass burning, dust, sea salt, vehicular emission, ship emission and coal combustion, could be identified.

Future research of online source apportionment using SPAMS should include converting particle count to mass for total fine particle mass and major compositions, including sulfate, nitrate, ammonium, EC and OC, and establishing local source profiles using SPAMS and protocols for evaluating source apportionment results with this technique.

4.3 Integration of Different Source Apportionment Methods

From the above discussion, it can be seen that many source apportionment studies in China are primarily based on receptor-oriented methods.[14] To evaluate the accuracy or uncertainty associated with source apportionment results, validation protocols should be better applied.[27] Zheng *et al.*[80] summarized studies of source profiles and the authors pointed out the need to establish more local source profiles in China. To ensure source apportionment from different cities across the country that can be comparable, it is very important to establish a monitoring network in China that has a common protocol for PM speciation, and receptor models should be used in a harmonized way.

As national and local emission inventories of fine PM and gaseous precursors are gradually established in China,[81,82] there is a need to integrate the emission inventory, source-oriented air quality model, and receptor-oriented model to better understand major sources of fine PM and to identify the local and regional source contributions. Uncertainty can be reduced through the application of multiple methods. It is still very challenging to effectively combine these methods.

Acknowledgements

We would like to thank Tian Zhou, Yanjun Zhang, and Xiaoshuang Guo for their helpful assistance. This research is supported by the National Natural Science Foundation of China (41571130033).

References

1. S. Guo, M. Hu, M. L. Zamora, J. F. Peng, D. J. Shang, J. Zheng, Z. F. Du, Z. J. Wu, M. Shao, L. M. Zeng, M. J. Molina and R. Y. Zhang, Elucidating severe urban haze formation in China, *Proc. Natl. Acad. Sci. U. S. A.*, 2014, **111**(49), 17373–17378.

2. Y. S. Wang, L. Yao, L. L. Wang, Z. R. Liu, D. S. Ji, G. Q. Tang, J. K. Zhang, Y. Sun, B. Hu and J. Y. Xin, Mechanism for the formation of the January 2013 heavy haze pollution episode over central and eastern China, *Earth Sci.*, 2014, **57**(1), 14–25.

3. G. J. Zheng, F. K. Duan, H. Su, Y. L. Ma, Y. Cheng, B. Zheng, Q. Zhang, T. Huang, T. Kimoto, D. Chang, U. Pöschl, Y. Cheng and K. B. He, Exploring the severe winter haze in Beijing: the impact of synoptic weather, regional transport and heterogeneous reactions, *Atmos. Chem. Phys.*, 2015, **15**, 2969–2983.

4. H. P. Chen and H. J. Wang, Haze days in North China and the associated atmospheric circulations based on daily visibility data from 1960 to 2012, *J. Geophys. Res.: Atmos.*, 2015, **120**, 5895–5909.

5. H. Z. Che, X. Y. Zhang, Y. Li, Z. J. Zhou, J. J. Qu and X. J. Hao, Haze trends over the capital cities of 31 provinces in China, 1981-2005, *Theor. Appl. Climatol.*, 2009, **97**(3–4), 235–242.

6. X. Zhang, L. Wang, W. Wang, D. Gao, X. Wang and D. Ye, Long-term trend and spatiotemporal variations of haze over China by satellite observations from 1979 to 2013, *Atmos. Environ.*, 2015, **119**, 362–373.

7. Z. Cheng, S. X. Wang, J. K. Jiang, Q. Y. Fu, C. H. Chen, B. Y. Xu, J. Q. Yu, X. Fu and J. M. Hao, Long-term trend of haze pollution and impact of particulate matter in the Yangtze River Delta, China, *Environ. Pollut.*, 2013, **182**, 101–110.

8. H. Che, X. Zhang, Y. Li, Z. Zhou and J. J. Qu, Horizontal visibility trends in China 1981-2005, *Geophys. Res. Lett.*, 2007, **34**, L24706.

9. S. X. Wang and J. M. Hao, Air quality management in China: Issues, challenges, and options, *J. Environ. Sci.*, 2012, **24**(1), 2–13.

10. S. X. Wang, B. Zhao, S. Y. Cai, Z. Klimont, C. P. Nielsen, T. Morikawa, J. H. Woo, Y. Kim, X. Fu, J. Y. Xu, J. M. Hao and K. B. He, Emission trends and mitigation options for air pollutants in East Asia, *Atmos. Chem. Phys.*, 2014, **14**, 6571–6603.

11. L. L. Wang, N. Zhang, Z. R. Liu, Y. Sun, D. S. Ji and Y. S. Wang, The influence of climate factors, meteorological conditions, and boundary-layer structure on severe haze pollution in the Beijing-Tianjin-Hebei region during January 2013, *Adv. Meteorol.*, 2014, **685971**, 1–14.

12. H. Guo, M. Xu and Q. Hu, Changes in near-surface wind speed in China: 1969-2005, *Int. J. Climatol.*, 2010, **31**(3), 349–358.

13. C. G. Lin, K. Yang, J. Qin and R. Fu, Observed coherent trends of surface and upper-air wind speed over China since 1960, *J. Climate*, 2012, **26**, 2891–2903.

14. M. Zheng, Y. J. Zhang, C. Q. Yan, X. L. Zhu, J. J. Schauer and Y. H. Zhang, Review of PM$_{2.5}$ source apportionment methods in China, *Acta Sci. Nat. Univ. Pekin.*, 2014, **50**(6), 1141–1154, (in Chinese).

15. X. An, T. Zhu, Z. Wang, C. Li and Y. Wang, A modeling analysis of a heavy air pollution episode occurred in Beijing, *Atmos. Chem. Phys.*, 2007, **7**(12), 3103–3114.

16. K. Huang, Typical types and formation mechanisms of haze in an Eastern Asia megacity, Shanghai, *Atmos. Chem. Phys.*, 2012, **12**(1), 105–124.

17. W. J. Li, L. Y. Shao and P. R. Buseck, Haze types in Beijing and the influence of agricultural biomass burning, *Atmos. Chem. Phys.*, 2010, **10**(17), 8119–8130.

18. X. G. Liu, J. Li, Y. Qu, T. Han, L. Hou and J. Gu, Formation and evolution mechanism of regional haze: a case study in the megacity Beijing, China, *Atmos. Chem. Phys.*, 2013, **13**(9), 4501–4514.

19. J. Ma, Y. Chen, W. Wang, P. Yan, H. Liu, S. Yang, Z. Hu and J. Lelieveld, Strong air pollution causes widespread haze-clouds over China, *J. Geophys. Res.*, 2010, **115**, D18204.

20. Y. Sun, G. Zhuang, A. Tang, Y. Wang and Z. An, Chemical characteristics of PM$_{2.5}$ and PM10 in haze-fog episodes in Beijing, *Environ. Sci. Technol.*, 2006, **40**(10), 3148–3155.

21. M. Tao, L. Chen, L. Su and J. Tao, Satellite observation of regional haze pollution over the North China Plain, *J. Geophys. Res.*, 2012, **117**, D12203.
22. C. Wang, Y. Yang and Y. Li, Analysis on the meteorological condition and formation mechanism of serious pollution in south Hebei Province in January 2013, *Res. Environ. Sci.*, 2013, **26**, 695–702, (in Chinese).
23. X. J. Zhao, P. S. Zhao, J. Xu, W. Meng, W. W. Pu, F. Dong, D. He and Q. F. Shi, Analysis of a winter regional haze event and its formation mechanism in the North China Plain, *Atmos. Chem. Phys.*, 2013, **13**, 5685–5696.
24. X. Sun, M. Hu, S. Guo, K. Liu and L. Zhou, ^{14}C-Based source assessment of carbonaceous aerosols at a rural site, *Atmos. Environ.*, 2012, **50**, 36–40.
25. Y. Sun, Q. Jiang, Z. Wang, P. Fu, J. Li, T. Yang and Y. Yin, Investigation of the sources and evolution processes of severe haze pollution in Beijing in January 2013, *J. Geophys. Res.: Atmos.*, 2014, **119**, 4380–4398.
26. J. Quan, Q. Zhang, H. He, J. Liu, M. Huang and H. Jin, Analysis of the formation of fog and haze in North China Plain (NCP), *Atmos. Chem. Phys.*, 2011, **11**(15), 8205–8214.
27. Z. Cheng, S. Wang, X. Fu, J. G. Watson, J. Jiang, Q. Fu, C. Chen, B. Xu, J. Yu, J. C. Chow and J. Hao, Impact of biomass burning on haze pollution in the Yangtze River delta, China: a case study in summer 2011, *Atmos. Chem. Phys.*, 2014, **14**, 4573–4585.
28. Ministry of Environmental Protection, 2014. http://www.zhb.gov.cn/zhxx/hjyw/201404/t20140416_270592.html.
29. http://www.bjepb.gov.cn/bjepb/413526/331443/331937/333896/396191/index.html.
30. http://www.tjhb.gov.cn/news/news_headtitle/201410/t20141009_570.html.
31. http://www.sjzhb.gov.cn/cyportal2.3/template/site00_article@sjzhbj.jsp?article_id=297e62b94811181601484396d4dd3589&parent_id=402882e74038b0b401407682305a01dd&parentType=0&siteID=site00&f_channel_id=null&a1b2dd=7xaac.
32. http://www.njhb.gov.cn/43123/201504/t20150430_3289890.html.
33. http://nb.people.com.cn/n/2015/0618/c365604-25290044.html.
34. Z. M. Xiao, X. H. Bi and Y. C. Feng, Source apportionment of ambient PM10 and PM2. 5 in urban area of Ningbo City, *Res. Environ. Sci.*, 2012, **25**(5), 549–555, (in Chinese).
35. http://www.hzepb.gov.cn/hbzx/mtbd/201506/t20150612_43016.html.
36. http://www.sepb.gov.cn/fa/cms/shhj//shhj2272/shhj2159/2015/01/88461.html.
37. http://www.gzepb.gov.cn/yhxw/201504/t20150420_79442.html.
38. http://www.gdep.gov.cn/news/hbxw/201504/t20150428_201065.html.
39. Ministry of Environmental Protection, 2010. China vehicle emission control annual report.
40. Ministry of Environmental Protection, 2014. 2013 China environmental bulletin.

41. R. Zhang, J. Jing, J. Tao, S. C. Hsu, G. Wang, J. Cao, C. S. L. Lee, L. Zhu, Z. Chen, Y. Zhao and Z. Shen, Chemical characterization and source apportionment of PM$_{2.5}$ in Beijing: seasonal perspective, *Atmos. Chem. Phys.*, 2013, **13**, 7053–7074.
42. R. J. Huang, Y. Zhang, C. Bozzetti, K. F. Ho, J. J. Cao, Y. Han, K. R. Daellenbach, J. G. Slowik, S. M. Platt, F. Canonaco, P. Zotter, R. Wolf, S. M. Pieber, E. A. Bruns, M. Crippa, G. Ciarelli, A. Piazzalunga, M. Schwikowski, G. Abbaszade, J. Schnelle-Kreis, R. Zimmermann, Z. An, S. Szidat, U. Baltensperger, I. E. Haddad and A. S. H. Prévôt, High secondary aerosol contribution to particulate pollution during haze events in China, *Nature*, 2014, **514**, 218–222.
43. Q. Wang, M. Shao, Y. Zhang, Y. Wei, M. Hu and S. Guo, Source apportionment of fine organic aerosols in Beijing, *Atmos. Chem. Phys.*, 2009, **9**, 8573–8585.
44. M. Canagaratna, T. Onasch, E. Wood, S. Herndon, J. Jayne, E. Cross, R. Miake-Lye, C. Kolb and D. Worsnop, Evolution of Vehicle Exhaust Particles in the Atmosphere, *J. Air Waste Manage. Assoc.*, 2009, **60**, 1192–1203.
45. J. Quan, X. Tie, Q. Zhang, Q. Liu, X. Li and Y. Gao, Characteristics of heavy aerosol pollution during the 2012–2013 winter in Beijing, China, *Atmos. Environ.*, 2014, **88**(5), 83–89.
46. T. Han, X. Liu, Y. Zhang, Q. Yu, L. Zeng and H. Min, Role of secondary aerosols in haze formation in summer in the megacity Beijing, *J. Environ. Sci.*, 2015, **31**(5), 51–60.
47. Q. X. Ma, Y. C. Liu and H. He, Synergistic effect between NO$_2$ and SO$_2$ in their adsorption and reaction on γ-alumina, *J. Phys. Chem.*, 2008, **112**(29), 6630–6635.
48. S. R. Tong, L. Y. Wu, M. F. Ge, W. G. Wang and Z. F. Pu, Heterogeneous chemistry of monocarboxylic acids on α-Al$_2$O$_3$ at ambient condition, *Atmos. Chem. Phys.*, 2010, **10**(2), 3937–3974.
49. J. D. Wang, S. X. Wang, J. K. Jiang, A. J. Ding, M. Zheng, B. Zhao, D. C. Wong, W. Zhou, G. J. Zheng, L. Wang, J. E. Pleim and J. M. Hao, Impact of aerosol–meteorology interactions on fine particle pollution during China's severe haze episode in January 2013, *Environ. Res. Lett.*, 2014, **9**, 094002.
50. J. Quan, Y. Gao, Q. Zhang, X. Tie, J. Cao and S. Han, Evolution of planetary boundary layer under different weather conditions, and its impact on aerosol concentrations, *Particuology*, 2013, **11**(1), 34–40.
51. Y. Jia, K. A. Rahn, K. He, T. Wen and Y. Wang, A novel technique for quantifying the regional component of urban aerosol solely from its sawtooth cycles, *J. Geophys. Res.: Atmos.*, 2008, **113**, D21309.
52. L. T. Wang, Z. Wei, J. Yang, Y. Zhang, F. F. Zhang, J. Su, C. C. Meng and Q. Zhang, The 2013 severe haze over southern Hebei, China: model evaluation, source apportionment, and policy implications, *Atmos. Chem. Phys.*, 2014, **14**, 3151–3173.

53. Z. F. Wang, L. I. Jie, W. Zhe, W. Y. Yang, T. Xiao and G. E. Baozhu, Modeling study of regional severe hazes over mid-eastern China in January 2013 and its implications on pollution prevention and control, *Sci. China*, 2014, **57**(1), 3–13.

54. K. Yang, R. R. Dickerson, S. A. Carn, C. Ge and J. Wang, First observations of SO_2 from the satellite Suomi NPP OMPS: Widespread air pollution events over China, *Geophys. Res. Lett.*, 2013, **40**, 4957–4962.

55. Y. Song, X. Y. Tang, S. D. Xie, Y. H. Zhang, Y. J. Wei, M. S. Zhang, L. M. Zeng and S. H. Lu, Source apportionment of PM2.5 in Beijing in 2004, *J. Hazard. Mater.*, 2007, **146**, 124–130.

56. Y. Song, S. D. Xie, Y. H. Zhang, L. M. Zeng, L. G. Salmon and M. Zheng, Source apportionment of $PM_{2.5}$ in Beijing using principal component analysis/absolute principal component scores and UNMIX, *Sci. Total Environ.*, 2006, **372**, 278–286.

57. Y. Song, Y. H. Zhang, S. D. Xie, L. M. Zeng, M. Zheng, L. G. Salmon, M. Shao and S. Slanina, Source apportionment of $PM_{2.5}$ in Beijing by positive matrix factorization, *Atmos. Environ.*, 2006, **40**, 1526–1537.

58. S. Guo, M. Hu, Q. F. Guo, X. Zhang, M. Zheng, J. Zheng, C. C. Chang, J. J. Schauer and R. Y. Zhang, Primary sources and secondary formation of organic aerosols in Beijing, China, *Environ. Sci. Technol.*, 2012, **46**, 9846–9853.

59. S. Guo, M. Hu, Q. Guo, X. Zhang, J. J. Schauer and R. Zhang, Quantitative evaluation of emission controls on primary and secondary organic aerosol sources during Beijing 2008 Olympics, *Atmos. Chem. Phys.*, 2013, **13**, 8303–8314.

60. M. Zheng, L. G. Salmon, J. J. Schauer, L. M. Zeng, C. S. Kiang, Y. H. Zhang and G. R. Cass, Seasonal trends in $PM_{2.5}$ source contributions in Beijing, China, *Atmos. Environ.*, 2005, **39**(22), 3967–3976.

61. B. L. Lv, B. Zhang and Y. Q. Bai, A systematic analysis of $PM_{2.5}$ in Beijing and its sources from 2000 to 2012, *Atmos. Environ.*, 2016, **124**, 98–108.

62. S. L. Tian, Y. P. Pan and Y. S. Wang, Size-resolved source apportionment of particulate matter in urban Beijing during haze and non-haze episodes, *Atmos. Chem. Phys. Discuss.*, 2015, **15**, 9405–9443.

63. L. L. Wang, Z. R. Liu, Y. Sun, D. S. Ji and Y. S. Wang, Long-range transport and regional sources of $PM_{2.5}$ in Beijing based on long-term observations from 2005-2010, *Atmos. Res.*, 2015, **157**, 37–48.

64. Q. Wang, D. W. Zhang, B. X. Liu, T. Chen, Q. Wei, J. X. Li and Y. P. Liang, Spatial and temporal variations of ambient $PM_{2.5}$ source contributions using positive matrix factorization, *China Environ. Sci.*, 2015, **35**, 2917–2924.

65. Y. Y. Yang, J. X. Li, Y. P. Liang, T. Chen, B. X. Liu, F. Sun, G. Cheng, J. P. Su and D. W. Zhang, Source apportionment of PM2.5 in Beijing by the chemical mass balance, *Acta Sci. Circumstantiae*, 2015, **35**, 2693–2700.

66. W. H. Zhao, Q. Xu, L. J. Li, L. Jiang, D. W. Zhang and T. Chen, Estimation of air pollutant emissions from coal burning in the semi-rural areas of Beijing plain, *Res. Environ. Sci.*, 2015, **28**(6), 869–876.

67. R. Delfino, N. Staimer, T. Tjoa, A. Polidori, M. Arhami, D. Gillen, M. Kleinman, N. Vaziri, J. Longhurst, F. Zaldivar and C. Sioutas, Circulating biomarkers of inflammation, antioxidant activity, and platelet activation are associated with primary combustion aerosols in subjects with coronary artery disease, *Environ. Health Perspect.*, 2008, **116**, 898–906.

68. B. Ritz, F. Yu, S. Fruin, G. Chapa, G. Shaw and J. A. Harris, Ambient air pollution and risk of birth defects in southern California, *Am. J. Epidemiol.*, 2002, **155**, 17–25.

69. M. Wilhelm and B. Ritz, Residential proximity to traffic and adverse birth outcomes in Los Angeles County, California, 1994-1996, *Environ. Health Perspect.*, 2003, **111**, 207–216.

70. M. Jerrett, M. Buzzelli, R. Burnett and P. DeLuca, Particulate air pollution, social confounders, and mortality in small areas of an industrial city, *Soc. Sci. Med.*, 2005, **60**, 2845–2863.

71. L. Liu, L. Kauri, M. Mahmuda, S. Weichenthalb, S. Cakmak, R. Shutt, H. You, E. Thomson, R. Vincent, P. Kumarathasan, G. Broad and R. Dales, Exposure to air pollution near a steel plant and effects on cardiovascular physiology: A randomized crossover study, *Int. J. Hyg. Environ. Health*, 2014, **217**, 279–286.

72. T. Pouliou, P. Kanaroglou, S. Elliott and L. Pengelly, Assessing the health impacts of air pollution: a re-analysis of the Hamilton children's cohort data using a spatial analytic approach, *Int. J. Environ. Health Res.*, 2008, **18**, 17–35.

73. S. Wu, F. Deng, H. Wei, J. Huang, X. Wang, Y. Hao, C. Zheng, Y. Qin, H. Lv, M. Shima and X. Guo, Association of cardiopulmonary health effects with source appointed ambient fine particulate in Beijing, China: A Combined Analysis from the Healthy Volunteer Natural Relocation (HVNR) Study, *Environ. Sci. Technol.*, 2014, **48**, 3438–3448.

74. T. Hohaus, D. Trimborn, A. Kiendler-Scharr, I. Gensch, W. Laumer, B. Kammer, S. Andres, H. Boudries, K. A. Smith, D. R. Worsnop and J. T. Jayne, A new aerosol collector for quasi on-line analysis of particulate organic matter: the Aerosol Collection Module (ACM) and first applications with a GC/MS-FID, *Atmos. Chem. Phys.*, 2010, **3**, 1423–1436.

75. M. Dall'Osto, X. Querol, F. Amato, A. Karanasiou, F. Lucarelli, S. Nava, G. Calzolai and M. Chiari, Hourly elemental concentrations in PM$_{2.5}$ aerosols sampled simultaneously at urban background and road site during SAPUSS-diurnal variations and PMF receptor modeling, *Atmos. Chem. Phys.*, 2013, **13**, 4375–4392.

76. J. Gao, X. Peng, G. Chen, J. Xu, G. L. Shi, Y. C. Zhang and Y. C. Feng, Insights into the chemical characterization and sources of PM$_{2.5}$ in Beijing at a 1-h time resolution, *Sci. Total Environ.*, 2016, **542**, 162–171.

77. S. S. Park, S. A. Jung, B. J. Gong, S. Y. Cho and S. J. Lee, Measurements at an air pollution monitoring supersite in Korea, *Aerosol Air Qual. Res.*, 2013, **13**, 957–976.

78. P. Prati, A. Zucchiatti, S. Tonus, F. Lucarelli, P. A. Mandò and V. Ariola, A testing technique of streaker aerosol samplers via PIXE analysis, *Nucl. Instrum. Methods Phys. Res., Sect. B*, 1998, **136–138**, 986–989.
79. A. Richard, N. Bukowiecki, P. Lienemann, M. Furger, M. Fierz, M. C. Minguillón, B. Weideli, R. Figi, U. Flechsig, K. Appel, A. S. H. Prévôt and U. Baltensperger, Quantitative sampling and analysis of trace elements in atmospheric aerosols: impactor characterization and syschrotron-XRF mass calibration, *Atmos. Meas. Tech.*, 2010, **3**, 1473–1485.
80. M. Zheng, Y. J. Zhang, C. Q. Yan, H. Y. Fu, H. Y. Niu, K. Huang, M. Hu, L. M. Zeng, Q. Z. Liu, B. Pei and Q. Y. Fu, Establishing PM$_{2.5}$ industrial source profiles in Shanghai, *China Environ. Sci.*, 2013, **33**(8), 1354–1359.
81. Y. Lei, Q. Zhang, K. B. He and D. G. Streets, Primary anthropogenic aerosol emission trends for China, 1990–2005, *Atmos. Chem. Phys.*, 2011, **11**, 931–954.
82. J. Y. Zheng, S. S. Yin, D. W. Kang, W. W. Che and L. J. Zhong, Development and uncertainty analysis of a high-resolution NH$_3$ emissions inventory and its implications with precipitation over the Pearl River Delta region, China, *Atmos. Chem. Phys.*, 2012, **12**, 7041–7058.

Case Studies of Source Apportionment from the Indian Sub-continent

MUKESH KHARE* AND ISHA KHANNA

ABSTRACT

The chapter reviews the studies on source apportionment conducted in the Indian sub-continent focussing mainly on respirable particulate matter including their sources and characteristics. The receptor models used in identification of major sources of respirable particulate matter and their fractions are also comprehensively described. Among the countries in the Indian sub-continent, India has carried out the greatest number of studies, including spatio-temporal variations, characterization and apportionment of the particulate matter sources. However, in Pakistan, the primary focus has been given towards apportionment of particulate matter, which has mainly been carried out in two mega cities, *i.e.* Lahore and Karachi. In Sri Lanka, Bhutan, Nepal and the Maldives, a limited number of studies have been carried out specifically targeting the apportionment of particulate matter. However, the existence of climate observatories in Nepal and Maldives is an added advantage towards possibilities of planning and conducting comprehensive studies investigating the particulate matter characteristics and their sources.

1   Introduction

Urban air pollution is one of the major problems affecting public health and the environment around the globe. Owing to urbanisation and fast-paced

---

*Corresponding author.

Issues in Environmental Science and Technology No. 42
Airborne Particulate Matter: Sources, Atmospheric Processes and Health
Edited by R.E. Hester, R.M. Harrison and X. Querol

development, the problem is more complex and severe in developing countries when compared to the developed world, particularly in terms of health impacts.[1] It has been reported that the majority of the air-pollution related health burden has been on low- and middle-income countries in the South-East Asia and Western Pacific Regions. In the year 2012, more than five million deaths were caused by indoor and outdoor air pollution in South Asian countries.[1] The major causes are unplanned growth of cities and exponential increases in population and number of motorized vehicles. These ultimately lead to increased air pollutants, in particular, exceedance of Particulate Matter (PM) levels in the ambient atmosphere. It has been reported that the coarser and finer fractions of PM *i.e.* PM_{10} and $PM_{2.5}$, respectively frequently violate the prescribed standard norms.[2] In a recent study conducted in Beijing, China, the nine-year average PM_{10} and $PM_{2.5}$ concentrations were observed to be 138.5 ± 92.9 µg m^{-3} (annual standard: 70 µg m^{-3}) and 72.3 ± 54.4 µg m^{-3} (annual standard: 35 µg m^{-3}), respectively.[3] However, in an another study in Delhi, India, the annual average PM_{10} and $PM_{2.5}$ concentrations were reported to be 232.1 ± 131.1 µg m^{-3} (annual standard: 60 µg m^{-3}) and 118.3 ± 81.7 µg m^{-3} (annual standard: 40 µg m^{-3}).[4] These values are much higher than the prescribed annual ambient average standards of Indian air quality.[5] In most of the Asian countries, the National Ambient Air Quality Standards (NAAQS) for PM_{10} are more or less similar; however, for $PM_{2.5}$, there are variations in respective countries (Table 1). Pakistan and Bangladesh have the most stringent standards for annual $PM_{2.5}$, which is comparable to Interim Target 3 of the WHO guidelines (Table 1).

The Indian sub-continent, also known as South Asia, refers to the group of countries that lie on the Indian tectonic plate. These include India, Pakistan, Bangladesh, Nepal, Bhutan, Sri Lanka and the Maldives (Figure 1). These countries together cover an area of about 4.4 million km^2 (1.7 million mi^2), which is 10% of the Asian continent or 3.3% of the world's land surface area. The Indian sub-continent accounts for more than forty percent of Asia's population and one-fourth of the world's population.[6] The climate of the Indian sub-continent varies considerably as per the Koppen climate classification.[7] As per the classification, the region has four major types of climate zones: (i) dry subtropical continental climate in India and Pakistan; (ii) equatorial climate in South India and Sri Lanka; (iii) tropical climate in the peninsula and (iv) alpine climate in the the Himalayas. Two monsoon systems majorly affect the climate in this sub-continent: the *summer monsoon* and the *winter monsoon*. In Sri Lanka and the Maldives, *winter monsoon* is dominant, however, for the rest of the region, the majority of the annual precipitation is provided by the *summer monsoon*. The difference between the two types is the direction of incoming winds, which is southwest for *summer monsoon* and northeast for *winter monsoon*.

51% of the Indian sub-continent's 1.4 billion people are exposed to $PM_{2.5}$ concentrations exceeding the Interim Target as given by WHO *i.e.* 35 µg m^{-3}.[8,9] Further, India and the Southern Asian countries are among the

Table 1 Annual and 24-hour average standards for PM$_{10}$ and PM$_{2.5}$.

Parameter	Averaging Time	Units	India[a]	Pakistan[b]	Bangladesh[c]	Bhutan[d]	Sri Lanka[e]	China[f]	Nepal[g]	USEPA[h]	UK[i]	WHO guidelines[j]
PM$_{10}$	24-hour	μg m^{-3}	100	150	150	100	100	150	120	150	50	50
	Annual		60	120	—	60	50	70	—	—	40	20
PM$_{2.5}$	24-hour		60	35	65	—	50	75	—	35	25	25
	Annual		40	15	15	—	25	35	—	12[k]	—	10

[a]Source: http://www.cpcb.nic.in/National_Ambient_Air_Quality_Standards.php.
[b]Source: http://www.mocc.gov.pk/gop/index.php?q=aHR0cDovLzE5Mi4xNjguNzAuMTM2L21vY2xjL3VzZXJmaWxlcEVzZmlssZS9NT0MvTmF0aW9uYW9uYWwlMjBFbnZppcm 9ubWVudCUyMFF1YWxpdHklMjBTdGFuZGFyZHMvTkVRUyUyMGZvciUyMEFYbmllbnQlMjBBaXIucGRm
[c]Source: http://www.indiaenvironmentportal.org.in/files/Bangladesh.pdf
[d]Source: http://www.nec.gov.bt/nec1/wp-content/uploads/2012/10/Air-Quality-Mgt-Strategy.pdf
[e]Source: http://www.investsrilanka.com/images/publications/pdf/environmental_norms.pdf
[f]Source: http://transportpolicy.net/index.php?title=China:_Air_Quality_Standards
[g]Source: http://www.bpc.com.np/uploads/file/eia/annexes/Annex-7.pdf
[h]Source: http://www.epa.gov/ttn/naaqs/criteria.html
[i]Source: http://uk-air.defra.gov.uk/assets/documents/National_air_quality_objectives.pdf
[j]Source: http://apps.who.int/iris/bitstream/10665/69477/1/WHO_SDE_PHE_OEH_06.02_eng.pdf
[k]Primary PM$_{2.5}$ only.

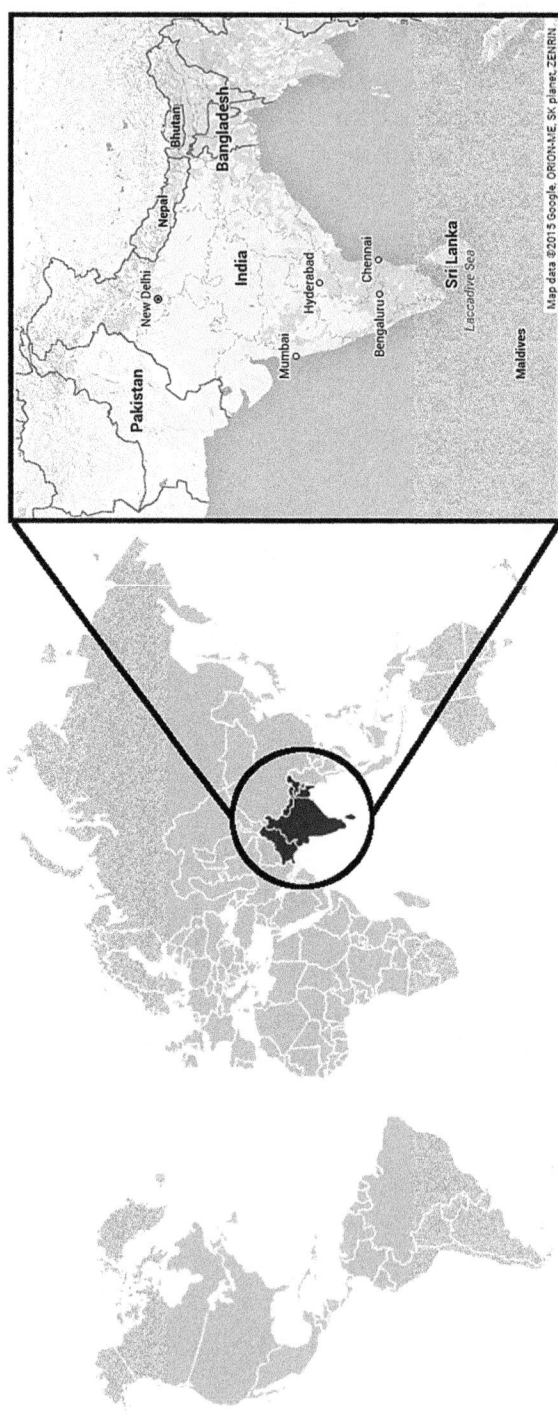

Figure 1 Location of Indian sub-continent and its countries.

top regions following the Northern and Southern China regions with the highest average urban $PM_{2.5}$ concentrations as they are emerging economies.[10] However, the air quality management in these South Asian countries has not developed at a similar pace to meet the evolving challenges of increasing air pollution, thus creating an imbalance. Besides, there is a large variation in the physical and chemical characteristics of PM in the sub-continent, which are based on the difference in intensities of the sources in different regions. It has been reported that globally 25% of urban ambient air pollution from $PM_{2.5}$ is contributed by traffic, 15% by industrial activities, 20% by domestic fuel burning, 18% from natural dust and salt and the rest from unspecified anthropogenic sources.[10] The source categories also have significantly diverse characteristics in terms of their chemical components. Therefore, a comprehensive spatio-temporal source apportionment (SA) study is required including detailed physico-chemical analysis of the PM as a function of their size.

SA studies *quantify the contribution of individual sources to particulate mass loadings based on source profiles and receptor characteristics with the nature of pollutants.* This may be accomplished using a tracer-based receptor model or an emission inventory coupled with dispersion models or a hybrid model.[11–13] Receptor models apportion the pollutant concentrations based on the mass balance of measured ambient air data with the composition of the contributing sources, *i.e.* source profiles based on the tracer species. Receptor models have been used for identification of sources and their respective contributions to airborne PM.[14–19] In the last four years in India (2010–2014), the most commonly used models for SA are Principal Component Analysis (PCA) followed by Enrichment Factor (EF), Positive Matrix Factorization (PMF), Chemical Mass Balance (CMB) and UNMIX. Except CMB, the other models are statistical based and do not require local source profiles as one of their inputs.[20] There have been numerous SA studies globally with varying particle sizes, chemical constituents analysed and receptor models being used for the apportionment. Initially, the studies have focused on apportionment of size fractions including coarser particles like SPM. In 1987–2005, it has been observed that 45% of the studies conducted in Europe targeted PM_{10} and 33% $PM_{2.5}$.[21] However, there has been a change in trend since 2006 with more focus on finer particles such as $PM_{2.5}$ and PM_1.[22] Another report has reviewed the studies conducted in Asian countries (Japan, China, Vietnam, India, Indonesia, Taiwan and Korea). The authors reported that the majority of these studies are limited to analysing coarser particles *i.e.* TSP and PM_{10}.[23] In India, it has been observed that the majority of the research studies on SA have taken place after 2005.[20] This shows that the focus of Indian researchers is gradually diverting towards undertaking advanced research in the area of SA for ensuring better air quality. The most dominant target metric for these studies has been PM_{10} (41%) followed by $PM_{2.5}$ (26%) and SPM (22%). However, in the recent past it has been observed that the research focus of Indian researchers has broadened by including finer fractions like $PM_{2.5}$ and PM_1.[24] This might be owing to recent findings regarding adverse effects of fine PM on human health and the environment.[25,26]

2 Source Signatures from Indian Sub-continent

The mass balance based receptor models, like CMB, require source profiles for estimation of the percentage contribution of each source towards the ambient concentrations. The statistical models, like PMF and PCA, do not require *a priori* knowledge of the source profiles; however characteristic compounds, also known as *tracers* or *source signatures*, should be known for each source type to be able to categorise the factors obtained from these models. These source profiles are the chemical composition of the emissions, with each chemical species expressed as a mass fraction of the total mass emitted. The source profiles required for the modelling need to be prepared locally as there is a significant variation in the source emission characteristics globally. For instance, Indian coal has a higher sulfur content than the coal available in the UK. Within India, coal from the north-eastern region contains high concentrations of Te owing to its marine origin, whereas from western region it contains As.[15,27] The metals and ions can be used as tracers to identify the sources based on their characteristic ratios. However, the same can also be emitted from other sources, which might lead to ambiguity in source identification. Therefore, organic molecular markers may be more reliable tracers as they may represent a particular source type only. Some common molecular marker compound classes include alkanes, cycloalkanes, polycyclic aromatic hydrocarbons (PAHs), steranes, fatty acids, sterols and methoxyphenols.

Selection of tracer species for conducting SA studies varies widely depending upon the source type (Figure 2). For instance, the majorly used tracers for crustal sources for PM are Al, Ca, Fe, Mg, Ba, Si, Mg, K, Na, Ni, Mn, Pb, Cu and Zn. However, road dust, soil and re-suspended dust may also be considered as crustal sources.[17,28,29] Owing to the presence of quartzite rocks, Fe has been used as a tracer for crustal sources in Delhi.[24] However, the presence of Fe along with OC may be attributed to road dust, while the presence of Cl^- and SO_4^{2-} with OC may be attributed to soil dust.[30] Similarly, the traffic source is usually characterised by the presence of elements Pb, Cu, Zn, V, Mn and Co. However, the tracers may also include emissions from fuel additives (Pb, Mn, Ba); tyre-wear and brake-wear (Zn, Cu, Al, Fe); and oil combustion (Ni). Apart from these, molecular markers like hopane, sterane and low molecular weight *n*-alkane are characteristic of vehicular emissions. In addition, PAHs are used to differentiate between gasoline (coronene, benzo(ghi)perylene, indeno(1,2,3-cd)pyrene) and diesel (chrysene, fluoranthene, pyrene) emissions.[31–33]

In the case of industrial emissions, there is a large variation in the tracers owing to a number of industries with different raw materials, fuels and processes involved. For instance, Cu, Mn and Ni are tracers of ferrous and steel processing; Cr and Zn from metal manufacturing; Pb, Cd and V from

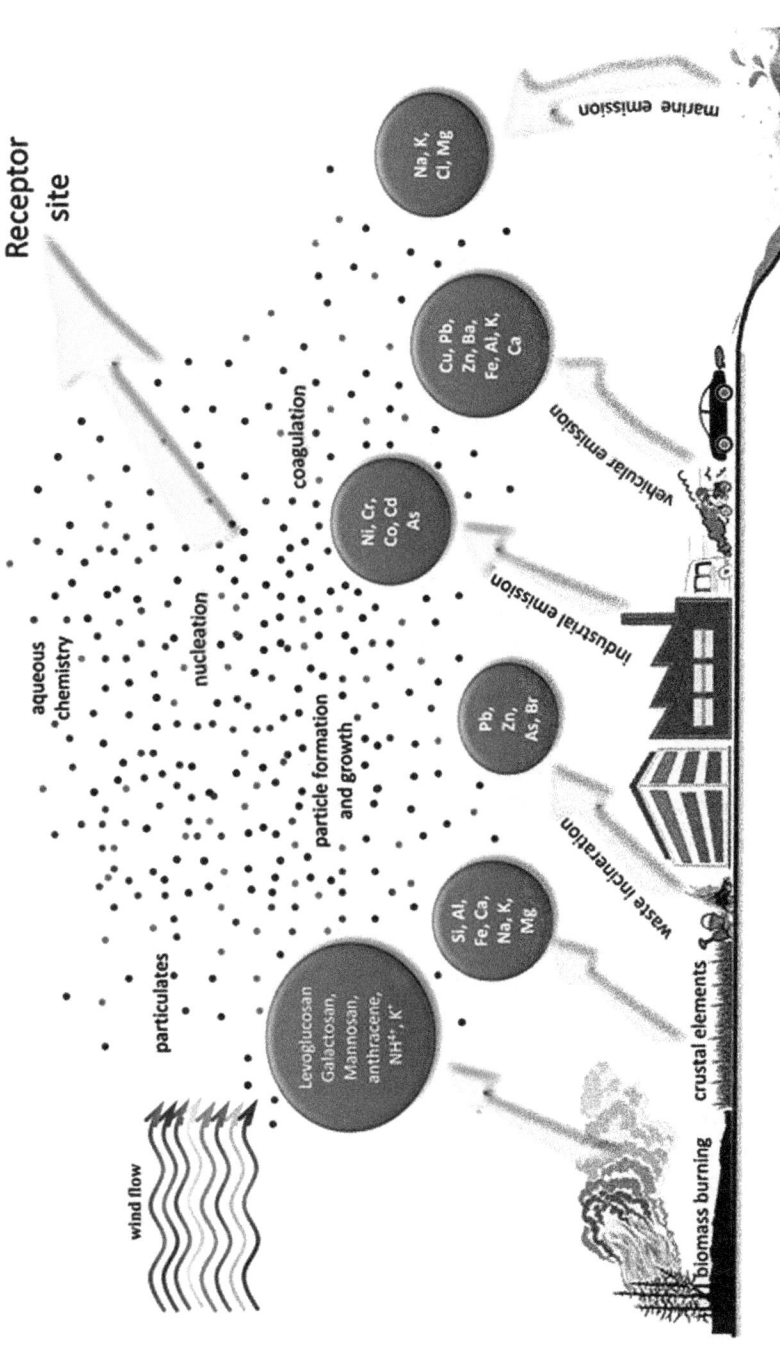

Figure 2 Evolution of elementary and organic signature molecules (reprinted from T. Banerjee *et al.* Source apportionment of airborne particulates through receptor modelling: Indian scenario, *Atmosp. Res.*, **164**, 167–187, Copyright (2015) with permission from Elsevier[113]).

battery repair; Zn, Cu and Ni from metallurgical industries.[34–37] Emissions from biomass burning consist of two major chemical components: organic carbon (OC), which primarily scatters solar radiation, and black carbon (BC), which primarily absorbs solar radiation. Additionally, wood burning forms a major source of biomass burning, which emits K in large quantities, thus making it a potential tracer for biomass burning.[38] Degradation products from biopolymers (*e.g.* anhydrosaccharides like mannosan, galactosan and levoglucosan from cellulose; methoxyphenols from lignin) are also frequently used tracers of biomass burning.[38,39] K is used as a tracer for crustal dust in the coarse range of PM and soluble K for biomass burning in the fine range.[27,40–42] Another source is coal combustion, which is an important constituent causing emissions from thermal power plants. Usually, Se, As and Te are used as tracers of coal combustion; however, in the Indian sub-continent, the most prominent tracers are Cr, Cu, Ni, V, K, Cd, Se, As, Pb, Cl^-, and PAHs such as phenanthrene, picene, benzo(*b*)fluoranthene, benzo(*a*)pyrene and anthracene.[18,24,33,41,43] Key markers for refuse burning includes Zn, Cr, and Ni.[43,44] It has been observed that refuse burning contributes significantly to the organic fraction of PM with tracers like naphthalene, fluoranthene, and dibenz(*a,h*)anthracene.[45–47] Inorganic sea salt constitutes dominant mass fraction of coarser aerosols and is identified by the presence of Na^+, K^+, Cl^-, Mg^{2+} and HCO_3^-.[15,44,48–50] However, constituents may undergo chemical transformations by reaction with SO_4^{2-} and NO_3^-, which may subsequently release Cl^- and Br^- to the atmosphere.[51] Also, NO_3^- may be formed by condensation of HNO_3 in $PM_{2.5}$.[51,52] However, the selection of some tracers is somewhat confusing and ambiguous. As a result, there are ambiguities in the source signatures selected in these studies, which may alter their interpretation and thus, their source contribution estimates.

3   Case Studies from the Indian Sub-continent

3.1   India

India is the seventh-largest country by area in the world and is bounded by water bodies on three sides—the Indian Ocean, the Arabian Sea and the Bay of Bengal in the south, south-west, and south-east, respectively. India borders seven other countries, namely Pakistan, China, Nepal, Bhutan, Myanmar and Bangladesh. It is also in close vicinity of Sri Lanka and the Maldives, which lie in the Indian Ocean. It has a population of more than 1.2 billion, making it the second-most populous country. The majority of the SA studies in India have been conducted in Delhi, its capital, followed by Mumbai, Chennai, Kanpur, and Kolkata. However, it has been observed that the ambient PM indicates spatio-temporal physical and chemical heterogeneity.[12,53] For instance, two coastal cities in India, Mumbai (situated on Western peninsula) and Chennai (situated on Southern Peninsula), have observed high contributions of marine sources to PM_{10} and $PM_{2.5}$; Delhi, the capital of India, observed dominance of contribution of crustal and vehicular sources in PM_{10} and $PM_{2.5}$, respectively; Kanpur, situated in the Indo-Gangetic basin, has observed 20–25% contribution of secondary particles to PM_{10}.[20] The Central Pollution Control Board (CPCB) has recently released the *Six-City* SA analysis report, which includes air quality monitoring, emission inventory, chemical speciation and SA of PM_{10}, including evaluation of control options using dispersion modeling to evolve city-specific action plans.[54] The study has reported that re-suspension of road dust was one of the major contributing sources for PM_{10} in residential, kerbside and industrial locations in all the six cities. Construction activities (22%) have also been observed to be another major source contributing to higher PM_{10} levels in Delhi. However, enough emphasis has not been given to $PM_{2.5}$ in the country. In India, 27% of SA studies have been conducted with the use of EFs and FA. Nearly 58% of SA studies that have been conducted in India have used FA technique for conducting SA. Among these, PCA has been the most dominant (34%), followed by PMF (9%) and UNMIX (1%). During 2005–2014, more than twenty-five percent of SA studies were conducted using CMB in India.[20]

Delhi, the capital of India and one of the six metropolitan cities, is bounded by the Indo-Gangetic alluvial plains in the north and east, the Thar Desert in the west and Aravalli hills in the south. Being the capital of the country, it has been studied most extensively for chemical composition as well as the SA of PM.[11,39,40,55,56] The first study in Delhi city has been performed in 1980 investigating trace elements in TSP. The study has reported that most of the TSP mass is associated with natural soil elements.[57] The particle size distribution has been studied for three sites in Delhi—residential, commercial and industrial. Three sub-fractions of PM_{10} have been studied: (i) >5.8 μm; (ii) 5.8–2.1 μm; and (iii) <2.1 μm. They have observed that at all three sites, up to 80% of PM_{10} is formed of particles with size 5.8 μm or below.[58] In another study, seasonal variability in PM_{10} mass concentration has been observed between 301–350 μg m^{-3} in the winter season and 51–100 μg m^{-3} in the monsoon season.[40] Later, in one of the studies, it was reported that 140 tonnes per day of PM_{10} mass is emitted out of which total carbon (TC) emissions and primary SO_4^{2-} and NO_3^{-} emissions account for 20% and 4.9%, respectively.[59] Vehicular emissions, road dust and coal combustion have been reported as the major sources of PM in the majority of the studies.[43,58,60,61] In one of the studies apportioning PM_{10} using PCA, it has been reported that vehicular sources are one of the major sources (53.9%), followed by industrial sources (19.4%) and crustal re-suspension (15.7%).[58] The source and origin of the ambient $PM_{2.5}$ concentrations in Delhi has been studied using the persistence analysis and nonparametric regression technique and it has been reported that the contribution by local sources is lead by the power plant, industries and dust coming from nearby Thar desert.[62] However, in a later study, it was reported that soil crustal dust contributes minimal SPM concentration.[55,61]

Mumbai, a coastal city in the western peninsula, is the most populous city in India with a population of over 12 million. It is bound by the Arabian Sea on the west and sits on Salsette Island in the Konkan coastal region. The first study conducted in Mumbai to quantify the source contribution at two traffic intersections used the FA-MR model and observed five major sources: road dust, vehicular emissions, marine aerosols, metal industries and coal combustion.[15] Further, a significant correlation was observed between TSP and PM_{10}, which may indicate the influence of common sources.[63] It has been observed that the thermal power plants are responsible for 19% SPM emissions.[64] Marine aerosols have been identified as significant sources of PM_{10} owing to Mumbai's coastal location.[15,44,65] However, another study identified crustal/road dust (69.41%), industrial emissions (11.76%) and fuel oil combustion (6.52%) as the key sources.[66]

Kolkata is another coastal city located on the banks of the Hooghly River over the Bengal Basin. It has a very limited number of studies, which have reported the PM fractions to be significantly affected by marine sources and meteorological conditions.[67] Like Mumbai, in Kolkata $PM_{2.5}$ has been found to be highly correlated with PM_{10} indicating common sources at local and regional level.[29,30] CMB has been used for apportioning PM_{10}, which estimated major contributions from coal combustion and vehicular emissions at the residential and industrial sites, respectively.[30] Other studies have included wood combustion, solid waste, tyre wear, soil dust and road dust as the significant sources of $PM_{2.5}$ mass loadings.

Chennai is located in the southern peninsula along the coast of the Bay of Bengal. Being the capital of Tamil Nadu, and owing to its proximity to the sea coast, the city's air is significantly influenced by the natural aerosols transported from the sea and the anthropogenic emissions owing to vehicular and industrial emissions. In most of the studies conducted in Chennai, it has been observed that both PM_{10} and $PM_{2.5}$ had a maxima in winter followed by monsoon and summer.[51,68–70] The marine aerosols have a more significant contribution in SPM mass loadings during summer as compared to monsoon (19%) and winter (20%).[70] Another study reported that $PM_{2.5}$ and PM_1 contributed 56% and 44% to PM_{10}, respectively. They also observed that $PM_{2.5}$ is comprised of 81% of PM_1.[69] SO_4^{2-}, NO_3^- and Mg^{2+} are reported as dominant species in fine and coarse PM.[69] The studies indicate that diesel exhausts (43–52% in PM_{10} and 44–65% in $PM_{2.5}$) and gasoline exhausts (6–16% in PM_{10} and 3–8% in $PM_{2.5}$) were found to be the most dominant sources. The other sources include paved road dust, brake lining and brake pad wear dust, marine aerosols and cooking.[51,68]

Kanpur, an industrial city in Uttar Pradesh, is a hub for many small-, medium- and large-scale industries. Additionally, the city is located in the Indo-Gangetic plain where agriculture is predominantly practised in almost all the seasons. For the above-mentioned reasons, the city's air is influenced by a number of combustion sources, including agricultural-waste burning post harvesting of crops, fossil-fuel combustion, and brick kilns.[71] Further, vehicular emissions have been reported as one of the most dominant sources (16%), followed by domestic fuel uses (16%), paved and unpaved roads (14%) and industrial sources (7%) for PM_{10}.[72] Crustal elements have also been reported as major sources owing to the dry texture of the soil in the region. High variability has been observed for PM_{10} during summer and PM_1 during winter.[73,74] The contribution of long range transport of mineral dust during summer and biomass burning during winter is reported to contribute significantly to both PM_{10} and $PM_{2.5}$.[75] Further, it has been reported that the inorganic secondary particles contribute 15–25% to the total PM_{10} mass.[75,76]

Agra, located in the north-central part of India, is the home of the one of the New Seven Wonders of the World, the Taj Mahal. Since the site has significant archaeological importance, reducing air pollution and investigating the causes of the same has gained utmost importance. Further, the city is closely located to the industrial towns and clusters which has become a source for a wide range of pollutants for which apportionment studies are essential. The majority of the SA studies have been conducted using trace element markers.[77–81] It has been found that there is a significant correlation between the PM mass and the meteorological parameters in this region, for instance, inverse proportionality with temperature and wind speed leading to maximum concentrations in winter and minimum in summer for both PM_{10} and $PM_{2.5}$.[77,78] Among the ionic species, Ca^{2+} has been observed to be most dominant, followed by SO_4^{2-}, NO_3^- and NH_4^+.[78,80] Further, soluble K^+ is also significantly correlated with OC due to biomass and waste burning.[81] The contribution of $PM_{2.5}$ to PM_{10} also varies from 30–70% in an urban site to 40–60% in a rural site.[77,82] Scanning Electron Microscopy has been used to differentiate between three types of particles based on their morphology and chemical constituents: C and O rich; Si, Na and Al rich; and S, Fe, K, and Co rich.[82] In 1995, key sources contributing to atmospheric aerosols in Agra were identified as crustal sources, industrial emissions, wood burning and coal combustion in brick kilns.[83] However, in 2009, resuspended dust, solid waste incineration and industrial emissions were identified to be key contributors to the metallic fraction of PM at an urban site.[77] Besides, re-suspension, construction activities, industrial emissions and biomass burning of agricultural waste were found to be the key sources of PM in the rural locations. Figures 3, 4 and 5 show comparative percent source contributions for SPM, PM_{10} and $PM_{2.5}$, respectively in different cities as reported in various studies.

3.2 Pakistan

In Pakistan, the majority of the studies have been conducted in major metropolitan areas like Karachi, Lahore, Islamabad and Faisalabad. Initial SA studies are from Karachi, the largest and most populous city of Pakistan. It is located on the coast of Arabian Sea and thus receives significant emissions in terms of marine aerosols. The studies have revealed a high TSP concentration of approximately 240 µg m^{-3} in the months of March–June with the highest contributions from soil and cement components (48%) and sea salt (12%).[84] These studies have focussed mainly on metallic and ionic components of the PM, which are followed by inclusion of OC and EC carbon

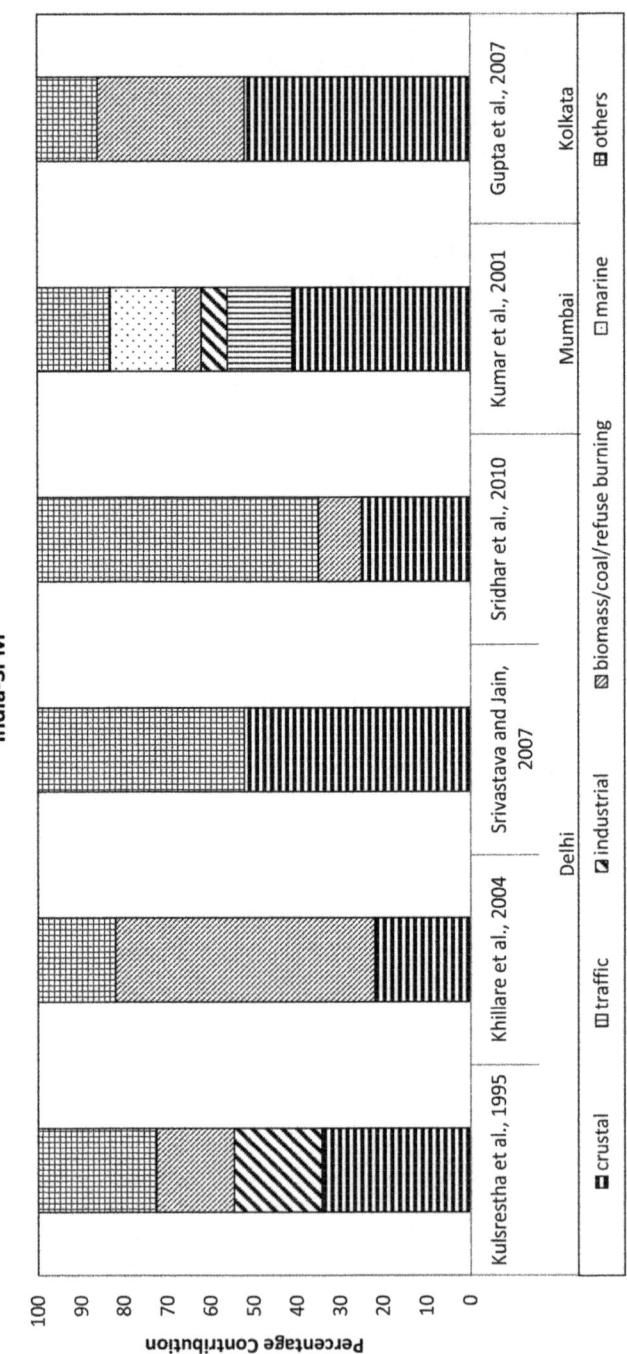

Figure 3   Percent source contribution for SPM in India.

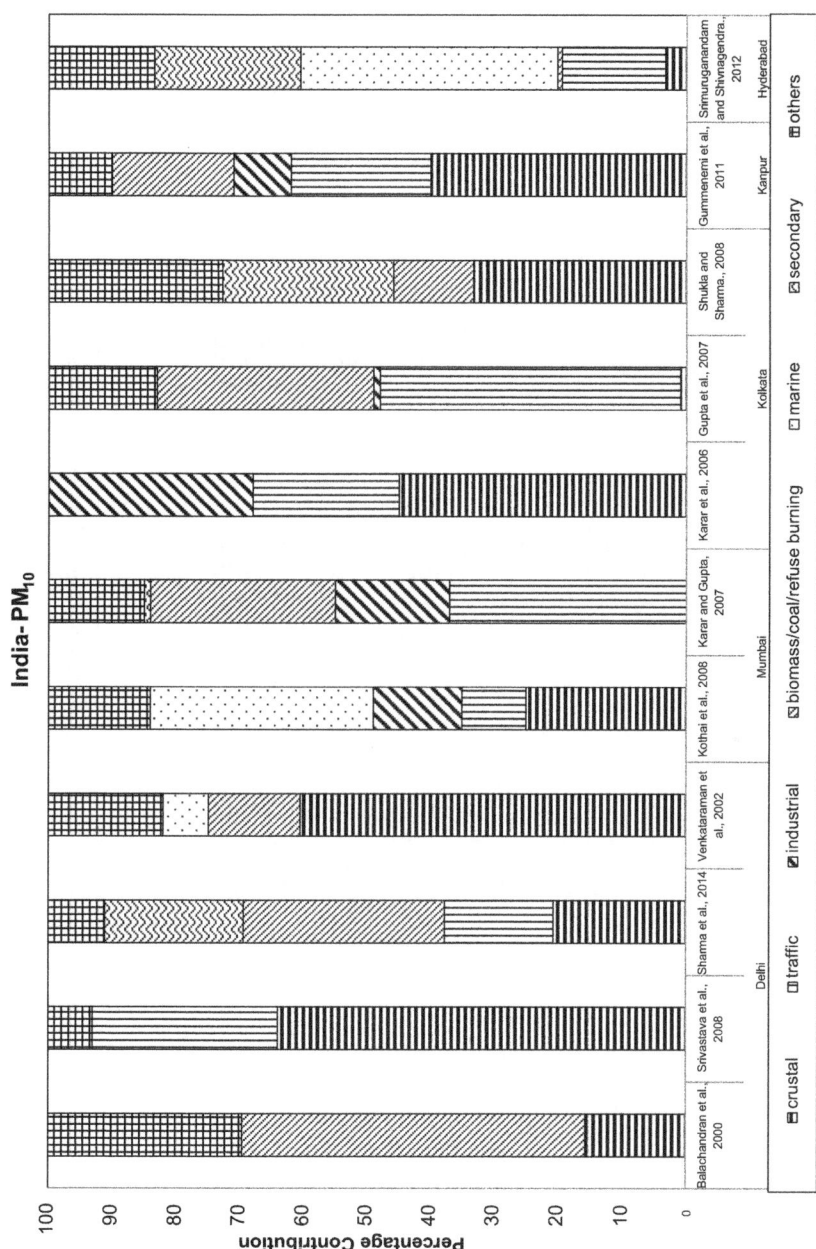

Figure 4 Percent source contribution for PM_{10} in India.

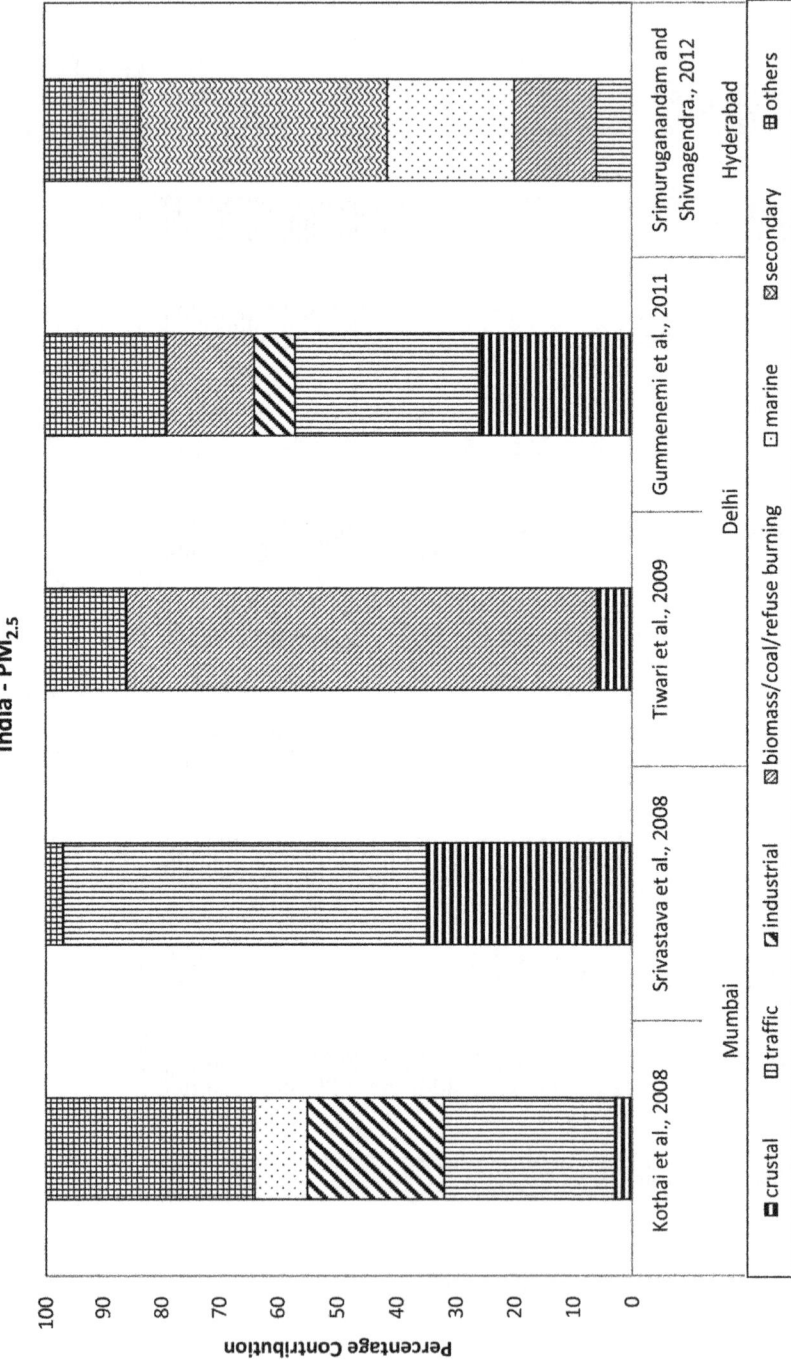

Figure 5 Percent source contribution for PM$_{2.5}$ in India.

fractions and PAHs.[14,37] The ambient PAHs present in PM have been found to be of the same order of magnitude as several Indian cities. However, they have been observed to be an order higher than those obtained in Birmingham in the same year.[37] Another study has been performed regarding particle size distributions for nine fractions ($PM_{<1.0}$, $PM_{1.0-2.5}$, $PM_{2.5-5}$, PM_{5-10}, PM_{10-15}, PM_{15-25}, PM_{25-50}, PM_{50-100} and $PM_{>100}$) and they found $PM_{5.0-10}$ to be most abundant, followed by $PM_{2.5-5.0}$ and PM_{15-25}.[85] The trace metals were also observed to be highly correlated with finer fractions up to PM_{10-15}. PM_{10}/TSP concentrations and their chemical components for Lahore have been compared with those of Birmingham and Coimbra and a remarkable difference was found in the values owing to their geographical differences.[14] The authors concluded that although statistical models may not efficiently discriminate between a large number of sources, PCA and multi-linear regression may provide useful results for semi-quantitative analysis. Five sources have been observed to be majorly contributing to $PM_{2.5}$ in Karachi: vehicular/industrial oil burning (39.7%), vehicles (18.5%), road dust (16.1%), steel industry (13.3%) and secondary aerosols (12.4%).[86] Figures 6 and 7 describe the percentage source contributions for SPM and PM_{10}/$PM_{2.5}$, respectively.

3.3 Bangladesh

Bangladesh is the world's eighth-most populous country, with over 160 million people. It is one of the most densely populated countries. The majority of studies in Bangladesh have been conducted in Dhaka, the capital and largest city of Bangladesh, which has a population of over 14 million. The majority of the studies have been conducted investigating $PM_{2.2}$ and $PM_{2.2-10}$ concentrations.[16,87,88] However, in one of the exclusive studies, the average $PM_{2.5}$ and PM_{10} concentrations for Dhaka from 1996–2011 were 36.7 ± 25.5 µg m^{-3} and 97.7 ± 68.6 µg m^{-3}, respectively which are higher than the Bangladesh NAAQS.[50] One of the first SA studies for PM from Dhaka has compared the chemical constituents of TSP with other cities from India and Pakistan. The study concludes that there is a considerable difference in EC/TC and K/EC ratios, showing a difference in the characteristics of biomass burned.[89] Further, the researchers reported that around 76% of the TSP is from soil-type material, 18% of carbonaceous material, and 6% of soluble ions and 0.3% of trace elements using reconstruction of aerosol mass. In another study at a background site at Bhola Island, it was observed that the site is affected by emissions from urbanized regions of Southeast Asia using a back-trajectory method.[90] Later, in one of the studies, the source contribution of two sites *i.e.* Dhaka and Rajshahi, was compared using PMF.[28] The study reported six major sources for $PM_{2.5}$, *i.e.* soil dust, road dust, cement, sea salt, motor vehicles and biomass burning. In Dhaka, about half of the $PM_{2.2}$ concentrations can be attributed to motor vehicles, whereas in Rajshahi, it is biomass burning. Similar studies have been conducted in Dhaka showing comparable results for $PM_{2.2}$ and $PM_{2.2-10}$ using CMB

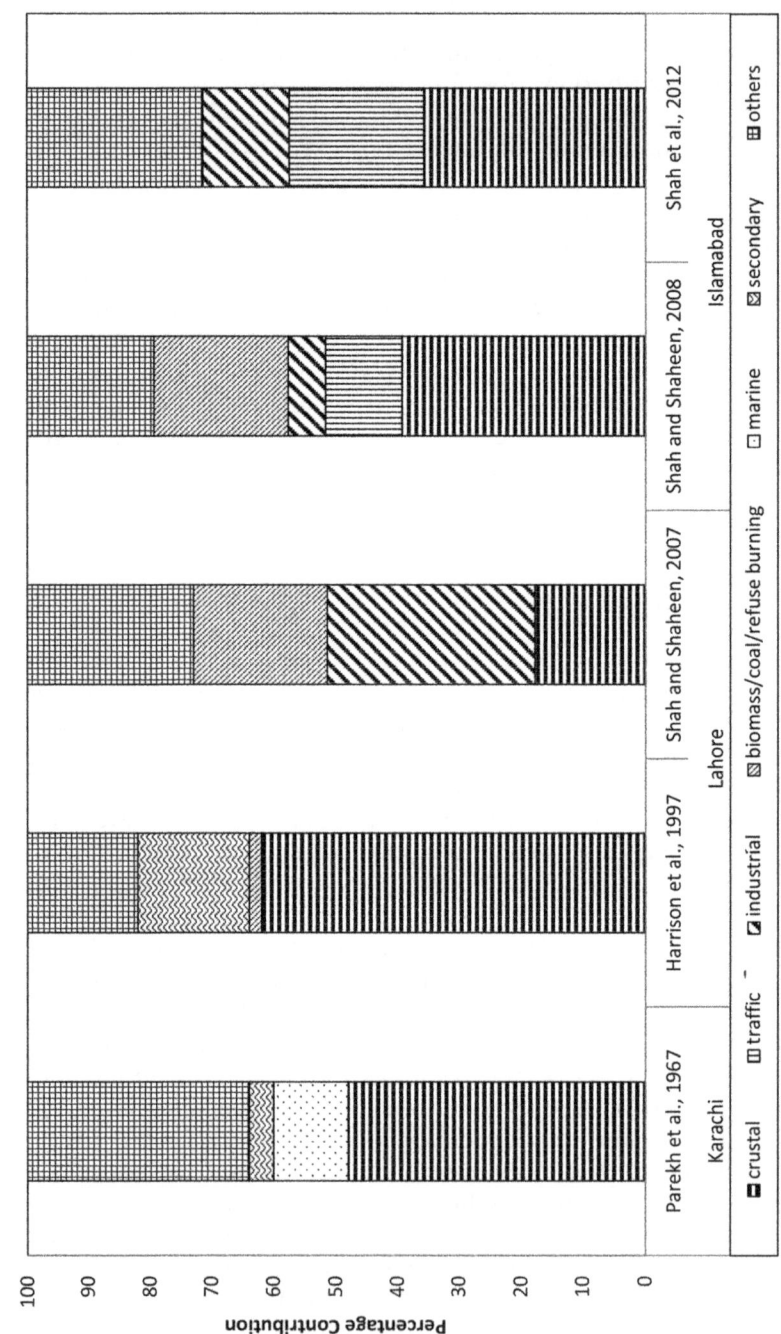

Figure 6 Percent source contribution for SPM in Pakistan.

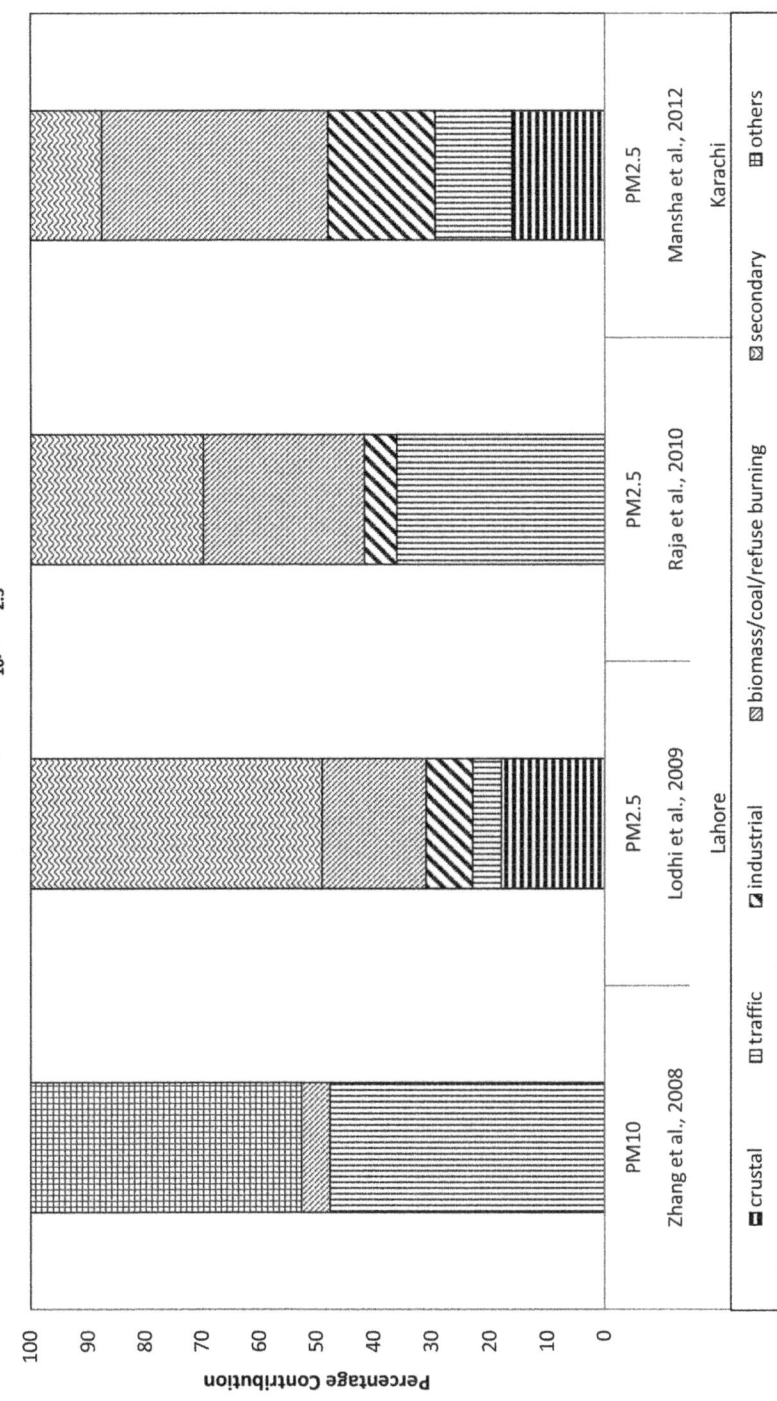

Figure 7　Percent source contribution for PM_{10} and $PM_{2.5}$ in Pakistan.

and PMF.[49,88] The trace metal concentrations for $PM_{2.5}$ and PM_{10} were observed to be higher than those of European and East Asian countries but lower than those of other South Asian countries.[91] Further, motor vehicles were the most dominant sources for $PM_{2.2}$ (54.4%) in 2001–2005 and 2001–2002 followed by brick kilns, road and soil dust (Figures 8 and 9).[16]

3.4   Nepal

Nepal is located in the Himalayas and bordered to the north by China and to the south, east, and west by India. It has an area of 147 181 km^2 (56 827 mi^2) and a population of approximately 27 million. The northern part of Nepal has 8 of the world's 10 tallest mountains, including Mount Everest, which is the highest point on Earth. They have established the Nepal Climate Observatory-Pyramid (NCO-P) for continuous monitoring of air mass and meteorological conditions, which is located at 5079 m above sea level on the southern foothills of Mt. Everest.[92] The NCO-P was set up in February 2006.[93] There have been various studies conducted at the NCOP to study the characteristics of the aerosols–PM_{10}, $PM_{2.5}$, PM_1, BC, O_3.[92,94] It has been observed that the NCO-P does not have any major anthropogenic sources in its vicinity and the annual average particle number concentrations for 10 nm–10 µm particles for the period 2006–2008 were 860 ± 55 cm^{-3}.[95] The majority of the studies have been conducted in the Himalayas of Nepal before setting up of NCO-P observatory.[96–98] There has been a predominance of particulate organic material followed by SO_4^{2-}, NH_4^+, and NO_3^- which are tracers of combustion sources. At a later stage, a study derived a logistic growth model to explain the growth of $PM_{2.5}$ in winter in the Kathmandu Valley.[99] The characterisation and SA of the aerosols revealed OC (64–68%) to be the most dominant species followed by ionic species (24–26%) and the major source to be vehicles (38.74%), followed by secondary particles (23%), aqueous processing (11.43%) and dust (9.05%).[18,100,101] The study also reported a lack of influence of biomass burning in atmospheric aerosols in wintertime in Kathmandu owing to a small concentration of K^+, a small value of $\delta^{15}N$ and a K^+/EC ratio of 0.03. The toxic equivalent quantity of PAHs in the aerosols of the Kathmandu valley ranged between 2.74 and 81.5 ng TEQ m^{-3}, which is considerably higher than those reported in other South Asian cities.[102] Recent studies regarding the characterization of PM_{10} and $PM_{2.5}$ have shown that there is a domination of anthropogenic sources in the southern side of the Himalayas in Nepal. These aerosols are responsible for constituting light-absorbing aerosol hazes, which are also known as the *Asian brown cloud*.[94] The brown clouds are the wide polluted tropospheric layers characterised by anthropogenic aerosol optical depth (AOD) greater than 0.3 and absorbing AOD greater than 0.03.[103] Further, a 3-year study has shown that the top of the atmosphere forcing values are significantly greater than the values reported by IPCC for greenhouse gases.[104] These results are based on optical properties of aerosols, *i.e.* aerosol absorption, scattering co-efficients and radiative forcing. However, another study has indicated that

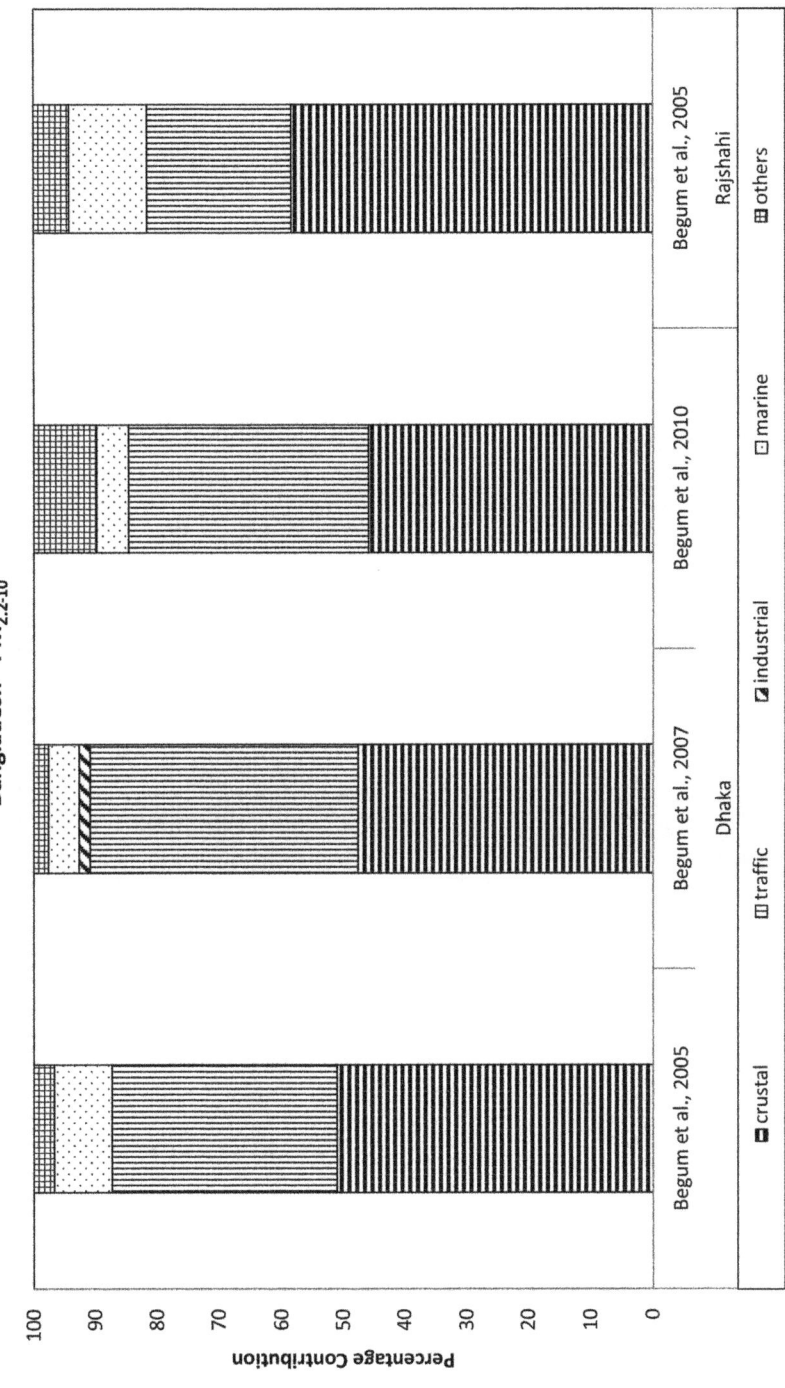

Figure 8　Percent source contribution for PM$_{2.2-10}$ in Bangladesh.

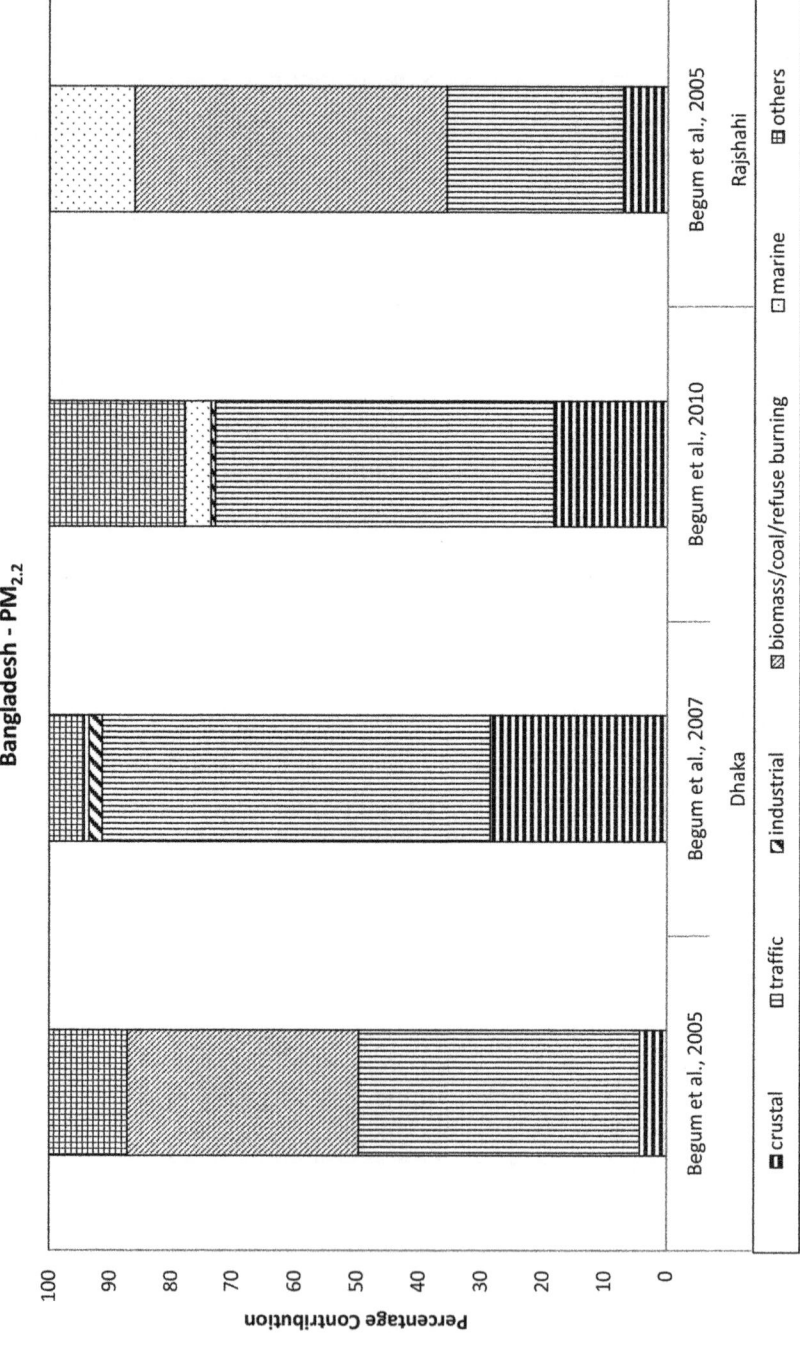

Figure 9 Percent source contribution for $PM_{2.2}$ in Bangladesh.

there is higher variability of the monsoon signal rather than variations in the intensity of emissions by studying black carbon refractory values from a Mera Peak ice core for ten years data.[105]

3.5 Sri Lanka

Sri Lanka is an island country in South Asia near south-east India that has maritime borders with India to the northwest and the Maldives to the southwest. It has an area of 65 610 km^2 and a population of over 20 million. A very limited number of studies have been conducted in Sri Lanka regarding the SA of PM. The PM$_{2.5}$ annual average for the period 2000–2005 has been found to be 29.0 ± 15.0 µg m^{-3} and the major contributors to be motor vehicles (48%), road dust (27%), biomass burning (21%) and sea salt (4%).[106] The study also identified soil dust and smoke trans-boundary events as contributors to PM$_{2.5}$ using the back-trajectory model HYSPLIT. The concentrations of 16 prioritized PAHs in PM$_{2.5}$ ranged from 57.43 to 1246.12 ng m^{-3}, with 695.94 ng m^{-3} in urban heavy traffic locations, 337.45 ng m^{-3} in suburban heavy traffic stations and 192.48 ng m^{-3} in rural high firewood burning areas.[19] The study revealed that the predominant source of PAHs is automobile emissions in the urban and suburban areas whereas in rural areas, domestic firewood burning is the major source.

3.6 Bhutan

Bhutan is a landlocked country in South Asia that is bordered by China to the north and by India to the east, west and south. It lies on the eastern end of the Himalayas. The land mainly consists of mountainous regions that are crisscrossed by a number of rivers. The studies performed in Bhutan for air pollution are very limited in number. For Thimphu, the capital of Bhutan, the PM$_{10}$ concentrations for the years 2010, 2011 and 2012 were 56 ± 47 µg m^{-3}, 35 ± 16 µg m^{-3} and 43 ± 19 µg m^{-3} respectively.[107] Another study reported mean PM$_{10}$ and PM$_{2.5}$ concentrations to be 27 and 13 for the wet season, and 36 and 29 for the dry season, respectively, for the Kanglung region in 2015.[108] Another study analysed the on-road air pollution for different segments on road on the East-West Highway in Bhutan, which showed that highest PM$_{10}$ concentrations were observed on the final segment of the road to the capital at 149 µg m^{-3} and particle number of 5.74×10^4 particles cm^{-3}. They also observed that the major sources for on-road air pollution are vehicle emissions, road works, unpaved roads and roadside combustion activities.[109]

3.7 The Maldives

The Maldives is a sovereign island country and archipelago in the Indian Ocean. It is located southwest of India and Sri Lanka in the Laccadive Sea. The capital and largest city is Malé. Encompassing a territory spread over roughly 90 000 km^2 (35 000 mi^2), the Maldives is one of the world's most geographically dispersed countries. It is the smallest Asian country in terms of both land area and population. Techniques like *natural abundance radiocarbon* (Δ14C) analysis have been used in studies to differentiate between fossil fuel combustion sources and biomass burning sources quantitatively. Additionally, studies have also reported the use of a *dual carbon isotope approach* for analysing the sources and the atmospheric processing of EC and water-soluble organic carbon (WSOC), which can be used for reducing the uncertainties in the aerosols in the domain of the study.[110–112] Further, it has also been observed that the cyclic nature of the Indian Monsoon largely influences the presence of BC and other anthropogenic aerosols over the Indian Ocean. Most of the studies reported from the Maldives are part of Atmospheric Brown Cloud (ABC) project as the MCOH Climate Observatory is located on Hanimaadhoo Island. In the first study conducted at the observatory built in the Republic of Maldives for the long-term monitoring of climate, a sampling campaign was conducted to investigate the influence of the shifting monsoon seasons on aerosols and climate change in 2006. It was observed that with the arrival of the dry monsoon, there is a large increase in the aerosol radiative forcing of the region.[113] Another study reported biofuel burning and fossil fuel combustion to be the most prominent sources of EC during the dry season in 2006.[114] The first study for characterization of organosulfates (OS) in Asia was performed in 2012, which has shown that average estimates of OS account for less than 1% of PM$_{2.5}$ mass, 2.3% of organic carbon, and 3.8% of total sulfate.[115] Organic molecular tracer analysis of PAHs and *n*-alkanes has been conducted using Monte Carlo SA (MCSA) techniques on winter data comparing the Island of Hanimaadhoo to a mountain top near Sinhagad in India.[31] The study showed that a higher input from coal combustion was observed at the Hanimaadhoo site (32–43 $\pm$ 21%) than at the Sinhagad site (24–25 $\pm$ 18%). In Hanimaadhoo, the MCSA results from this study indicate 34 $\pm$ 20% biomass burning contribution, which is lower than the value of 41 $\pm$ 5% found by radiocarbon apportionment of EC.[116] The local contribution to the load of PM$_{2.5}$ in Malé has been assessed using the MCOH laboratory to represent the local background.[117] This study showed that in the dry/winter season 90 $\pm$ 11% of PM$_{2.5}$ levels in Malé could be from long-range transport with only 8 $\pm$ 11% from local emissions. The contribution of TC to bulk mass PM$_{2.5}$ has been observed to be 17% in Malé and 13% at MCOH, suggesting larger contributions from incomplete combustion practices in the Malé local region.

4 Concluding Remarks

It is evident that SA studies can successfully be used in identification of PM sources and their detailed chemical profiles. It can be concluded that India seems to be the leading nation in the sub-continent as it has a large number of studies related to PM characterization and apportionment to its credit. At micro level investigations too, that include spatio-temporal variations of different fractions of PM, it has been observed that the most studies have been carried out in India. On the other hand, in Sri Lanka and Bhutan, limited published literature is available on apportionment of PM. However, the presence of climate observatories in Nepal and the Maldives monitoring atmospheric aerosols of different size fractions provides opportunities to carry out comprehensive studies on PM in the region investigating its apportionment and characterization. It has been observed that the trend regarding the target metric has been shifting from coarser fractions (TSP and PM_{10}) to finer fractions ($PM_{2.5}$ and PM_1). This may be owing to the availability and affordability of sophisticated equipment and also because they are major threats to the environment and to human health.

It has been shown that there are considerable differences in the results when inter-comparison of these studies is carried out. This may be due to different sampling locations based on topographic, territorial and urban planning based factors; sampling methodology; seasonal variations; and variations in analytical instrumentation and methodology used. Apart from these, collinearity between sources; non-apportioned regional background concentrations; non-inclusion of other sources *e.g.* biomass burning, brick kilns emissions, construction activities; and the inability to differentiate between sources, such as DG sets and diesel-fuelled vehicles, vehicular emissions and re-suspension, may also significantly contribute towards the differences among SA studies.

Further, the analysis of the chemical constituents of the PM has also seen an evident change from analysing only metallic and ionic components to detailed characterization including carbon fractions, organic molecular markers and other significant constituents. However, attention needs to be paid to *secondary* formation. Besides, it has been observed that many source tracers/signatures used in this region differ significantly when compared with source tracers/signatures used in Western countries.

The receptor models that have been mainly used are statistically based, like FA, PCA, and PMF, in contrast to mass-balance based, *i.e.* CMB. This is because statistical models do not require source profiles as their input data; however, the number of samples considered for multivariate models must be kept sufficiently high for efficient results. The majority of the studies have used USEPA SPECIATE source profiles for the CMB model owing to the unavailability of locally prepared source profiles. This situation has improved owing to the development of region-specific source profiles for the majority of the sources. However, these need to be further strengthened to capture the variability in different regions within the Indian Sub-Continent. Therefore, a comprehensive spatio-temporal SA study is needed, which may capture detailed physico-chemical characteristics of the PM as a function of their size. This information will further assist in planning interventions and policy decisions to improve ambient air quality for sustainable development and growth.

Acknowledgements

The authors would like to thankfully acknowledge the University Grant Commission (UGC), Government of India for providing a doctoral fellowship to Ms. Isha Khanna, Senior Research Scholar, Department of Civil Engineering, IIT-Delhi who is Co-Author of the chapter.

References

1. http://www.who.int/mediacentre/news/releases/2014/air-pollution/en/ (last accessed September 2015).
2. S. Gulia, S. M. Shiva Nagendra, M. Khare and I. Khanna, *Atmos. Pollut. Res.*, 2015, **6**, 286–304.
3. Z. Liu, B. Hu, L. Wang, F. Wu, W. Gao and Y. Wang, *Environ. Sci. Pollut. Res. Int.*, 2015, **22**, 627–642.
4. S. Tiwari, P. K. Hopke, A. S. Pipal, A. K. Srivastava, D. S. Bisht, S. Tiwari, A. K. Singh, V. K. Soni and S. D. Attri, *Atmos. Res.*, 2015, **166**, 223–232.
5. http://www.cpcb.nic.in/National_Ambient_Air_Quality_Standards.php (last accessed September 2015).
6. D. Levinson and K. Christensen, *Encyclopedia of Modern Asia*, Charles Scribner's Sons, 2002, vol. 6.
7. http://koeppen-geiger.vu-wien.ac.at/ (last accessed September 2015).
8. S. Dey, L. Di Girolamo, A. van Donkelaar, S. N. Tripathi, T. Gupta and M. Mohan, *Remote Sens. Environ.*, 2012, **127**, 153–161.
9. World Health Organization, *WHO Air Quality Guidelines for Particulate Matter, Ozone, Nitrogen Dioxide and Sulfur Dioxide*, 2006, 1–22.
10. F. Karagulian, C. F. C. Dora, A. M. Prüss-Ustün, S. Bonjour, H. Adair-Rohani and M. Amann, *Atmos. Environ.*, 2015, **120**, 475–483.
11. A. Srivastava and V. K. Jain, *J. Hazard. Mater.*, 2007, **144**, 283–291.
12. T. Banerjee, S. C. Barman and R. K. Srivastava, *Environ. Pollut.*, 2011, **159**, 865–875.
13. S. E. Haupt, G. S. Young and C. T. Allen, *J. Appl. Meteorol. Climatol.*, 2006, **45**, 476–490.
14. R. M. Harrison, D. J. T. Smith, C. A. Pio and L. M. Castro, *Atmos. Environ.*, 1997, **31**, 3309–3321.
15. A. V. Kumar, R. S. Patil and K. S. V. Nambi, *Atmos. Environ.*, 2001, **35**, 4245–4251.
16. B. A. Begum, S. K. Biswas, A. Markwitz and P. K. Hopke, *Aerosol Air Qual. Res.*, 2010, **10**, 345–353.
17. E. Stone, J. Schauer, T. A. Quraishi and A. Mahmood, *Atmos. Environ.*, 2010, **44**, 1062–1070.
18. R. K. Sharma, B. K. Bhattarai, B. K. Sapkota, M. B. Gewali, B. Kjeldstad, H. Lee and R. Pokhrel, *Res. J. Chem. Sci.*, 2013, **3**, 88–96.
19. A. P. Wickramasinghe, D. G. G. P. Karunaratne and R. Sivakanesan, *Atmos. Environ.*, 2011, **45**, 2642–2650.

20. T. Banerjee, V. Murari, M. Kumar and M. P. Raju, *Atmos. Res.*, 2015, **164–165**, 167–187.
21. M. Viana, T. a J. Kuhlbusch, X. Querol, a Alastuey, R. M. Harrison, P. K. Hopke, W. Winiwarter, M. Vallius, S. Szidat, a. S. H. Prévôt, C. Hueglin, H. Bloemen, P. Wåhlin, R. Vecchi, a. I. Miranda, a. Kasper-Giebl, W. Maenhaut and R. Hitzenberger, *J. Aerosol Sci.*, 2008, **39**, 827–849.
22. R. Vecchi, M. Chiari, A. D'Alessandro, P. Fermo, F. Lucarelli, F. Mazzei, S. Nava, A. Piazzalunga, P. Prati, F. Silvani and G. Valli, *Atmos. Environ.*, 2008, **42**, 2240–2253.
23. G. C. Fang, Y. S. Wu, S. H. Huang and J. Y. Rau, *Atmos. Environ.*, 2005, **39**, 3003–3013.
24. P. Pant and R. M. Harrison, *Atmos. Environ.*, 2012, **49**, 1–12.
25. A. Gurung and M. L. Bell, *Environ. Res.*, 2013, **124**, 54–64.
26. C. A. Pope and D. W. Dockery, *J. Air Waste Manage. Assoc.*, 2006, **56**, 709–742.
27. P. Khare and B. P. Baruah, *Atmos. Res.*, 2010, **98**, 148–162.
28. B. Begum, E. Kim, S. K. Biswas and P. K. Hopke, *Atmos. Environ.*, 2004, **38**, 3025–3038.
29. K. Karar and A. K. Gupta, *Atmos. Res.*, 2007, **84**, 30–41.
30. A. K. Gupta, K. Karar and A. Srivastava, *J. Hazard. Mater.*, 2007, **142**, 279–287.
31. R. J. Sheesley, A. Andersson and Ö. Gustafsson, *Atmos. Environ.*, 2011, **45**, 3874–3881.
32. L. N. Suvarapu, Y.-K. Seo, Y.-C. Cha and S.-O. Baek, *Asian J. Atmos. Environ.*, 2012, **6**, 169–191.
33. S. Gupta, K. Kumar, A. Srivastava, A. Srivastava and V. K. Jain, *Sci. Total Environ.*, 2011, **409**, 4674–4680.
34. B. Giri, K. S. Patel, N. K. Jaiswal, S. Sharma, B. Ambade, W. Wang, S. L. M. Simonich and B. R. T. Simoneit, *Atmos. Res.*, 2013, **120–121**, 312–324.
35. K. K. Shandilya, M. Khare and A. B. Gupta, *Environ. Monit. Assess.*, 2007, **128**, 431–445.
36. A. K. Pathak, S. Yadav, P. Kumar and R. Kumar, *Sci. Total Environ.*, 2013, **443**, 662–672.
37. D. J. T. Smith, R. M. Harrison, L. Luhana, C. A. Pio, L. M. Castro, M. N. Tariq, S. Hayat and T. Quraishi, *Atmos. Environ.*, 1996, **30**, 4031–4040.
38. B. R. Simoneit, *Environ. Sci. Pollut. Res. Int.*, 1999, **6**, 159–169.
39. P. Pant, A. Shukla, S. D. Kohl, J. C. Chow, J. G. Watson and R. M. Harrison, *Atmos. Environ.*, 2015, **109**, 178–189.
40. S. K. Sharma, T. K. Mandal, M. Saxena, Rashmi, A. Sharma, A. Datta and T. Saud, *J. Atmos. Sol.-Terr. Phys.*, 2014, **113**, 10–22.
41. Z. Chowdhury, M. Zheng, J. J. Schauer, R. J. Sheesley, L. G. Salmon, G. R. Cass and A. G. Russell, *J. Geophys. Res. Atmos.*, 2007, **112**, D15303.
42. P. G. Satsangi and S. Yadav, *Int. J. Environ. Sci. Technol.*, 2014, **11**, 217–232.

43. A. B. Chelani, D. G. Gajghate, C. V. Chalapatirao and S. Devotta, *Bull. Environ. Contam. Toxicol.*, 2010, **85**, 22–27.
44. G. G. Pandit and V. D. Puranik, *Aerosol Air Qual. Res.*, 2008, **8**, 423–436.
45. B. Negi, S. Sadasivan and U. Mishra, *Atmos. Environ.*, 1967, **21**, 1259–1266.
46. H. K. Bandhu, S. Puri, M. L. Garg, B. Singh, J. S. Shahi, D. Mehta, E. Swietlicki, D. K. Dhawan, P. C. Mangal and N. Singh, *Nucl. Instrum. Methods Phys. Res., Sect. B*, 2000, **160**, 126–138.
47. S. Gummeneni, Y. Bin Yusup, M. Chavali and S. Z. Samadi, *Atmos. Res.*, 2011, **101**, 752–764.
48. R. Sunder Raman and S. Ramachandran, *Atmos. Environ.*, 2010, **44**, 1200–1208.
49. B. A. Begum, S. K. Biswas, E. Kim, P. K. Hopke and M. Khaliquzzaman, *J. Air Waste Manage. Assoc.*, 2005, **55**, 227–240.
50. B. A. Begum, P. K. Hopke and A. Markwitz, *Atmos. Pollut. Res.*, 2013, **4**, 75–86.
51. B. Srimuruganandam and S. M. S. Nagendra, *Chemosphere*, 2012, **88**, 120–130.
52. A. K. Azad and T. Kitada, *Atmos. Environ.*, 1998, **32**, 1991–2005.
53. R. Kumar, A. Elizabeth and A. G. Gawane, *Aerosol Sci. Technol.*, 2006, **40**, 477–489.
54. Central Pollution Control Board, *Central Pollution Control Board Environmental Standards: National Ambient Air Quality Standards*, 2011.
55. A. Srivastava and V. K. Jain, *Transp. Res. Part D Transp. Environ.*, 2008, **13**, 59–63.
56. I. Khanna, M. Khare and P. Gargava, *J. Geosci. Environ. Prot.*, 2015, **3**, 72–77.
57. L. T. Khemani, M. S. Naik, G. A. Momin, R. Kumar, R. N. Chatterjee, G. Singh and B. V. Ramana Murty, *J. Atmos. Chem.*, 1985, **2**, 273–285.
58. S. Balachandran, B. R. Meena and P. S. Khillare, *Environ. Int.*, 2000, **26**, 49–54.
59. P. Gargava, J. C. Chow, J. G. Watson and D. H. Lowenthal, *Aerosol Air Qual. Res.*, 2014, **14**, 1515–1526.
60. A. Srivastava, S. Gupta and V. K. Jain, *Atmos. Res.*, 2009, **92**, 88–99.
61. V. Shridhar, P. S. Khillare, T. Agarwal and S. Ray, *J. Hazard. Mater.*, 2010, **175**, 600–607.
62. A. B. Chelani, *Aerosol Air Qual. Res.*, 2013, **13**, 1768–1778.
63. A. K. Gupta, R. S. Patil and S. K. Gupta, *Environ. Monit. Assess.*, 2004, **95**, 295–309.
64. A. D. Bhanarkar, P. S. Rao, D. G. Gajghate and P. Nema, *Atmos. Environ.*, 2005, **39**, 3851–3864.
65. A. Srivastava, *Atmos. Environ.*, 2004, **38**, 6829–6843.
66. R. M. Tripathi, A. Vinod Kumar, S. T. Manikandan, S. Bhalke, T. N. Mahadevan and V. D. Puranik, *Atmos. Environ.*, 2004, **38**, 135–146.
67. S. Kar, J. P. Maity, A. C. Samal and S. C. Santra, *Environ. Monit. Assess.*, 2010, **168**, 561–574.

68. B. Srimuruganandam and S. M. S. Nagendra, *Sci. Total Environ.*, 2012, **433**, 8–19.
69. B. Srimuruganandam and S. M. S. Nagendra, *Sci. Total Environ.*, 2011, **409**, 3144–3157.
70. V. S. Chithra and S. M. S. Nagendra, *Atmos. Environ.*, 2013, **77**, 579–587.
71. K. Ram, M. M. Sarin and S. N. Tripathi, *J. Geophys. Res.*, 2010, **115**, D24313.
72. S. N. Behera, M. Sharma, O. Dikshit and S. P. Shukla, *Water, Air, Soil Pollut.*, 2010, **218**, 423–436.
73. A. Chakraborty and T. Gupta, *Int. J. Civ. Environ. Eng.*, 2009, **1**, 87–90.
74. S. P. Shukla and M. Sharma, *Environ. Eng. Sci.*, 2008, **25**, 849–862.
75. K. Ram, M. M. Sarin and S. N. Tripathi, *Environ. Sci. Technol.*, 2012, **46**, 686–695.
76. S. P. Shukla and M. Sharma, *Environ. Eng. Sci.*, 2008, **25**, 849–862.
77. A. Kulshrestha, P. G. Satsangi, J. Masih and A. Taneja, *Sci. Total Environ.*, 2009, **407**, 6196–6204.
78. R. Singh and B. S. Sharma, *Environ. Monit. Assess.*, 2012, **184**, 5945–5956.
79. M. Habil, D. D. Massey and A. Taneja, *Air Qual., Atmos. Health*, 2013, **6**, 575–587.
80. A. Satsangi, T. Pachauri, V. Singla, A. Lakhani and K. M. Kumari, *Aerosol Air Qual. Res.*, 2013, **13**, 1877–1889.
81. T. Pachauri, V. Singla, A. Satsangi, A. Lakhani and K. M. Kumari, *Aerosol Air Qual. Res.*, 2013, **13**, 523–536.
82. A. S. Pipal, A. Kulshrestha and A. Taneja, *Atmos. Environ.*, 2011, **45**, 3621–3630.
83. U. C. Kulshrestha, N. Kumar, A. Saxena, K. M. Kumari and S. S. Srivastava, *Environ. Monit. Assess.*, 1995, **34**, 1–11.
84. P. P. Parekh, B. Ghauri, Z. R. Siddiqi and L. Husain, *Atmos. Environ.*, 1987, **21**, 1267–1274.
85. M. H. Shah and N. Shaheen, *J. Hazard. Mater.*, 2007, **147**, 759–767.
86. M. Mansha, B. Ghauri, S. Rahman and A. Amman, *Sci. Total Environ.*, 2012, **425**, 176–183.
87. B. A. Begum, S. K. Biswas and P. K. Hopke, *Air Qual. Atmos. Health*, 2008, **1**, 125–133.
88. B. a Begum, S. K. Biswas and P. K. Hopke, *Aerosol Air Qual. Res.*, 2007, **7**, 446–468.
89. A. Salam, H. Bauer, K. Kassin, S. M. Ullah and H. Puxbaum, *Atmos. Environ.*, 2003, **37**, 2517–2528.
90. A. Salam, H. Bauer, K. Kassin, S. M. Ullah and H. Puxbaum, *J. Environ. Monit.*, 2003, **5**, 483–490.
91. A. Salam, T. Hossain, M. N. A. Siddique and A. M. S. Alam, *Air Qual. Atmos. Health*, 2008, **1**, 101–109.
92. P. Bonasoni, P. Laj, A. Marinoni, M. Sprenger, F. Angelini, J. Arduini, U. Bonafè, F. Calzolari, T. Colombo, S. Decesari, C. Di Biagio, A. G. di

Sarra, F. Evangelisti, R. Duchi, M. Facchini, S. Fuzzi, G. P. Gobbi, M. Maione, A. Panday, F. Roccato, K. Sellegri, H. Venzac, G. Verza, P. Villani, E. Vuillermoz and P. Cristofanelli, *Atmos. Chem. Phys.*, 2010, **10**, 7515–7531.

93. P. Bonasoni, P. Laj, F. Angelini, J. Arduini, U. Bonafè, F. Calzolari, P. Cristofanelli, S. Decesari, M. C. Facchini, S. Fuzzi, G. P. Gobbi, M. Maione, A. Marinoni, A. Petzold, F. Roccato, J. C. Roger, K. Sellegri, M. Sprenger, H. Venzac, G. P. Verza, P. Villani and E. Vuillermoz, *Sci. Total Environ.*, 2008, **391**, 252–261.

94. S. Decesari, M. C. Facchini, C. Carbone, L. Giulianelli, M. Rinaldi, E. Finessi, S. Fuzzi, A. Marinoni, P. Cristofanelli, R. Duchi, P. Bonasoni, E. Vuillermoz, J. Cozic, J. L. Jaffrezo and P. Laj, *Atmos. Chem. Phys.*, 2010, **9**, 25487–25522.

95. K. Sellegri, P. Laj, H. Venzac, J. Boulon, D. Picard, P. Villani, P. Bonasoni, a. Marinoni, P. Cristofanelli and E. Vuillermoz, *Atmos. Chem. Phys.*, 2010, **10**, 10679–10690.

96. A. B. Shrestha, C. P. Wake, J. E. Dibb, P. a. Mayewski, S. I. Whitlow, G. R. Carmichael and M. Ferm, *Atmos. Environ.*, 2000, **34**, 3349–3363.

97. E. E. Hindman and B. P. Upadhyay, *Atmos. Environ.*, 2002, **36**, 727–739.

98. C. M. Carrico, M. H. Bergin, A. B. Shrestha, J. E. Dibb, L. Gomes and J. M. Harris, *Atmos. Environ.*, 2003, **37**, 2811–2824.

99. R. K. Aryal, B. K. Lee, R. Karki, A. Gurung, B. Baral and S. H. Byeon, *J. Hazard. Mater.*, 2009, **168**, 732–738.

100. P. Shrestha, A. P. Barros and A. Khlystov, *Atmos. Chem. Phys.*, 2010, **10**, 11605–11621.

101. K. M. Shakya, L. D. Ziemba and R. J. Griffin, *Aerosol Air Qual. Res.*, 2010, **10**, 219–230.

102. P. Chen, S. Kang, C. Li, M. Rupakheti, F. Yan, Q. Li, Z. Ji, Q. Zhang, W. Luo and M. Sillanpää, *Sci. Total Environ.*, 2015, **538**, 86–92.

103. A. Marinoni, P. Cristofanelli, P. Laj, R. Duchi, F. Calzolari, S. Decesari, K. Sellegri, E. Vuillermoz, G. P. Verza, P. Villani and P. Bonasoni, *Atmos. Chem. Phys.*, 2010, **10**, 8551–8562.

104. S. Marcq, P. Laj, J. C. Roger, P. Villani, K. Sellegri, P. Bonasoni, A. Marinoni, P. Cristofanelli, G. P. Verza and M. Bergin, *Atmos. Chem. Phys.*, 2010, **10**, 5859–5872.

105. P. Ginot, M. Dumont, S. Lim, N. Patris, J.-D. Taupin, P. Wagnon, A. Gilbert, Y. Arnaud, A. Marinoni, P. Bonasoni and P. Laj, *Cryosphere*, 2014, **8**, 1479–1496.

106. M. C. S. Seneviratne, V. A. Waduge, L. Hadagiripathira, S. Sanjeewani, T. Attanayake, N. Jayaratne and P. K. Hopke, *Atmos. Pollut. Res.*, 2011, **2**, 207–212.

107. National Statistics Bureau Royal Government of Bhutan, 2013, 1–297.

108. T. Wangchuk, C. He, M. R. Dudzinska and L. Morawska, *Atmos. Environ.*, 2015, **113**, 151–158.

109. T. Wangchuk, L. D. Knibbs, C. He and L. Morawska, *Atmos. Environ.*, 2015, **118**, 98–106.

110. E. N. Kirillova, A. Andersson, R. J. Sheesley, M. Kruså, P. S. Praveen, K. Budhavant, P. D. Safai, P. S. P. Rao and Ö. Gustafsson, *J. Geophys. Res. Atmos.*, 2013, **118**, 614–626.
111. C. Bosch, A. Andersson, E. N. Kirillova, K. Budhavant, S. Tiwari, P. S. Praveen, L. M. Russell, N. D. Beres, V. Ramanathan and Ö. Gustafsson, *J. Geophys. Res. Atmos.*, 2014, **119**, 11743–11759.
112. K. Budhavant, A. Andersson, C. Bosch, M. Kruså, E. N. Kirillova, R. J. Sheesley, P. D. Safai, P. S. P. Rao and Ö. Gustafsson, *Environ. Res. Lett.*, 2015, **10**, 064004.
113. C. E. Corrigan, V. Ramanathan and J. J. Schauer, *J. Geophys. Res. Atmos.*, 2006, **111**, 1–15.
114. E. A. Stone, G. C. Lough, J. J. Schauer, P. S. Praveen, C. E. Corrigan and V. Ramanathan, *J. Geophys. Res.*, 2007, **112**, D22S23.
115. E. a. Stone, L. Yang, L. E. Yu and M. Rupakheti, *Atmos. Environ.*, 2012, **47**, 323–329.
116. O. Gustafsson, M. Kruså, Z. Zencak, R. J. Sheesley, L. Granat, E. Engström, P. S. Praveen, P. S. P. Rao, C. Leck and H. Rodhe, *Science*, 2009, **323**, 495–498.
117. K. Budhavant, A. Andersson, C. Bosch, M. Kruså, A. Murthaza and Ö. Gustafsson, *Sci. Total Environ.*, 2015, **536**, 72–78.

Health Effects of Airborne Particles in Relation to Composition, Size and Source

FRANK J. KELLY* AND JULIA C. FUSSELL

ABSTRACT

Uncertainty regarding the sources and chemical/physical properties of particular matter (PM) responsible for adverse health effects remains, despite momentous research efforts. The ambitious 10 year US NPACT initiative is deemed to have made a valuable contribution to the policy arena by demonstrating that no particle components can as yet be conclusively ruled out as *not* having an effect on public health. Upon focusing on studies conducted in different regions of world, within air sheds that vary with respect to a PM composition, size and source the very complex issue of differential toxicity is reaffirmed. Not only are individual PM characteristics and sources associated with certain effects in some locations and not in others but also, strengths of associations between effects and individual chemical components of the ambient mix vary from one effect to another. To further our understanding so that we can definitively conclude, or otherwise, that additional indicators have a role in protecting public health more effectively than the targeting total PM mass, comparison and synthesis of existing data through systematic reviews and quantitative meta-analysis must continue. Future studies should embrace refined modeling techniques and PM speciation data, enhance individual and population indoor/outdoor exposure, incorporate specific disease categories and better define susceptible individuals. Regionally specific studies are also needed to

---

*Corresponding author.

Issues in Environmental Science and Technology No. 42
Airborne Particulate Matter: Sources, Atmospheric Processes and Health
Edited by R.E. Hester, R.M. Harrison and X. Querol
© The Royal Society of Chemistry 2016
Published by the Royal Society of Chemistry, www.rsc.org

predict the impact of effective and sustainable control strategies. Owing to future population growth and increased ambient $PM_{2.5}$ concentrations, mortality from air pollution has been estimated to double by 2050—a statistic that calls for global air quality control measures, informed by sophisticated developments in research, and interpretation of the latter into region-specific clean air policies.

1   Introduction

Large parts of the world's population are exposed to indoor and/or outdoor airborne particulate matter (PM) that does not meet national standards let alone the health-based World Health Organisation (WHO) Air Quality Guidelines (AQG).[1-3] Ambient $PM_{2.5}$ (PM less than 2.5 µm in diameter) pollution is now the world's largest single environmental health risk, contributing to approximately 7 million deaths per year worldwide, predominantly in the Western Pacific and Southeast Asia (see Figure 1).[2,3] Particulate air pollution has thus been the focus of a global research effort for several decades and, as a consequence, the evidence base for the association between short- (day to day) and long-term or chronic (years to decades) exposure to PM and cardiopulmonary mortality is now fully acknowledged.[5-9] Furthermore, epidemiological studies in the developed world provide no evidence of a threshold within the studied range of ambient concentrations, with recent studies showing associations between $PM_{2.5}$ and mortality at levels below current annual national standards and WHO AQG.[10-12] A strong scientific consensus also exists that short-term and chronic exposures have cardiovascular effects and lead to respiratory disease, with evidence particularly strong for ischaemic heart disease (IHD),[13] reduced lung function and heightened severity of symptoms in individuals with asthma and chronic obstructive lung disease (COPD).[14] That ambient PM is having an impact on cardiorespiratory endpoints is evident from emerging data linking long-term exposure with atherosclerosis,[15-18] and a host of childhood respiratory conditions including an increased susceptibility to infection,[19-21] symptoms of asthma,[22] and low lung function.[23,24] A growing number of studies have also investigated its potential to negatively influence a broader number of disease outcomes. For example, evidence exists linking long-term exposure to $PM_{2.5}$ with adverse birth outcomes, whilst emerging data suggest possible effects on diabetes,[25-27] neurodevelopment,[28] and cognitive function.[29]

Health studies describing robust associations between ambient PM and a diverse list of adverse effects have over the years contributed to the WHO AGQ and various national air quality standards that use the mass concentration of $PM_{2.5}$ or PM_{10} (particles with a diameter of 10 µm or less) as the metric, thus treating all particles as equally toxic, without regard to their source and chemical composition. It is unlikely, however, that every component within the overall ambient PM mix is equally harmful to the exposed population. As a consequence, there has been an enormous drive to identify which

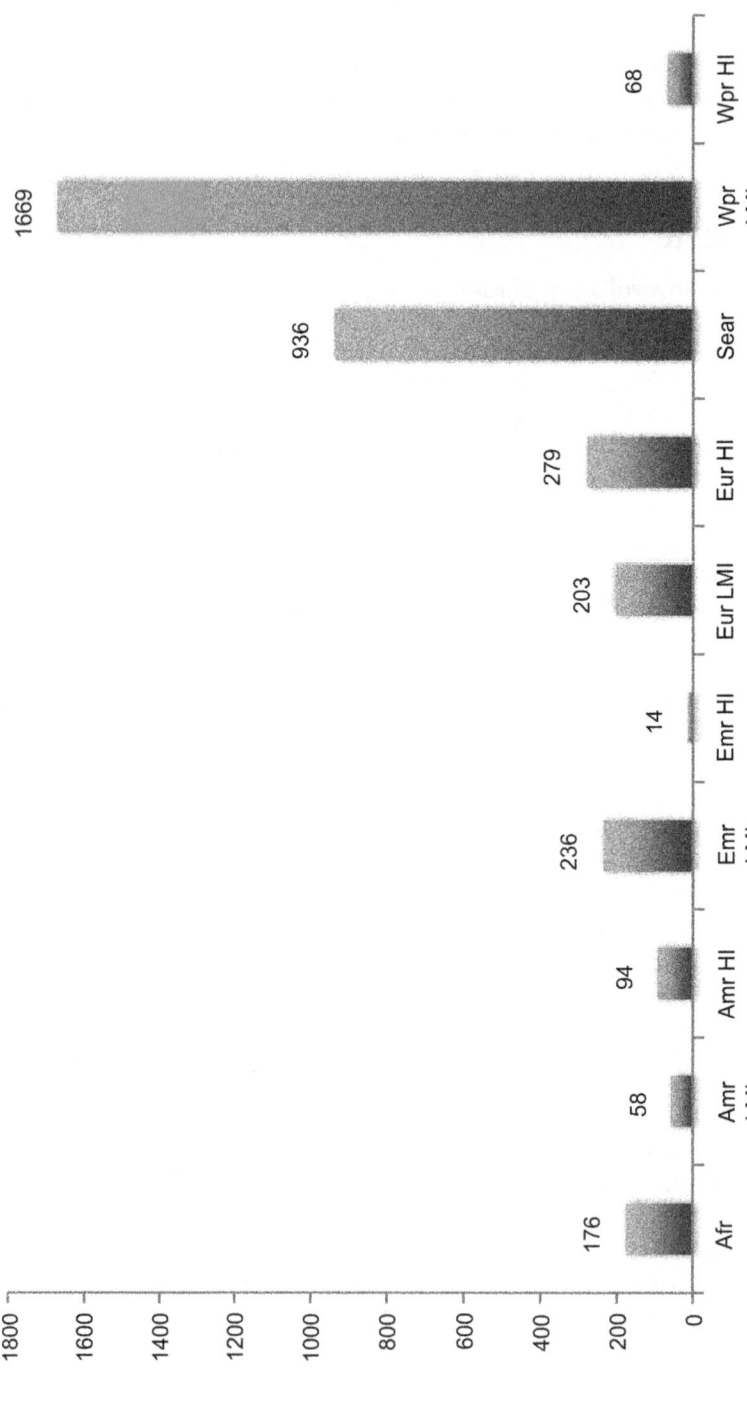

Figure 1   Total deaths ('000) attributable to ambient air pollution in 2012, by region. Amr: America, Afr: Africa, Emr: Eastern Mediterranean, Sear: Southeast Asia, Wpr: Western Pacific, LMI: low and middle income, HI: high income. Reprinted from World Health Organisation, Burden of disease from air pollution, 2014, http://www.who.int/phe/health_topics/outdoorair/databases/FINAL_HAP_AAP_BoD_24March2014.pdf. Accessed 12 May 2016.

component(s)/source(s) of ambient PM, and/or which of their physical and chemical characteristic(s) are most harmful to health. Information gleaned on the most toxic constituents could facilitate a re-appraisal of ambient air quality standards/guidelines and prioritisation of targeted PM control strategies to more effectively protect public health. However, whilst epidemiological and toxicological research findings have shown that PM mass comprises fractions and sources with varying types and degrees of health effects,[30] the question of differential PM toxicity represents one of the most challenging areas of environmental health research. Rather than constituting a single entity, ambient particulate pollution is a complex, heterogeneous mixture that can exist in the atmosphere as solids or liquids. Primary PM is directly emitted from source (road transport, biomass burning, industrial activities, land and sea) into the atmosphere, whilst secondary particles are formed within the atmosphere following chemical reactions with other pollutants. The mix includes emissions from man-made activities as well as natural sources. These particles vary not only in chemical composition, mass, size, number, shape and surface area, but also source, solubility and reactivity. Furthermore, the chemical constituents of PM can be internal or upon the particulate surface, with a core and a shell of different compositions. Attempts to identify PM-specific effects are complicated further since PM can vary in space and time as a consequence of atmospheric chemistry and weather conditions, as well as complex interactions with gaseous air pollutants (*e.g.* ozone [O_3] and nitrogen dioxide [NO_2]) that share biologically plausible associations with various health endpoints. In fact, some of the strongest epidemiological data on associations between PM and ill health highlights the complexities of unraveling the independent effects of pollutants owing to a common source and high degrees of inter-correlation.[31] There is no shortage of comprehensive reviews providing critical perspectives on the physical and chemical characteristics that determine PM toxicity.[32–35] So as not to duplicate previous work, this review adopts a different approach in an attempt to augment the existing literature. We begin with a summary of the latest scientific evidence and current conclusions, drawing upon international projects,[36,37] Health Effects Institute (HEI) research reports,[38–42] authoritative reviews,[13] and important individual publications on the contribution to PM-induced adverse health effects played by chemical constituents, size and source. Then, given the focus of this current issue on airborne PM in different regions of the world, we endeavour to review the differential toxicity of PM by summarising key findings from large-scale, where available, epidemiological efforts conducted in North America, Europe, China and India and, as such, within air sheds that vary with respect to PM composition, size and source.

2 Current Evidence on Differential Toxicity

2.1 Black Carbon and Organic Carbon

A large fraction of ambient PM in many areas is derived from combustion (burning of fossil fuels and biomass) and, as such, contains considerable amounts of carbon-laden material.[43] Added to this, combustion-derived particles are believed to cause more harm to health than those generated by other means.[44] Carbonaceous particles also originate from biological sources (*e.g.* viruses, pollen grains, plant debris) and contain a secondary organic fraction formed from the oxidation of natural and anthropogenic volatile organic compounds.

Non-volatile carbon is termed black carbon (BC) (when measured by testing optical absorption) or elemental carbon (EC) (when measured through thermo-optical methods). Both metrics are highly correlated and used interchangeably by epidemiologists as they are proxies for the same emissions. Systematic reviews into the health effects of BC particles have confirmed that sufficient epidemiological evidence exists to link daily variations in BC particles with all-cause and cardiovascular mortality and cardiopulmonary hospital admissions.[36,37,45,46] Evidence is also conclusive that annual BC exposure is associated with all-cause and cardiopulmonary mortality, whilst Janssen and colleagues estimated that the increase in life expectancy associated with a hypothetical traffic abatement measure was 4–9 times higher when expressed in BC particles compared with an equivalent change in $PM_{2.5}$ mass.[45] Toxicological research has not however identified distinct mechanistic effects, suggesting that BC may not be a directly toxic component of fine PM but rather, may operate as a universal carrier of combustion-derived chemicals (semi-volatile organic fractions, transition metals) of varying toxicity not only to the lungs but also to the systemic circulation and beyond.[36] Nevertheless, considering that short-term studies show that health effect associations with BC particles were more robust than those with $PM_{2.5}$ or PM_{10}, although BC particles may not constitute a causal agent, it is the opinion that they may well constitute a relevant indicator of harmful particulate substances from primary traffic-related, wood smoke or other combustion particles compared to undifferentiated PM mass.[36]

Organic carbon (OC) is a highly complex and heterogeneous mixture, made from combustion and non-combustion, anthropogenic and natural, and primary and secondary compounds. It can also co-exist with BC and, as a consequence, it is a huge challenge to identify the potential toxicity of specific OC constituents. While it is possible to measure BC particles with sufficient accuracy, the measurement of primary and secondary organic aerosols requires further refinement before being introduced into larger epidemiological studies. Increasing amounts of data are however accumulating on associations between total OC and a variety of health effects including short-term perturbations in both respiratory[47,48] and cardiovascular endpoints,[49–53] as well as long-term effects on all-cause, cardiopulmonary and ischaemic heart disease mortality.[54] In two of the cohort studies included in the HEI's systematic and multidisciplinary National Particle Component Toxicity (NPACT) initiative, OC was associated with effects in one, whilst in another it was not.[40,41]

2.2   Metals

Metallic constituents of ambient PM, especially those within the fine fraction, are often cited as those most likely to exert toxicity. They originate from metallurgical processes, oil combustion and non-exhaust emissions (engine/brake/tyre wear, road dust). Epidemiological studies invariably focus on transition metals, such as iron (Fe), vanadium (V), nickel (Ni), chromium (Cr) and copper (Cu), based on their potential to produce reactive oxygen species in biological tissues.[50,54–60] Non-redox active metals (zinc [Zn], aluminium [Al] and lead [Pb]) have also deserved attention owing to their ability to influence the toxic effects of transition metals, either exacerbating or lessening the production of free radicals. Whilst study design inconsistencies (variability in hypotheses, locations, number of metals included in analyses, lower limits of detection, outcomes studied) hinder the ability to compare the relative toxicity of different elements, recent evidence, including output from the NPACT project, points to associations between Ni and V and cardiovascular hospital admissions.[39,50,59] A recent qualitative review and meta-analysis of 19 time series studies that included metals as a particle metric suggests that, in the main, Zn, Ni and V are associated with increased mortality.[46]

2.3   Inorganic Secondary Aerosols

The major components in inorganic secondary aerosols are sulfates and nitrates, formed in the atmosphere from the oxidation of sulfur dioxide (SO_2) and NO_2 into acids, which are then neutralised by atmospheric ammonia derived largely from agricultural sources. These aerosols may exist as pure aqueous or solid particles, or as a surface layer on other solid particles such as carbon. Alternatively, the surface of a sulfate particle may be covered by potentially toxic components such as transition metals or organics. Owing to their high solubility, abundance in biological systems and historically a lack of toxicological data, the general consensus has been that alone these aerosols have little or no biological potency at environmentally relevant levels.[61] Despite this, epidemiological evidence continues to accumulate on the short-term effects of sulfate on cardiovascular mortality as well as respiratory and cardiovascular hospital admissions.[50,51] Data have also emerged on associations between daily increments in ambient sulfate and physiological changes to the cardiovasculature, namely ventricular arrhythmias,[62] and markers of endothelial dysfunction.[63] Reasons for the apparent lack of coherence in results between toxicology and epidemiology are unclear. One explanation is that the secondary inorganic aerosol interacts with other ambient pollutant particles and/or gases so as to increase the toxicity of the latter and/or form new products with potential damage to health.[64–67]

2.4   Size

The behaviour of particles in the atmosphere and within the human respiratory system is primarily dictated by their physical properties, which have a strong dependence on size. Particle diameter varies from a few nanometres to tens of micrometres, constituting ultrafine (UFP; diameter of 0.1 μm or less [$PM_{0.1}$]), fine ($PM_{2.5}$) and coarse (diameter between 2.5 and 10 μm [$PM_{2.5-10}$]) particles. As a rough guide, particles less than about 2.5 μm penetrate to the alveoli and terminal bronchioles, larger particles of up to 10 μm will deposit predominantly in the primary bronchi and much larger ones (up to 100 μm) will deposit in the nasopharynx. PM_{10} are sometimes termed "thoracic" particles in that they can escape the initial defenses of the nose and throat and penetrate beyond the larynx to deposit along the airways in the thorax.

2.4.1   Ultrafine Particles.   The greatest number of particles belong to the ultrafine size range and whilst they do not contribute large quantities to PM mass, they dominate the surface area of particulate pollution. They originate largely from primary combustion (vehicles, industry, power stations) emissions and secondary particles produced by gas-to-particle conversion processes. As such, they are dominated by OC, EC, sulfates and nitrates. UFPs possess properties that have led scientists to hypothesise that this size fraction may present a particular threat to health relative to fine or coarse PM. Apart from the relationship between particle diameter and penetration within the lung and possible onward passage across the air–blood barrier, their large surface area creates a high capacity to adsorb and deliver toxic chemicals. In addition, finer particles have a greater propensity to penetrate indoor environments, be suspended in the atmosphere for longer periods and be transported over large distances. However, whilst toxicological studies have clearly demonstrated the potential of this size fraction to adopt differential patterns of deposition, clearance and translocation,[68] epidemiological data only provides suggestive rather than strong and consistent evidence of independent adverse effects of UFPs.[69,70] Indeed, a recent HEI review panel charged with increasing our understanding of UFP toxicity noted 'The current evidence does not support a conclusion that exposure to UFPs alone can account in substantial ways for the adverse effects of $PM_{2.5}$.[42]

2.4.2   Coarse Particulate Matter.   Coarse PM include mechanically generated particles (including metals from brake pad and disc and tyre), soil, dust, large salt particles from sea spray, pollen, mould and spores. Accumulating evidence from epidemiological studies suggests that short-term exposures to this size fraction are associated with effects on adverse cardiorespiratory health, including premature mortality.[71–75] Overall conclusions, following several systematic reviews and assessments, vary as to whether effect estimates are higher or lower than those for $PM_{2.5}$.[32] Long-term studies of coarse PM are limited and report no or little evidence of an effect on mortality or cardiovascular health.[76,77] Output from toxicological studies comparing coarse and fine PM report that coarse particles can be as toxic as $PM_{2.5}$ on a mass basis.[78,79] However, the data are scarce and difficult to interpret owing inherent differences in the inhalable dose and deposition efficiency of varying size fractions.

2.5 Source

Ambient PM is derived from a variety of sources (coal combustion, shipping, power generation, the metal industry, biomass combustion, desert dust episodes and road transport), each containing a mixture of pollutants that are associated with different types of health effects.[37,80] Studies that have adopted accurate characterisation of exposure of subjects to different local pollutants are fairly consistent in identifying emissions from vehicle traffic (and in some locations, oil combustion and other industrial activities) as causing more harm than coal-fired power emissions or secondary organic aerosols.[33] A critical review of the literature on the health effects of traffic-related air pollution concluded that sufficient evidence had accumulated to support a causal relationship between exposure to traffic-related air pollution and exacerbation of asthma. Further evidence was found to be suggestive of a causal relationship with onset of childhood asthma, non-asthma respiratory symptoms, impaired lung function, total and cardiovascular mortality, and cardiovascular morbidity.[38] The PM components from road traffic include engine emissions, plus non-exhaust sources that are often characterised by elevated concentrations of transition metals Cu, antimony, Zn, Fe) from brake wear, tyre abrasion and dust from road surfaces. The largest single source is derived from diesel exhaust (DE). Indeed, owing to the increased domestic market penetration of diesel engines, the fuel powering the majority of our buses and taxis in many industrialised countries and the fact that they generate up to 100 times more particles than comparable gasoline engines with 3-way catalytic convertors,[81] diesel exhaust particles (DEPs) contribute significantly to the air shed in many of the world's largest cities. DEPs have also been shown to have substantial toxicological capacity, facilitated by size (80% by mass have an aerodynamic diameter of <1 μm) and surface chemistry characteristics. DEPs have a highly adsorptive carbon core that acts as a vector for the delivery, deep into the lung, of redox active metals, polyaromatic hydrocarbons and quinones. It is therefore not surprising that significant health impacts appear to be associated with proximity to roads carrying a high proportion of diesel powered heavy and light good vehicles.[82,83] In 2012, the International Agency for Research on Cancer classified particulates in diesel fumes as carcinogenic to humans based on sufficient evidence that it is linked to an increased risk of lung cancer, as well as limited evidence linking it to an increased risk of bladder cancer.[84] Although most studies into the toxicity and health consequences of roadside PM have focused on DEPs, the non-exhaust sources are attracting increasing interest.[85] Indeed, contributions from brake/tyre wear and road surface abrasion, particularly in individuals living near major roads, will assume greater importance with progressive reductions in exhaust emissions. Of note, their potential to elicit health effects is largely ignored at the regulatory level despite links with cardiopulmonary toxicity.[86–89]

3   Overall Conclusions on Differential Toxicity

Whilst the scientific evidence of PM toxicity is rapidly expanding, reaffirming and strengthening previously reported associations as well as revealing new health outcomes, the issue of differential toxicity is complicated and clearly not fully resolved. The current database of experimental and epidemiologic studies does not allow individual characteristics or sources to be definitely identified as being closely related to specific health effects.[37,76,90] It appears that the strengths of associations between effects and individual chemical components of the ambient aerosol vary from effect to effect and that the situation is further complicated by components being associated with certain effects in some locations, but not in others. The ambitious US NPACT studies—that used coordinated toxicology, epidemiology, and exposure research to examine and compare the toxicity of PM components, gases, and sources—concluded that 'the studies do not provide compelling evidence that any specific source, component, or size class of PM may be excluded as a possible contributor to PM toxicity.'[90] Another illustration of the variety of results reported in the literature is a recent systematic review of the findings of epidemiology, controlled human exposure and toxicology studies that used apportionment methods to relate sources of PM with human health outcomes.[91] Among the 29 studies reviewed, soil, sea salt, local SO_2, secondary sulfate, motor vehicle emissions, coal burning, wood smoke, biomass combustion, Cu smelter emissions, residual oil combustion, and incinerator emissions were found to be associated with health outcomes. Another noteworthy quote on the subject from Krall and colleagues states 'Associations with a given $PM_{2.5}$ chemical component should be considered as potentially indicative of associations with another component or set of components with similar sources'.[92] In fact, it is a belief of various commentators in the field that the literature suggests that various complex mixtures may be involved, and that the capability of PM to induce disease may be the result of multiple components acting on different physiological mechanisms.[93,94]

4   Global Variation in the Composition and Toxicity of Particulate Matter

Ambient PM is location- and season-dependent, producing individual air sheds in different parts of the world. These are dictated by location-specific characteristics (*e.g.* prevalence of diesel vehicles, population density, meteorology/climate), which in turn are reflected by a region's economic and cultural make-up and ultimately result in different exposure patterns. It is therefore entirely feasible that a variation in both the type and degree of

PM-induced health effects around the world reflects a given local PM pollution signature. For example, effect estimates of long-term exposure to $PM_{2.5}$ and PM_{10} vary among different geographical locations, with elevated risks in some areas of Europe and the United States,[95-99] but no or little association in others.[77,95-100] There are several possible reasons that may explain this and, in addition to a true variation in PM composition, one must also consider methodological differences, chance findings and population susceptibility. Differential toxicity of PM sources and mixtures was a suggested explanation for global heterogeneity in effect estimates in a systematic review and meta-analysis of 110 peer-reviewed time series studies assessing evidence for associations between $PM_{2.5}$ and daily mortality and hospital admissions for a range of diseases.[101] To determine consistency of evidence worldwide (rather than for comparison), the researchers stratified their analyses by WHO geographical region (America region A, America region B, Europe, Western Pacific region A, Western Pacific region B, SE Asia region D). For all-age, all-cause mortality, a 10 μg m^{-3} increment in $PM_{2.5}$ was associated with a 1.04% (95% CI 0.52, 1.56) increase in the risk of death but worldwide there was a substantial regional variation (statistic for heterogeneity $[I^2] = 93\%$) from 0.25% (Western Pacific region B) to 2.08% (America region B) (see Table 1). Positive associations with $PM_{2.5}$ were also heterogeneous between WHO regions for cardiovascular deaths and increases in risk of admission for respiratory diseases. The researchers discussed that the reasons for the heterogeneity require explanation and further work, as they may be relevant to the formulation of policy measures—an issue that will be revisited in our discussion. So as to tie in with the other contributions to this issue that focus on findings of source apportionment studies conducted in North America, Southern Europe, North Western Europe, China and the Indian Sub-Continent, we now review the health effects of airborne particles (in relation wherever possible to composition, size and source) in these regions.

4.1   North America

Studies initially done by Pope and colleagues and more recently by Correia and colleagues have provided evidence that a decline in $PM_{2.5}$ air pollution in the United States during the 1980s and 1990s, and again from 2000 to 2007, was associated with increased life expectancy.[102,103] Of particular relevance to our review, the chemical constituents of $PM_{2.5}$ that were the main drivers of the observed associations have now been examined.[104] The study estimated associations between temporal changes in seven major components of $PM_{2.5}$ (sulfate, nitrate, EC, OC, Al, sodium [Na], silicon [Si]) that make up over 80% of $PM_{2.5}$ mass, and temporal changes in life expectancy in 95 urban and non-urban US counties that had adequate chemical components of $PM_{2.5}$ mass data across all seasons between 2002 and 2007. In multiple pollutant models, in which each individual pollutant was also adjusted for the other six constituents, it was found that a reduction

Table 1 Meta-analysis results for all-age, all-cause mortality and cause-specific mortality by WHO region.[101,a,b]

WHO region	All[c] (SC/MC)	Selected[d] (SC/MC)	RE (95% CI)[e]	I[2,f] (%)
All cause				
AMR A	13/12	5/2	0.94 (0.73 to 1.16)	93
AMR B	4/0	2/0	2.08 (1.60 to 2.56)	
EUR A	12/1	9/1	1.23 (0.45 to 2.01)	
WPR A	0/1	0/1	0.90 (−0.70 to 2.53)	
WPR B	5/0	3/0	0.25 (0.06 to 0.44)	
Summary[g]	—	4/4	1.04 (0.52 to 1.56)	
Cardiovascular				
AMR A	10/3	6/1	0.84 (0.47 to 1.20)	76
AMR B	3/0	2/0	0.13 (−0.71 to 0.98)	
EUR A	6/1	6/1	2.26 (1.23 to 3.29)	
WPR B	4/0	2/0	0.56 (0.31 to 0.81)	
Summary[g]	—	4/2	0.84 (0.41 to 1.28)	
Respiratory				
AMR A	4/5	4/1	1.39 (0.62 to 2.16)	0
AMR B	3/0	2/0	0.88 (−1.88 to 3.71)	
EUR A	7/0	7/0	3.81 (0.57 to 7.16)	
WPR B	4/0	2/0	1.49 (0.04 to 2.96)	
Summary[g]	—	4/1	1.51 (1.01 to 2.01)	

[a]AMR, Region of the Americas; EUR, European Region; WPR/SEAR, South East Asian Region.
[b]Reprinted with permission. © (2014). BMJ. All rights reserved.
[c]Numbers of single-city (SC)/multicity (MC) estimates available from all studies.
[d]Numbers of single-city (SC)/multicity (MC) estimates selected for meta-analysis (see estimate selection protocol in Methods section).
[e]Random effects summary estimate (95% CI) per 10 µg m^{-3}.
[f]I^2 statistic for heterogeneity.
[g]Estimate numbers for 'Summary' refers to the number of pooled (from single-city estimates) and multicity estimates used to calculate the overall summary estimate across WHO regions.

in sulfates (power generation, industrial combustion, agriculture) was associated with an increase in life expectancy in all 95 counties (see Figure 2). In addition, decreases in ammonium (fertilizers, waste disposal) and Na (industry and agriculture) were associated with increases in life expectancy in the non-urban counties alone. That exposure to sulfate affects mortality is consistent with the results of a retrospective analysis by Pope and colleagues that found a 2.5 µg m^{-3} decrease in sulfate particle concentrations observed during an eight month smelters strike was associated with a small (1.5–4.0%) reduction in the number of deaths in the region.[105] The long-term effects of several PM$_{2.5}$ components (EC, OC, sulfates, nitrate, Si, Fe, K, Zn) on mortality were also examined in the California Teachers Study where no statistically significant associations between all-cause mortality and any of the measured constituents were reported.[54,106] However, significant hazard ratios (HR) for cardiopulmonary mortality were found for sulfates (HR = 1.14, 95% CI 1.01, 1.29 per interquartile range [IQR], 2.2 µg m^{-3}), nitrate (HR = 1.11, 95% CI 1.03, 1.19 per IQR 3.2 µg m^{-3}) and Si (HR = 1.05,

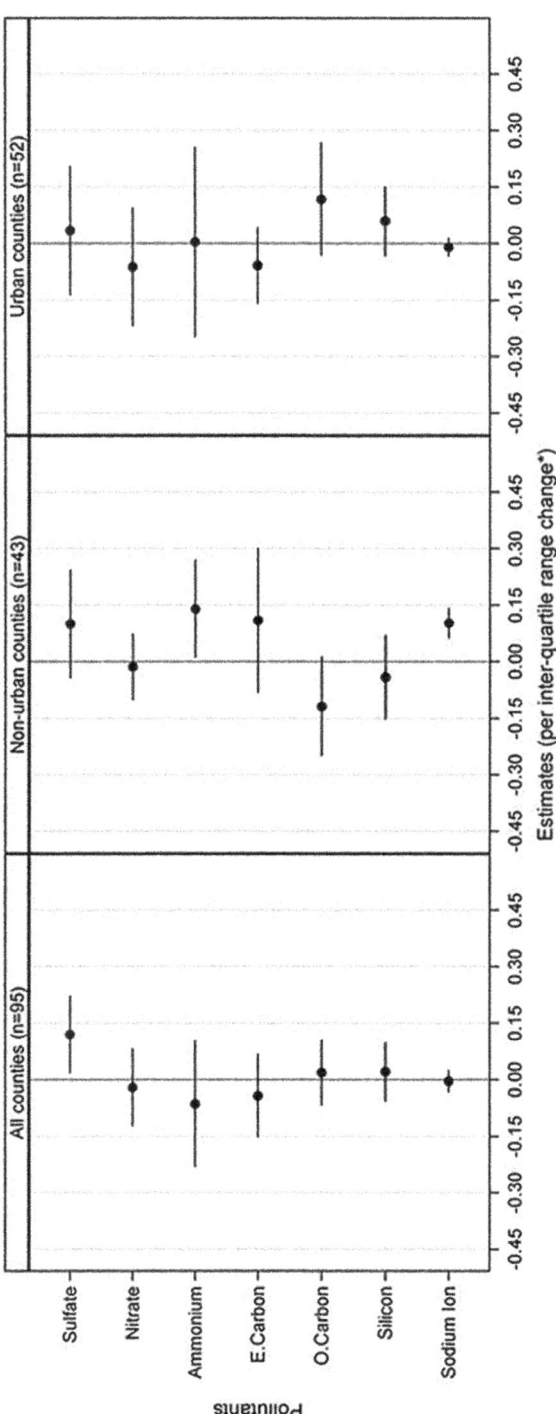

Inter-quartile range (μg/m3): PM2.5=2.20, Sulfate=0.32, Nitrate=0.33, Ammonium=0.24, Elemental Carbon=0.16, Organic Carbon=1.12, Silicon=0.04, and Sodium=0.04.

Figure 2 Point estimates and 95% confidence intervals for the association across counties between a decrease (between 2002 and 2007) in an IQR μg m^{-3} of the PM$_{2.5}$ total mass or in each of the seven chemical components and an increase (between 2002 and 2007) in life expectancy (years). Results are reported for all of the 95 counties (left panel), 43 non-urban counties (middle panel) and 52 urban counties (right panel). These estimates were obtained under an 'adjusted multi-pollutant model' that includes changes in all the seven components as exposure predictors and the additional confounders that represent changes in demographics, socioeconomic, population, lung cancer, and COPD variables.
Reprinted with permission from F. Dominici, Y. Wang, A. W. Correia, M. Ezzati, C. A. Pope III and D. W. Dockery, *Epidemiology*, 2015, **26**, 556.[104]

95% CI 1.00, 1.10 per IQR 0.03 μg m^{-3}), whereas HRs were elevated but non-significant for K, Fe and Zn. All components were associated with mortality from IHD, whereas none of the pollutants were associated with pulmonary mortality.[106] A higher risk for all-cause mortality in association with sulfur (S) (HR = 1.09 per 200 ng m^{-3}) as well as for IHD mortality in association with increased levels of PM$_{2.5}$ Fe, Ni, and Zn has been reported in the NPACT Study.[41] NPACT also used data from the Women's Health Initiative–Observational Study cohort to examine the association with cardiovascular mortality and (fatal and nonfatal) cardiovascular events.[40] Here an association was not found with all cardiovascular deaths and S (HR = 1.01, 95% CI: 0.92, 1.12 per 0.25 μg m^{-3}), but the association with cardiovascular events was statistically significant (HR = 1.09; 95% CI 1.05, 1.14 per 0.25 μg m^{-3}).

Among recent national level time series analyses, Krall and colleagues estimated short-term associations between non-accidental mortality and seven (EC, OC, sulfate, nitrate, ammonium, Si, Na) major chemical constituents of PM$_{2.5}$ across 72 urban counties from 2000 to 2005.[92] They found that daily changes in EC, OC, and Si and Na were associated with a daily change in mortality with IQR increases associated with estimated increases in mortality of 0.39%, 0.22%, 0.17% and 0.16%, respectively. In another nationwide time-series study, Dai and colleagues estimated the effects of PM$_{2.5}$ species on daily mortality (covering approximately 4.5 million deaths) across 75 US cities from 2000 to 2006.[107] They found that Si and calcium (Ca) (crustal elements), often elevated near roads, were associated with all-cause mortality whilst sulfates (reflecting exposure to power plant emissions as well as regional pollution) were also associated with higher risks for all-cause but particularly respiratory mortality. Apart from national initiatives, local and regional time-series studies have also reported estimated effects for PM$_{2.5}$ constituents on mortality but the cited responsible component for particle toxicity (EC, OC, sulfates, nitrate, Si, ammonium) varies among these investigations.[50,60,108–110] With respect to short-term effects on morbidity, a national study on cardiovascular and respiratory hospitalisations reported significant associations with EC and OC,[111] whilst others have found that PM$_{2.5}$ mass higher in either EC, Ni, V or Ni and OC significantly increased its effect on hospital admissions.[39,55,59] Regional studies have reported positive associations of sulfates and water soluble metal concentrations with preterm birth,[112] and EC, Si, Al, Zn, V, Ni with low birth weight.[113]

A differential effect of particle size (PM$_{2.5}$ versus PM$_{2.5-10}$) on mortality (all cause, cardiovascular disease, myocardial infarction, stroke, respiratory disease) has been investigated in a national, multi-city time series study.[53] For PM$_{2.5}$, significant associations were found with all analysed causes of death, with the highest effect for stroke (1.78% increase) and respiratory mortality (1.68% increase) for a 10 μg m^{-3} increase in 2 day averaged PM$_{2.5}$ (and a 0.98% increase [95% CI 0.75, 1.22] in total mortality). For an equivalent increase in PM$_{2.5-10}$, significant but smaller increases were found whereby effect sizes per unit of mass were about half those for PM$_{2.5}$.

4.2   Europe

The discussion on evidence of the differential toxicity of PM in European countries will primarily focus on two multi-city air pollution epidemiological efforts, namely the European Study of Cohorts for Air Pollution Effects (ESCAPE) and the MED-PARTICLES project.

The multi-centre ESCAPE project is a European Union wide (conducted across 36 study areas) study investigating the health effects of long-term exposure to particle components (see Figure 3). The initiative has utilised existing cohorts to which a harmonized exposure assessment, derived from land use regression (LUR models), is applied. Analysis of natural-cause mortality data from the 19 European cohorts revealed positive HRs for residential exposure of almost all of the eight *a priori*-selected PM components representing major sources (Cu, Fe, K, Ni, S, Si, V, Zn) and statistically significant for $PM_{2.5}$ S (HR $= 1.14$, 95% CI 1.06, 1.23 per 200 ng m^{-3}). In a two-pollutant model, the association with $PM_{2.5}$ S was robust to

Figure 3   Cohort Locations in which elements were measured.
Reproduced from *Environ. Health Perspect.*[114]

Table 2 Results from random-effects meta-analyses from single-pollutant and two-pollutant models for association with natural-cause mortality (using main model 3) [HR (95% CI)].[a,b]

Exposure	Adjusted for	Single-pollutant	Two-pollutant
$PM_{2.5}$ S[c]	$PM_{2.5}$	1.15 (1.06, 1.24)	1.13 (1.03, 1.24)
$PM_{2.5}$ S[d]	PM_{10} Ni	1.14 (1.04, 1.25)	1.14 (1.04, 1.25)
$PM_{2.5}$ S[e]	$PM_{2.5}$ Si	1.14 (1.05, 1.23)	1.13 (1.04, 1.22)
$PM_{2.5}$ S[f]	PM_{10} K	1.16 (1.06, 1.27)	1.15 (1.05, 1.26)
$PM_{2.5}$[c]	$PM_{2.5}$ S	1.07 (1.02, 1.13)	1.02 (0.96, 1.09)
PM_{210} Ni[d]	$PM_{2.5}$ S	1.09 (0.98, 1.22)	1.06 (0.95, 1.18)
$PM_{2.5}$ Si[d]	$PM_{2.5}$ S	1.09 (0.98, 1.21)	1.08 (0.97, 1.20)
$PM_{2.5}$ K[f]	$PM_{2.5}$ S	1.09 (0.99, 1.08)	1.02 (0.98, 1.06)

[a]Limited to studies for which correlation between two pollutants was <0.7. HRs are presented for the following increments: 200 ng m^{-3} $PM_{2.5}$ S, 5 µg m^{-3} $PM_{2.5}$, 2 ng m^{-3} PM_{10} Ni, 100 ng m^{-3} $PM_{2.5}$ Si, 100 ng m^{-3} PM_{10} K.
[b]Reproduced from *Environ. Health Perspect.*[114]
[c]FINRISK and SAPALDIA not included.
[d]HUBRO, SALIA, and SAPALDIA not included.
[e]HUBRO, SAPALDIA, and EPIC-Athens not included.
[f]FINRISK, HURBO, and SIDRIA-Rome not included.

adjustment for $PM_{2.5}$ mass, whereas the association with $PM_{2.5}$ mass was reduced (see Table 2).[114] The estimated effect of $PM_{2.5}$ S on natural-cause mortality within the study population corresponds to an HR of 1.24 (95% CI 1.10, 1.41) per 1 µg m^{-3} sulfate. In contrast, no statistically significant association between long-term exposure to 8 elemental constituents and total cardiovascular mortality was found.[115] Cited reasons for this null effect included errors in exposure measurement, exposure contrast, studied outcomes, the application of a current LUR to assess historic exposures or a true absence of an effect. Elevated risks, though not statistically significant, were found for cardiovascular disease mortality and $PM_{2.5}$ Si and S in $PM_{2.5}$ and PM_{10}, respectively.

Since systemic inflammation is a plausible biological mechanism behind the effects of long-term exposure to ambient PM on mortality from cardiovascular and respiratory disorders, Hampel and colleagues investigated possible links between long-term residential exposure to elemental components of PM and inflammatory blood markers.[116] A 5 ng m^{-3} increase in $PM_{2.5}$ Cu and a 500 ng m^{-3} increase in PM_{10} Fe were associated with a 6.3% (95% CI 0.7, 12.3) and 3.6% (95% CI 0.3, 7.1) increase in high-sensitivity C-reactive protein (hsCRP), respectively. Associations between the elemental components and fibrinogen levels were slightly weaker, with a 10 ng m^{-3} increase in $PM_{2.5}$ Zn associated with a 1.2% (95% CI 0.1, 2.4) increase in this inflammatory marker. Despite the fact that the detected associations were small, that Cu and Fe are correlated and might represent more than one source, and that the findings are estimated rather than measured, these results may be indicative that exposure to transition metals within ambient PM is related to chronic systemic inflammation providing a link to long-term health effects of PM.

ESCAPE has also conducted a multi-centre study using 5 European birth cohorts to investigate associations between the eight fine PM elements and lung function at the age of 6 or 8 years.[117] Estimations of PM_{10} Ni and PM_{10} S at the residential address were associated with decreases in forced expiratory volume in the first second of 1.6% (95% CI 0.4, 2.7) and 2.3% (95% CI 0.1, 4.6) per increase in exposure of 2 and 200 ng m^{-3}, respectively. Associations remained unchanged after adjusting for PM mass; however, associations were not evident in all cohorts and heterogeneity of associations with exposure to PM components was larger than for exposure to PM mass. When assessing associations between the PM elements at the birth address and early-life pneumonia in seven birth cohort studies, substantial within- and between-cohort variability in element concentrations was reported.[118] In an adjusted meta-analysis, pneumonia was weakly associated with Zn derived from PM_{10} (OR: 1.47 [95% CI 0.99, 2.18] per 20 ng m^{-3} increase) but associations with the other elements were consistently observed.

MED-PARTICLES has characterised particulate pollution and its health effects across fourteen cities in Mediterranean area (three cities in Spain, one in France, eight in Italy, two in Greece). It was initiated owing to scarce epidemiological data in a region that differs from other European countries in having mild meteorological conditions, encouraging outdoor activities throughout the year and experiencing enhanced formation of secondary pollutants owing to intense solar radiation. Southern Europe is also densely populated, with intense traffic congestion and a high proportion of diesel cars. The region also sees elevated sea traffic and a high frequency of Saharan dust episodes, especially during the spring and summer.[119] The project's investigation on short-term effects of PM_{10}, $PM_{2.5}$ and $PM_{2.5-10}$ on mortality outcomes was conducted in ten European Mediterranean metropolitan areas.[120] A 10 μg m^{-2} increase in $PM_{2.5}$ on the same day and the previous day was associated with a 0.55% (95% CI 0.27, 0.84) increase in all-cause mortality and a 0.57% (95% CI 0.07, 1.08) increase in cardiovascular mortality. For respiratory mortality, an increase of 1.91% (95% CI 0.71, 3.12) in association with cumulative exposures over the same day and 5 previous days suggested more delayed effects. Positive associations with $PM_{2.5-10}$ were not significant, whilst associations with PM_{10} seemed to be driven by $PM_{2.5}$. The stronger impact of fine particles is consistent with US findings.[53] The specific causes of death associated have also been investigated.[121] Positive associations between a 10 μg m^{-3} increase in $PM_{2.5}$, with deaths due to diabetes (1.23%, 95% CI −1.63, 4.17) after immediate exposure and cardiac causes (1.33%, 95% CI 0.27, 2.40), LRTI (1.37%, 95% CI −1.94, 4.78) and COPD (2.53%, 95% CI −0.01, 5.14) after prolonged exposure were reported. The only statistically significant association was with cardiac deaths. To a lesser extent, this size fraction was also associated with deaths due to cerebrovascular diseases. Coarse particles displayed positive, if not statistically significant, associations with mortality due to diabetes and cardiac causes that showed greater variance depending on exposure period, co-pollutant and seasonality adjustment. Associations between short-term effects of fine and coarse particles on hospitalisation in eight Southern European cities were similar across PM size metrics, although stronger albeit more heterogeneous for respiratory admissions.[122] Specifically, increases of 10 µg m^{-3} in $PM_{2.5}$, 6.3 µg m^{-3} in $PM_{2.5-10}$, and 14.4 µg m^{-3} in PM_{10} (lag 0–1 days) were associated with increases in cardiovascular admissions of 0.51% (95% CI 0.12, 0.90), 0.46% (95% CI 0.10, 0.82), and 0.53% (95% CI 0.06, 1.00), respectively. For respiratory hospitalisations, associations ranged from 1.15% (95% CI 0.21, 2.11) for PM_{10} to 1.36% (95% CI 0.23, 2.49) for $PM_{2.5}$ (lag 0–5 days). All MED-PARTICLE studies investigating health effects of different size fractions have found stronger associations during warmer months, consistent with US findings.[53] This may be explained by increased exposure owing to increased outdoor activities and open windows or by a more toxic air pollution mixture (not least owing to desert dust episodes, discussed below) compared to winter periods.[123,124] In addition to particle size, MED-PARTICLES has investigated short-term effects of PM constituents on daily hospitalisations and mortality.[125,126] In five southern European cities, the majority of the elements studied (EC, sulfate, Si, Ca, Fe, Zn, Cu, Ti, Mn, V, Ni) showed increased percent changes in cardiovascular and/or respiratory hospitalisations. The percent increase by one IQR change ranged from 0.69% to 3.29% and highest effects were found for EC and Ni. After adjustment for total PM levels, only associations for Mn, Zn and Ni remained significant. Although positive associations were identified for total mortality (Fe and Ti), cardiovascular mortality (EC and Mg) and respiratory mortality (sulfate, K), for which the greatest effect was observed, the patterns were less clear. In a separate study conducted in two southern Mediterranean cities (Barcelona, Spain and Athens, Greece), associations were found between daily concentrations of BC and total, cardiovascular and respiratory mortality.[126] Notably, the pooled estimate for all-cause mortality, of 2.5% (95% CI 0.7, 4.3) per µg m^{-3}, is about 10 times that of $PM_{2.5}$ on a per unit basis.

Outside of the ESCAPE and MED-PARTICLES projects, research in London has examined the urban atmospheric environment, patterns of exposure of the population to traffic pollution and their relationships to total and cause-specific mortality and hospital admissions.[127,128] Specifically, workers assembled a database of a large number of daily, measured and modelled pollutants, characterising air pollution in the city between 2011 and 2012. Then, based upon analyses of temporal and seasonal patterns and correlations between the metrics, knowledge of local emission sources and reference to the existing literature they selected, *a priori*, markers of traffic pollution (NO_x [general traffic pollution]; EC/BC and carbon monoxide [markers of diesel and petrol exhaust respectively]; Cu and Zn [brake/tyre wear]; Al [mineral dust]) for daily mortality and hospital admissions time series analyses. Positive associations were found for EC adjusted for particle mass (2.66% [95% CI, 0.11, 5.28]) and BC adjusted for particle mass (2.72% [95% CI, 0.09, 5.42]) and respiratory mortality per IQR.[127] Consistent associations were also found, again predominately with

EC/BC (0.56% to 1.65% increase per IQR change) and also to a lesser degree with carbon monoxide and PM_{10} Al content, with cardiovascular (for those below 65 years) and paediatric respiratory hospital admissions.[128] The specific effect estimates for hospitalisations were variable to the adjustment of other pollutants, but remained largely consistent in direction. All mortality and hospital admission associations were higher during the warmer months of the year.

Southern Europe is frequently affected by dust outbreaks from the Sahara desert, which globally produces approximately half of the annual atmospheric mineral dust.[129] The Mediterranean countries are particularly vulnerable to these episodes owing not only to proximity to the Sahara (as well as to the Arabian Peninsula—another major source of dust particles) but also to low precipitation in the Mediterranean basin that promotes a long residence time of atmospheric PM. As a consequence, the region has recurrent air quality problems, with Saharan dust events contributing to exceedances of the airborne PM_{10} daily EU limit of 50 µg m^{-3}.

Several epidemiological studies conducted in single cities of Southern Europe have evaluated whether outbreaks of Saharan dust exacerbate the effects of man-made pollution ($PM_{2.5}$, $PM_{2.5-10}$, PM_{10}) on daily mortality and cardiopulmonary hospitalisation. A review of the literature by Karanasiou and colleagues concluded that although there is no association of $PM_{2.5}$ with total or cause specific daily mortality during Saharan dust intrusions, the results for PM_{10} and $PM_{2.5-10}$ were deemed to be more inconsistent;[130] some published studies found no evidence of increased mortality due to PM_{10} during Sahara dust days while other studies did report higher associations of PM_{10} and $PM_{2.5-10}$ with mortality and morbidity on dust *versus* non-dust days.[131–135] More recent studies provide additional support for the negative health effects of dust episodes in Southern Europe. Findings not only report that local PM_{10} mass has stronger effects on cardiovascular and respiratory mortality during Saharan dust outbreaks than on other days,[136,137] but also that PM_{10} originating from desert sources was positively associated with cardiorespiratory mortality and hospital admissions (see Table 3).[138] Potential reasons to explain these discrepancies are multi-fold: differences in PM size fraction studied, ways used to calculate $PM_{2.5-10}$, time periods examined and methodologies used to define an episodic day. In addition, inconsistent results may relate to the different sections of the Sahara desert that affected the studied areas and transport processes and air mass routes followed to western and eastern Mediterranean countries. Indeed, different regions of the Sahara desert have individual mineralogical properties that may affect the PM toxicological potential.[139] Moreover, the air path followed will influence the chemical composition of the mineral dust as it combines with anthropogenic emissions of industrial regions (North African countries), thereby becoming carriers of industrial pollutants.[140] The absorption and interaction of such air pollutants throughout its long journey across many industrial cities has been cited as a possible explanation behind the

Table 3 Estimated percent increase (95% CI) in risk of mortality and hospital admissions associated with 10 μg m^{-3} increase in non-desert and desert PM$_{10}$.[a,b,c]

Outcome	Lag	PM$_{10}$			Non-desert PM$_{10}$			Desert PM$_{10}$		
		% IR (95% CI)	I^2	Het. p value	% IR (95% CI)	I^2	Het. p value	% IR (95% CI)	I^2	Het. p value
Mortality										
Natural	0–1	0.51 (0.27, 0.75)	22	0.23	0.55 (0.24, 0.87)	32	0.15	0.65 (0.24, 1.06)	0	0.75
CV	0–5	0.66 (−0.02, 1.34)	40	0.08	0.49 (−0.31, 1.29)	46	0.04	1.10 (0.16, 2.06)	0	0.77
Resp	0–5	2.01 (0.92, 3.12)	31	0.15	2.46 (0.96, 3.98)	41	0.07	1.28 (−0.42, 3.01)	0	1.00
Hospital admissions										
CV, age 15 +	0–1	0.29 (0.00, 0.58)	41	0.10	0.37 (−0.04, 0.78)	59	0.02	0.32 (−0.24, 0.89)	0	0.50
Resp, age 15 +	0–5	0.69 (0.20, 1.19)	32	0.17	0.62 (0.03, 1.21)	21	0.27	0.70 (−0.45, 1.87)	10	0.35
Resp, age 0–14	0–5	1.66 (0.93, 2.39)	0	0.47	1.82 (0.77, 2.88)	24	0.23	2.47 (0.22, 4.77)	9	0.36

[a]I^2 statistics represents the amount (%) of heterogeneity among city-specific estimates; heterogeneity p-value is calculated from the X^2 test on the Cochran's Q statistic.

[b]The estimates for non-desert and desert PM$_{10}$ are obtained from two-pollutant models adjusted for the other PM source in turn, while the estimates for PM$_{10}$ are from single-pollutant models.

[c]Reproduced from *Environ. Health Perspect.*[138]

increased toxicity of airborne particles during dust storms.[140] Saharan dust has also been shown to carry large amounts of biogenic factors, such as fungi, bacteria and viruses,[141,142] and may also contain toxic allergens or irritants,[143,144] offering further plausible explanations of the increased health risk associated with dust exposure. The condensation of secondary components from gaseous precursors onto the surface of dust particles may also trigger heightened toxicity of local PM during Saharan dust days.[137] Furthermore, as has been demonstrated over Barcelona, increasing intensity of dust outbreaks is associated with a progressive lowering of the mixing layer height (MLH). This enhances the accumulation of local and regional atmospheric pollution from all sources and importantly, as the MLH reduced, the risk of mortality associated with the same concentration of particulate matter was found to increase.[145]

4.3   Western Pacific and Southeast Asia

Although air pollution health research has largely been dominated by work conducted in developed countries of the western world, studies now emerging from regions such as Western Pacific and Southeast Asia confirm in broad terms the spectrum of adverse health effects reported in the West. They also illustrate that exposure and its subsequent toll on health, is the most severe (by far) in these regions.[3] It is a disheartening fact that the developing regions with the worst air quality in 1990 invariably experienced further deteriorations of air quality by 2010, with emissions of several pollutants predicted to increase further.[114,146] In stark contrast, western regions with better air quality in 1990 have experienced ongoing improvements owing to the implementation of clean air policies.[102,103] In 2012, around 5.1 million people died as a result of ambient and indoor air pollution exposure in the WHO Western Pacific and Southeast Asia Regions, where China and India are the main contributors, respectively.

Of the 4.3 million global deaths each year attributed to indoor or household air pollution, 3.3 million occur in Western Pacific and Southeast Asia.[3] After high blood pressure, tobacco and alcohol, household cooking with solid fuels is the greatest health risk in the world,[147] with more people dying from smoke from home fires than from malaria, tuberculosis and HIV/AIDS combined. In agreement, recent projections using a global atmospheric chemistry model found that emissions from residential energy use such as heating and cooking, prevalent in India and China, have the largest impact on premature mortality globally.[4] The heating of homes and daily cooking over traditional, often inefficient, coal, and biomass (wood, animal dung, agricultural waste and charcoal) stoves in unventilated spaces creates mean 24 hour indoor concentrations of $PM_{2.5}$ of 337 μg m^{-3}—ten times the WHO indoor air quality guidelines.[148] Although this practice is more prevalent in rural households, it remains commonplace in many cities and in fact globally, because the world's population is rising so quickly, the number of people using (around 3 billion) solid fuels for cooking today is greater than at any time in human history.[149,150] Estimations for 2000 were that residential coal and biomass use contributed to 86% of BC emissions in both India and China whilst for OC, the proportion was 96% in India and 97% in China.[151] China, India, and many other countries with emerging economies also persistently experience some of the highest outdoor air pollutant levels globally.[152] Moreover, this is a feature not confined to densely populated megacities, but also a problem in smaller cities and interurban areas owing to unprecedented increases in vehicle traffic and industrial activity as well as dependence on coal for electricity.[153,154]

4.3.1 China. Research in China on the adverse health effects of specific types or sources of ambient particles is limited due to a lack of monitoring data and few city-wide mortality/morbidity reporting systems. Instead, the great majority of studies focused on the relation between daily mortality and total mass of a given size of particles.[155–157] The relative effects of ambient PM have however been the focus of some single-city, time series analyses, undertaken in highly populous and developing megacities with extraordinarily high concentrations of $PM_{2.5}$ and excessive traffic densities. Li and co-workers explored the short-term effects of different PM size fractions in Beijing.[158] After controlling for potential confounding factors, a 10 μg m^{-3} increase in $PM_{2.5}$ levels was associated with a 0.65% (95% CI 0.29, 0.80), 0.63% (95% CI 0.25, 0.83), and 1.38% (95% CI 0.51, 1.71) increase in non-accidental mortality, respiratory mortality, and circulatory mortality, respectively. A 10 μg m^{-3} increase in PM_{10} was similarly associated with increases of 0.15% (95% CI 0.04, 0.22%), 0.08% (95% CI 0.01, 0.18%), and 0.44% (95% CI, 0.12, 0.63), whilst no significant effect of $PM_{2.5-10}$ on daily mortality outcomes was reported.[158] Kan and colleagues (2007) also found significant associations of daily mortality with $PM_{2.5}$, but not with $PM_{2.5-10}$ in Shanghai. A 10 μg m^{-3} increase in the 2 day moving average (lag01) concentration of $PM_{2.5}$ corresponded to a 0.36% (95% CI 0.1, 0.61), 0.41% (95% CI 0.01, 0.82) and 0.95% (95% CI 0.16, 1.73) increase of total, cardiovascular and respiratory mortality, respectively.[156] An investigation in Beijing into the role of key chemical constituents reported that for a 3 day lag, the non-accident mortality increased by 1.52, 0.19, 1.03, 0.56, 0.42, and 0.32% for particulate matter $PM_{2.5}$, PM_{10}, K, sulfate, Ca, and nitrate based on IQR ranges of 36.00, 64.00, 0.41, 8.75, 1.43, and 2.24 μg m^{-3}, respectively.[159] The authors concluded that the most important $PM_{2.5}$ chemical components and source categories affecting public health in Beijing were secondary aerosols transformed from coal combustion, biomass burning and resuspended road dust. $PM_{2.5}$ mass, as well as several constituents originating from fossil fuel combustion, were also associated with daily total non-accidental and cardiopulmonary mortality in Xi'an, another heavily polluted Chinese city (see Figure 4).[110] Ammonium, nitrate, chloride ion, chloride, OC, EC, and Ni all maintained significant positive associations with mortality outcomes after adjusting for $PM_{2.5}$, whilst nitrate demonstrated stronger

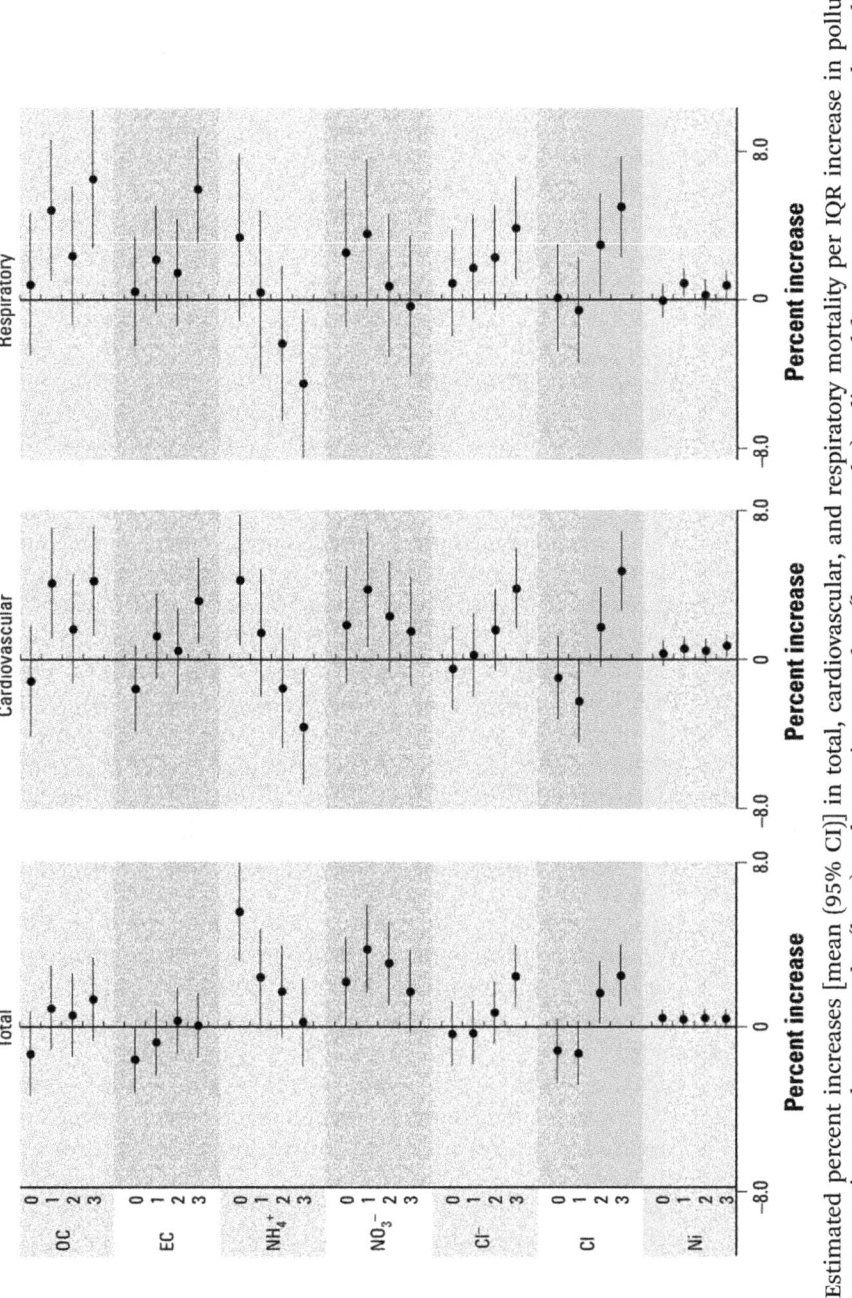

Figure 4   Estimated percent increases [mean (95% CI)] in total, cardiovascular, and respiratory mortality per IQR increase in pollutant concentrations on the current day (lag 0) or the previous 1–3 days (lags 1, 2, and 3), adjusted for $PM_{2.5}$ mass, temporal trend, day of the week, temperature, relative humidity, and SO_2 and NO_2 concentrations. Reproduced from *Environ. Health Perspect.*[110]

associations with total and cardiovascular mortality than $PM_{2.5}$ mass. For a 1 day lag, IQR increases in $PM_{2.5}$ mass and nitrate (114.9 and 15.4 µg m^{-3}, respectively) were associated with 1.8% (95% CI, 0.8, 2.8) and 3.8% (95% CI 1.7, 5.9) increases in total mortality, respectively. Independent associations have been reported between BC and mortality outcomes in Shanghai, where major sources are road traffic, coal burning, shipping emissions and industrial activity.[160,161] An IQR increase (2.7 µg m^{-3}) of BC corresponded to a 2.3% (95% CI 0.6, 4.1), 3.2% (95% CI 0.6, 5.7), and 0.6% (95% CI −4.5, 5.7) increase in total, cardiovascular and respiratory mortality, respectively.[162] When adjusted for $PM_{2.5}$, the effects increased and remained statistically significant but in contrast, statistically significant associations between $PM_{2.5}$ and daily mortality did not remain when adjusted for BC. Studies have also linked short-term exposure to several PM constituents with increased hospital admissions. In Hong Kong, Pun and colleagues have linked PM_{10} from vehicle exhaust, regional combustion, residual oil and secondary particles and aged sea salt with increased cause-specific emergency hospital admissions,[163,164] whilst findings in Shanghai provide evidence to link daily emergency room visits with $PM_{2.5}$ constituents from fossil fuel combustion, such as OC and EC.[165] The Healthy Volunteer Natural Relocation (HVNR) study followed 40 healthy university students in Beijing who relocated from a suburban to an urban campus with different air pollution concentrations, presenting the opportunity to examine the relationship between $PM_{2.5}$ chemical constituents and health effects. Potential links between many different components and a host of endpoints including pulmonary function, blood pressure, and circulatory biomarkers for inflammation, coagulation and oxidative stress were identified.[166–169]

Of note, associations per 10 µg m^{-3} of $PM_{2.5}$ with total, cardiovascular and respiratory mortality in Shanghai (0.36%, 0.41%, 0.95% increases, respectively) and Xi'an (0.20%, 0.3% and 0.4% increases, respectively) are lower compared with associations reported for populations in developed countries.[110,156] The multi-city time series analysis in the US found that a 10 µg m^{-3} increase in $PM_{2.5}$ was associated with a 0.98% increase in total mortality, a 0.85% increase in cardiovascular mortality, and a 1.68% increase in respiratory mortality,[53] whilst a Canadian meta-analysis reported a 1.2% increase of $PM_{2.5}$ and total mortality.[170] Qiao and colleagues also reported smaller (0.16%) effect estimates for Shanghai emergency room visits per unit increase of $PM_{2.5}$ compared with studies of $PM_{2.5}$ and emergency-room visit in developed countries where $PM_{2.5}$ levels are much lower.[165,171,172] Explanations for the potentially weaker associations between health outcomes and unit increases in air pollution exposures in China than in developed countries are multi-fold. In addition to a possible flatter exposure–response curve at higher concentrations,[173] differences in the composition and toxicity of PM, local concentrations of PM, transport of pollutants from outdoor air to indoor and population characteristics (*e.g.* sensitivity and age structure) all have the potential to affect risk per unit increase of ambient pollutants.

4.3.2 India. A lack of research precludes an overview of studies examining how different types and sources of ambient particles may impact public health in India. Unsurprisingly, however, a large number of studies have examined the health effects associated with biomass use for cooking since it represents such a significant factor behind the daunting air pollution challenges that India faces. Despite heterogeneity among published studies, the use of proxy measurements of exposure and the failure to adjust for important confounders (socioeconomic status, smoking, fuel types and age), available evidence suggests that exposure to solid fuel smoke presents a three-fold risk for COPD in adults and a two-fold risk for acute lower respiratory infections in children.[174] The vulnerability of women and young children is a consequence of spending more time at home undertaking the cooking and heating and results in an incremental and ongoing (three to seven hours a day through a person's lifetime) inhalation of small particles.[175] Comparative assessments of respiratory morbidity among women have shown a doubled risk of airflow obstruction in those cooking with biomass fuels compared to liquified petroleum gas (LPG) users. Furthermore, that different indices of lung function were significantly lower by 16–25 years suggests that biomass smoke during childhood may impair lung growth.[176] In connection with these effects, an interesting hypothesis has been put forth linking chronic PM exposure to a developing lung during childhood with an ensuing decline in lung function and increased vulnerability to COPD in adulthood in a chain of events involving oxidative stress, impaired innate immunity, and subsequent bacterial infection.[177] The two most prevalent biomass fuels in India are wood and cow dung and to explore the mechanisms underlying the pulmonary response, toxicology studies have compared the relative impacts of PM collected from homes in rural India during cooking with biomass in animal models and *in vitro* systems.[178–180] Acute exposure to mice resulted in pronounced neutrophilic inflammation, proinflammatory cytokine production, airway resistance, and hyperresponsiveness, which were significantly higher in cow dung PM-exposed mice (see Figure 5). Sub-chronic exposures, however, induced eosinophilic inflammation, PM-specific antibody responses, and alveolar destruction, which was greatest in wood PM-exposed mice.[180] It is perhaps not surprising that the oxidative potential of inhalable particles emitted during the burning of mixed biomass and wood has also been demonstrated to be considerable and in fact greater than that of diesel, gasoline and residual oil fly ash particles.[178,179]

Whilst adverse respiratory health effects of biomass exposure are well established, much less is known about its impact on cardiovascular diseases, which are a major and growing contributor to mortality and morbidity in South Asia.[147,148] Cross-sectional studies are however providing evidence to suggest positive associations between the use of traditional biomass fuels and hypertension, cardiovascular disease mortality and several relevant biomarkers (P-selectin, platelet aggregation, depleted superoxide dismutase,

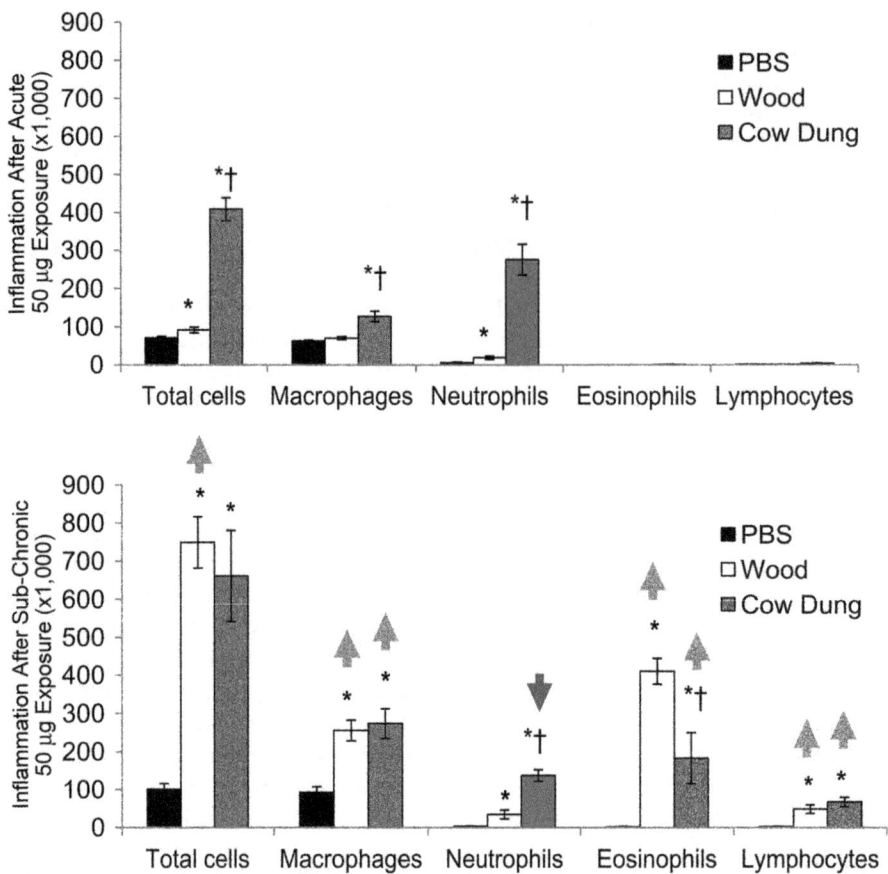

Figure 5   Subchronic exposure to biomass PM elicits innate and adaptive immune responses. BAL inflammation 24 hours after a single (top panel) or 8-week (bottom panel) exposure to 50 μg biomass PM. Arrows represent significant changes between acute and subchronic exposures.[180]
Reprinted with permission of the American Thoracic Society. Copyright © 2015 American Thoracic Society. Sussan *et al.*, 2015, Impact of Biomass Fuels On Pregnancy Outcomes In Central East India, *Am. J. Resp. Cell Mol. Biol.*, **50**, 538–548. The *American Journal of Respiratory Cell and Molecular Biology* is an official journal of the American Thoracic Society.

oxidised low-density lipoprotein, anti-cardiolipin antibodies) linked with adverse cardiovascular outcomes.[181] In a healthy volunteer study, acute exposure to dilute wood smoke (mean particle concentration of 314 ± 38 μg m^{-3} for 3 hours with intermittent exercise) as a model of exposure to biomass combustion was associated with an immediate increase in central arterial stiffness and a simultaneous reduction in heart rate variability.[182] These findings are consistent with previous studies reporting vascular impairment after exposure to other particulate air pollutants such as diesel exhaust.[183] Again, in line with data emerging elsewhere in the world linking long-term exposure to ambient $PM_{2.5}$ with adverse birth outcomes, a growing body of literature suggests that smoke from biomass burning is linked to stillbirth occurrence, preterm delivery, reduced birth weight and neonatal death[184–186] (see Table 4).

This section has deliberately focused on the health effects of biomass smoke owing to its enormous toll on morbidity and mortality; however, emissions from coal based power stations are also having a devastating impact. India is one of the world's largest producers of energy, with production planned to increase in the coming years to meet growing demands. Despite this, the country lacks effective pollutant controls such that it is estimated that emissions from the country's 111 coal fired power plants resulted in 80 000 to 115 000 premature deaths, 20.9 million cases of asthma attacks, 90 0000 emergency room visits, and 160 million days of restricted activity in 2011–2012.[187] The ubiquitous and interrelated nature of household and ambient particulate air pollution in India, plus the commonality of health effects associated with these sources, has led to a call for cohort studies that focus on how joint exposure to ambient and household air pollution may interact to produce long-term adverse health effects. The Tamil Nadu Air Pollution and Health Effects study, likened to ESCAPE in Europe and the American Cancer Society, Multi-Ethnic Study of Atherosclerosis and Air Pollution & Nurses Health in USA, thus aims to understand the complexity of the rural–urban continuums in exposure. It will also focus on the differential impacts on maternal/child *versus* adult outcomes in an attempt to facilitate health impact assessments and inform policy actions in India and other parts of the world where such exposures coexist.[188]

5 Discussion

The uncertainty regarding which chemical and physical properties of PM are responsible for adverse health effects remains, despite momentous research efforts. Indeed, the ambitious 10 year US NPACT initiative is deemed to have made a valuable contribution to the policy arena by demonstrating that no particle components can as yet be conclusively ruled out as *not* having an effect on public health. The remit of this review was therefore not to make a definitive ranking of individual PM properties in terms of biological activity and, in turn, a threat to public health. Rather, in an attempt to augment the wealth of existing text on this subject, we chose to examine the differential health risks of the ambient aerosol, by focusing on recent, large-scale (where available) epidemiological projects conducted in different regions of world, within air sheds that vary with respect to PM composition, size and source.

Data from the large North American and European cohort studies support consistent associations between $PM_{2.5}$ sulfate and mortality, whilst links with nitrate, Si, Fe, Ni and Zn have also been cited.[54,104,113] The list of components associated with mortality endpoints in time series studies include EC, OC, Si, Na, Ca and sulfates in North America and BC in

Table 4 Pregnancy outcomes comparing women cooking with wood *versus* gas, unadjusted and adjusted analysis. [a,b]

	Birth weight		Small for gestational age (birth weight <10%)	Stillbirth [c]	Preterm delivery (<37 weeks)
	Mean birth weight (g)	Low birth weight (<2500 g)			
Gas	2736 ± 409	48/253 (19.0%)	20/244 (8.2%)	0/253 (0%)	33/245 (13.5%)
Wood	2623 ± 429	286/1199 (23.9%)	71/1190 (6.0%)	50/1255 (4.0%)	390/1194 (32.7%)
Effect size (wood *versus* gas), unadjusted (95% CI)	− 112 (−170, −55)	1.33 (0.95, 1.88)	0.71 (0.42, 1.19)	2.71 (0.99, ∞)	3.11 (2.12, 4.59)
Adjusted effect size (95% CI)	− 14 (−93, 66) [d]	0.95 (0.58, 1.57) [e]	0.53 (0.23, 1.19) [f]	2.06 (0.08, ∞) [g]	2.29 (1.24, 4.21) [h]

[a] Reproduced from *Environ. Health.* [186]

[b] For birth weight outcomes, analyses limited to singleton live births with recorded birth weights. For stillbirths, all singleton births included. For preterm delivery, analyses limited to singleton live births with recorded Ballard examinations. Values represent n (%) or mean ± STD.

[c] ORs and lower confidence interval estimated using exact logistic regression.

[d] Adjusted for propensity score, cohort (Jharkhand *versus* Chhattisgarh), maternal age, body mass index, squared body mass index, gravidity, hypertension at delivery, hemoglobin at delivery, and time spent cooking.

[e] Adjusted for propensity score, cohort (Jharkhand *versus* Chhattisgarh), maternal age, body mass index, gravidity, hemoglobin at delivery, and time spent cooking.

[f] Adjusted for propensity score, cohort (Jharkhand *versus* Chhattisgarh), gravidity, hemoglobin at delivery, fever in week prior to delivery and time spent cooking.

[g] **Adjusted for propensity score alone.**

[h] Adjusted for propensity score, cohort (Jharkhand *versus* Chhattisgarh), maternal age, body mass index, gravidity, hypertension at delivery, hemoglobin at delivery, presence of windows, and time spent cooking.

Europe.[92,107,126,127] In rapidly developing China with high concentrations of particulate air pollution, constituents originating from fossil fuel combustion, biomass burning, secondary particles transformed from coal combustion and resuspended road dust have all been identified as influencing daily mortality outcomes.[110,159,162] With respect to size, a stronger impact of $PM_{2.5}$ (*versus* PM_{10} and $PM_{2.5-10}$) on mortality appears consistent in North America, Europe and China,[53,119,156,158] whilst there is more evidence to suggest that the exacerbation by Saharan dust of PM-induced health effects manifests through PM_{10} rather than the fine fraction.[136–138] PM components most consistently linked to short-term increments in hospitalisations include EC, OC, Ni, V and Zn.[39,59,111,125,128,163–165] This brief summary reaffirms that the complex issue of differential toxicity is clearly not fully resolved. Not only are some individual PM characteristics and sources associated with certain effects in some locations and not in others, but also strengths of associations between effects and individual chemical components of the ambient mix vary from one effect to another. To progress this area so that we can definitively conclude, or otherwise, that additional indicators have a role in protecting public health more effectively than the targeting total PM mass, a number of steps are needed.

Comparing and synthesising existing data through systematic reviews and quantitative meta-analysis must continue. Difficulties will however remain in analysing the relative effects of particle characteristics owing to a host of limitations within a given set of studies under examination: range of components (*e.g.* metals) investigated, unavailability of daily measurements or those below detection limits, exposure measurement error, inconsistency in statistical methods and a narrow range of health end points and/or particle components in the same population limiting within-study comparison. Future studies should rely on better refined modeling techniques and PM speciation data. With the aim of overcoming inevitable challenges, we should also begin to design studies to explore the role of mixtures of pollutants that may result in independent, synergistic or even antagonistic effects. This may be particularly relevant in the light of persistent epidemiological evidence linking PM sulfates with health effects,[145,189] recent findings of the significant the contribution that agricultural sources have on mortality in the eastern United States and Europe,[4] and work showing the association of MLH with daily mortality on Saharan dust outbreaks.[145] There is also a need to enhance individual and population indoor/outdoor exposure estimates—an area in which approaches are expanding rapidly owing to the availability of remote sensing tools, innovative personal samplers, and emerging mobile apps and technology. This is particularly relevant for emissions such as EC and other markers of vehicle emissions that vary locally. In addition, studies should incorporate specific disease categories to further our understanding of the populations at risk and on this point we must better define susceptible individuals based on our ever-growing knowledge of epigenetic regulation of genes, genotype–phenotype associations and inheritable gene–environment interactions.

Since PM composition and sources differ dramatically by location, regionally specific studies, particularly in developing countries, are needed to understand the impact of the local environment and atmospheric transformations of air pollutants within specific socioeconomic groups, and thus predict the impact of effective and sustainable control strategies. It is assumed that the exposure–response functions for particulate pollution determined *via* Western epidemiological studies incorporated with sophisticated relative risk models are relevant to the less developed world.[190] There have however been rather few studies to cast light on this question, and as yet we do not have sufficient evidence to identify the shape of the concentration–response function relating long-term exposure to air pollution and mortality at the high ambient concentrations observed in many places in the world. As a consequence of a combination of population growth and increased ambient $PM_{2.5}$ concentrations, mortality from air pollution has been estimated to double by 2050.[4] Predictions also support the need for global air quality control measures if such projections are to be avoided.[191] This will undoubtedly necessitate sophisticated developments in research to better our understanding of the relative toxicity of particles from various sources and interpretation and synthesis of the latter into effective and region specific clean air policies.

Acknowledgements

This work was supported by the UK's cross-research council Environmental Exposures and Health Initiative (NE/I007806/1) and the National Institute for Health Research Health Protection Research Unit (NIHR HPRU) in Health Impact of Environmental Hazards at King's College London in partnership with Public Health England (PHE). The views expressed are those of the authors and not necessarily those of the NHS, the NIHR, the Department of Health or Public Health England.

References

1. World Health Organisation (WHO) Regional Office for Europe, Air quality guidelines global update 2005: Particulate matter, ozone, nitrogen dioxide and sulfur dioxide. Copenhagen, 2006, http://www.euro.who.int/__data/assets/pdf_file/0005/78638/E90038.pdf. Accessed 12 May 2016.
2. European Environment Agency (EEA), Air quality in Europe, 2014, http://www.eea.europa.eu/publications/air-quality-in-europe-2014. Accessed 21 September 2015.
3. World Health Organisation (WHO), Burden of disease from air pollution, 2014, http://www.who.int/phe/health_topics/outdoorair/databases/FINAL_HAP_AAP_BoD_24March2014.pdf. Accessed 12 May 2016.
4. J. Lelieveld, J. S. Evans, M. Fnais, D. Giannadaki and A. Pozzer, *Nature*, 2015, **525**, 367–371.

5. J. Schwartz and D. W. Dockery, *Am. Rev. Respir. Dis.*, 1992, **145**, 600.
6. D. W. Dockery, C. A. Pope, X. Xu, J. D. Spengler, J. H. Ware, M. E. Fay, B. G. Ferris and F. E. Speizer, *N. Engl. J. Med.*, 1993, **329**, 1753.
7. B. Ostro, L. Chestnut, N. Vichit-Vadakan and A. Laixuthai, *J. Air Waste Manage. Assoc.*, 1999, **49**, 100.
8. K. Katsouyanni, G. Touloumi, E. Samoli, A. Gryparis, A. Le Tertre, Y. Monopolis *et al.*, *Epidemiology*, 2001, **12**, 521.
9. L. Filleul, V. Rondeau, S. Vandentorren, N. Le Moual, A. Cantagrel, I. Annesi-Maesano *et al.*, *Occup. Environ. Med.*, 2005, **62**, 453.
10. D. L. Crouse, P. A. Peters, A. van Donkelaar, M. S. Goldberg, P. J. Villeneuve, O. Brion *et al.*, *Environ. Health Perspect.*, 2012, **120**, 708.
11. R. Beelen, O. Raaschou-Nielsen, M. Stafoggia, Z. J. Andersen, G. Weinmayr, B. Hoffmann *et al.*, *Lancet*, 2014, **383**, 785.
12. L. Shi, A. Zanobetti, I. Kloog, B. A. Coull, P. Koutrakis, S. J. Melly and J. D. Schwartz, *Environ. Health Perspect.*, 2015, **124**, 46.
13. R. D. Brook, S. Rajagopalan, C. A. Pope III, R. J. Brook, A. Bhatnagar, A. V. Diez-Roux *et al.*, *Circulation*, 2010, **121**, 2331.
14. F. J. Kelly and J. C. Fussell, *Clin. Exp. Allergy*, 2011, **41**, 1059.
15. N. Künzli, M. Jerrett, W. J. Mack, B. Beckerman, L. LaBree, F. Gilliland *et al.*, *Environ. Health Perspect.*, 2005, **113**, 201.
16. B. Hoffmann, S. Moebus, S. Möhlenkamp, A. Stang, N. Lehmann, N. Dragano, A. Schmermund, M. Memmesheimer and K. Mann, *Circulation*, 2007, **116**, 489.
17. M. Bauer, S. Moebus, S. Möhlenkamp, N. Dragano, M. Nonnemacher, M. Fuchsluger, C. Kessler, H. Jakobs, M. Memmesheimer, R. Erbel, K.-H. Jöckel and B. Hoffmann on behalf of the HNR Study Investigative Group, *J. Am. Coll. Cardiol.*, 2010, **56**, 1803.
18. N. Künzli, M. Jerrett, R. Garcia-Esteban, X. Basagaña, B. Beckermann, F. Gilliland *et al.*, *PLoS One*, 2010, **5**, e9096.
19. W. A. Jedrychowski, F. P. Perera, J. D. Spengler, E. Mroz, L. Stigter, E. Flak *et al.*, *Int. J. Hyg. Environ. Health*, 2013, **216**, 395.
20. E. A. Macintyre, U. Gehring, A. Molter, E. Fuertes, C. Klümper, U. Krämer *et al.*, *Environ. Health Perspect.*, 2013, **22**, 107.
21. S. Mehta, H. Shin, R. Burnett, T. North and A. J. Cohen, *Air Qual., Atmos. Health*, 2013, **6**, 69.
22. U. Gehring, A. H. Wijga, M. Brauer, P. Fischer, J. C. de Jongste, M. Kerkhof, M. Oldenwening, H. A. Smit and B. Brunekreef, *Am. J. Respir. Crit. Care Med.*, 2010, **181**, 596.
23. P. Latzin, M. Röösli, A. Huss, C. E. Kuehni and U. Frey, *Eur. Respir. J.*, 2009, **33**, 594.
24. E. Eenhuizen, U. Gehring, A. H. Wijga, H. A. Smit, P. H. Fischer, M. Brauer *et al.*, *Eur. Respir. J.*, 2013, **41**, 1257.
25. U. Kramer, C. Herder, D. Sugiri, K. Strassburger, T. Schikowski, U. Ranft *et al.*, *Environ. Health Perspect.*, 2010, **118**, 1273.
26. R. C. Puett, J. E. Hart, J. Schwartz, F. B. Hu, A. D. Liese and F. Laden, *Environ. Health Perspect.*, 2011, **119**, 384.

27. L. F. Coogan, M. White, R. D. Jerrett, J. G. Brook, E. Su, R. Seto, J. R. Burnett, Palmer and L. Rosenberg, *Circulation*, 2012, **125**, 767.

28. M. Guxens and J. Sunyer, *Swiss Med. Wkly.*, 2012, **141**, w13322.

29. U. Ranft, T. Schikowski, D. Sugiri, J. Krutmann and U. Krämer, *Environ. Res.*, 2009, **109**, 1004.

30. F. J. Kelly and J. C. Fussell, *Atmos. Environ.*, 2012, **60**, 504.

31. W. J. Gauderman, E. Avol, F. Gilliland, H. Vora, D. Thomas, K. Berhane, R. McConnell, N. Kuenzli, F. Lurmann, E. Rappaport, H. Margolis, D. Bates and J. Peters, *N. Engl. J. Med.*, 2004, **351**, 1057.

32. B. Brunekreef and B. Forsberg, *Eur. Respir. J.*, 2005, **26**, 309.

33. T. J. Grahame and R. B. Schlesinger, *Inhalation Toxicol.*, 2007, **19**, 457.

34. J. L. Mauderly and J. C. Chow, *Inhalation Toxicol.*, 2008, **20**, 257.

35. L. C. Chen and M. Lippmann, *Inhalation Toxicol.*, 2009, **21**, 1.

36. World Health Organisation (WHO) Regional Office for Europe, Health effects of black carbon, 2012, http://www.euro.who.int/__data/assets/pdf_file/0004/162535/e96541.pdf?ua=1. Accessed 29 September 2015.

37. World Health Organisation (WHO) Regional Office for Europe. (2013). Review of evidence on health aspects of air pollution—REVIHAAP project, technical report, 2013, http://www.euro.who.int/__data/assets/pdf_file/0004/193108/REVIHAAP-Final-technical-report-final-version.pdf. Accessed 29 September 2015.

38. Health Effects Institute (HEI), Special Report 17, Traffic-related air pollution: a critical review of the literature on emissions, exposure and health effects, Panel on the Health Effects of Traffic-Related Air Pollution, Boston MA: Health Effects Institute, 2010, http://pubs.healtheffects.org/getfile.php?u=553. Accessed 29 September 2015.

39. Health Effects Institute (HEI), Research Report 161. Assessment of the Health Impacts of Particulate Matter Characteristics. M. L. Bell, Boston MA: Health Effects Institute. 2012. http://pubs.healtheffects.org/getfile.php?u=685. Accessed 30 September 2015.

40. Health Effects Institute (HEI) Research Report 178, National particle component toxicity (NPACT) initiative report on cardiovascular effects. S. Vedal, M. J. Campen, J. D. McDonald, J. D. Kaufman, T. V. Larson, P. D. Sampson, L. Sheppard, C. D. Simpson, A. A. Szpiro, Boston MA: Health Effects Institute, 2013. http://pubs.healtheffects.org/getfile.php?u=946. Accessed 30 September 2015.

41. Health Effects Institute (HEI), Research Report 177, National particle component toxicity (NPACT) initiative: integrated epidemiological and toxicological studies of the health effects of particulate matter components. M. Lippmann, L.-C. Chen, T. Gordon, K. Ito, G. Thurston, Boston MA: Health Effects Institute, 2013. http://pubs.healtheffects.org/getfile.php?u=934. Accessed 30 September 2015.

42. Health Effects Institute (HEI) Understanding the health effects of ambient ultrafine particles, Review Panel on Ultrafine Particles, MA:

Health Effects Institute, 2013, http://pubs.healtheffects.org/getfile.php?u=893. Accessed 16 February 2015.

43. J. C. Chow and J. G. Watson, *Aerosol Air Qual. Res.*, 2007, **7**, 121.
44. World Health Organisation (WHO) Regional Office for Europe, Health Effects of Transport-Related Air Pollution, 2005, http://www.euro.who.int/__data/assets/pdf_file/0006/74715/E86650.pdf. Accessed 30 September 2015.
45. N. A. Janssen, G. Hoek, M. Simic-Lawson, P. Fischer, L. van Bree, H. ten Brink *et al.*, *Environ. Health Perspect.*, 2011, **119**, 1691.
46. R. Atkinson, I. Mills, H. Walton and H. Anderson, *J. Exposure Sci. Environ. Epidemiol.*, 2015, **25**, 208.
47. J. J. Kim, K. Huen, S. Adams, S. Smorodinsky, A. Hoats, B. Malig *et al.*, *Environ. Health Perspect.*, 2008, **116**, 1274.
48. K. Hildebrandt, R. Rückerl, W. Koenig, A. Schneider, M. Pitz, J. Heinrich *et al.*, *Part. Fibre Toxicol.*, 2009, **6**, 25.
49. R. J. Delfino, N. Staimer, T. Tjoa, M. Arhami, A. Polidori, D. L. Gillen, S. C. George, M. M. Shafer, J. J. Schauer and C. Sioutas, *Epidemiology*, 2010, **21**, 396.
50. K. Ito, R. Mathes, Z. Ross, A. Nadas, G. Thurston and T. Matte, *Environ. Health Perspect.*, 2011, **119**, 467.
51. S. Y. Kim, J. L. Peel, M. P. Hannigan, S. J. Dutton, L. Sheppard, M. L. Clark *et al.*, *Environ. Health Perspect.*, 2012, **120**, 1094.
52. J. Y. Son, J. T. Lee, K. H. Kim, K. Jung and M. L. Bell, *Environ. Health Perspect.*, 2012, **120**, 872.
53. A. Zanobetti and J. Schwartz, *Environ. Health Perspect.*, 2009, **117**, 898.
54. B. Ostro, M. Lipsett, P. Reynolds, D. Goldberg, A. Hertz, C. Garcia, K. D. Henderson and L. Bernstein, *Environ. Health Perspect.*, 2010, **118**, 363.
55. M. L. Bell, K. Ebisu, R. D. Peng, J. M. Samet and F. Dominici, *Am. J. Respir. Crit. Care Med.*, 2009, **179**, 1115.
56. J. J. De Hartog, J. G. Ayres, A. Karakatsani, A. Analitis, H. T. Brink, K. Hameri *et al.*, *Environ. Health Perspect.*, 2009, **117**, 105.
57. B. Ostro, L. Roth, B. Malig and M. Marty, *Environ. Health Perspect.*, 2009, **117**, 475.
58. H. H. Suh, A. Zanobetti, J. Schwartz and B. A. Coull, *Environ. Health Perspect.*, 2011, **119**, 1421.
59. A. Zanobetti, M. Franklin, P. Koutraki and J. Schwartz, *Environ. Health*, 2009, **8**, 58.
60. J. Zhou, K. Ito, R. Lall, M. Lippman and G. Thurston, *Environ. Health Perspect.*, 2011, **119**, 461.
61. R. B. Schlesinger and F. Cassee, *Inhalation. Toxicol.*, 2003, **15**, 197.
62. H. R. Anderson, B. Armstrong, S. Hajat, R. Harrison, V. Monk, J. Poloniecki *et al.*, *Epidemiology*, 2010, **21**, 405.
63. M. A. Bind, A. Baccarelli, A. Zanobetti, L. Tarantini, H. Suh, P. Vokonas *et al.*, *Epidemiology*, 2012, **23**, 332.

64. G. Rubasinghege, R. W. Lentz, M. M. Scherer and V. H. Grassian, *Proc. Natl. Acad. Sci. U. S. A.*, 2010, **107**, 6628.
65. W. Li, L. Shao, R. Shen, S. Yang, Z. Wang and W. Tang, *J. Air Waste Manage. Assoc.*, 2011, **61**, 1166.
66. O. B. Popovicheva, N. M. Persiantseva, E. D. Kireeva, T. D. Khokhlova and N. K. Shonija, *J. Phys. Chem. A*, 2011, **115**, 298.
67. M. Oakes, E. D. Ingall, B. Lai, M. M. Shafer, M. D. Hays, Z. G. Liu *et al.*, *Environ. Sci. Technol.*, 2012, **46**, 6637.
68. W. G. Kreyling, S. Hirn and C. Schleh, *Nat. Biotechnol.*, 2010, **28**, 1275.
69. R. Rückerl, A. Schneider, S. Breitner, J. Cyrys and A. Peters, *Inhalation Toxicol.*, 2011, **23**, 55.
70. S. Weichenthal, *Environ. Res.*, 2012, **115**, 26.
71. R. D. Peng, H. H. Chang, M. L. Bell, A. McDermott, S. Zeger, J. M. Samet *et al.*, *JAMA, J. Am. Med. Assoc.*, 2008, **299**, 2172.
72. R. W. Atkinson, G. W. Fuller, H. R. Anderson, R. H. Harrison and B. Armstrong, *Epidemiology*, 2010, **21**, 501.
73. J. K. Mann, J. R. Balmes, T. A. Bruckner, K. M. Mortimer, H. G. Margolis, B. Pratt *et al.*, *Environ. Health Perspect.*, 2010, **118**, 1497.
74. K. Meister, C. Johansson and B. Forsberg, *Environ. Health Perspect.*, 2012, **20**, 431.
75. H. Qiu, I. T. Yu, L. Tian, X. Wang, L. A. Tse, W. Tam *et al.*, *Environ. Health Perspect.*, 2012, **120**, 572.
76. R. C. Puett, J. E. Hart, J. D. Yanosky, C. Paciorek, J. Schwartz, H. Suh *et al.*, *Environ. Health Perspect.*, 2009, **117**, 1697.
77. R. C. Puett, J. E. Hart, H. Suh, M. Mittleman and F. Laden, *Environ. Health Perspect.*, 2011, **119**, 1130.
78. D. W. Graff, W. E. Cascio, A. Rappold, A. Zhou, Y. C. Huang and R. B. Devlin, *Environ. Health Perspect.*, 2009, **117**, 1089.
79. T. C. Wegesser, K. E. Pinkerton and J. A. Last, *Environ. Health Perspect.*, 2009, **117**, 893.
80. Environmental Protection Agency (EPA), Integrated science assessment for particulate matter (final report). Washington, DC: United States Environmental Protection Agency, 2009, http://cfpub.epa.gov/ncea/cfm/recordisplay.cfm?deid=216546. Accessed 21 September 2015.
81. Quality of urban air review group, Airborne particulate matter in the UK, 1996, http://uk-air.defra.gov.uk/assets/documents/reports/empire/quarg/quarg_11.pdf. Accessed 29 September 2015.
82. N. A. Janssen, B. Brunekreef, P. van Vliet, F. Aarts, K. Meliefste, H. Harssema and P. Fischer, *Environ. Health Perspect.*, 2003, **111**, 1512.
83. A. M. Gowers, P. Cullinan, J. G. Ayres, H. R. Anderson, D. P. Strachan, S. T. Holgate, I. C. Mills and R. L. Maynard, *Respirology*, 2012, **17**, 887.
84. International Agency for Research on Cancer (IARC), Diesel engine exhaust carcinogenic, 2012, http://www.iarc.fr/en/media-centre/pr/2012/pdfs/pr213_E.pdf. Accessed 29 September 2015.

85. H. A. van der Gon, M. E. Gerlofs-Nijland, R. Gehrig, M. Gustafsson, N. Janssen, R. M. Harrison *et al.*, *J. Air Waste Manage. Assoc.*, 2013, **63**, 136.
86. M. Riediker, R. B. Devlin, T. R. Griggs, M. C. Herbst, P. A. Bromberg, R. W. Williams *et al.*, *Part. Fibre Toxicol.*, 2004, **1**, 2.
87. R. R. Gottipolu, E. R. Landa, M. C. Schladweiler, J. K. McGee, A. D. Ledbetter, J. Richards, G. J. Wallenborn and U. P. Kodavanti, *Inhalation Toxicol.*, 2008, **20**, 473.
88. M. Gasser, M. Riediker, L. Mueller, A. Perrenoud, F. Blank, P. Gehr and B. Rothen-Rutishauser, *Part. Fibre Toxicol.*, 2009, **6**, 30.
89. P. Mantecca, G. Sancini, E. Moschini, F. Farina, M. Gualtieri, A. Rohr *et al.*, *Toxicol. Lett.*, 2009, **189**, 206.
90. Health Effects Institute (HEI) National Particle Component Toxicity (NPACT) Initiative, EXECUTIVE SUMMARY, HEI NPACT Review Panel, Boston MA: Health Effects Institute, 2013, http://www.healtheffects.org/Pubs/NPACT-ExecutiveSummary.pdf. Accessed 30 September 2015.
91. L. W. Stanek, J. D. Sacks, S. J. Dutton and J. J. B. Dubois, *Atmos. Environ.*, 2011, **45**, 5655.
92. J. R. Krall, G. B. Anderson, F. Dominici, M. L. Bell and R. D. Peng, *Environ. Health Perspect.*, 2013, **121**, 1148.
93. R. M. Harrison and J. Yin, *Atmos. Environ.*, 2008, **42**, 1413.
94. R. B. Schlesinger, N. Kunzli, G. M. Hidy, T. Gotschi and M. Jerrett, *Inhalation Toxicol.*, 2006, **18**, 95.
95. C. A. Pope III, R. T. Burnett, M. J. Thun, E. E. Calle, D. Krewski, K. Ito *et al.*, *JAMA, J. Am. Med. Assoc.*, 2002, **287**, 1132.
96. F. Laden, J. Schwartz, F. E. Speizer and D. W. Dockery, *Am. J. Respir. Crit. Care Med.*, 2006, **173**, 667.
97. M. J. Lipsett, B. D. Ostro, P. Reynolds, D. Goldberg, A. Hertz, M. Jerrett *et al.*, *Am. J. Respir. Crit. Care Med.*, 2011, **184**, 828.
98. J. Lepeule, F. Laden, D. Dockery and J. Schwartz, *Environ. Health Perspect.*, 2012, **120**, 965.
99. G. Cesaroni, C. Badaloni, C. Gariazzo, M. Stafoggia, R. Sozzi, M. Davoli *et al.*, *Environ. Health Perspect.*, 2013, **121**, 324.
100. R. Beelen, G. Hoek, P. A. van den Brandt, R. A. Goldbohm, P. Fischer, L. J. Schouten, M. Jerrett, E. Hughes, B. Armstrong and B. Brunekreef, *Environ. Health Perspect.*, 2008, **116**, 196.
101. R. W. Atkinson, S. Kang, H. R. Anderson, I. C. Mills and H. A. Walton, *Thorax*, 2014, **69**, 660.
102. C. A. Pope III, R. T. Burnett, D. Krewski, M. Jerrett, Y. Shi, E. E. Calle *et al.*, *Circulation*, 2009, **120**, 941.
103. A. W. Correia, C. A. Pope III, D. W. Dockery, Y. Wang, M. Ezzati and F. Dominici, *Epidemiology*, 2013, **24**, 23.
104. F. Dominici, Y. Wang, A. W. Correia, M. Ezzati, C. A. Pope III and D. W. Dockery, *Epidemiology*, 2015, **26**, 556.
105. C. A. Pope III, D. L. Rodermund and M. M. Gee, *Environ. Health Perspect.*, 2007, **115**, 679.

106. B. Ostro, P. Reynolds, D. Goldberg, A. Hertz, R. T. Burnett, H. Shin, E. Hughes and C. Garcia, *Environ. Health Perspect.*, 2011, **119**, A242.
107. L. Dai, A. Zanobetti, P. Koutrakis and J. D. Schwartz, *Environ. Health Perspect.*, 2014, **122**, 837.
108. B. Ostro, W.-Y. Feng, R. Broadwin, S. Green and M. Lipsett, *Environ. Health Perspect.*, 2007, **115**, 13.
109. C. Cakmak, R. Dale and C. B. Vida, *Int. J. Occup. Environ. Health*, 2009, **15**, 152.
110. J. Cao, H. Xu, Q. Xu, B. Chen and H. Kan, *Environ. Health Perspect.*, 2012, **120**, 373.
111. R. Peng, M. Bell, A. Geyh, A. McDermott, S. Zeger, J. Samet *et al.*, *Environ. Health Perspect.*, 2009, **117**, 957.
112. L. A. Darrow, M. Klein, W. D. Flanders, L. A. Waller, A. Correa, M. Marcus, J. A. Mulholland, A. G. Russell and P. E. Tolbert, *Epidemiology*, 2009, **20**, 689.
113. M. L. Bell, K. Belanger, K. Ebisua, J. F. Gent, H. J. Lee, P. Koutrakisc and B. P. Leaderer, *Epidemiology*, 2010, **21**, 884.
114. R. Beelen, G. Hoek, O. Raaschou-Nielsen, M. Stafoggia, Z. J. Andersen, G. Weinmayr *et al.*, *Environ. Health Perspect.*, 2015, **123**, 525.
115. S. X. Wang, R. Beelen, M. Stafoggia, O. Raaschou-Nielsen, Z. J. Andersen and B. Hoffmann, *Atmos. Chem. Phys.*, 2014, **14**, 6571.
116. R. Hampel, A. Peters, R. Beelen, B. Brunekreef, J. Cyrys, U. de Faire *et al.*, for the ESCAPE TRANSPHORM study groups, *Environ. Int.*, 2015, **82**, 76.
117. M. Eeftens, G. Hoek, O. Gruzieva, A. Mölter, R. Agius, R. Beelen *et al.*, *Epidemiology*, 2014, **25**, 648.
118. E. Fuertes, E. MacIntyre, R. Agius, R. Beelen, B. Brunekreef, S. Bucci *et al.*, *Int. J. Hyg. Environ. Health*, 2014, **217**, 819.
119. X. Querol, A. Alastuey, J. Pey, M. Cusack, N. Perez, N. Mihalopoulos, C. Theodosi, E. Gerasopoulos, N. Kubilay and M. Koçak, *Atmos. Chem. Phys.*, 2009, **9**, 4575.
120. E. Samoli, M. Stafoggia, S. Rodopoulou, B. Ostro, C. Declercq, E. Alessandrini *et al.* for the MED-PARTICLES Study Group, *Environ. Health Perspect.*, 2013, **121**, 932.
121. E. Samoli, M. Stafoggia, S. Rodopoulou, B. Ostro, E. Alessandrini, X. Basagaña *et al.*, MED-PARTICLES Study group, *Environ. Int.*, 2014, **67**, 54.
122. M. Stafoggia, E. Samoli, E. Alessandrini, E. Cadum, B. Ostro, G. Berti *et al.*, *Environ. Health Perspect.*, 2013, **121**, 1026.
123. R. D. Peng, F. Dominici, R. Pastor-Barriuso, S. L. Zeger and J. M. Samet, *Am. J. Epidemiol.*, 2005, **161**, 585.
124. M. Stafoggia, J. Schwartz, F. Forastiere, C. A. Perucci and SISTI Group, *Am. J. Epidemiol.*, 2008, **167**, 1476.

125. X. Basagaña, B. Jacquemin, A. Karanasiou, B. Ostro, X. Querol, D. Agis *et al.*, on behalf of the MED-PARTICLES Study group, *Environ. Int.*, 2015, **75**, 151.

126. B. Ostro, A. Tobias, A. Karanasiou, E. Samoli, X. Querol, S. Rodopoulou *et al.* and the MED-PARTICLES Study Group, *Occup. Environ. Med.*, 2015, **72**, 12.

127. R. Atkinson, A. Analitis, E. Samoli, G. Fuller, D. Green, I. Mudway *et al.*, *J. Exposure Sci. Environ. Epidemiol.*, 2016, **26**, 125.

128. E. Samoli, R. W. Atkinson, A. Ananlitis, G. W. Fuller, D. C. Green, I. Mudawy *et al.*, *Occup. Environ. Med.*, 2016, **73**, 300.

129. J. Pey, X. Querol, A. Alastuey, F. Forastiere and M. Stafoggia, *Atmos. Chem. Phys.*, 2013, **13**, 1395.

130. A. Karanasiou, N. Moreno, T. Moreno, M. Viana, F. de Leeuw and X. Querol, *Environ. Int.*, 2012, **47**, 107.

131. E. Samoli, P. T. Nastos, A. G. Paliatsos, K. Katsouyanni and K. N. Priftis, *Environ. Res.*, 2011, **111**, 418.

132. S. Sajani, R. Miglio, P. Bonasoni, P. Cristofanelli, A. Marinoni, C. Sartini *et al.*, *Occup. Environ. Med.*, 2011, **68**, 446.

133. N. Middleton, P. Yiallouros, S. Kleanthous, O. Kolokotroni, J. Schwartz, D. W. Dockery *et al.*, *Environ. Health*, 2008, **22**, 7.

134. L. Pérez, A. Tobias, X. Querol, N. Künzli, J. Pey, A. Alastuey *et al.*, *Epidemiology*, 2008, **19**, 800.

135. E. Samoli, E. Kougea, P. Kassomenos, A. Analitis and K. Katsouyanni, *Sci. Total Environ.*, 2011, **409**, 2049.

136. L. Pérez, A. Tobias, J. Pey, N. Pérez, A. Alastuey, J. Sunyer *et al.*, *Epidemiology*, 2012a, **23**, 768.

137. L. Pérez, A. Tobías, X. Querol, J. Pey, A. Alastuey, J. Díaz *et al.*, *Environ. Int.*, 2012b, **48**, 150.

138. M. Stafoggia, S. Zauli-Sajani, J. Pey, E. Samoli, E. Alessandrini, X. Basagaña, A. Cernigliaro, M. Chiusolo, M. Demaria and the MED-PARTICLES Study Group, *Environ. Health Perspect.*, 2016, **124**, 413.

139. T. Moreno, X. Querol, S. Castillo, A. Alastuey, E. Cuevas, L. Herrmann *et al.*, *Chemosphere*, 2006, **65**, 261.

140. S. Rodríguez, X. Querol, A. Alastuey, G. Kallos and O. Kakaliagou, *Atmos. Environ.*, 2001, **35**, 243.

141. D. Griffin, C. Kellogg and E. Shinn, *Global Change Hum. Health*, 2001, **2**, 2.

142. D. W. Griffin, *Clin. Microbiol. Rev.*, 2007, **20**, 45.

143. V. H. Garrison, W. T. Foreman, S. Genualdi, D. W. Griffin, C. A. Kellogg, M. S. Majewski *et al.*, *Rev. Biol. Trop.*, 2006, **54**(S3), 9.

144. P. N. Polymenakou, M. Mandalakis, E. G. Stephanou and A. Tselepides, *Environ. Health Perspect.*, 2008, **116**, 292.

145. M. Pandolfi, A. Tobias, A. Alastuey, J. Sunyer, J. Schwartz, J. Lorente *et al.*, *Sci. Total Environ.*, 2014, **283**, 494.

146. J. Baumgartner, Y. Zhang, J. J. Schauer, W. Huang, Y. Wang and M. Ezzati, *Proc. Natl. Acad. Sci. U. S. A.*, 2014, **111**, 13229.

147. S. S. Lim, T. Vos, A. D. Flaxman, G. Danaei, K. Shibaya, H. Adair-Rahani *et al.*, *Lancet*, 2012, **380**, 2224.

148. K. R. Smith, N. Bruce, K. Balakrishnan, H. Adair-Rohani, J. Balmes, Z. Chafe *et al.*, *Annu. Rev. Public Health*, 2014, **35**, 185.

149. Health Effects Institute (HEI) Outdoor air pollution and health in the developing countries of Asia: a comprehensive review. Boston MA: Health Effects Institute, 2010, http://pubs.healtheffects.org/getfile. php?u=602. Accessed 30 September 2015.

150. S. Bonjour, H. Adair-Rohani, J. Wolf, N. G. Bruce, S. Mehta, A. Prüss-Ustün, M. Lahiff, E. A. Rehfuess, V. Mishra and K. R. Smith, *Environ. Health Perspect.*, 2013, **121**, 784.

151. T. Ohara, H. Akimoto, J. Kurokawa, N. Horii, K. Yamaji, X. Yan and T. Hayasaka, *Atmos. Chem. Phys.*, 2007, **7**, 4419.

152. K. F. Chung, J. Zhang and N. Zhong, *Respirology*, 2011, **16**, 1023–1026.

153. M. Brauer, M. Amann, R. T. Burnett, A. Cohen, F. Dentener, M. Ezzati *et al.* on behalf of the Outdoor Air Pollution Expert Working Group of the Global Burden of Disease Project, *Environ. Sci. Technol.*, 2012, **46**, 652.

154. T. C. Su, S. Y. Chen and C. C. Chan, *Prog. Cardiovasc. Dis.*, 2011, **53**, 369.

155. S. A. Venners, B. Wang, Z. Xu, Y. Schlatter, L. Wang and X. Xu, *Environ. Health Perspect.*, 2003, **111**, 562.

156. H. Kan, S. J. London, G. Chen, Y. Zhang, G. Song, N. Zhao *et al.*, *Environ. Int.*, 2007, **33**(3), 376.

157. Y. Ma, R. Chen, G. Pan, X. Xu, W. Song, B. Chen and H. Kan, *Sci. Total Environ.*, 2011, **409**, 2473.

158. P. Li, J. Xin, Y. Wang, S. Wang, G. Li, X. Pan *et al.*, *Environ. Sci. Pollut. Res.*, 2013, **20**, 6433.

159. P. Li, J. Xin, Y. Wang, S. Wang, K. Shang, Z. Liu, G. Li, X. Pan, L. Wei and M. Wang, *Atmos. Environ.*, 2013, **81**, 253.

160. D. G. Streets, S. Gupta, S. T. Waldhoff, M. Q. Wang, T. C. Bond and Y. Bo, Black carbon emissions in China, *Atmos. Environ.*, 2001, **35**, 4281–4296.

161. F. Yang, K. He, B. Ye, X. Chen, L. Cha, S. H. Cadle, T. Chan and P. A. Mulawa, *Atmos. Chem. Phys. Discuss.*, 2005, **5**, 217.

162. F. Geng, J. Hua, Z. Mu, L. Peng, X. Xu, R. Chen and H. Kan, *Environ. Res.*, 2013, **120**, 27.

163. V. C. Pun, I. T. S. Yu, H. Qiu, K. F. Ho, Z. Sun, P. K. K. Louie, T. W. Wong and L. Tian, *Am. J. Epidemiol.*, 2014, **179**, 1086.

164. V. C. Pun, L. Tian, I. T. S. Yu, M. A. Kioumourtzoglou and H. Qiu, *Environ. Sci. Technol.*, 2015, **49**, 3830.

165. L. Qiao, J. Cai, H. Wang, W. Wang, M. Zhou, S. Lou, R. Chen, H. Dai, C. Chen and H. Kan, *Environ. Sci. Technol.*, 2014, **48**, 10406.

166. S. Wu, F. Deng, H. Wei, J. Huang, H. Wang, M. Shima *et al.*, *Part. Fibre Toxicol.*, 2012, **9**, 49.

167. S. Wu, F. Deng, J. Huang, H. Wang, M. Shima, X. Wang *et al.*, *Environ. Health Perspect.*, 2013, **121**, 6.

168. S. Wu, F. Deng, Y. Hao, M. Shima, X. Wang, C. Zheng *et al.*, *J. Hazard. Mater.*, 2013, **260**, 183.

169. S. Wu, D. Yang, H. Wei, B. Wang, J. Huang, H. Li, M. Shima, F. Deng and X. Guo, *Chemosphere*, 2015, **135**, 347.
170. R. T. Burnett, J. Brook, T. Dann, C. Delocla, O. Philips, S. Cakmak, R. Vincent, M. S. Goldberg and D. Krewski, *Inhalation Toxicol.*, 2000, **12**, 15.
171. J. L. Peel, P. E. Tolbert, M. Klein, K. B. Metzger, W. D. Flanders, K. Todd, J. A. Mulholland, P. B. Ryan and H. Frumkin, *Epidemiology*, 2005, **16**, 164.
172. M. Neuberger, H. Moshammer and D. Rabczenko, *Int. J. Environ. Res. Public Health*, 2013, **10**, 4728.
173. C. A. Pope III, M. Ezzati and D. W. Dockery, *N. Engl. J. Med.*, 2009, **360**, 376.
174. O. P. Kurmi, K. B. Lam and J. G. Ayres, *Eur. Respir. J.*, 2012, **40**, 239.
175. United Nations Millennium Project, Investing in development. A practical plan to achieve the Millennium Development Goals, 2015, London, Sterling (VA): Earthscan and United Nations http://www.unmillenniumproject.org/documents/MainReportComplete-lowres.pdf. Accessed 30 September 2015.
176. O. P. Kurmi, G. S. Devereux, W. C. Smith, S. Semple, M. F. Steiner, P. Simkhada, K. B. Lam and J. G. Ayres, *Eur. Respir. J.*, 2013, **41**, 25.
177. J. Grigg, *Proc. Am. Thorac. Soc.*, 2009, **6**, 564.
178. I. S. Mudway, S. T. Duggan, C. Venkataraman, G. Habib, F. J. Kelly and J. Grigg, *Part. Fibre Toxicol.*, 2005, **2**, 6.
179. O. P. Kurmi, C. Dunster, J. G. Ayres and F. J. Kelly, Oxidative potential of smoke from burning wood and missed biomass fuels, *Free Radical Res.*, 2013, **47**, 829.
180. T. E. Sussan, V. Ingole, J.-H. Kim, S. McCormick, J. Negherbon, J. Fallica, J. Akulian, L. Yarmus and D. Feller-Kopman, *Am. J. Respir. Cell Mol. Biol.*, 2014, **50**, 538.
181. S. S. Yamamotoa, R. Phalkeya and A. A. Malik, *Int. J. Hyg. Environ. Health*, 2014, **217**, 13.
182. J. Unosso, A. Blomberg, T. Sandström, A. Mual, C. Boman, R. Nyström, R. Westerholm, N. L. Mills and D. E. Newby, *Part. Fibre Toxicol.*, 2013, **10**, 20.
183. M. Lundbäck, N. L. Mills, A. Lucking, S. Barath, K. Donaldson, D. E. Newby *et al.*, *Part. Fibre Toxicol.*, 2009, **6**, 7.
184. M. B. Epstein, M. N. Bates, N. K. Aror, K. Balakrishnan, D. W. Jack and K. R. Smith, *Int. J. Hyg. Environ. Health*, 2013, **216**, 523.
185. P. V. Lakshmi, N. K. Virdi, A. Sharma, J. P. Tripathy, K. R. Smith, M. N. Bates and R. Kumar, *Environ. Res.*, 2013, **121**, 17.
186. B. J. Wylie, B. A. Coull, D. H. Hamer, M. P. Singh, D. Jack, K. Yeboah-Antwi, L. Sabin, N. Singh and W. B. MacLeod, *Environ. Health*, 2014, **13**, 1.
187. Conservation Action Trust, Urban Emissions, Greenpeace India, Coal kills - an assessment of death and disease caused by India's dirtiest energy source, 2013. http://www.greenpeace.org/india/Global/india/report/Coal_Kills.pdf. Accessed 28 September 2015.
188. K. Balakrishnan, S. Sambandam, P. Ramaswamy, S. Ghosh, V. Venkatesan, G. Thangavel, K. Mukhopadhyay, P. Johnson, S. Paul,

N. Puttaswamy, R. S. Dhaliwal, D. K. Shukla and SRU-CAR Team, *BMJ Open*, 2015, **5**, e008090.

189. F. R. Cassee, M. E. Héroux, M. E. Gerlofs-Nijland and F. J. Kelly, *Inhalation Toxicol.*, 2013, **25**, 802.

190. R. T. Burnett, C. A. Pope III, M. Ezzati, C. Olives, S. S. Lim, S. Mehta *et al.*, *Environ. Health Perspect.*, 2014, **122**, 397.

191. J. S. Apte, J. D. Marshall, A. J. Cohen and M. Brauer, *Environ. Sci. Technol.*, 2015, **49**, 8057.

Subject Index